# Biochemistry

## for Physiotherapy and
## Allied Health Sciences Students

# Biochemistry
## for Physiotherapy and Allied Health Sciences Students

**Beena V Shetty** PhD
Professor
Kasturba Medical College, Mangalore
Karnataka, India

**Nandini M** PhD
Associate Professor
Kasturba Medical College, Mangalore
Karnataka, India

**Vinitha Ramanath Pai** PhD
Associate Professor
Yenapoya Medical & Dental College, Mangalore
Karnataka, India

JAYPEE BROTHERS MEDICAL PUBLISHERS (P) LTD

New Delhi • Ahmedabad • Bengaluru • Chennai • Hyderabad
Kochi • Kolkata • Lucknow • Mumbai • Nagpur

*Published by*

Jitendar P Vij
**Jaypee Brothers Medical Publishers (P) Ltd**
B-3, EMCA House, 23/23B Ansari Road, Daryaganj
**New Delhi** 110 002, India
Phones: +91-11-23272143, +91-11-23272703, +91-11-23282021, +91-11-23245672, Rel: +91-11-32558559
Fax: +91-11-23276490, +91-11-23245683 e-mail: jaypee@jaypeebrothers.com
Visit our website: www.jaypeebrothers.com

**Branches**

- 2/B, Akruti Society, Jodhpur Gam Road Satellite
  **Ahmedabad** 380 015 Phones: +91-79-26926233, Rel: +91-79-32988717, Fax: +91-79-26927094
  e-mail: ahmedabad@jaypeebrothers.com
- 202 Batavia Chambers, 8 Kumara Krupa Road, Kumara Park East
  **Bengaluru** 560 001 Phones: +91-80-22285971, +91-80-22382956, +91-80-22372664, Rel: +91-80-32714073
  Fax: +91-80-22281761 e-mail: bangalore@jaypeebrothers.com
- 282 IIIrd Floor, Khaleel Shirazi Estate, Fountain Plaza, Pantheon Road
  **Chennai** 600 008 Phones: +91-44-28193265, +91-44-28194897, Rel: +91-44-32972089
  Fax: +91-44-28193231 e-mail: chennai@jaypeebrothers.com
- 4-2-1067/1-3, 1st Floor, Balaji Building, Ramkote Cross Road
  **Hyderabad** 500 095 Phones: +91-40-66610020, +91-40-24758498, Rel:+91-40-32940929
  Fax:+91-40-24758499, e-mail: hyderabad@jaypeebrothers.com
- No. 41/3098, B & B1, Kuruvi Building, St. Vincent Road
  **Kochi** 682 018, Kerala Phones: +91-484-4036109, +91-484-2395739, +91-484-2395740
  e-mail: kochi@jaypeebrothers.com
- 1-A Indian Mirror Street, Wellington Square
  **Kolkata** 700 013 Phones: +91-33-22651926, +91-33-22276404, +91-33-22276415, Rel: +91-33-32901926
  Fax: +91-33-22656075, e-mail: kolkata@jaypeebrothers.com
- Lekhraj Market III, B-2, Sector-4, Faizabad Road, Indira Nagar
  **Lucknow** 226 016 Phones: +91-522-3040553, +91-522-3040554 e-mail: lucknow@jaypeebrothers.com
- 106 Amit Industrial Estate, 61 Dr SS Rao Road, Near MGM Hospital, Parel
  **Mumbai** 400012 Phones: +91-22-24124863, +91-22-24104532, Rel: +91-22-32926896
  Fax: +91-22-24160828, e-mail: mumbai@jaypeebrothers.com
- "KAMALPUSHPA" 38, Reshimbag, Opp. Mohota Science College, Umred Road
  **Nagpur** 440 009 (MS) Phone: Rel: +91-712-3245220, Fax: +91-712-2704275 e-mail: nagpur@jaypeebrothers.com

***Biochemistry for Physiotherapy and Allied Health Sciences Students***



*First Edition*: 2008
Reprint : **2009**
ISBN 978-81-8448-338-3
*Typeset at* JPBMP typesetting unit
*Printed at* Rajkamal Electric Press, Plot No. 2, Phase-IV, Kundli, Haryana.

# Foreword

I have great pleasure in writing a brief introduction to the *Biochemistry for Physiotherapy and Allied Health Sciences Students* authored by Dr. Beena Shetty, Professor of Biochemistry, Dr M Nandini and Dr Vinitha Ramanath Pai, Associate Professors of Biochemistry, from Kasturba Medical College, Mangalore and Yenapoya Medical College, Mangalore. They are experienced teachers and have been teaching Biochemistry for the medical, nursing and physiotherapy students for over two decades. I had the pleasure of working with them as a senior colleague for a long time. Hence I can vouch for their competence, dedication to teaching and commitment to work.

Physiotherapy and allied sciences taught in the Health Universities in our country encompass almost all the basic science subjects of the undergraduate medical students. Though it is said that an elementary knowledge of these subjects is adequate for the allied science students, the curricula presented by the respective boards of studies are ambitious and sometimes vague. The existing textbooks (both Indian and Western) are voluminous. The students appearing at the University examinations in these subjects are confused and lost in the maze of information. Information offering fundamental knowledge is the need of the hour. I recall Dr R Raghunandana Rao (retired Professor of Biochemistry in various medical colleges including the Kasturba Medical College, Mangalore), one of the brilliant teachers I have known, often saying that if a student is taught to appreciate fundamentals, details will eventually take care of themselves. Good fundamentals will inspire a student to look into the "Future of Science" that is rapidly developing in biological complexity, with greater inspiration.

The highly complicated molecule DNA is the basis for all life on earth. As DNA replicates (makes copies of itself), there would have been occasional errors in the order of nucleotides along the spiral (of DNA) over the billions of years since life originated. To dwell into the complexity of mutation with an enquiring mind is awe-inspiring. The 9-membered purines and the 6-membered pyrimidines are the key fundaments in the search. One who is at ease with their chemical nature and sequence in DNA will be advantageously placed to fathom the biological complexity. Further, rapid developments in the diagnostic criteria for liver and renal functions need to be understood clearly with the existing fundamental knowledge.

The authors have taken pains in laying foundation for the fundamentals of all aspects of Biochemistry without overdose. The book, at the same time, is student-friendly and will be useful from the examination point of view, which, after all, is the passport for grading.

I am sure the efforts of Drs. Beena Shetty, Nandini and Vinitha Ramanath Pai will be successful and the students will be benefited.

**Dr S Narasimha Rao,** PhD
Retd, Prof and HOD, Dept. of Biochemistry
Kasturba Medical College, Mangalore and
Manipal College of Medical Sciences
Pokhara (Nepal)

# Preface

Biochemistry is a fascinating subject which is growing at a very fast pace. An understanding of the biochemical concepts is very essential for the medical and nutritional aspects of human welfare too. Hence, there is a need for a sound basic knowledge of the subject to be imparted to young students who are on the first step for a career in health sciences of any type.

This textbook is an extension of our experiences in teaching undergraduate courses of various allied sciences like medical, dental, nursing and physiotherapy. It has been designed keeping in mind the prescribed university syllabus of MAHE and RGUHS and is intended to impart a simple, concise knowledge of the subject the students need to know. Relevant materials on the chemistry, structure, function and metabolism of the various biomolecules have all been discussed under one heading to give a clearer understanding of the topic. We believe this approach is essential for proper understanding and easy reference.

**Beena V Shetty**
**Nandini M**
**Vinitha Ramanath Pai**

# Acknowledgements

We are highly indebted to *Dr. Ashalatha Rao,* Prof. Dept. of Biochemistry, Fr. Muller's Medical College, Mangalore, who helped us with constructive criticism and valuable suggestions. *Prof. S Narasimha Rao's* ideas have also been an inspiration towards making this book and we are thankful to him. We are grateful to *Dr CV Raghuveer,* Dean, KMC, Mangalore, *Dr BH Sripathi Rao,* Principal, YDC, Mangalore, *Dr A Ramachandran*, Principal, YMC, Mangalore and *Dr Vivian D'Souza,* HOD Dept. of Biochemistry, KMC, Mangalore for their constant encouragement.

# Contents

# 1 Cell Biology

The cell is the structural and functional unit of life. The human body is composed of specialized tissues which in turn consist of vast clusters of cells. Two general types of cells are recognized in nature. They are prokaryotes and eukaryotes.

- The word *Prokaryote* is derived from a Greek word (Pro – before; karyon – nucleus). They lack a well defined nucleus and possess simple structure. Present mainly in the bacteria, blue green algae and certain types of microorganisms.
- The word *Eukaryote* is also derived from the Greek word (Eu – true ; karyon – nucleus). They possess a well defined nucleus and are more complex in their structure and function. Present in all higher organisms, (both plant and animal) as well as in fungi, protozoa, etc.

The differences between a prokaryotic and eukaryotic cell are given in Table 1.1.

## EUKARYOTIC CELL

In humans, eukaryotic cells exist in a variety of sizes and shapes that allow adaptation to their special functions. Despite their diversity in function they have certain features in common. The nucleus is separated from the rest of the cell constituents by an envelope called the *nuclear membrane* and contains the DNA (Fig. 1.1).

- The cell units are separated from their surroundings by a plasma membrane. The plasma membrane is 7-10 nm thick.
- It is primarily composed of a lipid bilayer, in which is dispersed a considerable amount of protein. A small amount of carbohydrate is present in membranes as glycoproteins and glycolipids.
- Both protein and lipid molecules have freedom of movement within the framework of the membrane which is made possible by the fluid nature of the bilayer.

**Table 1.1:** Comparison between a prokaryotic and eukaryotic cell

| | *Characteristic* | *Prokaryotic Cell* | *Eukaryotic Cell* |
|---|---|---|---|
| 1. | Size | Small (1-10 μm) | Large (10-100 μm) |
| 2. | Subcellular organelles | Absent | Distinct organelles present, e.g. nucleus, mitochondria, lysosomes |
| 3. | Nucleus | Not well defined | Well defined nucleus surrounded by a membrane |
| 4. | Cytoplasm | Organelles and cytoskeleton absent | Present |
| 5. | Cell division | Fission | Mitosis |

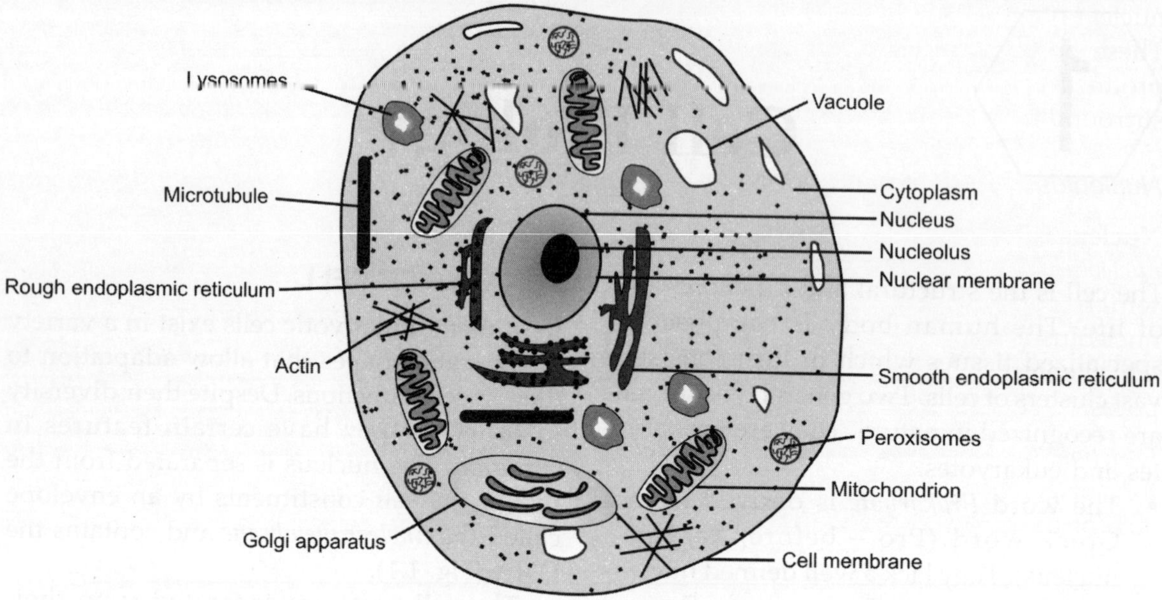


**Fig. 1.1:** Structure of an eukaryotic cell
*(For colour version see Plate 1)*

- Movement of proteins is maintained in a lateral direction. This important property of fluidity facilitates the conformational changes that occur in the protein molecules of the membrane during their functional activity.
- The plasma membrane performs the following functions.
  — It holds the cells together and makes it a functional unit.
  — Serves as a selective barrier—permitting the entrance of essential nutrients into the cell.
  — Secretes the waste products and thus keeps out toxic materials.
  — Binds certain regulatory substances, particularly hormones in a location that is most advantageous to the performance of their function.

## SUBCELLULAR ORGANELLES

The various subcellular organelles present in the cell include the following :

i. Nucleus
ii. Mitochondria
iii. Endoplasmic reticulum
iv. Golgi apparatus
v. Lysosomes
vi. Peroxisomes
vii. Cytosol and cytoskeleton

The subcellular organelles are also surrounded by membranes. A brief note on the structure and function of various subcellular organelles are mentioned.

### Nucleus

It is the largest cellular organelle surrounded by a double layered membrane known as nuclear envelope. The ground material present in the nucleus is known as *nucleoplasm*. Each of the membranes is a typical bilayer composed of phospholipids and cholesterol. The outer membrane is usually covered with ribosomes and is continuous with the endoplasmic reticulum. At certain intervals the two nuclear membranes have nuclear pores with a diameter of about 90 nm. These pores permit the free passage of the products synthesized in the nucleus into the surrounding cytoplasm.

### Nucleolus

Nucleus contains a dense body known as the *nucleolus*. This consists of that portion of DNA which codes for ribosomal RNA. The nucleolus is therefore rich in *ribosomal RNA* which enters the cytosol through the nuclear pores.

*Function:* Nucleus contains DNA, the repository of genetic information. Approximately 90% of the cells DNA is found concentrated in the organelle as part of structures called *chromosomes.*

- Human cells contain the genetic information distributed among 46 such chromosomes.
- Eukaryotic DNA is associated with basic proteins (histones) in the ratio of 1:1 to form *nucleosomes*. An assembly of nucleosomes constitutes the *chromatin*-fibres of the chromosomes.
- The nucleoplasm is rich in enzymes such as DNA polymerases and RNA polymerases which are involved in the synthesis of DNA and RNA and also possess the enzymes for DNA repair.

## Mitochondria

Are the second largest organelles in the cell measuring 0.5-1 µm in diameter and 1-10 µm in length. There are usually 1-1000 mitochondria per cell depending on the nature and function of the cell. They are composed of a double membrane system. The outer membrane is smooth and composed of a phospholipid and cholesterol rich bilayer with an equal amount of protein. The inner membrane is invaginated for forming shelves or *cristae* which extends into the matrix.

*Function:* Mitochondria are the site of respiration and concomitant production of ATP which is the energy currency of the cell.

- ATP generated in the mitochondria is exported to all the parts of the cell to provide energy for cellular work, such as motility, active transport of nutrients and certain energy requiring metabolic reactions. Mitochondria are often termed *power houses of the cell.*
- Their number and size may reflect the need for energy and the nature of metabolic activity occurring in the tissue. For example, in cardiac muscle, where metabolism is essentially aerobic, the mitochondria are large and numerous, whereas in skeletal muscle which can function anaerobically as well, the mitochondria are smaller and fewer.
- The components of electron transport chain (ETC) and oxidative phosphorylation concerned with ATP production are buried in the inner mitochondrial membrane.
- The matrix contains several enzymes concerned with catabolism of carbohydrates, lipids and amino acids. The matrix enzymes also participate in the synthesis of heme and urea.
- Mitochondria also contain a certain amount of DNA and ribosomes and are capable of synthesizing some of its own proteins.

Eukaryotic cells possess an extensive network of internal membranes pervading the cytoplasm. They are termed the cytomembranes and are usually subdivided into endoplasmic reticulum and Golgi apparatus.

## Endoplasmic Reticulum

Consists of a complicated array of vesicular spaces separated from the cytosol by a system of membrane (7nm). The endoplasmic reticulum is divided into rough and smooth types.

### *Rough Endoplasmic Reticulum (RER)*

Is lined on the cytoplasmic surface with multitude particles called **ribosomes.** These are responsible for the rough appearance of this type of endoplasmic reticulum under the electron microscope. Cells that produce proteins for secretion or export, e.g., pancreatic acinar cells and intestinal goblet cells, are rich in membrane bound ribosomes.

*Function:* The rough endoplasmic reticulum is the site of protein synthesis.

### *Smooth Endoplasmic Reticulum (SER)*

Is in continuation with the rough endoplasmic reticulum. It is distinguishable from the rough by the absence of ribosomes.

*Function:*

- Mainly concerned with the synthesis of lipids.
- It is also the location of glycosyl transferases involved in the synthesis of glycoproteins and glycolipids.
- It is responsible for the detoxification of xenobiotics.

## Golgi Apparatus

Appear as flattened sacs or vesicles and are continuous with the endoplasmic reticulum. They function in modifying and sorting out the proteins.

- The newly synthesized proteins are transported and temporarily stored in the Golgi apparatus (e.g., zymogen granules found in the pancreatic acinar cells prior to the release of proteins from the cells).
- Secretory proteins as well as proteins of lysosomes and plasma membrane are synthesized on the membrane bound polyribosomes. They are then transported to the Golgi apparatus where they are suitably modified and sent to appropriate destinations.
- Secretory proteins are enveloped in vesicles. Secretion involves the fusion of the vesicles with the plasma membrane followed by discharge of the contents into the extracellular space. This process is called *exocytosis.*
- In doing so they must successively pass through each of the following cytomembrane systems.

RER → SER → Golgi Apparatus → Vesicles → ECF

*Function:* Primarily concerned with the modification (sorting) and export of proteins.

## Lysosomes

Are spherical vesicles enveloped by a single membrane (0.4–0.9 µm in diameter). They contain a variety of hydrolytic degradation enzymes and are regarded as the digestive tract of the cell. The enzymes attack the various substrates and are as follows:

| | | |
|---|---|---|
| Glycosidase | - | Glycogen |
| Arylsulfatase | - | Glycolipids |
| Lipase | - | Lipids |
| Phospholipase | - | Phospholipids |
| Cathepsins | - | Proteins |
| Phosphatase | - | Phosphorylated substrate |

- The lysosomes *(primary lysosomes)* fuse with membrane bound vesicles containing the substrates to be partially or wholly digested. The fused lysosome is called *secondary lysosome.*
- Digestive products such as amino acids leave the secondary lysosome and enter the cytosol. When undigested or indigestible substances accumulate they are called *residual bodies,* e.g. the cells of the proximal tubule.

- The pH of the lysosomal matrix is more acidic (pH < 5) than the cytosol (pH > 7) and this facilitates the degradation of different compounds.
- The digestive enzymes of the cell are confined to the lysosomes in the best interest of the cell. Escape of these enzymes into the cytosol will destroy the functional macromolecules of the cell and result in many complications.
- Occurrence of diseases such as arthritis and some of the muscular disorders have been partly attributed to the release of lysosomal enzymes.

*Function:* The lysosomal enzymes are responsible for maintaining the cellular compounds in a dynamic state by their degradation and recycling.

### Peroxisomes

Are also known as microbodies and are spherical with a diameter of 0.8–1.5 μm. They are distinguishable by their content of *catalase,* an enzyme that degrades hydrogen peroxide ($H_2O_2$).

*Function:*

- Catalase protects the cell from the toxic effects of $H_2O_2$ by converting it to $H_2O$ and $O_2$.

$$2\,H_2O_2 \xrightarrow{\textit{Catalase}} 2\,H_2O + O_2$$

- Peroxisomes are also involved in the oxidation of long chain fatty acids ( $>C_{18}$ ) and synthesis of plasmalogens and glycolipids.

### Cytosol

The extranuclear cell content that possesses both soluble and insoluble constituents is termed as cytosol. The cytosol makes up 70-75% of the cellular weight. Many proteins, enzymes, metabolic intermediates and inorganic salts are found in this gel like medium.

*Function:*

- Important metabolic processes such as glycolysis, gluconeogenesis, HMP shunt pathway and fatty acid synthesis occur in the cytosol.
- Recent studies indicate that cytoplasm contains a complex network of protein filaments spread throughout and constitutes the cytoskeleton. The cytoplasmic filaments are of three types, *viz*
  - — Microfilaments
  - — Actin filaments
  - — Intermediate filaments.

The filaments are polymers of proteins which are responsible for the structure, shape, and intracellular movement of organelles and also provide support to the cell.

# 2 Carbohydrates

Carbohydrates are also referred to as saccharides (*saccharon* in Greek means sugar). Being one of the three bulk elements of our food, they act as main source of energy for the body. They perform several functions in the body, mainly in energy metabolism and as structural components. They are primarily made up of carbon, hydrogen and oxygen.

## BIOMEDICAL IMPORTANCE

- Glucose is the *major metabolic fuel* of the body and is a precursor for the synthesis of all carbohydrates in the body.
- The most important and common disorder associated with carbohydrate metabolism is *diabetes mellitus*, leading to hypertension, nephropathy, neuropathy and blindness.
- *Lactose intolerance* which is very common among the elderly orientals occurs due to deficiency of lactase.
- Deficiency of enzymes of glycolysis may lead to *haemolytic anaemia or fatigue* (in which muscle is affected).
- *Lactic acidosis*, accumulation of end product of anaerobic glycolysis, may result due to several causes.
- Genetic deficiency of the enzymes of glycogen metabolism, may lead to *glycogen storage diseases*, resulting in poor mobilization and deposition of abnormal forms of glycogen in the liver or muscle.

## CHEMISTRY

### Definition

Carbohydrates may be defined as *polyhydroxy aldehydes or ketones* or compounds which yield them on hydrolysis.

### Functions of Carbohydrates

1. *In energy metabolism*
   a. Most abundant dietary source of energy of the body. Dietary carbohydrates are digested to monosaccharides, mainly glucose, which provides more than half of the energy required by the body.
   b. Serve as storage forms of energy, e.g., glycogen in animals and starch in plants.
2. *As structural components*
   a. As glycoproteins and glycolipids carbohydrates participate in the structure of cell membrane and other cellular functions, such as adhesion and recognition of molecules.
   b. Matrix of connective tissue is made up of glycosaminoglycans, which are heteropolysaccharides.
   c. Fibres of plants are made up of cellulose.
   d. Plant gums, corn cobs, seed husks are all made up of carbohydrates.

e. Bacterial cell wall, shells of crustaceans are made up of carbohydrates.

### Classification

Carbohydrates are broadly classified into 3 groups:

- *Monosaccharides,* which cannot be hydrolyzed into simpler forms.
- *Oligosaccharides,* which on hydrolysis yield two to ten monosaccharide units.
- *Polysaccharides,* yield more than ten monosaccharide units on hydrolysis.

This classification is based on the number of monosaccharide units. The mono and oligosaccharides are sweet to taste, crystalline in character and soluble in water. Hence they are commonly called as *sugars.*

Monosaccharides are called *"simple sugars"* or *"monosaccharoses"* or *"monoses,"* e.g., glucose and fructose.

General formula $C_n(H_2O)_n$.

Monosaccharides can be divided into different categories based on:

i. their functional group or
ii. the number of carbon atoms present.

#### *Based on the Functional Group (Fig. 2.1)*

a. *Aldoses*: When the functional group in monosaccharides is an aldehyde, they are known as *"aldoses"*, e.g., glyceraldehyde, glucose.
b. *Ketoses*: When the functional group is a keto group, they are referred to as *ketoses,* e.g., dihydroxyacetone, fructose.

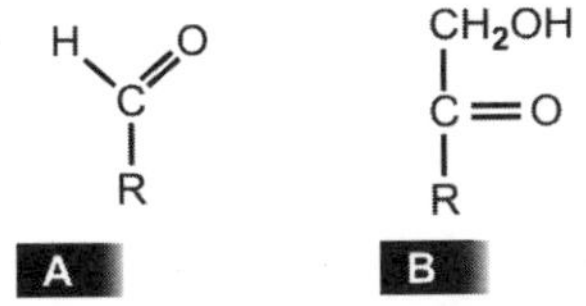


**Figs 2.1:** General structure of (A) Aldose (B) Ketose

#### *Based on the Number of Carbon Atoms (Table 2.1)*

- Comprise both aldoses and ketoses.

### ISOMERISM

Carbohydrates exhibit two main types of isomerisms.

- Functional isomerism
- Stereoisomerism.

### Functional Isomerism

Occurs due to the presence of functional groups, i.e., aldose or ketose group.

Aldoses are those which possess an aldehyde group, e.g., glucose and ketoses are those which possess a keto group, e.g., fructose (Refer page 9 and 10).

### Stereoisomerism

This type of isomerism occurs due to difference in spatial arrangement of constituents in compounds having the same structural formula.

Many of the carbohydrates contain the same number of atoms and the same kind of groups, yet are definitely different substances. For example, the formula $C_6H_{12}O_6$ represents 16 different simple sugars all possessing the structure $CH_2OH$-CHOH-CHOH-CHOH-CHOH-CHO. This is due to different arrangements of the constituent groups of the molecule in space. The phenomenon represents what is called *"stereoisomerism"* and these sugars are referred to as *"stereoisomers."*

There are two types of stereoisomerisms.

- Optical isomerism
- Geometric or cis-trans isomerism: Occurs due to difference in arrangement of groups around double bond (dealt under lipids).

**Table 2.1:** Classification of monosaccharides with examples

| *Subclass* | *Example* | *Importance* |
|---|---|---|
| *Trioses: 3 carbon sugars* | | |
| Aldotriose | D-Glyceraldehyde 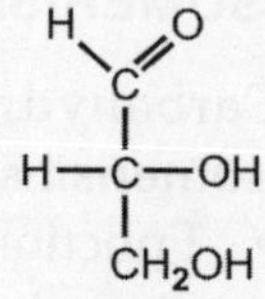 | Glyceraldehyde-3-phosphate is an intermediate of glycolysis |
| Ketotriose | Dihydroxyacetone 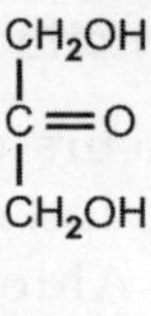 | As dihydroxyacetone-1-phosphate is an intermediate of glycolysis |
| *Tetroses: four carbon sugars* | | |
| Aldotetrose | D-Erythrose 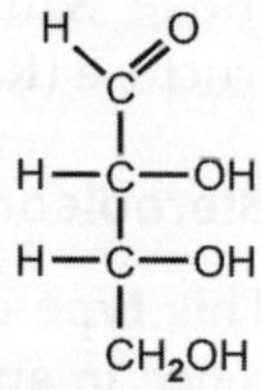 | As erythrose-4-phosphate is an intermediate of the HMP shunt pathway |
| *Pentoses: five carbon sugars* | | |
| Aldopentose | D-Ribose 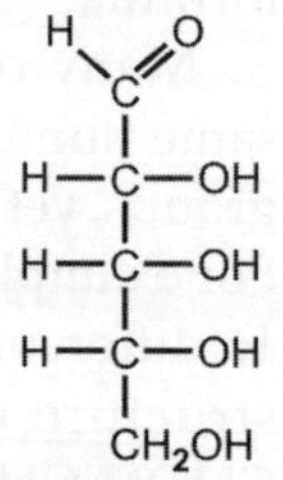 | Ribose-5-phosphate is an intermediate of HMP shunt pathway. It is also a constituent of RNA and nucleotide coenzymes like ATP, $NAD^+$, $NADP^+$. |
| | D-Xylose 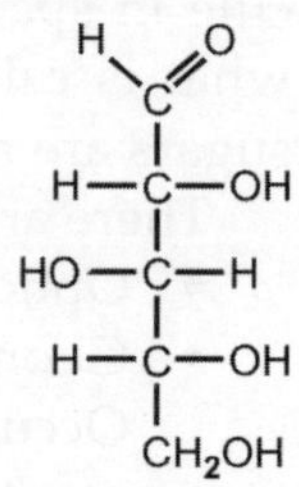 | Constituent of glycoproteins and gums |

*Contd...*

*Table 2.1 contd..*

| *Subclass* | *Example* | *Importance* |
|---|---|---|
| Ketopentose | D- Ribulose 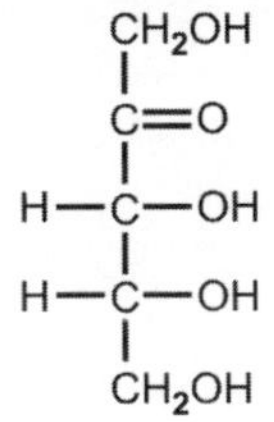 | D-Ribulose-5-phosphate is an intermediate of HMP shunt pathway |
| | D-Xylulose 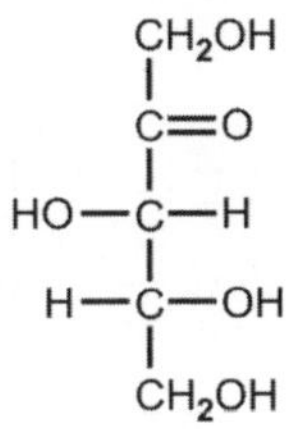 | D-xylulose-5-phosphate is an intermediate of HMP shunt pathway. Both D and L-xylulose-5-phosphate are intermediates of uronic acid pathway |
| *Hexoses: 6 carbon sugars* | | |
| Aldohexoses | D-Glucose 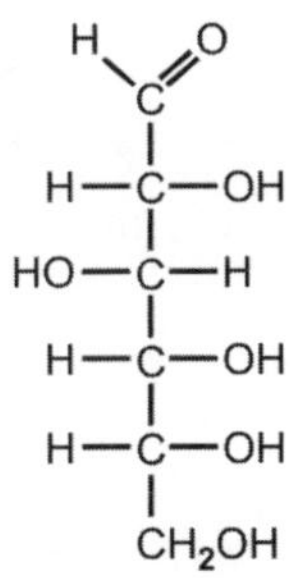 | Most abundant carbohydrate. Glucose is the major sugar present in blood and other body fluids. It is a fuel of life. Glucose is oxidized preferentially by all tissues of the body to provide energy. Ordinarily more than half the energy is provided by the oxidation of glucose. It is an important constituent of disaccharides-maltose, lactose and sucrose; polysaccharides-starch, glycogen and cellulose. |
| | D-Galactose 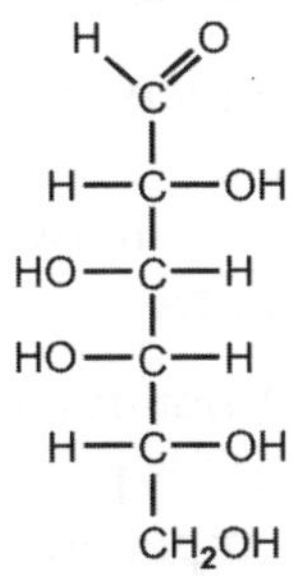 | Constituent of milk sugar lactose. Also is a constituent of glycoproteins and glycolipids |

Contd...

*Table 2.1 contd..*

| *Subclass* | *Example* | *Importance* |
|---|---|---|
| | D-Mannose 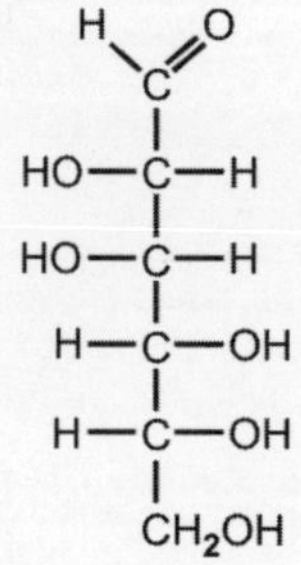 | Constituent of glycoproteins |
| Ketohexoses | D-Fructose 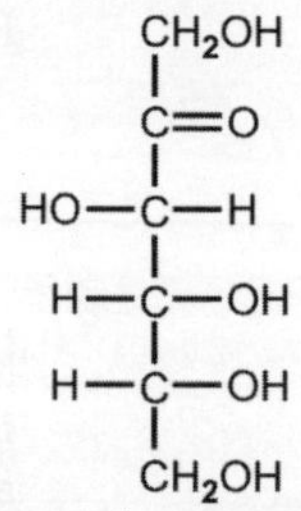 | Present in fruits and honey. It is a part of disaccharide sucrose and polysaccharide inulin. Its 6-phosphate derivative is an intermediate of glycolysis and HMP shunt pathway |
| *Heptoses: 7 carbon sugars* | | |
| Ketoheptose | D-Sedoheptulose 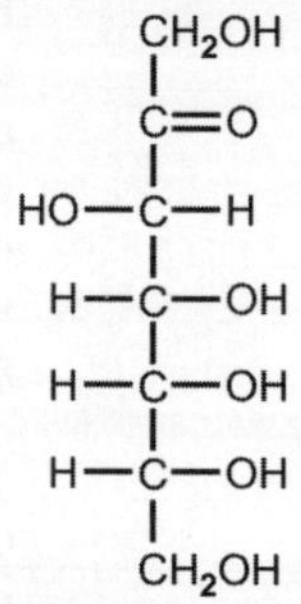 | Sedoheptulose-7-phosphate is an intermediate of HMP shunt pathway. |

### Optical Isomerism

The carbohydrates exhibit the property of optical activity and exist as optical isomers. The optical isomerism is due to the presence of an "*asymmetric carbon*" atom.

An asymmetric carbon atom is one to which four different constituent groups are attached. The number of asymmetric carbon atoms (n) determines the possible number of isomers of a given compound, which is equal to $2^n$.

Glucose which has four asymmetric 'C' atoms has 16 isomers.

*Optical activity*: When a beam of plane polarized light is passed through a solution containing a compound with an asymmetric carbon atom, the beam is rotated towards the right or the left. If it is rotated towards the right, the compound is dextrorotatory and is designated as '*d*' or '+'. If rotated towards the left, it is levorotatory designated '*l*' or '–'.

### D and L isomers

Based on the configuration of groups around the asymmetric carbon atom next to the primary alcoholic group, termed the *penultimate* carbon atom, the monosaccharides are termed either D or L sugars.

Glyceraldehyde is the simplest monosaccharide with an asymmetric carbon atom. It exists as two isomers.

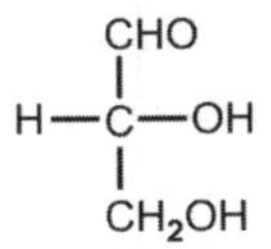

D-glyceraldehyde

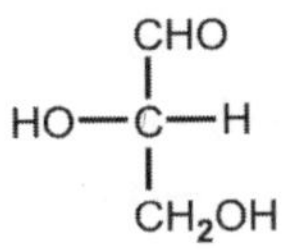

L-glyceraldehyde

If the –OH group is on the right side on the penultimate carbon atom, the sugar belongs to *D* series. If it is on the left side, it belongs to the '*L*' series.

All D-aldose sugars are considered as derivatives of 'D'-glyceraldehyde and L-aldoses derivatives of L-glyceraldehyde. There will be two tetroses derived from D-glyceraldehyde and four pentoses—two each derived from D-erythrose and D-threose and eight hexoses. Similarly, there are two tetroses, four pentoses and eight hexoses derived from the L-series (Fig. 2.2).

D and L refer to only configuration and not to the nature of optical activity. So an optical isomer may be designated D(+), D(-), L(+) or L(-) based on its structural relation to glyceraldehyde and optical activity.

### Enantiomers

Are stereoisomers in which one isomer is the mirror image of the other and require the presence of an asymmetric carbon atom. The D and L monosaccharides of the same name are *"enantiomers"* (Fig. 2.3).

If a solution contains a mixture of D and L forms of a sugar (enantiomers), it is called *racemic mixture*. All the sugars in the body exist in D forms. The enzymes present in the body can act only on D sugars.

### Diastereoisomers

Diastereoisomers are stereoisomers which are not mirror images of each other. For example, D-erythrose and D-threose, D-ribose, D-xylose and D-arabinose (Fig. 2.2).

### Epimers

Epimers are stereoisomers which differ in configuration of groups around a single carbon other than the first (Fig. 2.4).

For example, glucose and mannose — $C_2$ epimers.

Glucose and galactose — $C_4$ epimers.

### Anomerism

*Hemiacetal formation*:

- Lower sugars exist as straight chain forms, whereas, pentoses and hexoses exist in the ring form.
- The hydroxyl groups of a monosaccharide can react with its own aldehyde or keto functional group to form *"hemiacetal"* or *"hemiketal."*
- The aldehyde group of D-glucose at $C_1$ reacts with the alcohol group at $C_5$ to form two types of cyclic hemiacetals: the α and the β forms (Fig. 2.5).
- The first carbon atom is called the *anomeric carbon atom.*
- These two isomers which arise as a result of the difference in arrangement of groups around the first carbon atom are called *anomers.*

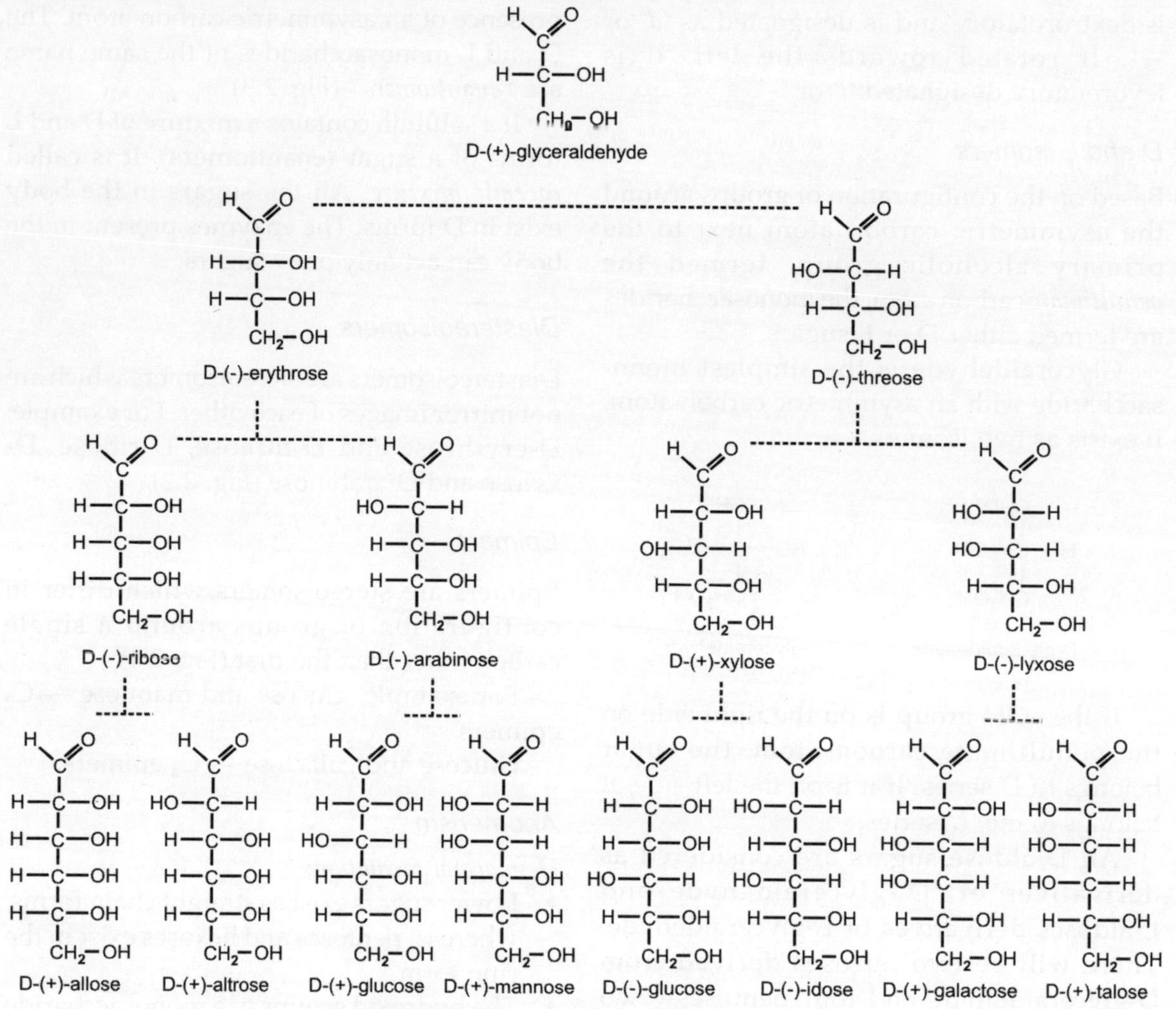


**Fig. 2.2:** Configuration of the aldoses

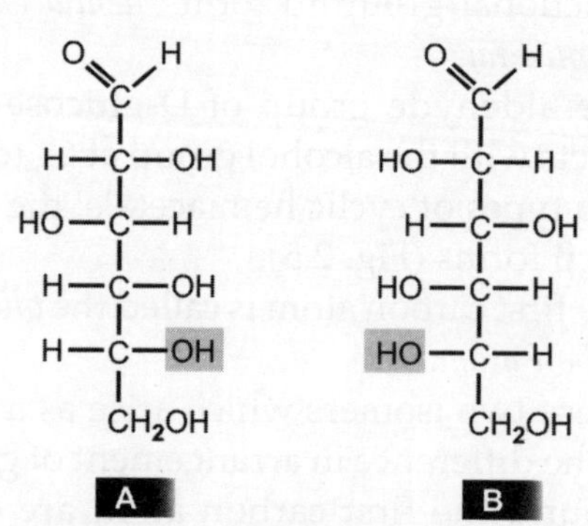


**Fig. 2.3:** Example for enantiomers (A) D-glucose (B) L-glucose

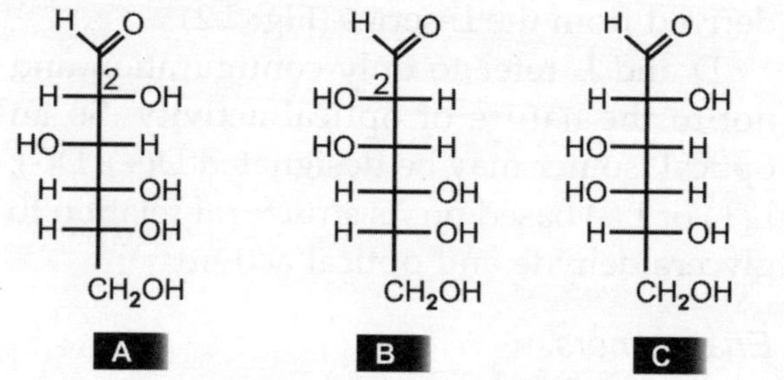


**Fig. 2.4:** Examples for epimers (A) D-glucose (B) D-mannose (C) D-galactose

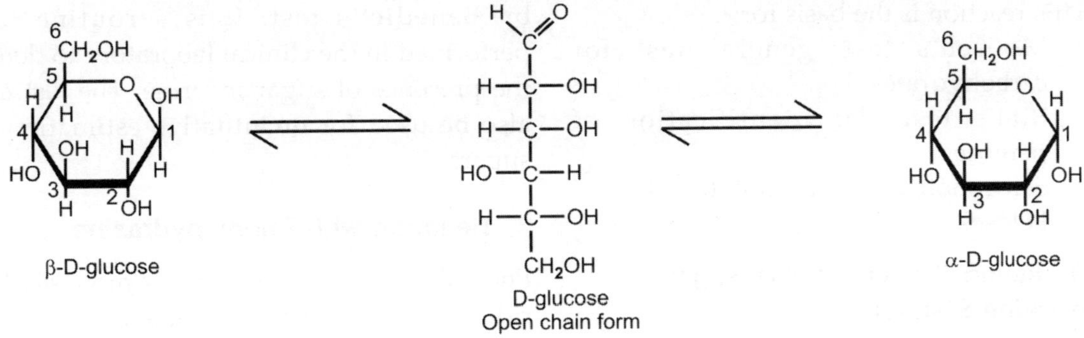


**Fig. 2.5:** Anomers of glucose

- Anomers are stereoisomers formed due to hemiacetal formation by aldehyde or ketone group.
- Haworth's projection formula is a better representation of the spatial relationship of the different constituents in a ring structure (Fig. 2.6).
- The sugars may exist as furanose or as pyranose ring structures. Pyranose form is the most stable form for glucose, whereas, furanose form is the most stable form for fructose.

### *Mutarotation*

The α and β anomers have different optical rotation.

α D-glucose — +112.2°
β D-glucose — +18.7°

When α-D-glucose solution or β-D-glucose solution is prepared in water, the optical rotation gradually changes to +52.7°. This is called *mutarotation.*

*Mutarotation* is defined as the change in specific optical rotation representing the interconversion of α and β forms of a sugar to an equilibrium mixture.

## IMPORTANT PROPERTIES

### 1. Action of Acids

Sugars are dehydrated in the presence of strong acids to give furfural or furfural derivatives which condense with phenols to give coloured compounds. The polysaccharides and oligosaccharides get hydrolyzed to their constituent monosaccharides and give the reaction.

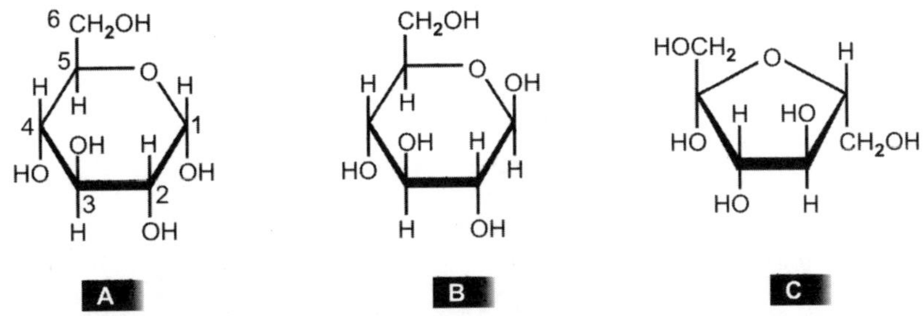


**Fig. 2.6:** Structures of (A) α-D-glucopyranose (B) β-D-glucopyranose and (C) β-D-fructofuranose

This reaction is the basis for:

a. Molisch's test—general test for carbohydrates.
b. Bial's test—for identification of pentoses.
c. Seliwanoff's test—for identification of ketoses.

## 2. Reducing Property of the Sugars in Alkaline Solution

Monosaccharides, both aldoses and ketoses, and polysaccharides containing a free functional group, tautomerize and form *enediol* in alkaline solution. These enediols are very unstable and reactive. The enediols break at the double bond liberating an aldehyde and other products (shorter chain acids). These compounds are *"reducing agents."* They reduce cupric ions to cuprous ions in solution.

This property is the basis of determining the presence of reducing sugars in a solution by Benedict's test. It is a routine test performed in the clinical laboratory to detect the presence of sugar in urine. The test can also be used for quantitative estimation of sugars.

## 3. Reaction with Phenylhydrazine

Phenylhydrazine reacts with the monosaccharides and other carbohydrates containing a free functional group to form hydrazones and osazones.

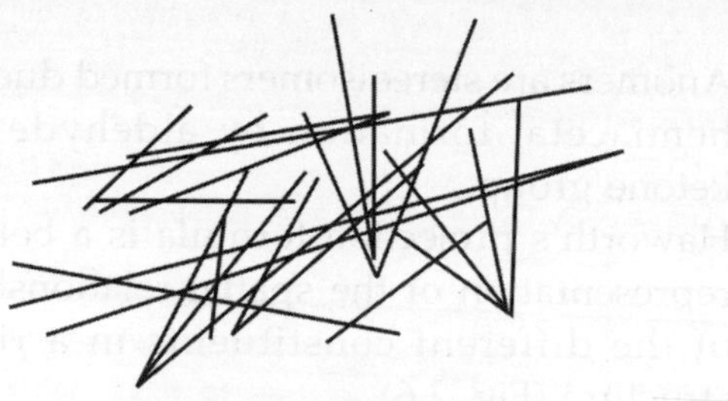

Needle shaped osazones

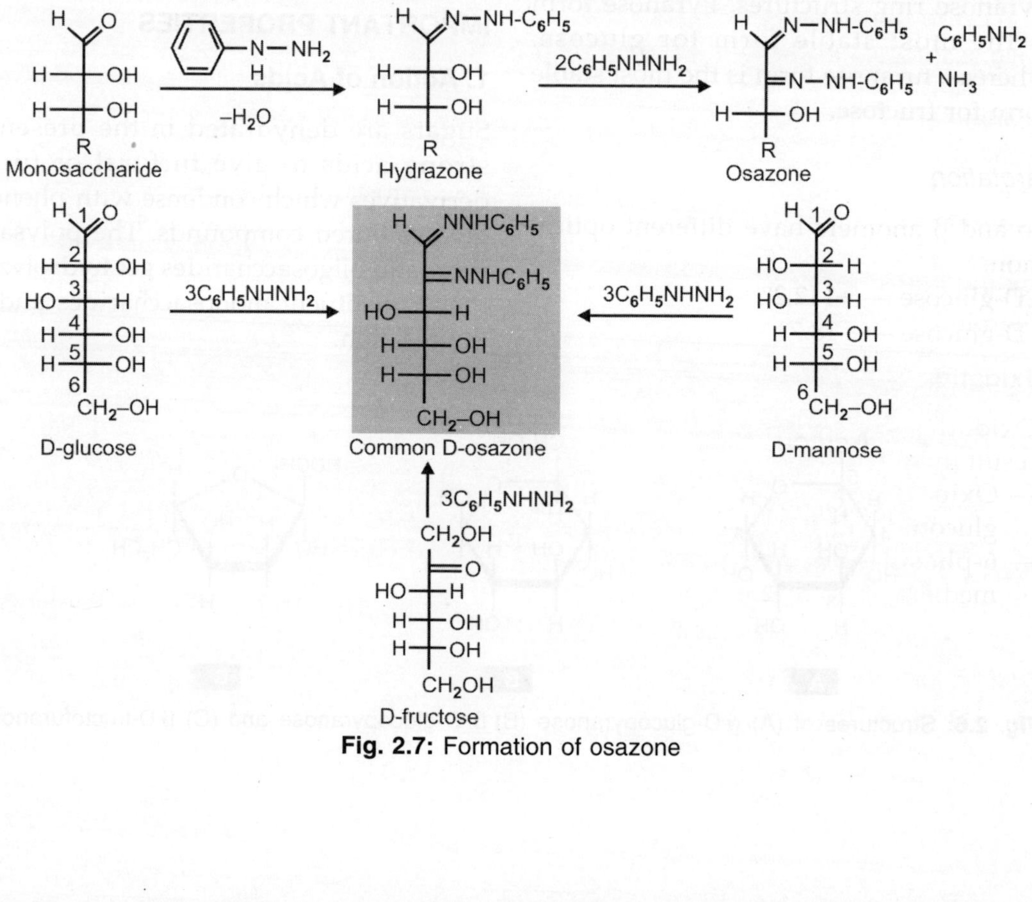


**Fig. 2.7:** Formation of osazone

The osazones are relatively insoluble and crystalline in beautiful characteristic forms for different sugars. Hence used in identification of sugars. The osazone formed from glucose, fructose and mannose are identical, i.e., needle shaped because these three sugars differ in their structure only in the first two carbon atoms (Fig. 2.7). These carbon atoms are involved in formation of osazones.

### 4. Reduction

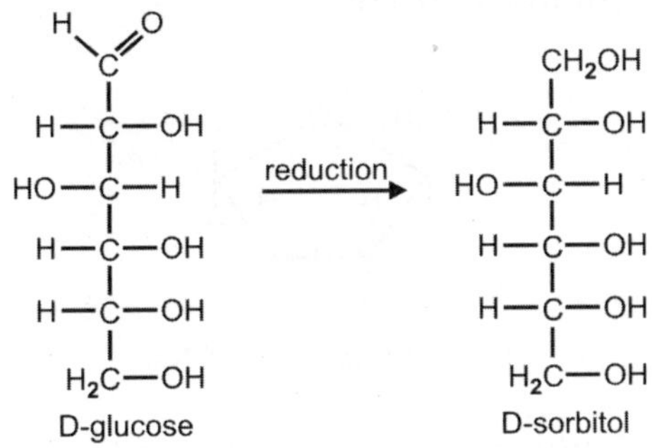


**Fig. 2.8:** Reduction of glucose to sorbitol

The aldehyde group of aldoses is reduced to give the corresponding alcohol.

Accumulation of sorbitol formed from glucose is responsible for diabetic cataract (Fig. 2.8).

Ribitol formed from ribose is a component of vitamin $B_2$, riboflavin.

### 5. Oxidation

- Oxidation of the aldehyde group will result in the formation *of aldonic* acid.
  — Oxidation of glucose gives rise to gluconic acid.
  — 6-phosphogluconic acid is an intermediate of HMP shunt pathway.
- Oxidation of only the primary alcoholic group will result in the formation of *"uronic acids."*
  — D-glucuronic acid is formed from glucose.
  — L-diuronic acid, an epimer of D-glucuronic acid, is a component of glycosaminoglycans.

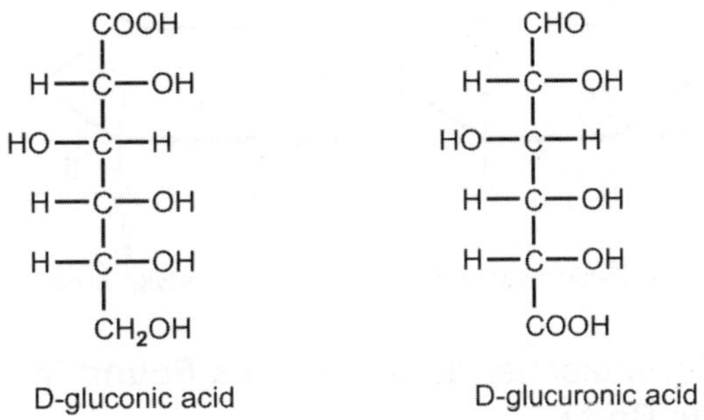


### 6. Ester Formation

The hydroxyl groups can react with acids like phosphoric, sulfuric and acetic acid to form esters. Phosphorylation of glucose is an obligatory step in its utilization. Phosphorylation

i. activates the sugar.
ii. prevents the diffusion of the sugar out of the cell; because membranes are impermeable to sugar phosphates. Intermediate of glycolysis and HMP shunt pathway are all phosphorylated.

Sulfated sugars are formed as components of glycosaminoglycans, glycolipids.

### 7. Formation of Amino Sugars

Amino groups may be substituted for various hydroxyl groups of the sugar to give *amino sugars*. Most common natural amines are D-glucosamine (2-amino, 2-deoxy, D-glucose) and or D-galactosamine (2-amino, 2-deoxy, D-galactose).

Both exist in ring form similar to glucose and galactose.

The amino sugar may be acetylated at the amino groups to give rise to N-acetyl glucosamine or N-acetyl galactosamine.

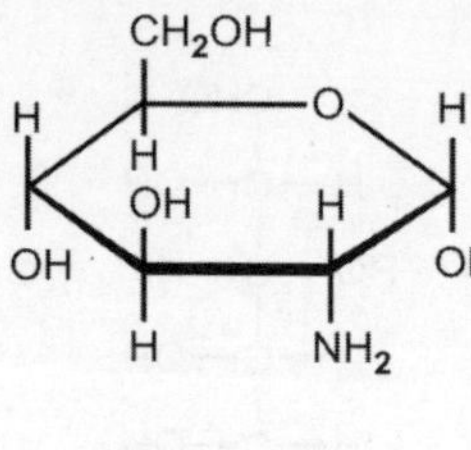


α-D-glucosamine

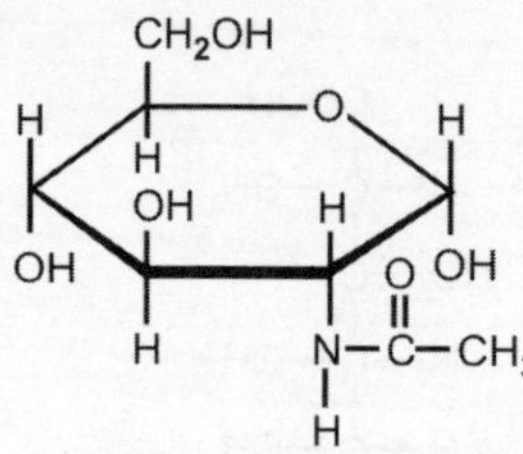


α-D-N-acetylglucosamine

### Monosaccharide Derivatives Found in the Body

Several derivatives can be formed from monosaccharides, which serve different functions in the body.

#### 1. Uronic Acids

i. *D-glucuronic acid,* formed from glucose by oxidation of primary alcohol group is:
   - An intermediate of uronic acid pathway.
   - Is a component of glycosamino-glycans.
   - Used in detoxification reactions.

ii. *L-iduronic acid,* 5-epimer of glucuronic acid is a constituent of glycosamino-glycans.

#### 2. Amino Sugars

Formed by substitution of various hydroxyl groups of the sugar by amino groups.

*i. D-glucosamine*

*ii. D-galactosamine*

Both are found as constituents of glyco-saminoglycans and glycoproteins.

#### 3. Acetylated Amino Sugars

The amino groups of the amino sugar are acetylated.

*i. N-acetyl D-glucosamine*

*ii. N-acetyl D-galactosamine*

Both occur as constituents of glycosamino-glycans and glycoproteins.

#### 4. Deoxy Sugars

One of the hydroxyl groups is replaced by a hydrogen atom.

*i. 2-deoxy D-ribose:*

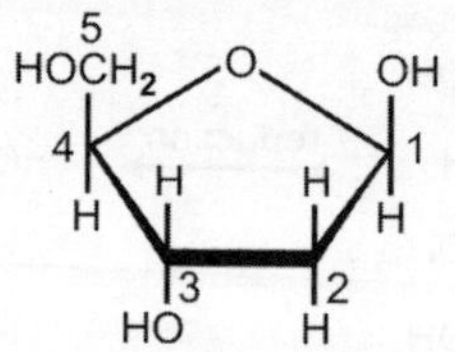


Component of DNA

*ii. L-fucose:* 6-deoxy L-galactose

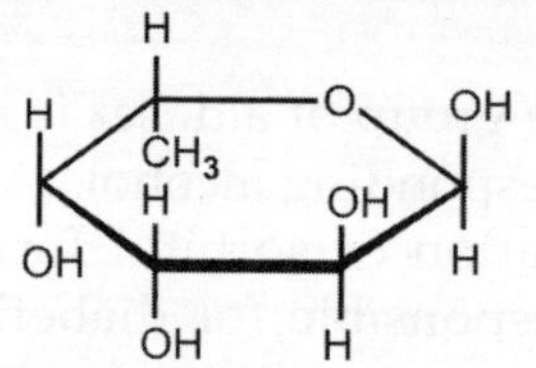


Present in glycoproteins and glycolipids.

#### 5. Sugar Alcohols

i. *Sorbitol* is formed from both D-glucose and D-fructose during conversion of fructose to glucose. Increased concentration observed during diabetes mellitus is responsible for causing cataract.

ii. *Mannitol* is formed from D-mannose. It is used clinically as a diuretic.

iii. *Galactitol* is formed from D-galactose. Increased concentrations are found in cataract.

*iv. Ribitol* is formed from D-ribose. It is a component of riboflavin, vitamin $B_2$.

*v. Glycerol* is formed from D-glyceraldehyde and dihydroxy acetone, it is the alcohol in triacyl glycerols.

## 6. Sialic Acids

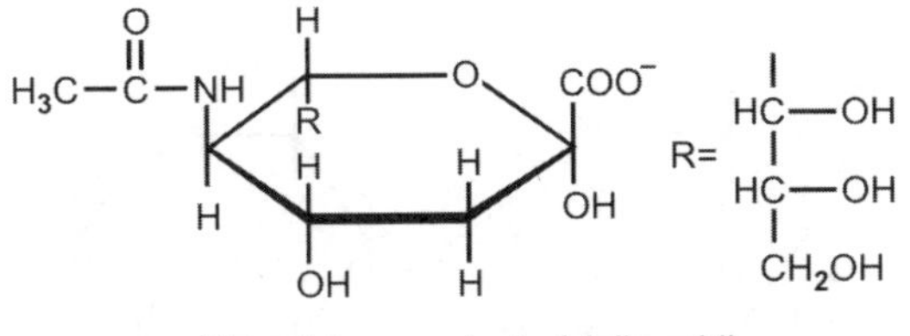


N-acetylneuraminate (sialic acid)

- *N-acetyl neuraminic acid (NANA)* is a derivative of N-acetyl mannosamine and pyruvic acid.
- It is an important component of glycoproteins and glycolipids.

## GLYCOSIDES

Glycosides are formed when the hydroxyl group (hemiacetal) of anomeric carbon atom of a monosaccharide reacts with a hydroxyl group of another carbohydrate or hydroxyl/amino group of a non-carbohydrate i.e., a protein, purine, pyrimidine, bilirubin, steroid or lipid. The bond so formed is known as a *glycosidic bond.*

The monosaccharides are held together by glycosidic bonds to result in the formation of *"oligo"* and *"polysaccharides."* The nomenclature of glycosidic bonds is based on the linkage between the carbon atoms and the status of the anomeric carbon (α or β). For example, maltose is formed by linkage between the first carbon of one α-D-glucose unit with the fourth carbon of another α-D-glucose molecule. The linkage in maltose is termed α-1, 4 glycosidic linkage.

The non-carbohydrate moiety (when present) is referred to as an *aglycone.* If the sugar is linked to the aglycone through its hydroxyl group, then the bond is termed an *O-glycosidic bond* and if it is linked through an amino group, then the bond is termed as *N-glycosidic bond.*

### Physiologically Important Glycosides

1. *Nucleotides*: Components of nucleic acids.
2. *Glycoproteins*: Components of receptors, antibodies.
3. *Glycolipids:* Components of nervous tissue.
4. *Cardiac glycosides* (steroidal glycosides): Digoxin and digitoxin have steroid as aglycone. Stimulate muscle contraction. Components of digitoxin are four galactose residues, xylose and digitonin which is a steroid.
5. *Streptomycin,* an antibiotic, is used in treatment of tuberculosis.

It is an aminoglycoside consisting of three sugar residues linked to amino groups.

## OLIGOSACCHARIDES

- Composed of 2-10 monosaccharide units.
- Most important oligosaccharides are the disaccharides maltose, lactose, sucrose, isomaltose and cellobiose.

### Maltose

- Maltose is made up of two glucose units held together by α-1, 4-glycosidic bond (Fig. 2.9). Generally not found in the free form in the body. It is formed during the course of digestion of starch by α-amylase.
- It is a reducing sugar because the aldehyde group present on the $C_1$ of the second glucose residue is free. Maltose can be hydrolyzed by dilute acid or the enzyme maltase to liberate two molecules of α-D-glucose.

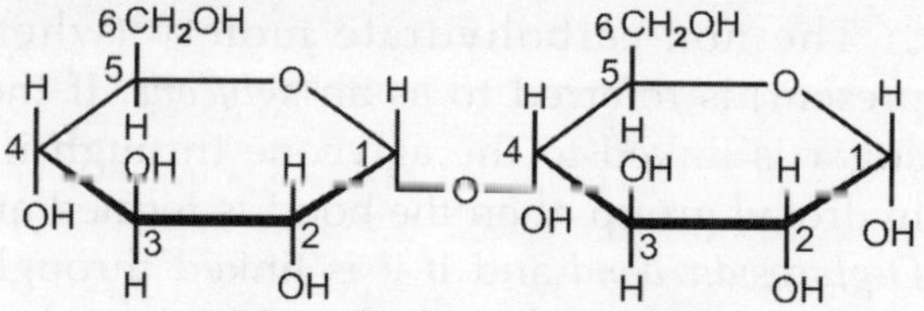

**Fig. 2.9:** Structure of maltose
(α-D-glucosyl 1,4 α-D-glucose)

- Gives sunflower shaped osazones.

Maltosazone

**Isomaltose**

- Has α-1, 6-glycosidic bond between two glucose units (Fig. 2.10).
- Formed during the course of digestion of starch by α-amylase.
- Also a reducing sugar.
- Digested by the enzyme oligo, 1. 6-glucosidase present in intestinal juice.

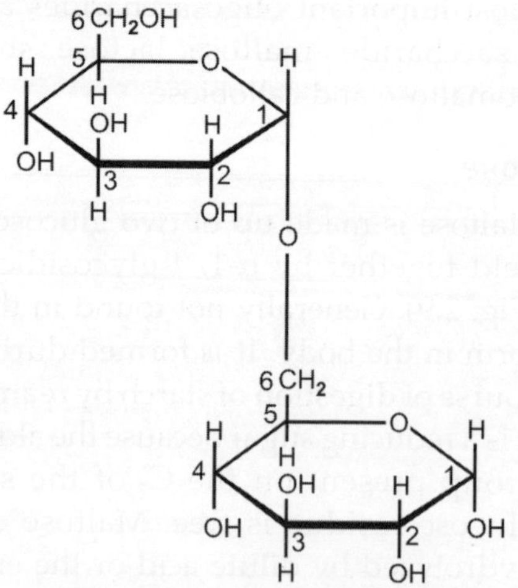

**Fig. 2.10:** Structure of isomaltose
(α-D-glucosyl 1,6 α-D-glucose)

**Lactose**

- Composed of a molecule of galactose and a molecule of glucose linked by β-1,4 glycosidic linkage (Fig. 2.11).

Lactosazone

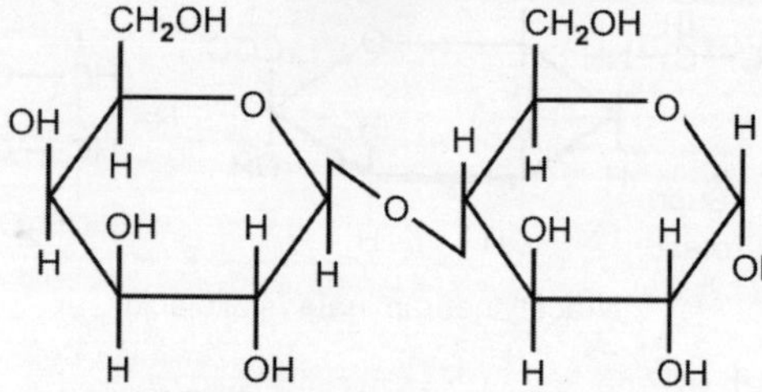

**Fig. 2.11:** Structure of lactose
(β-D-galactosyl 1,4 α-D-glucose)

- Commonly known as the milk sugar.
- The anomeric carbon of glucose residue is free, so it is a reducing sugar.
- Hydrolyzed by lactase to glucose and galactose.
- Gives dense badminton ball like osazones.

**Sucrose**

Sucrose is made up of a molecule of α-D-glucose and a molecule of β-D-fructose held together by (α β - 1, 2) glycosidic linkage between the $C_1$ of α-glucose and $C_2$ of β-fructose (Fig. 2.12).

- Is the sweetening agent found in sugar.
- Present in cane sugar, sugarbeets and also in fruits.

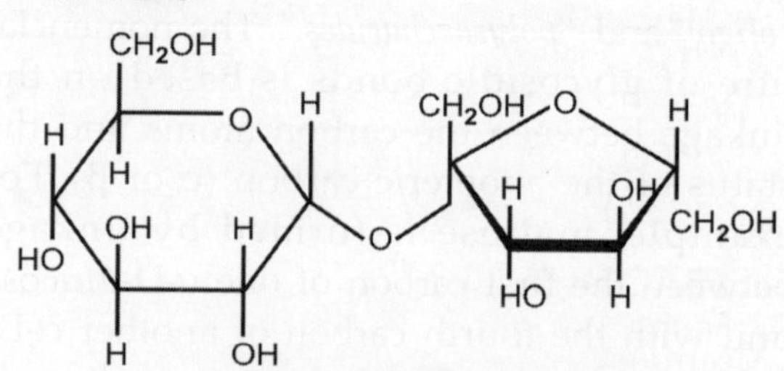

**Fig. 2.12:** Structure of sucrose
(α-D-glucosyl-1,2 β-D-fructose)

- Since reducing groups of both glucose and fructose are involved in glycosidic bond formation, sucrose is a non-reducing sugar.
- Sucrose may be hydrolyzed by dilute acids or the enzyme sucrase.
- Hydrolysis of sucrose which is dextrorotatory (+66.5°) produces one molecule of glucose (+52.5°) and one molecule of levorotatory fructose (–92°).

$$\text{Sucrose} \xrightarrow{H_2O} \text{Glucose} + \text{Fructose}$$

Optical rotation+66.5° + 52.5° – 92°
(net –28.2°)

- Net effect is that the solution containing the hydrolyzed product becomes levorotatory (–28.2°). So hydrolyzed mixture of sucrose is called *"invert sugar"* and the enzyme bringing about hydrolysis of sucrose, i.e., sucrase is called *invertase.*

### Cellobiose (Fig. 2.13)

- Composed of two molecules of glucose in β-1,4-glycosidic linkage.
- Is produced as a product of incomplete hydrolysis of cellulose, the plant polysaccharide.

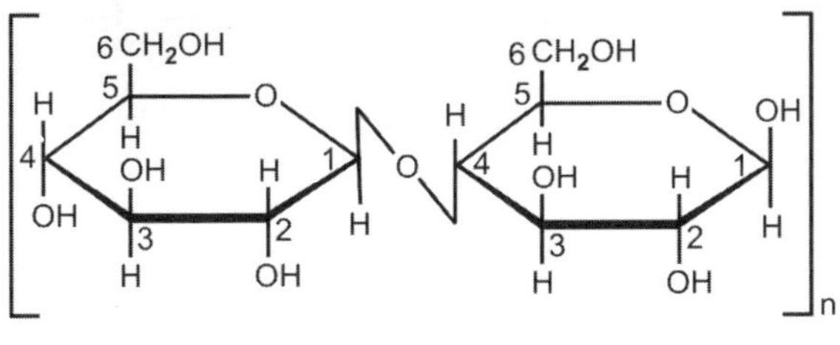


**Fig. 2.13:** Structure of cellobiose (β-D-glucosyl-1,4 β-D-glucose)

## POLYSACCHARIDES

- Composed of more than 10 molecules of monosaccharides or their derivatives held together by glycosidic bonds.
- May be linear or branched.
- Primarily concerned with two important functions i.e.
  — storage of energy
  — as structural components.

### Classification

There are two types of polysaccharides
- *Homopolysaccharides* and
- *Heteropolysaccharides.*

*Homopolysaccharides*: These are made up of similar monosaccharide units. When they are made up of only glucose units, they are termed *glucans*, e.g., starch, glycogen, cellulose. When they are made up of only fructose units, they are termed as *fructosans*, e.g., inulin.

*Heteropolysaccharides*: These are made up of more than one type of monosaccharide units or their derivatives. So, on hydrolysis, they yield a mixture of a few monosaccharides and their derivatives, e.g., glycosaminoglycans.

### Homopolysaccharides

#### *Starch*

- Storage carbohydrate in plants.
- Important dietary source of energy for higher animals including man.
- Found in cereals, roots, tubers and other food grains.
- Composed of glucose units held together by α-glycosidic bonds.

#### *Structure*

Consists of two types of polysaccharides:
- Water soluble *amylose* (15-20%).
- Water insoluble *amylopectin* (80-85%).

#### *Amylose*

- Amylose is a long, unbranched chain with 400-1000 glucose units held together by α-1, 4-glycosidic linkages (Fig. 2.14).

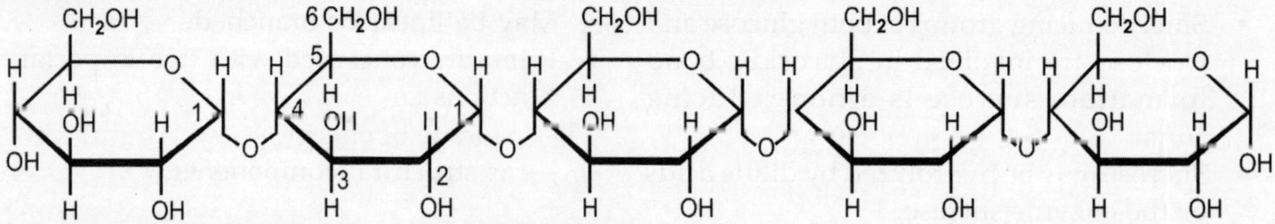

Fig. 2.14: Structure of amylose

- Molecular weight ranges from 5,000-200,000.
- In solution, it is a helically coiled structure.
- Gives blue colour with iodine solution because iodine gets trapped within the coils. On heating, amylose loses its helical structure and the blue colour disappears. On cooling, it regains the helical structure and the blue colour with iodine reappears.

*Amylopectin (Fig. 2.15)*

- Has branched structure. The straight chain portion has α-1, 4 linkages, whereas at the branches the glucose units are linked by α-1, 6 linkages.
- There are about 20-30 glucose units in the outer branches and there is a branch after every 8-9 glucose units in the interior.
- Molecular weight varies from 50,000-1 million.
- Absorbs water and swells to form a paste.
- Gives violet colour with iodine.

*Dextrins*

- Dextrins are breakdown products of starch formed by the action of the enzyme amylase or dilute acids.
- Dextrins are formed by sequential hydrolysis of starch (Table 2.2).

Depending on the number of reducing groups present in solution, the Benedicts test gives blue (no reducing groups) or green, yellow, orange or red (few or large number of reducing groups) colour.

- Dextrins are present in plant tissue during synthesis and degradation of starch.
- May be linear or branched.

*Glycogen*

- Termed as animal starch, because it is the storage polysaccharide in animals.

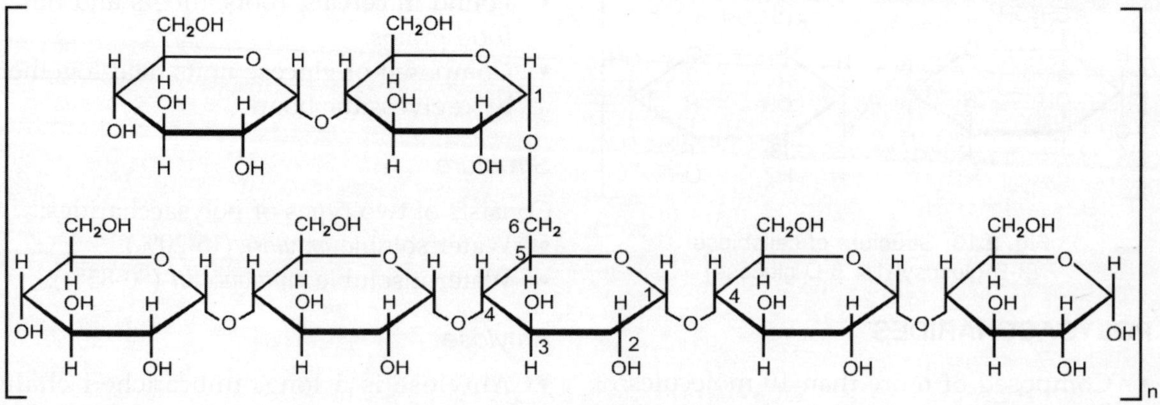

Fig. 2.15: Structure of amylopectin

**Table 2.2:** Sequential hydrolysis of starch

| *Compound* | *Reaction with $I_2$ solution* | *Benedict's test* |
|---|---|---|
| Starch ↓ | blue | Blue (no reduction) |
| Amylodextrin ↓ | violet | Blue (no reduction) |
| Erythrodextrin ↓ | red | Green (slight reduction) |
| Achrodextrin ↓ | no colour | Yellow or orange (greater reduction) |
| Maltose ↓ | no colour | Red (high reduction) |
| Glucose | no colour | Red (high reduction) |

- It is stored in the *liver* and *muscle* in animals. About 5% by weight of liver and 0.5% by weight of muscle is glycogen.
- It is a glucan.
- It has a branched structure resembling the structure of amylopectin of starch. It is more branched than amylopectin. That is, the outer branches have 6-7 glucose residues and there are only 3-4 glucose units between the branching points in the interior. The high degree of branching helps in rapid degradation, providing large amount of glucose when needed for maintaining glucose levels or energy production.

### *Cellulose*

- It is the most abundant carbohydrate.
- Occurs in plant cell wall. Cotton and flax are rich sources containing as much as 96-97% cellulose.
- It is a linear glucan, the glucose units linked by β-1, 4 linkages.
- The mammalian enzymes can act only on the α-1, 4 linkages, hence cannot digest cellulose. Therefore, cellulose provides roughage and adds bulk to stools.

### *Dextran*

- High molecular weight glucans produced by certain microorganisms.
- Glucose units are linked by different types of linkages i.e., α-1, 6; α-1, 4 and α-1, 3.
- Used
  1. clinically as plasma expanders during treatment of shock.
  2. in gel filtration, which is used for separation of compounds in solution based on molecular weight.

### *Inulin*

- Inulin is a fructosan. Consists of fructose units linked by β-1, 2 linkages.
- Reserve carbohydrate in various bulbs and tubers such as dahlia, onion, garlic, etc.
- Clinically used to find the renal clearance, to determine the glomerular filtration rate.

### *Chitin*

- Structural polysaccharide found in the exoskeleton of some invertebrates like crustaceans and insects.
- Composed of N-acetyl, D-glucosamine units held together by β-1,4 glycosidic linkages.

## Heteropolysaccharides

Those polysaccharides which yield more than one type of monosaccharide or derivatives of monosaccharides are heteropolysaccharides. Include the following:

i. *Glycosaminoglycans (GAGS) or mucopolysaccharides* present in the matrix of the connective tissue and
ii. Oligosaccharides of the *glycoproteins.*

## Glycosaminoglycans (GAG)

- Are present in the matrix of connective tissue.

- A GAG consists of unbranched heteropolysaccharide chains made up of repeating disaccharide units.
- The disaccharide units consist of one amino sugar (D-glucosamine or galactosamine or their derivatives) and one uronic acid (α-glucuronic acid or α-Iduronic acid). Many are sulfated.
- They are polyanionic because of the large number of carboxyl and sulfate groups.
- Because of the negative ions they hold high amount of cations and water.
- Highly viscous and elastic due to the large amount of water absorbed. They are mucilagenous, which led to their original naming as *mucopolysaccharides.*
- There are six classes differing in the nature of amino sugar and uronic acid, type of glycosidic linkage and tissue of occurrence (Table 2.3, Fig. 2.16).

## Funotionc

1. Structural components of the extracellular matrix.
2. Lubricants and shock absorbers.
3. May contribute to water—electrolyte balance.
4. Restrict movement of large particles in tissues.
5. Heparin is an anticoagulant.
6. Enzyme activator—heparin activates lipoprotein lipase.
7. Participate in cell—cell interaction and cell adhesion.
8. Have a structural role in sclera.
9. Have a role in corneal transparency.

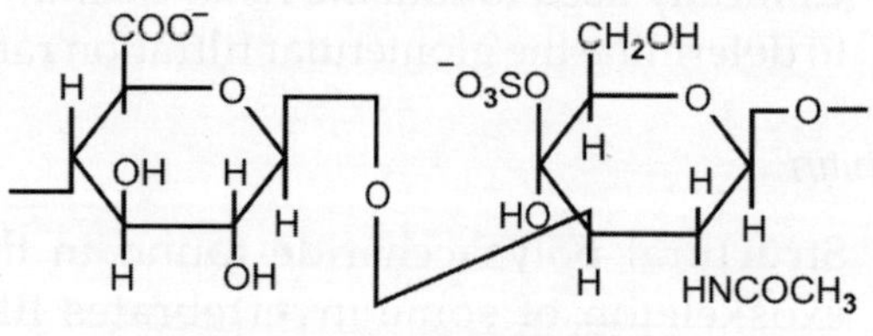

Repeating unit of chondroitin-4-sulphate

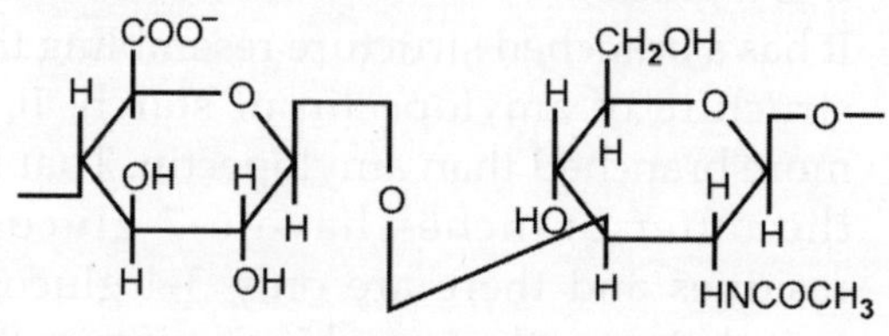

Repeating unit of hyaluronic acid

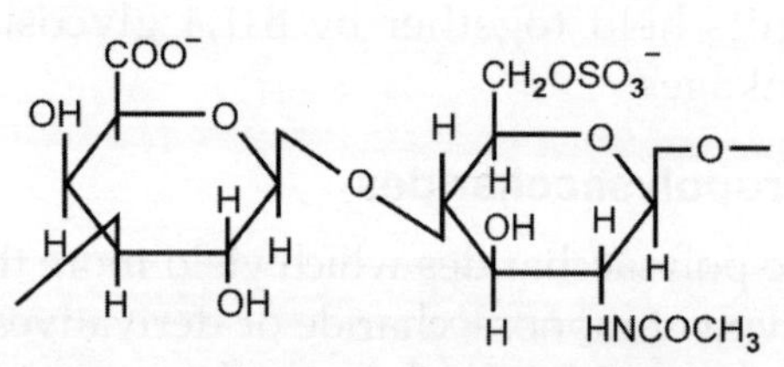

Repeat unit of keratin sulphate

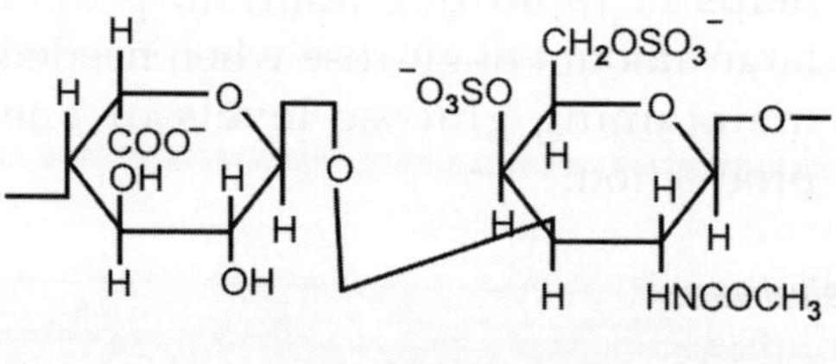

Repeat unit of dermatan sulphate

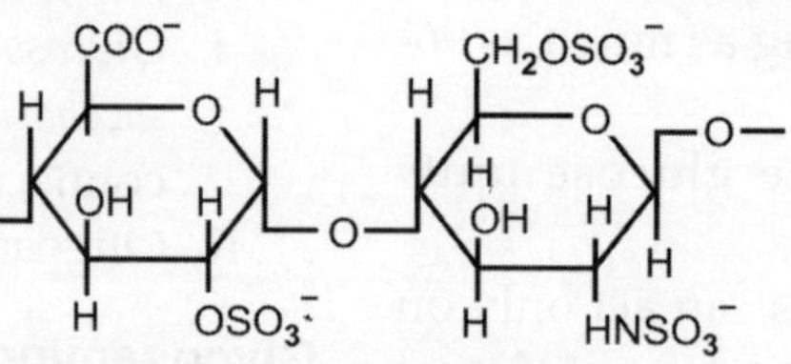

Repeat unit of heparin

**Fig. 2.16:** Structure of glycosaminoglycans

**Table 2.3:** Different types of GAGs

| | *Type* | *Repeating sugar units* | *Tissues in which present* |
|---|---|---|---|
| 1. | Hyaluronic acid<br>**Different from other GAGs in that it is not sulfated and not covalently linked to proteins* | N-acetyl glucosamine and glucuronic acid | Synovial fluid, vitreous humour, umbilical cord, loose connective tissue, skin, cartilage. |
| 2. | Chondroitin sulphate<br>**Most abundant GAG* | N-acetyl galactosamine and glucuronic acid, sulfated on $C_4$ or $C_6$ | Cartilage, bone, cornea, tendon, ligament. |
| 3. | Keratan sulphate<br>**No uronic acid* | N-acetyl glucosamine and galactose. $C_6$ of either sugar may be sulfated | Cartilage, cornea, intervertebral disc. |
| 4. | Heparin<br>**Only intracellular GAG* | 6–O sulfated glucosamine and L–iduronic sulphate, glucuronic acid, N-acetyl glucosamine in low amounts. Sulfate on $C_3$ or $C_6$ of glucosamine or $C_2$ of uronic acid | Mast cells that line arteries of skin, liver, lung, intestinal mucosa. |
| 5. | Heparan sulfate | Same as above except fewer sulphate groups. More N-acetyl glucosamine. Glucuronic acid more than iduronic acid | Extracellular GAG found in basement membrane or cell surfaces. |
| 6. | Dermatan sulfate | N-acetyl galactosamine and L–iduronic acid. | Wide distribution. Skin, arterial walls. |

*Glycoproteins* have oligosaccharide chains made up of different hexoses or their derivatives linked to the protein chain by covalent bonds, e.g., serum proteins, immunoglobulins, mucins, human chorionic gonadotropin.

## DIGESTION AND ABSORPTION

Digestion is the process involving hydrolysis of large and complex organic molecules of foodstuffs into smaller and preferably water soluble molecules which can be easily absorbed in the intestine for utilization by the organism.

The principal dietary carbohydrates include:

- polysaccharides—starch and glycogen
- disaccharides—lactose and sucrose and
- small amount of free monosaccharides–glucose and fructose.
- In addition to this, there are also cellulose, pectin and gums which cannot be digested by the human enzymes and which constitute the *fibre*. These provide roughage and prevent constipation hence decreasing the incidence of diseases like colon cancer and atherosclerosis.

When food is cooked, the starch granules, the major carbohydrates found in the diet, are broken down and partly solubilized.

## Digestion

The hydrolysis of the glycosidic bonds is carried out by a group of enzymes called *glycosidases* which are specific to the bond, structure and configuration of monosaccharide units. Digestion begins in the mouth, though most of it occurs in the intestine.

## Mouth

Food is masticated. During the process of mastication, the carbohydrates present in the food are acted upon by *α-amylase* (old name: ptyalin), produced by the salivary glands (Fig. 2.17).

### *α-amylase*

*Mode of action*: Randomly cleaves the α-1, 4 glycosidic bonds.

*Optimum pH*: 6.8

*Activator*: $Cl^-$

*Products formed*: Large molecular weight dextrins and small amounts of maltose.

There is limited digestion of food in the mouth because of its presence in the mouth for a short duration.

## Stomach

*pH:* Acidic due to the presence of HCl.

Salivary amylase is denatured when the food enters the stomach because of the high acidity of the stomach contents.

No glycosidases in the stomach.

Only limited hydrolysis of sucrose by HCl occurs in the stomach.

Sucrose ⟶ glucose + fructose

↓

Directly absorbed across the wall of the stomach into blood.

## Intestine

The acidic dietary contents from the stomach on reaching the small intestine are neutralized by bicarbonate of the pancreatic juice.

*The pancreatic juice* also contains an *α-amylase,*

*Optimum pH*: 7.1

*Activator:* $Cl^-$

*Mode of action*: Cleaves α-1, 4 glycosidic bonds randomly in starch and dextrin

*Products formed*: Maltose, isomaltose, branched and unbranched oligosaccharides.

*Intestinal juice:*

- Produced by glands of Brünner and Lieberkühn.
- Contains membrane bound enzymes.

### *Enzymes present in intestinal juice and their action*

1. *Oligosaccharidases*: Exoenzymes, cleave off one sugar unit at a time from oligosaccharides.
2. *Maltase*: Cleaves α-1,4 glycosidic bond.
   Maltose ⟶ 2 glucose (*Maltase*)
3. *Isomaltase*: Cleaves α-1,6 glycosidic bond.
   Isomaltose ⟶ 2 glucose (*Isomaltase*)
4. *Sucrase:*
   Sucrose ⟶ glucose + fructose (*Sucrase*)
5. *Lactase:*
   Lactose ⟶ glucose + galactose (*Lactase*)

## End products of digestion of carbohydrates

- Glucose
- Galactose
- Fructose
- Small amounts of pentoses and mannose.

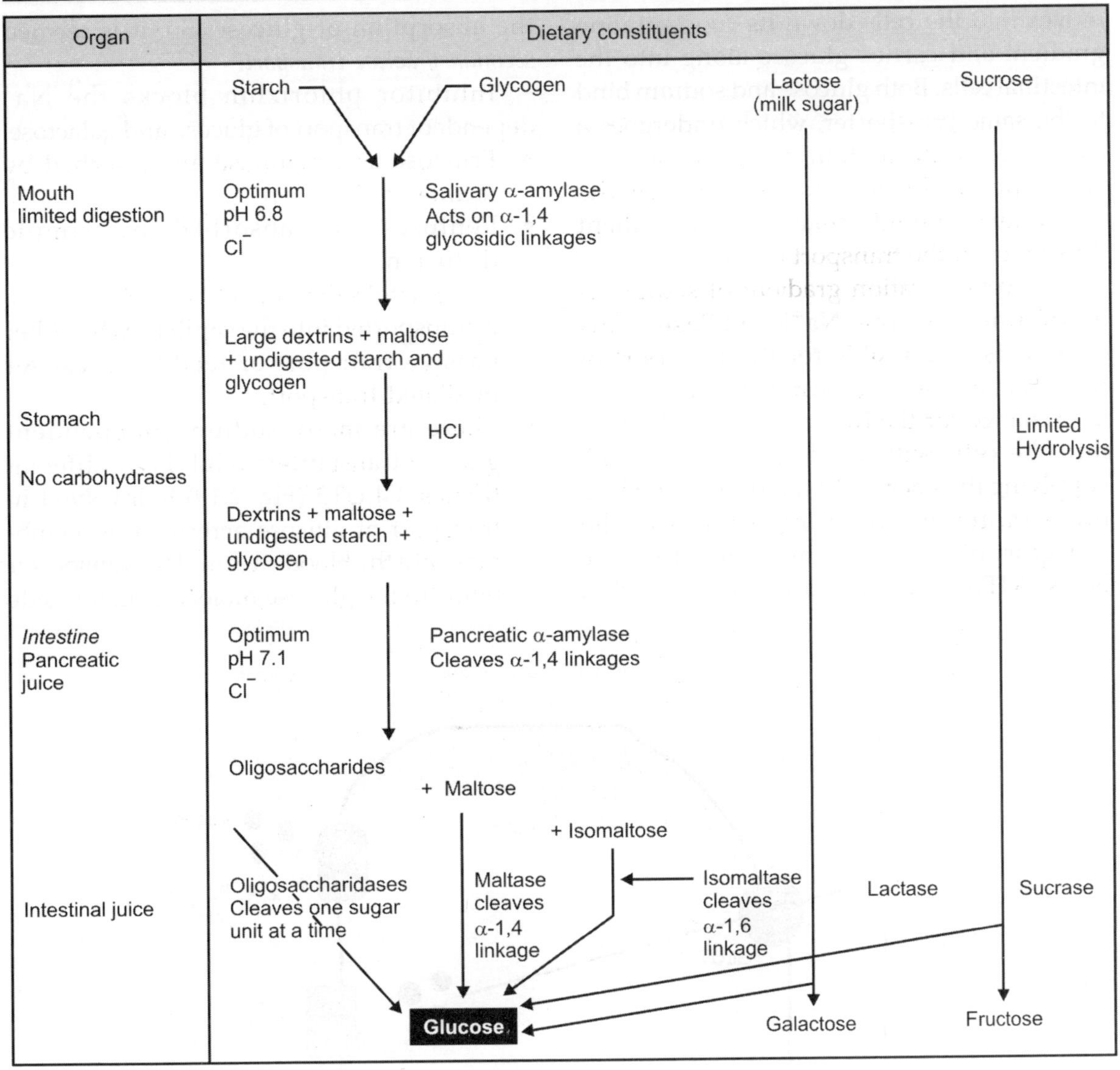


**Fig. 2.17:** Digestion of carbohydrates

## Absorption of Carbohydrates

*Site*: Duodenum and upper jejunum of the small intestine.

Glucose accounts for nearly 80% of the monosaccharides absorbed in the intestine.

## Transport of Glucose

- By sodium dependent glucose transporter (SGluT)
- It is a secondary active transport mechanism.

### *Mechanism (Fig. 2.18)*

Glucose and sodium share the same transport system present on the luminal membrane of the intestinal mucosal cells. The concentration of sodium is higher in the intestinal lumen compared to mucosal cells. Sodium, therefore, moves into the cells down its concentration gradient and carries glucose along into the intestinal cells. Both glucose and sodium bind to the same transporter, which undergoes a change in conformation, releasing sodium and glucose on the cytosolic side of the membrane. The intestinal sodium gradient thus helps in the transport of glucose.

The concentration gradient of sodium is maintained by the $Na^+K^+$ATPase. This enzyme is responsible for the transport of three $Na^+$ ions outside the mucosal membrane in exchange for the transport of two $K^+$ ions into the cell, with the hydrolysis of ATP supplying the energy for the transport. Since the sodium gradient required for the transport of glucose is maintained by the $Na^+K^+$ATPase, ATP is indirectly involved in the absorption of glucose. So, it is termed *secondary active transport*.

Inhibitor phlorizhin blocks the $Na^+$ dependent transport of glucose and galactose.

- Fructose and mannose are absorbed by facilitated diffusion.
- Pentoses are absorbed by simple diffusion.
- Once within the mucosal cell, the glucose is transported into the capillaries by either passive transport or facilitated carrier mediated transport.
- There are many sodium independent glucose transporters (GLUT) in different tissues. GLUT2 (Fig. 2.19) is involved in transport of glucose across basal membrane into the blood stream. The transporter binds to the glucose molecule on one side

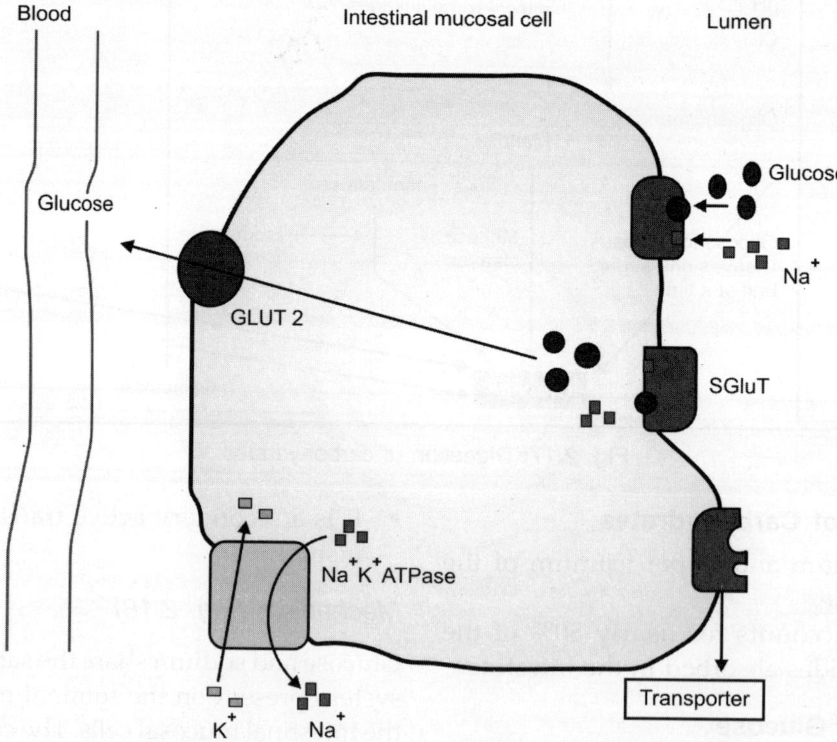


**Fig. 2.18:** Mechanism of absorption of glucose in the intestine

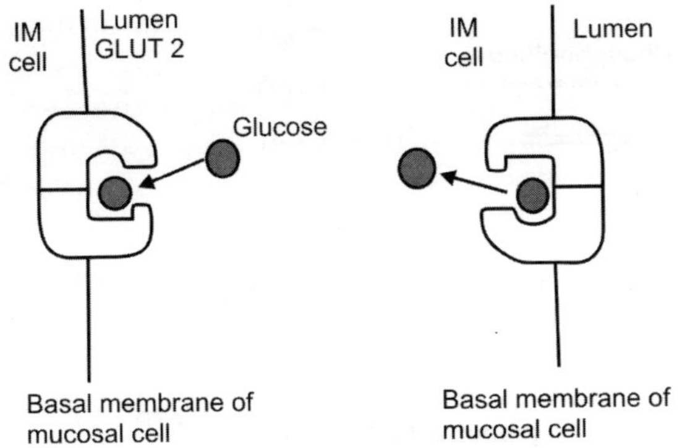


**Fig. 2.19:** Mechanism of glucose absorption by GLUT 2

of the membrane, undergoes a conformational change and releases the glucose on the other side of the membrane.

Di, oligo and polysaccharides that are not hydrolysed by α-amylase or the enzymes of the intestinal juice, cannot be absorbed. Therefore, when they reach the ileum, they are acted upon by bacteria. Bacteria can utilise these carbohydrates because they possess many more types of saccharidases than humans. The monosaccharides that are released are acted upon by intestinal bacterial enzymes forming products like lactate, hydrogen gas, methane and carbon dioxide. These compounds can cause fluid secretion into intestine and increase intestinal motility, thus causing distention of the gut and intestinal cramps.

## DISORDERS

### 1. Lactose Intolerance

*Cause*: Lactase deficiency

Lactose cannot be digested. It is acted upon by intestinal bacteria causing the formation of acids like lactic acid and gases like methane, carbon dioxide and hydrogen.

The acids and gases are osmotically active intestinal irritants.

So, ingestion of lactose in milk or milk products causes diarrhoea, flatulence and abdominal pain.

Since lactose and the products formed from it are osmotically active, there is decreased fluid reabsorption from the intestine causing diarrhoea.

*Types of lactose intolerance:*

a. *Inherited lactase deficiency*:
   - Very rare
   - Symptoms develop soon after birth.
   - Treatment: Lactose free diet i.e., no milk, but curd can be given, as it contains galactosidase activity, which hydrolyzes lactose.

b. *Primary low lactase activity*:
   - Relatively common, especially in Oriental people, mainly in the elderly.
   - There is gradual decrease in lactase activity.

c. *Secondary low lactase activity:*
   - Secondary to intestinal diseases like sprue, colitis, gastroenteritis, kwashiorkor.

### 2. Sucrase Isomaltase Deficiency

- Inherited.
- Symptoms same as in lactose intolerance.
- Occurs in childhood.

## GLYCOLYSIS OR EMBDEN–MEYERHOFF PATHWAY

Glucose that is present in the blood is taken up by the liver freely and in other tissues uptake is facilitated by insulin. The glucose that is taken up is oxidized mainly through the glycolytic pathway.

Glycolysis is the pathway by which glucose is degraded to pyruvate under aerobic (oxygen available) conditions and

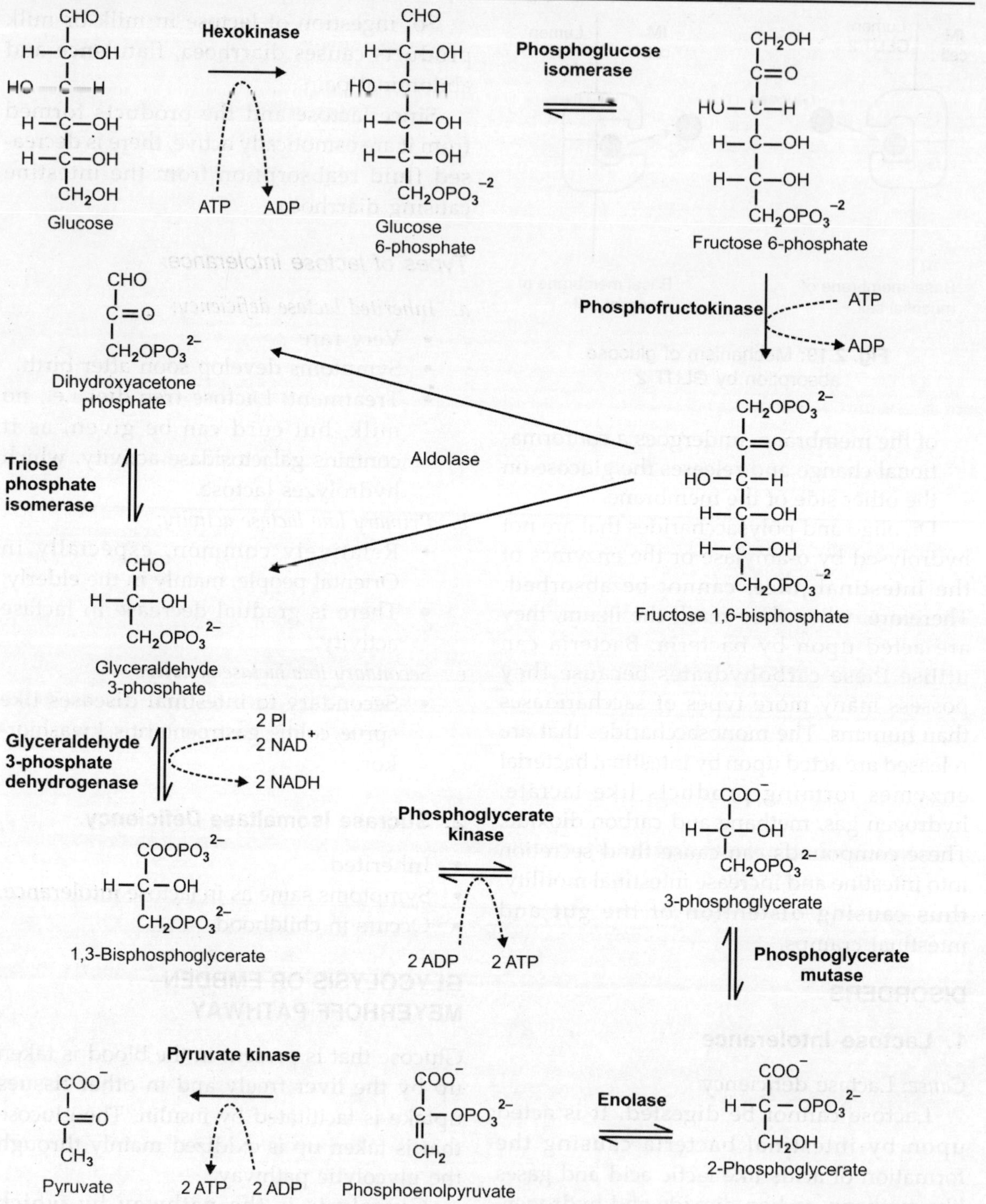


Fig. 2.20: The glycolytic pathway

lactate under anaerobic (oxygen not available) conditions with the production of energy.

**Site**: All living cells

**Subcellular Site**: Cytosol

## REACTIONS OF GLYCOLYSIS (FIG. 2.20)

Can be studied under three stages:

I. Priming stage
II. Splitting stage
III. Energy generation stage.

### I. Priming Stage

Conversion of glucose to fructose-1, 6-bisphosphate:

*Step 1*: Phosphorylation of glucose to glucose-6-phosphate.

*Enzyme: Hexokinase / glucokinase (Table 2.4).*

- Obligatory step for glucose utilisation
- Irreversible reaction
- Requires ATP and $Mg^{2+}$.

Due to low $K_m$ and high affinity glucose is utilised by hexokinase even at low concentrations, whereas glucokinase acts only when blood glucose concentration is high (above 100 mg/dl) i.e., after a meal.

The cell membrane is impermeable to glucose-6-phosphate. The glucose-6-phosphate formed has many fates. Depending on the energy status of the cells, it enters one of the many pathways given in Fig. 2.21.

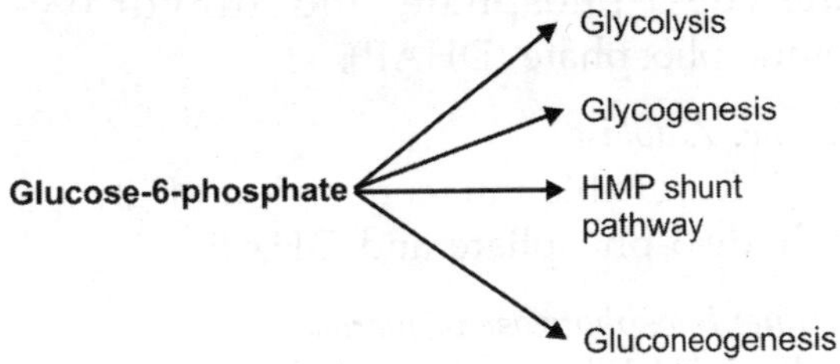


**Fig. 2.21:** Fate of glucose-6-phosphate

*Step 2*: Isomerization of glucose-6-phosphate to fructose-6-phosphate.

*Enzyme: Phosphohexose isomerase.*

- $Mg^{2+}$ is required.

*Step 3*: Phosphorylation of fructose-6-phosphate to fructose-1, 6-bisphosphate

*Enzyme: Phosphofructokinase 1 (PFK)*

- Committed step of the pathway
- Rate limiting step
- Irreversible reaction
- Requires ATP and $Mg^{2+}$.

*(Note: Phosphofructokinase 2 is involved in formation of fructose-2, 6-bisphosphate, involved in regulation of glycolysis).*

### II. Splitting Stage

Conversion of fructose-1,6-bisphosphate to glyceraldehyde-3-phosphate and dihydroxyacetone phosphate.

**Table 2.4:** Comparison of properties of glucokinase and hexokinase

| | *Hexokinase* | *Glucokinase* |
|---|---|---|
| 1. | Present in all tissues | Present only in liver |
| 2. | Has broad specificity—catalyzes phosphorylation of glucose as well as other sugars like fructose and mannose | Has absolute specificity for glucose |
| 3. | Has low $K_m$ for substrates (around 0.1 mM) | Has high $K_m$ for glucose (around 10 mM) |
| 4. | Inhibited by increased concentration of glucose-6-phosphate | Not inhibited by glucose-6-phosphate |
| 5. | Constitutive enzyme | Inducible enzyme |

*Step 4*: Six carbon fructose-1,6-bisphosphate is split into two, three carbon sugars, glyceraldehyde-3-phosphate and dihydroxyacetone phosphate (DHAP).

*Enzyme*: *Aldolase*

*Step 5*: Reversible interconversion of glyceraldehyde-3-phosphate and DHAP.

*Enzyme: Phosphotriose isomerase.*

Glyceraldehyde-3-phosphate is continuously removed by the next reaction of glycolysis. So, though it is a reversible reaction, the equilibrium shifts towards formation of glyceraldehyde-3-phosphate. Two molecules of glyceraldehydes-3-phosphate are formed from each molecule of glucose.

*Inhibitor*: Bromohydroxyacetone phosphate.

### III. Energy Generation Phase

Conversion of glyceraldehyde-3-phosphate to pyruvate.

*Step 6*: Glyceraldehyde-3-phosphate is converted to 1, 3-bisphosphoglycerate, a high energy compound.

*Enzyme*: *Glyceraldehyde-3-phosphate dehydrogenase.*

- Substrate is oxidized and phosphorylated.
- $NAD^+$ is reduced to $NADH+H^+$, which under aerobic conditions can be oxidized in the electron transport chain, generating 3 ATP, by oxidative phosphorylation.

*Inhibitor*: Iodoacetate.

*Step 7*: 1,3-bisphosphoglycerate is converted to 3-phosphoglycerate.

*Enzyme*: *3-phosphoglycerate kinase.*

- ADP is phosphorylated to generate ATP.

In this step, ATP is formed directly by transfer of a phosphate group from the substrate 1,3-bisphosphoglycerate to ADP. This is called *substrate level phosphorylation.*

*Step 8*: 3-phosphoglycerate is isomerised to 2- phosphoglycerate.

*Enzyme*: *Phosphoglycerate mutase.*

- isomerisation reaction involving transfer of phosphate group from position 3 to 2 in the substrate.
- requires $Mg^{2+}$

*Step 9*: 2-phosphoglycerate is converted to phosphoenolpyruvate, a high energy compound.

*Enzyme*: *Enolase.*

- Substrate undergoes dehydration.
- Requires $Mg^{2+}$ or $Mn^{2+}$.

*Inhibitor* : Fluoride

*Sodium fluoride is added during collection of blood for glucose estimation, to prevent glycolysis in blood cells, so that, spurious low values due to utilization of glucose by blood cells after blood collection may be avoided.*

*Step 10*: Phosphoenolpyruvate is converted to pyruvate.

*Enzyme*: *Pyruvate kinase*

- ADP is phosphorylated to ATP.
- Another example of *substrate level phosphorylation.*
- Irreversible reaction.
- Requires $K^+$ and $Mg^{2+}$ or $Mn^{2+}$.

Pyruvate is the end product of aerobic glycolysis. It is a key compound and has many fates, as shown in Figure 2.22.

Major fate of pyruvate produced during glycolysis depends on the availability of oxygen in the cells.

Under anaerobic conditions, pyruvate is reduced to lactate by $NADH+H^+$ produced during the glyceraldehyde-3-phosphate dehydrogenase step.

*Enzyme*: *Lactate dehydrogenase.*

*Importance*: Formation of lactate allows regeneration of $NAD^+$ which can be reused by glyceraldehyde-3-phosphate dehydrogenase, so that glycolysis can proceed even in the absence of oxygen to produce ATP.

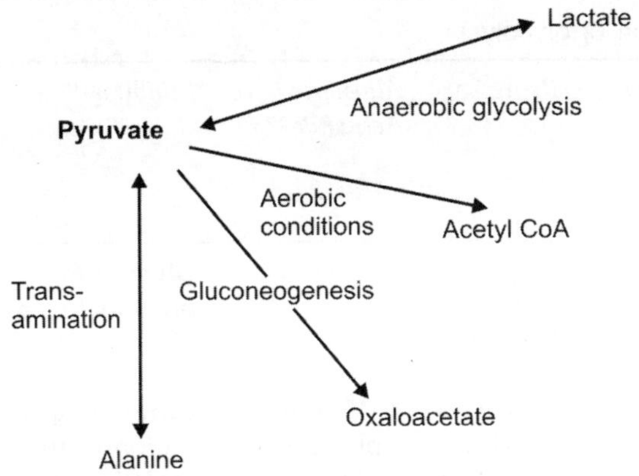


**Fig. 2.22:** Various fates of pyruvate

Under aerobic conditions, NADH+$H^+$ is oxidized by oxidative phosphorylation which requires presence of $O_2$.

Glycolysis in erythrocytes leads to lactate formation, because mitochondria required for further oxidation of pyruvate are not present in RBCs.

## Energetics

### *Aerobic Glycolysis*

| *Enzyme catalyzing the step* | *No. of ATP produced* |
|---|---|
| Glyceraldehyde-3-phosphate dehydrogenase (from NADH) | 3 |
| Phosphoglycerate kinase | 1 |
| Pyruvate kinase | 1 |
| | 5 |
| 2 molecules of glyceraldehyde-3-phosphate produced from one molecule of glucose | × 2 |
| Total | 10 |
| ATP used: | |
| Hexokinase | -1 |
| Phosphofructokinase | -1 |
| Net ATP production | 8 |

### *Anaerobic Glycolysis*

NADH+$H^+$ produced during glyceraldehyde- 3-phosphate dehydrogenase reaction is consumed during lactate dehydrogenase reaction. Not available for formation of ATP.

| ATP produced: | |
|---|---|
| Phosphoglycerate kinase | 2 |
| Pyruvate kinase | 2 |
| | 4 |
| ATP used (as in aerobic glycolysis) | –2 |
| Net production of ATP during anaerobic glycolysis | 2 |

## REGULATION

The three enzymes catalyzing the irreversible steps of glycolysis are regulatory. These are:

1. Hexokinase/glucokinase.
2. Phosphofructokinase-1, the most important regulatory enzyme, catalyzing rate limiting, committed step of the pathway and an allosteric enzyme.
3. Pyruvate kinase.

These enzymes may be subject to allosteric regulation, feedback regulation and/or may be under the influence of the hormones insulin and glucagon and/ epinephrine (Table 2.5).

Pyruvate kinase is also subject to regulation by reversible covalent modification i.e., phosphorylation and dephosphorylation (refer regulation of glycogen metabolism).

Dephosphorylated pyruvate kinase is in active form, phosphorylated pyruvate kinase is inactive.

## RAPOPORT-LEUBERING (RL) CYCLE OR DIPHOSPHOGLYCERATE SHUNT

It is a metabolic bypass occurring in the erythrocytes as a supplement to glycolysis.

**Table 2.5:** Regulation of glycolysis

| *Enzyme* | *Activity in* | | *Inducer* | *Repressor* | *Allosteric Activator* | *Inhibitor* |
|---|---|---|---|---|---|---|
| | *Carbo-hydrate feeding* | *Starvation, diabetes* | | | | |
| 1. Hexokinase | | | | | | Glucose-6-phosphate |
| 2. Glucokinase | ↑ | ↓ | Insulin | Glucagon (cAMP) | | |
| 3. Phosphofructo-kinase-1 | ↑ | ↓ | Insulin | Glucagon (cAMP) | AMP, Fructose 6-phosphate, Pi, Fructose 2, 6-bis phosphate | Citrate (concentration ↑ during oxidation of fatty acids and ketone bodies) ATP. |
| 4. Pyruvate kinase | ↑ | ↓ | Insulin | Glucagon (cAMP) | Fructose 1, 6-bisphosphate | ATP, alanine |

In the RL cycle, the step catalyzed by phosphoglycerate kinase where there is production of 1 ATP molecule, is bypassed.

1, 3 BPG is converted to 2, 3 BPG by *BPG mutase*.

2, 3 BPG is then hydrolysed by *2, 3 BPG phosphatase* to give 3-phosphoglycerate and Pi, without the formation of a high energy bond.

Both the enzyme activities are present on the same enzyme molecule (bifunctional enzyme).

The 3-phosphoglycerate is then processed through the glycolytic pathway (Fig. 2.23).

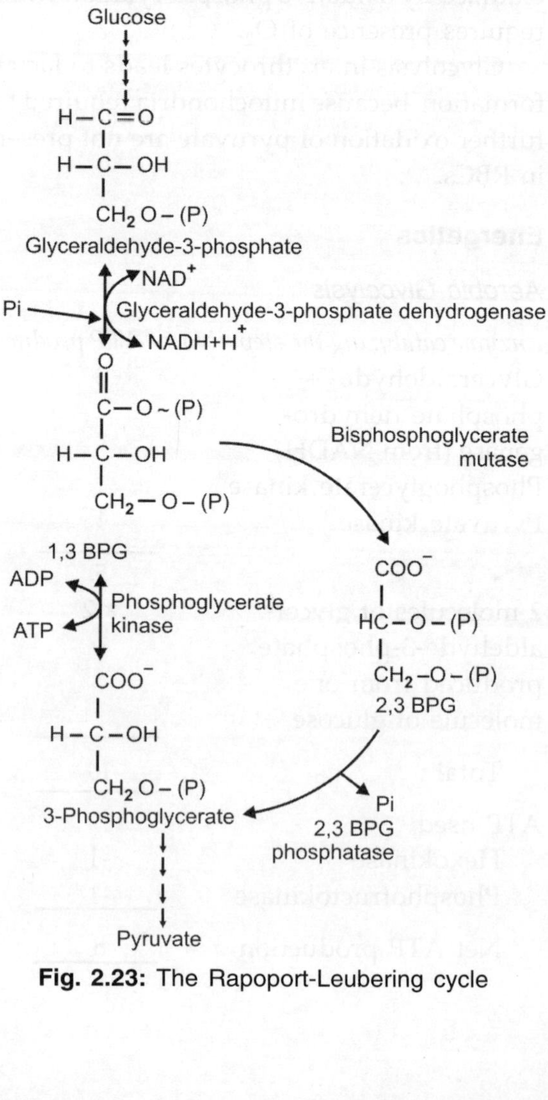


**Fig. 2.23:** The Rapoport-Leubering cycle

## Advantages

1. 2, 3 BPG produced, combines with β chain of haemoglobin, causing a decrease in affinity for oxygen and displacement of the oxyhaemoglobin dissociation curve to the right. Its presence in red cell helps oxy- haemoglobin to unload $O_2$. At high altitudes 2, 3 BPG concentration increases in RBCs and helps to release more $O_2$ from oxyhaemoglobin (Fig. 2.24).
2. There is no net production of ATP in the bypass step, which may be of advantage to the economy of the red cell, allowing

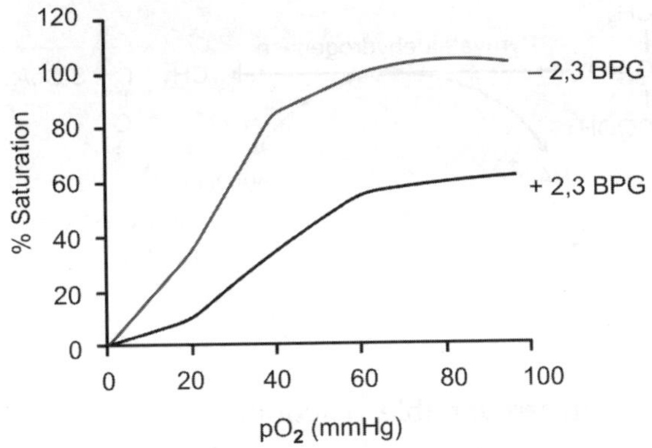


**Fig. 2.24:** Oxygen dissociation curve in the presence and absence of 2,3 BPG

glycolysis to occur when the need for ATP is minimal.

## IMPORTANCE OF GLYCOLYSIS

1. Energy yielding pathway especially in tissues lacking mitochondria, e.g., RBC, cornea, lens, regions of retina. Brain is also dependant on glucose for energy.
2. The ability of glycolysis to provide ATP in the absence of $O_2$ is important in muscle because it allows muscle to perform its function even when $O_2$ supply is insufficient.
3. Glycerol-3-phosphate required for triacylglycerol synthesis can be formed from DHAP which is an intermediate in glycolysis, mainly in adipose tissue and liver.
4. Nonessential amino acid alanine can be synthesized from pyruvate, end product of glycolysis and serine from 3-phosphoglycerate, intermediate of glycolysis.
5. 2,3 BPG formed during Rapoport-Leubering shunt helps to unload oxygen in RBCs.
6. Deficiency of glycolytic enzymes i.e., hexokinase and pyruvate kinase is manifested as haemolytic anaemia in RBCs; fatigue and intolerance to exercise in skeletal muscle.

## CORI'S CYCLE

Under anaerobic conditions lactate is produced from pyruvate. This is a dead end of metabolism and lactate produced, mainly by the active skeletal muscle, has to be converted back to glucose by gluconeogenesis.

This cannot be done in the muscle, because, muscle lacks the key enzymes of gluconeogenesis.

- The plasma membrane is freely permeable to lactate. Lactate is carried from the skeletal muscle, by the blood to the liver, where it is oxidized to pyruvate and the pyruvate formed is converted to glucose by gluconeogenesis. This glucose may then be transported back to muscle.
- The cycle involving the synthesis of glucose in liver from skeletal muscle lactate and the reuse of glucose thus synthesized by the muscle for energy purposes is called as *"Cori's Cycle"* (Fig. 2.25).

### Lactic Acidosis

#### *Causes*

1. Overproduction of lactic acid occurs due to oxygen depletion.

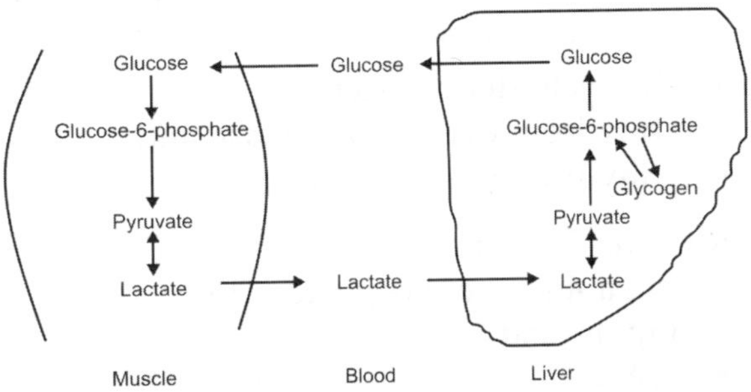


**Fig. 2.25:** Cori's cycle

2. Under utilisation of lactic acid occurs due to:
   a. Liver diseases
   b. Ethanol ingestion
   c. Ingestion of certain drugs.

pH is decreased causing metabolic acidosis.

*Treatment:*
- Bicarbonate administration
- Treatment of the underlying cause.

## FORMATION OF ACETYL COA FROM PYRUVATE

Major fate of pyruvate is it's conversion to acetyl CoA by *"Oxidative decarboxylation."* (One $NAD^+$ is reduced to $NADH+H^+$ and three carbon pyruvate is decarboxylated to two carbon acetyl CoA).

***Site*** : Mitochondria.

Major source of pyruvate is glycolysis in cytosol, but conversion of pyruvate to acetyl CoA occurs in the mitochondria. Transport of pyruvate from cytosol to mitochondria occurs by the pyruvate transporter along with $H^+$ utilising the inward proton gradient (Fig. 2.26).

Conversion of pyruvate to acetyl CoA is the central step linking glycolysis to TCA cycle.

***Enzyme***: *Pyruvate dehydrogenase (PDH complex) complex*. It is a multienzyme system comprising of three enzymes:
- Pyruvate dehydrogenase
- Dihydrolipoyl transacetylase and
- Dihydrolipoyl dehydrogenase.

Requires 5 coenzymes:
- Thiamin pyrophosphate (TPP)
- Lipoic acid
- NAD

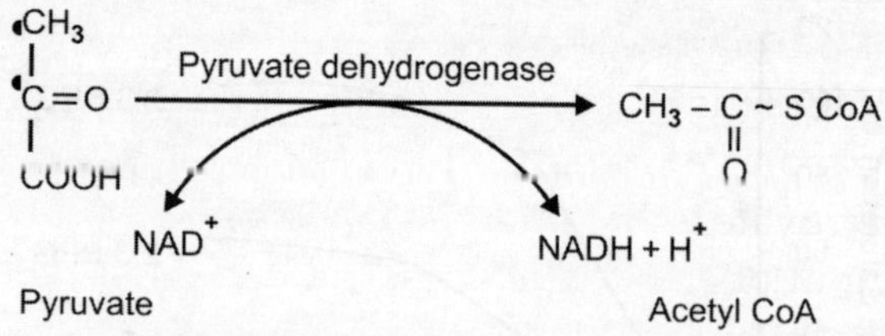


It is an irreversible reaction.

### Energetics

One $NADH + H^+$ - 3ATP
(2 pyruvate molecules produced from one glucose molecule) × 2

6 ATP

### Regulation

Pyruvate and acetyl CoA are important central compounds in metabolism. Acetyl CoA is formed not only from glucose but

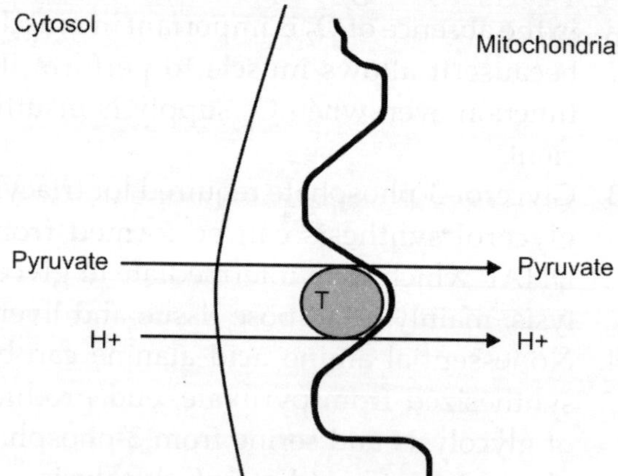


**Fig. 2.26:** Transport of pyruvate to mitochondria from cytosol

also from fatty acids and amino acids when glucose is in short supply. Because of this reason oxidative phosphorylation of pyruvate is an important control point. If the input of acetyl CoA from lipids is adequate, the pyruvate dehydrogenase is inhibited, preventing further input from glucose.

Control of pyruvate dehydrogenase occurs by insulin/glucagon dependent reversible dephosphorylation/phosphorylation (refer glycogen metabolism). Active form is dephosphorylated and inactive form is phosphorylated.

It is also an allosteric enzyme.

*Negative modulators*: NADH, acetyl CoA

*Positive modulators*: Pyruvate, ADP.

## CITRIC ACID CYCLE OR TRICARBOXYLIC ACID CYCLE (TCA CYCLE) OR KREB'S CYCLE (after Hans Krebs who postulated the cycle)

TCA cycle is the cyclic process of reactions in mitochondria that bring about catabolism

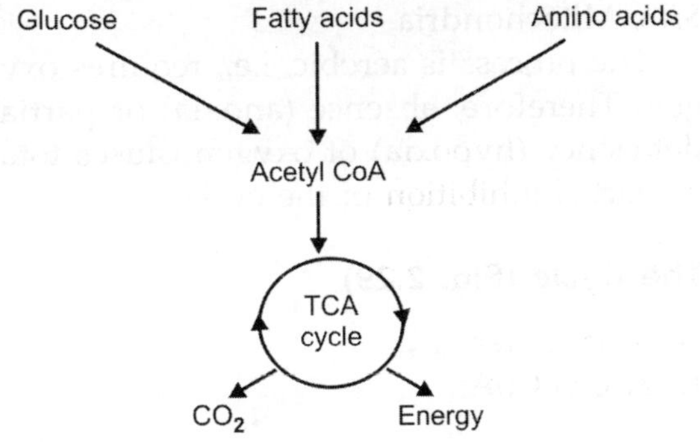


**Fig. 2.27:** TCA cycle as final common pathway

of two carbon acetyl residues, liberating two molecules of carbon dioxide and hydrogen equivalents which upon oxidation, lead to formation of ATP.

It is the final common pathway for oxidation of carbohydrates, lipids and proteins (Fig. 2.27). This is because, glucose, fatty acids and many amino acids are all metabolized to acetyl CoA or intermediates of TCA cycle.

The major fate of acetyl CoA is the complete oxidation to $CO_2$ and $H_2O$ through the TCA cycle (Fig. 2.28).

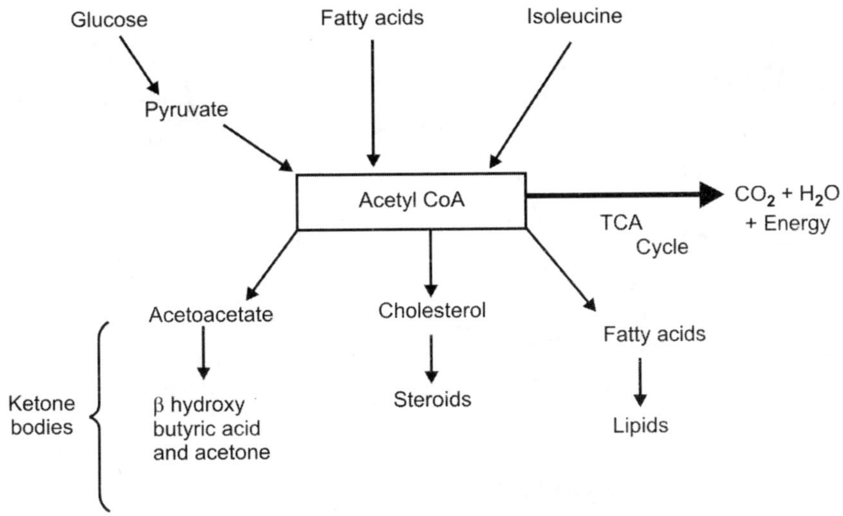


**Fig. 2.28:** Formation and fate of acetyl CoA

*Site*: Mitochondria.

The process is aerobic, i.e., requires oxygen. Therefore, absence (anoxia) or partial deficiency (hypoxia) of oxygen causes total or partial inhibition of the cycle.

### The Cycle (Fig. 2.29)

Starting material

1. Acetyl CoA
2. Oxaloacetate, a 4-carbon dicarboxylic acid, which is regenerated at the end of the cycle. It plays a catalytic role.

*Step 1*: Condensation of acetyl CoA with oxaloacetate to form citrate.

*Enzyme: Citrate synthase.*

- There is a C-C bond formed between methyl carbon of acetyl CoA and carbonyl carbon of oxaloacetate.

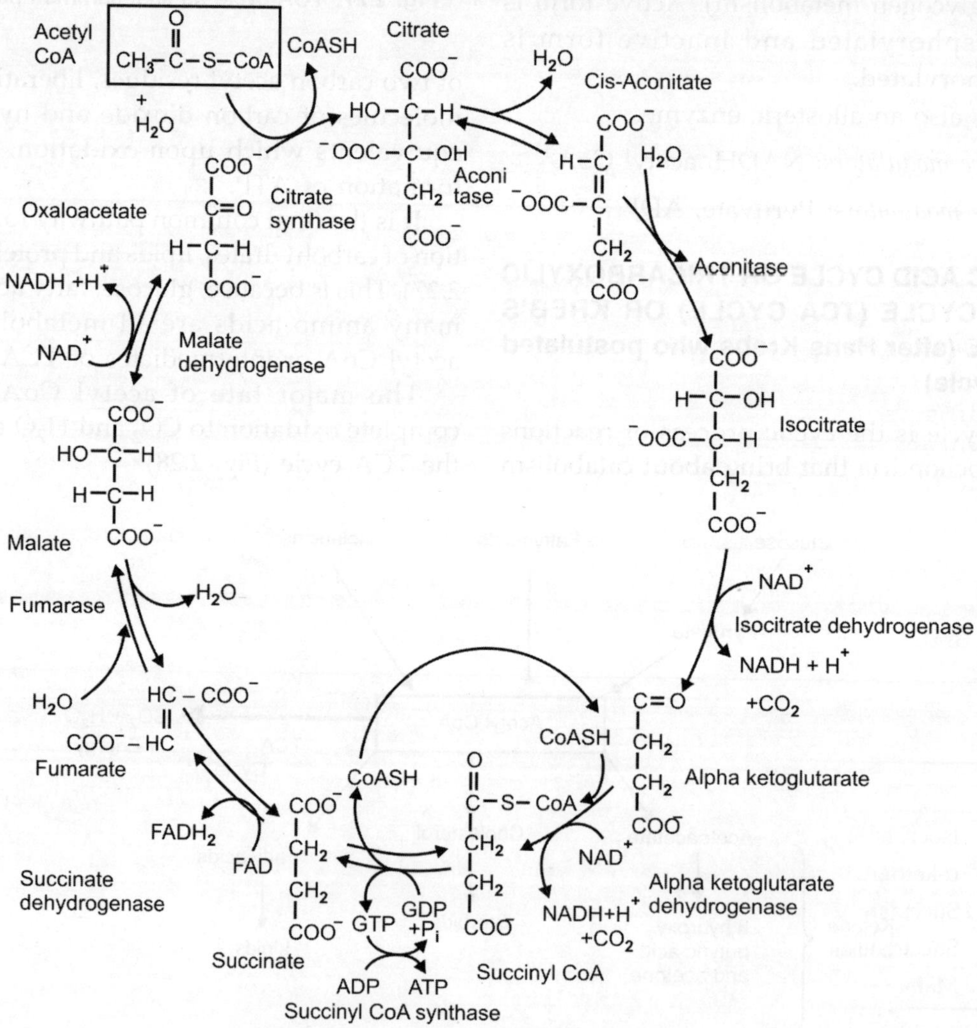


**Fig. 2.29:** Reactions of the citric acid cycle

- It is a highly exergonic reaction, where the energy is provided by the cleavage of the thioester bond.
- It is an irreversible reaction.

*Step 2*: Conversion of citrate to isocitrate.

*Enzyme*: *Aconitase,* an iron-sulfur protein, contains iron in $Fe^{2+}$ form.

*Intermediate*: Cis-aconitate.

Citrate is first dehydrated to cis-aconitate, which is then rehydrated to isocitrate.

*Inhibitor*: Fluoroacetate; forms fluoroacetyl CoA and reacts with oxaloacetate to form fluorocitrate, which inhibits the enzyme.

*Step 3*: Isocitrate is converted to α-ketoglutarate.

*Enzyme*: *Isocitrate dehydrogenase,* requires $Mg^{2+}$ or $Mn^{2+}$.

*Intermediate*: Oxalosuccinate.

Isocitrate is first oxidized, with the transfer of hydrogens to $NAD^+$ and formation of oxalosuccinate, which remains enzyme bound. Oxalosuccinate is then decarboxylated to give α-ketoglutarate.

*Step 4*: Oxidative decarboxylation of α ketoglutarate to succinyl CoA, a high energy compound.

*Enzyme: α-ketoglutarate dehydrogenase complex* which is similar to pyruvate dehydrogenase complex.

Consists of 3 enzymes:

- α-ketoglutarate dehydrogenase
- Dihydrolipoyl transacetylase and
- Dihydrolipoyl dehydrogenase.

Requires 5 coenzymes

- TPP
- $NAD^+$
- FAD
- Lipoic acid
- Coenzyme A.

A molecule of carbon dioxide is removed and one $NADH+H^+$ is generated.

*Inhibitor*: Arsenite

*Step 5*: Conversion of succinyl CoA to succinate.

*Enzyme*: *Succinate thiokinase or succinyl CoA synthetase.*

The high energy thioester bond of succinyl CoA is broken to form succinate and the energy liberated is used to phosphorylate GDP to GTP (substrate level phosphorylation).

GTP can be converted to ATP by the reaction.

**Table 2.6:** Energy generating steps of TCA cycle

| *Enzyme* | *Compound Formed* | *ATP generated* |
|---|---|---|
| Isocitrate dehydrogenase | 1 $NADH+H^+$ | 3 |
| α-ketoglutarate dehydrogenase | 1 $NADH+H^+$ | 3 |
| Succinate thiokinase | 1 GTP | 1 |
| Succinate dehydrogenase | 1 $FADH_2$ | 2 |
| Malate dehydrogenase | 1 $NADH + H^+$ | 3 |
| | Total | 12 |

$$GTP + ADP \xleftrightarrow[\text{diphoophato kinase}]{\text{Nucleoside}} ATP + GDP$$

*Step 7*: Fumarate is hydrated to form L-malate.

*Enzyme: Fumarase.*

Malate formed easily penetrates the mitochondrial membrane. So, it can move in either direction, i.e., either into the mitochondria or cytosol.

*Step 8*: Regeneration of oxaloacetate.

*Enzyme*: *Malate dehydrogenase.*

*Coenzyme*: $NAD^+$ reduced to $NADH+H^+$.

Stoichiometry:

Acetyl CoA + $3NAD^+$ + FAD + $2H_2O$ + GDP + Pi $\longrightarrow$ 2 $CO_2$ + 3NADH + $3H^+$ + $FADH_2$ + GTP + CoASH

## Energetics

Energy yield from complete oxidation of one molecule of glucose:

| | | |
|---|---|---|
| In aerobic glycolysis | : | 8 ATP |
| Pyruvate dehydrogenase reaction | : | 6 ATP |
| TCA cycle (12 × 2) (Table 2.6) | | 24 ATP |
| Total | | 38 ATP |

## REGULATION (TABLE 2.7)

In most tissues, the function of TCA cycle is to provide energy. So, the cycle is controlled by the energy status of the cell, i.e., [ATP]/[ADP] and [NADH]/[NAD] ratio. This is called as the *respiratory control*, because activity of several enzymes is dependent on availability of $NAD^+$ which in turn is dependent on availability of ADP for phosphorylation during transfer of reducing equivalents from $NADH+H^+$ and $FADH_2$ to oxygen through the electron transport chain.

## IMPORTANCE OF TCA CYCLE

1. Energy production.
2. It is an amphibolic pathway. Has both oxidative and synthetic functions (Fig. 2.30).
   - Oxidises acetyl CoA formed from glucose, fatty acids and amino acids.
   - TCA cycle intermediates are removed continuously for many essential reactions. For example:
     — Oxaloacetate for gluconeogenesis
     — α-ketoglutarate for transamination to glutamate.
     — Citrate, transferred to cytoplasm as a source of acetyl CoA for fatty acid synthesis
     — Succinyl CoA for heme biosynthesis.

**Table 2.7:** Regulation of TCA cycle

| *Enzyme* | *Activator* | *Inhibitor* | |
|---|---|---|---|
| 1. Citrate synthase | | ATP, long chain fatty acids | |
| 2. Isocitrate dehydrogenase | ADP | ATP, NADH | |
| 3. α-ketoglutarate dehydrogenase | ADP, pyruvate | NADH, acetyl CoA | Also subject to reversible phosphorylation dephosphorylation like PDH complex |
| 4. Succinate dehydrogenase | | Oxaloacetate | |
| 5. Malate dehydrogenase | $NAD^+$ | NADH | |

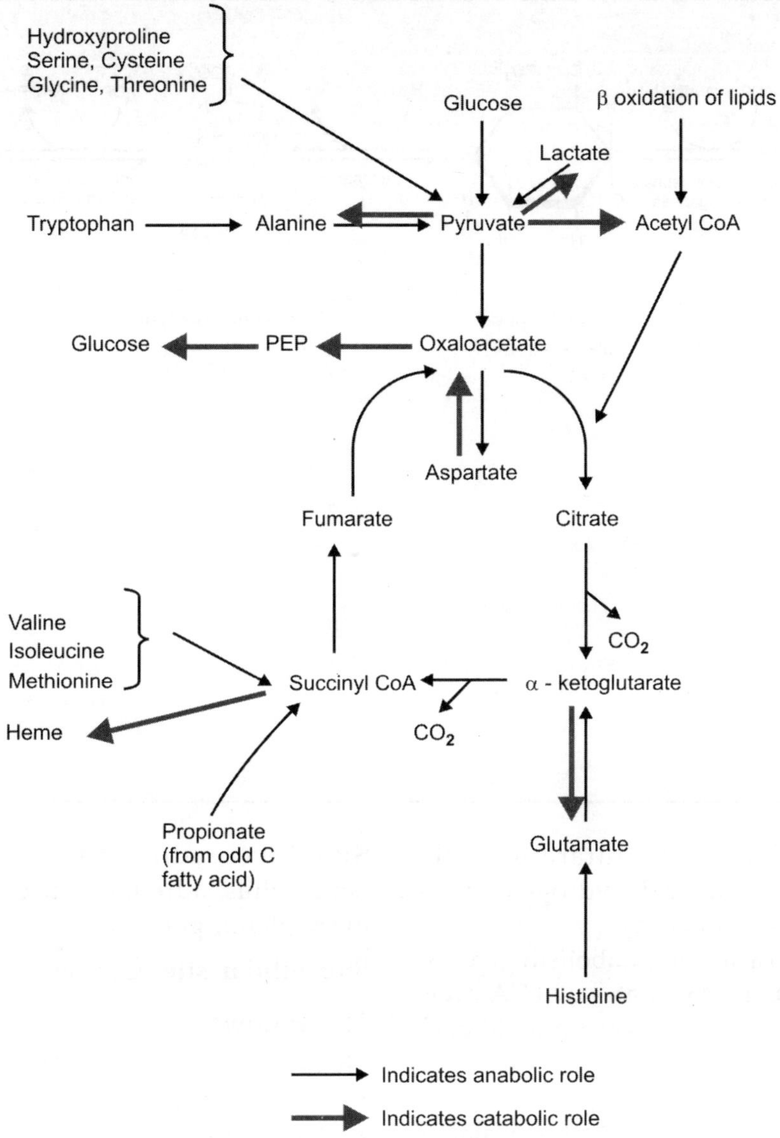


**Fig. 2.30:** Amphibolic role of TCA Cycle

## Anaplerosis

Reactions that lead to net synthesis of TCA cycle intermediates are called *anaplerotic reactions.*

1. Most important anaplerotic reaction is carboxylation of pyruvate to oxaloacetate in mitochondria.

*Enzyme: Pyruvate carboxylase.*
*Coenzymes*: ATP, biotin

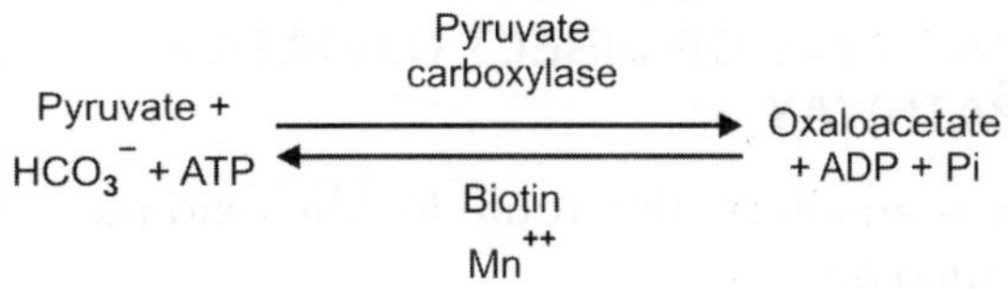

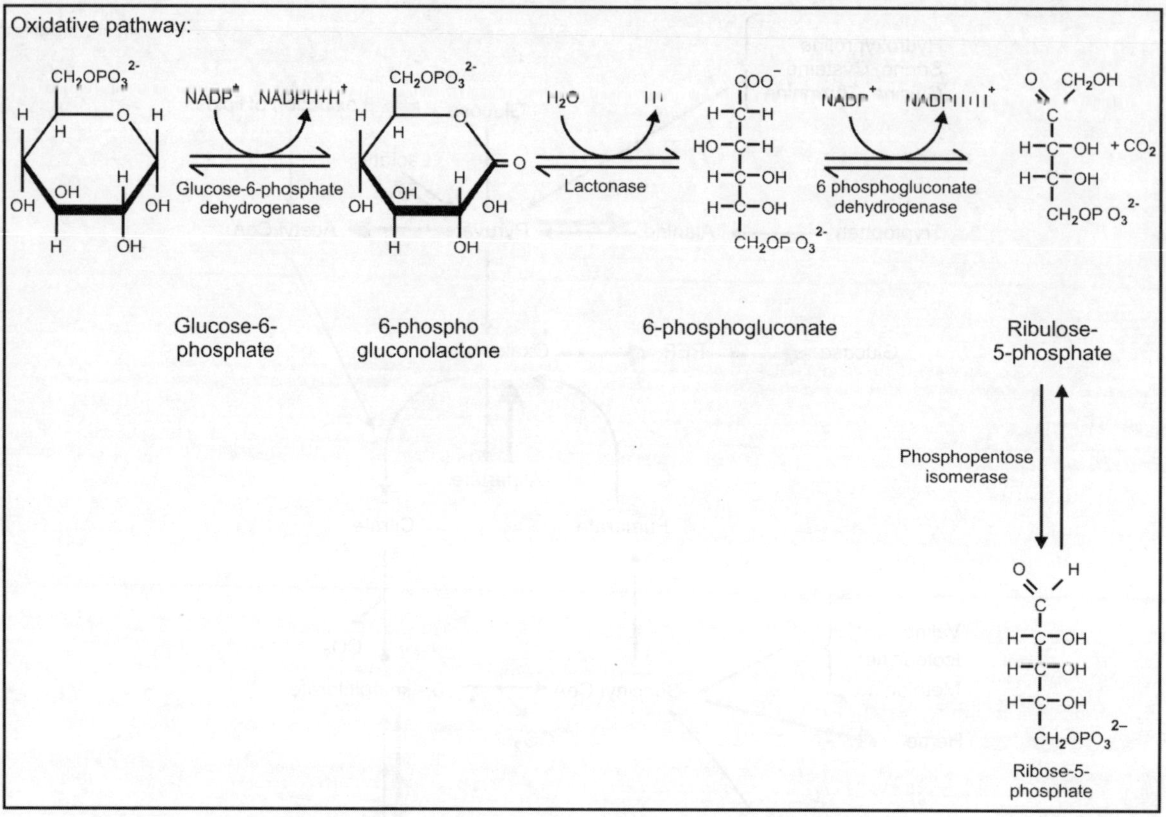


2. Formation of α-ketoglutarate by the action of L-glutamate dehydrogenase (see amino acid metabolism).
3. Many amino acids on catabolism give rise to one of the intermediates of TCA cycle.
4. Conversion of pyruvate to malate catalyzed by malic enzyme.

$$\text{Pyruvate} + CO_2 + NADPH + H^+ \rightleftharpoons \text{L – Malate} + NADP^+$$

*Malic enzyme*

## HEXOSE MONOPHOSPHATE (HMP) SHUNT OR PENTOSE PHOSPHATE PATHWAY OR DIRECT OXIDATIVE PATHWAY

It is an alternative route for the oxidation of glucose.

**Site:** Liver, adipose tissue, erythrocytes, neutrophils, adrenal cortex, lactating mammary gland, gonads.

**Subcellular site:** Cytosol.

### The Pathway

There are two phases in HMP shunt pathway,
- The oxidative pathway and
- The nonoxidative pathway.

### Oxidative Phase

Two NADPH + $H^+$ molecules are generated and a molecule of $CO_2$ is removed resulting in the formation of a pentose phosphate.

The first reaction is the oxidation of glucose-6-phosphate by glucose-6-phosphate dehydrogenase, an NADP dependent enzyme, forming 6-phosphogluconolactone.

The 6-phosphogluconolactone is then hydrolyzed by gluconolactone hydrolase forming 6-phosphogluconate.

In the next step, 6-phosphogluconate is again oxidized and decarboxylated by an NADP dependent 6-phosphogluconate dehydrogenase, resulting in the formation of ribulose-5-phosphate.

### Nonoxidative Pathway

The ribulose 5-phosphate may either be isomerized to the corresponding aldopentose, ribose-5-phosphate, by ribose-5-phosphate ketoisomerase or to its' 3-epimer, xylulose-5- phosphate, by ribulose 5-phosphate 3 epimerase.

These two sugars by repeated rearrangements may give rise to fructose-6-phosphate and glyceraldehyde-3-phosphate, both intermediates of glycolysis.

An overivew of the pathway is given in the box below:

**Importance of HMP shunt pathway:**

Two most important functions are the following:

1. Generation of NADPH + $H^+$ which is required for
   - The synthesis of fatty acids and cholesterol
   - Integrity of RBC membrane
   - Phagocytosis in WBC and
   - For hydroxylation by the cytochrome $P_{450}$ system.
2. Also provides ribose-5-phosphate required for the synthesis of nucleic acids and coenzymes like $NAD^+$ and ATP.

## GLUCONEOGENESIS

### Definition

Gluconeogenesis is the biosynthesis of glucose from noncarbohydrate sources.

The noncarbohydrate sources include:

- Pyruvate
- Lactate
- Glucogenic amino acids
- Propionate and
- Glycerol

**Site**: Liver and renal cortex

**Subcellular site**: Cytosol, mitochondria and endoplasmic reticulum (Table 2.8).

Gluconeogenesis and glycolysis are opposing pathways. All the reactions of glycolysis, except those catalyzed by hexokinase, phosphofructokinase and pyruvate kinase are reversible. These irreversible steps of glycolysis are bypassed by alternate enzymes during gluconeogenesis. These enzymes are called as the *key gluconeogenic enzymes*.

**Table 2.8:** Site of occurrence of key gluconeogenic enzymes

| | *Enzyme* | *Site of occurrence* |
|---|---|---|
| 1. | Pyruvate kinase | Mitochondria |
| 2. | Phosphoenolpyruvate carboxykinase | Cytosol and mitochondria |
| 3. | Fructose-1,6-bisphosphatase | Cytosol |
| 4. | Glucose-6-phosphatase | Endoplasmic reticulum |

The key gluconeogenic enzymes are –

1. *Pyruvate carboxylase:*
2. *Phosphoenolpyruvate carboxykinase (PEPCK):*

} bypass the reaction catalyzed by pyruvate kinase

3. *Fructose-1,6-bis-phosphatase:* Bypasses the phosphofructokinase step.
4. *Glucose-6-phosphatase:* Bypasses the hexokinase step.

| Reactants | Enzyme | Products |
|---|---|---|
| 1. Ribulose-5-phosphate | ⟷ Phosphopentose isomerase | Ribose-5-phosphate |
| 2. Ribulose-5-phosphate | ⟷ Phosphopentose epimerase | Xylulose-5-phosphate |
| 3. Xylulose-5-phosphate + Ribose-5-phosphate | ⟷ Transketolase | Glyceraldehyde-3-phosphate + Sedoheptulose-7-phosphate |
| 4. Sedoheptulose-7-phosphate + Glyceraldehyde-3-phosphate | ⟷ Transaldolase | Erythrose-4-phosphate + Fructose-6-phosphate |
| 5. Xylulose-5-phosphate + Erythrose-4-phosphate | ⟷ Transketolase | Glyceraldehyde-3-phosphate + Fructose-6-phosphate |

## REACTION SEQUENCE

### From Lactate and Pyruvate

- Lactate is converted to pyruvate by lactate dehydrogenase in the cytosol.
- Pyruvate is transported to mitochondria by pyruvate-proton symport system.
- In the mitochondria, pyruvate is first carboxylated to oxaloacetate (OA) using

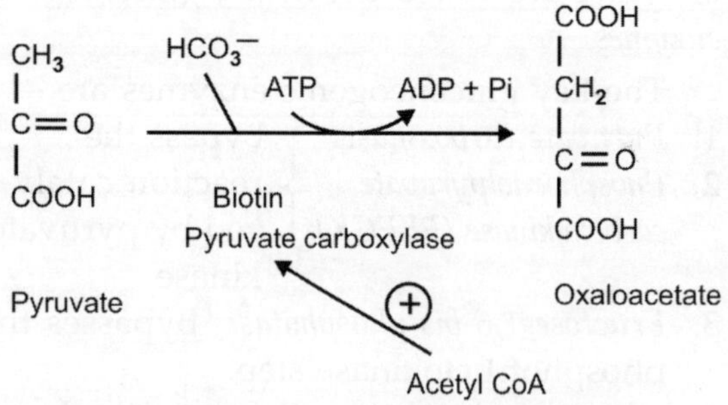


$HCO_3^-$, ATP, $Mg^{2+}$ or $Mn^{2+}$, by biotin dependent pyruvate carboxylase.

The oxaloacetate is then decarboxylated and phosphorylated to phosphoenol pyruvate (PEP) by phosphoenol pyruvate carboxykinase utilising GTP as the energy source to form phosphoenol pyruvate.

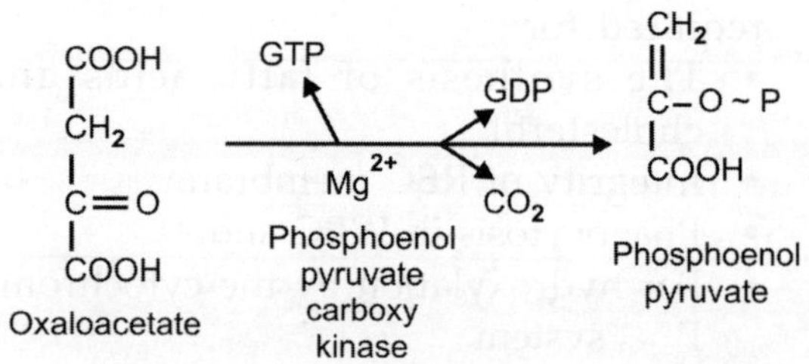


PEPCK is present both in mitochondria and cytosol.

If oxaloacetate is converted to PEP in the mitochondria, the PEP can easily traverse the mitochondrial membrane and be transported to the cytosol.

The mitochondrial membrane is impermeable to oxaloacetate. Oxaloacetate is transported to the cytosol by:

1. Reduction to malate by malate dehydrogenase. The malate can be transferred to cytosol by the malate—$PO_4$ antiport system. Malate is reconverted to OA in the cytosol.
2. OA may be transaminated to aspartate which is then transported to the cytosol by aspartate—glutamate transporter. Aspartate is then reconverted to OA by the reversal of transaminase reaction.

In both the cases the PEPCK then changes the OA to PEP with the help of GTP.

The glycolytic enzymes then convert PEP to fructose-1, 6-bis-phosphate.

Fructose-1, 6-bisphosphatase, another key gluconeogenic enzyme then converts fructose-1, 6-bis phosphate to fructose-6-phosphate.

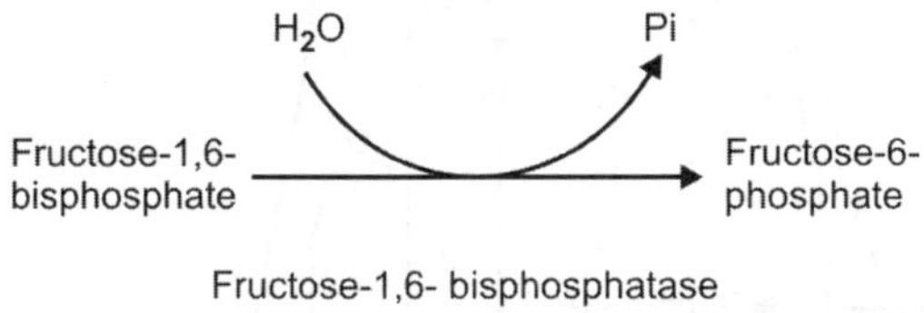


Fructose-6-phosphate is converted to glucose-6-phosphate by the reversal of the phosphohexoisomerase reaction.

The glucose-6-phosphate is then converted to glucose in the endoplasmic reticulum by the action of glucose-6-phosphatase another key gluconeogenic enzyme.

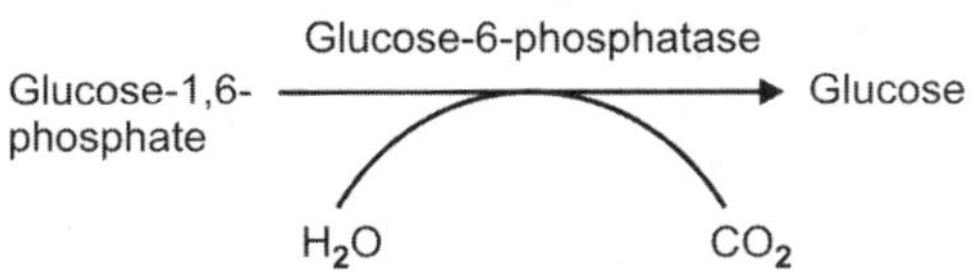


### Gluconeogenesis from Amino Acids

After losing their amino groups by transamination, most amino acids except leucine and lysine form pyruvate or intermediates of TCA cycle which can be converted to malate. Malate can be transferred to the cytosol and changed to glucose (refer Fig. 2.30, amphibolic role of TCA cycle). It is the major source of glucose during starvation; from breakdown of muscle proteins.

Of the amino acids transported from muscle to the liver during starvation, alanine predominates. *"Glucose-alanine cycle"*(Fig. 2.32) transports glucose from liver to muscle; forms pyruvate, followed by transamination to alanine. The alanine is transported to liver, transaminated and pyruvate formed undergoes gluconeogenesis (Fig. 2.31). So there is a net transfer of amino nitrogen from muscle to liver and free energy from liver to muscle.

### From Glycerol

Glycerol is the product of hydrolysis of glycerolipids in adipose tissue.

Glycerol is first converted to glycerol-3-phosphate by glycerokinase using ATP and $Mg^{2+}$ in hepatic and renal cytosol.

Cytosolic glycerol-3-phosphate dehydrogenase oxidizes glycerol-3-phosphate to DHAP, using $NAD^+$ as the electron acceptor. DHAP is an intermediate of glycolysis, can be converted to glucose.

### From Propionate

Propionate is formed as the end product of metabolism of the amino acids valine, isoleucine and methionine and fatty acids containing odd number of carbon atoms. Propionate is converted to succinyl CoA an intermediate of TCA cycle which is converted to malate and then to glucose (Fig. 2.33).

## IMPORTANCE OF GLUCONEOGENESIS

1. Certain tissues of the body, like the nervous system, renal medulla and erythrocytes require a continuous supply of glucose. Gluconeogenesis helps to keep

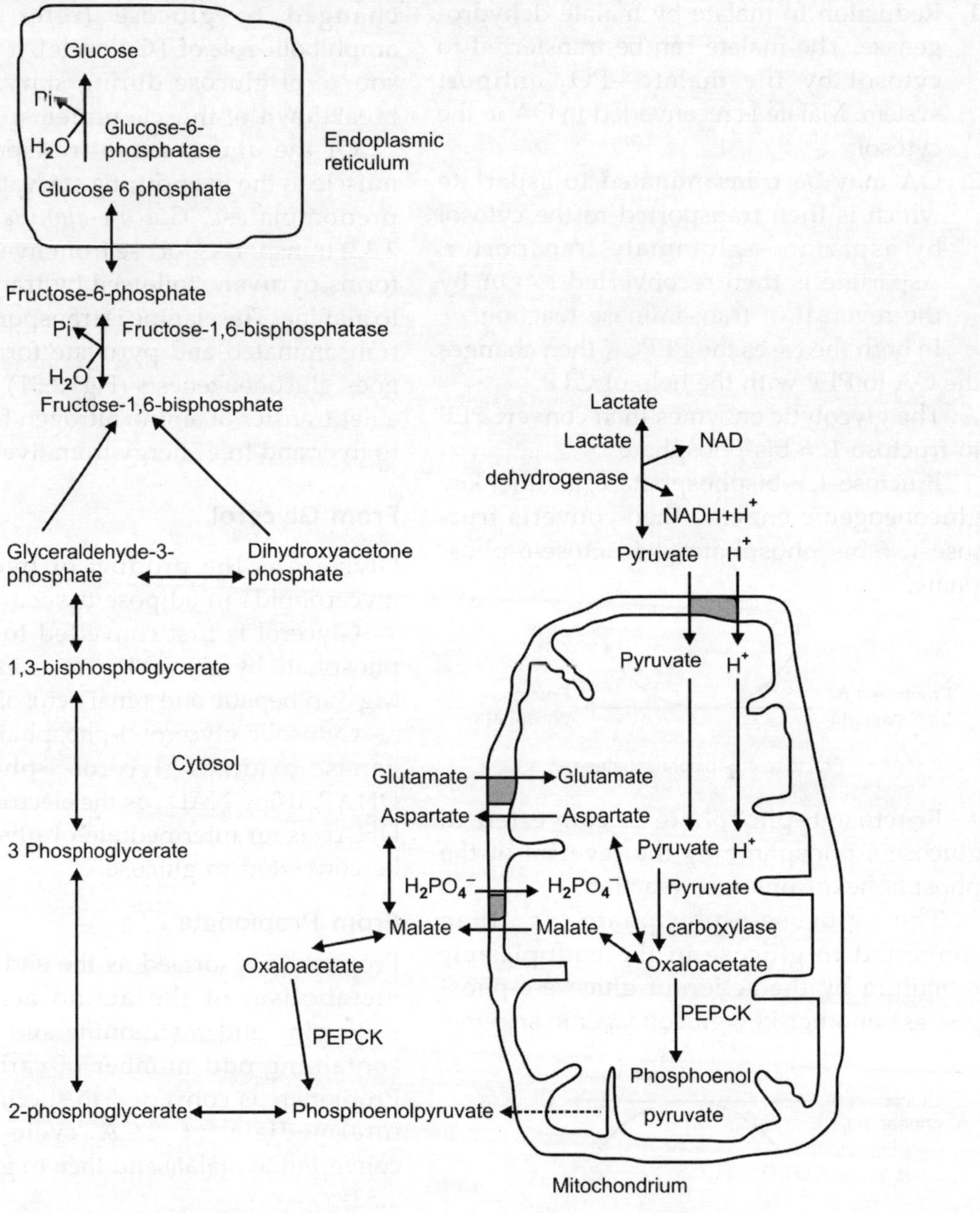


**Fig. 2.31:** Gluconeogenesis from lactate and pyruvate

the blood glucose level within the normal range during fasting conditions.

2. Plays a role in maintaining the levels of the intermediates of the citric acid cycle.
3. Gluconeogenesis is used to clear the end products of various metabolic pathways.
   - Lactate, end product of anaerobic glycolysis from muscle.

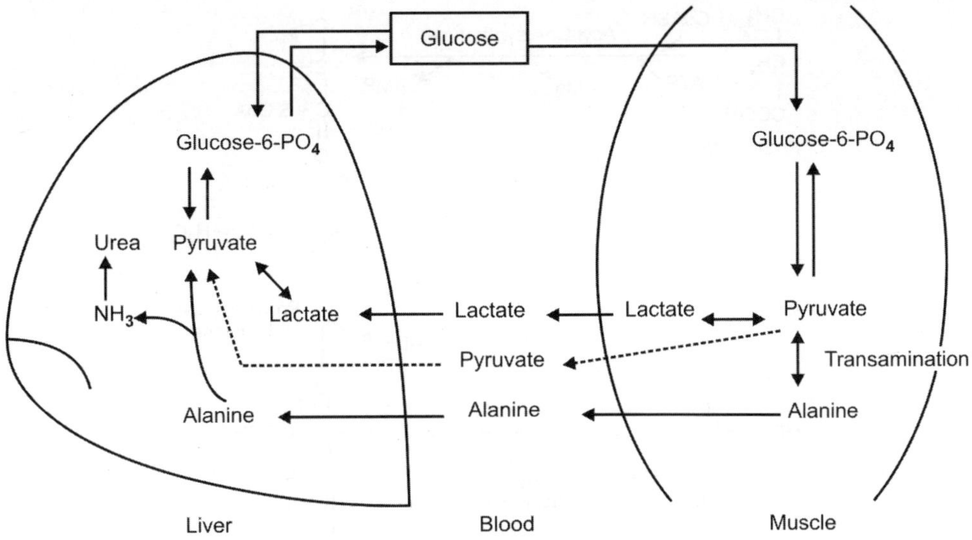


**Fig. 2.32:** Glucose-alanine cycle

- Glycerol, formed from glycerolipids in adipose tissue.
- Propionate, produced from metabolism of odd carbon fatty acids.
- Carbon skeletons of amino acids.

## GLYCOGEN METABOLISM

Glycogen is the storage polysaccharide of the animal body. So, it is often called as "*animal starch.*" It is stored in,

- The liver (upto 6% by weight)
- The muscle (<1% by weight).

## FUNCTION

In the muscle, it is a source of readily available glucose units for glycolysis, to provide energy for the contracting muscle.

In the liver, it helps in maintaining blood glucose level. Excess glucose present after a meal is taken up by the liver and converted to glycogen and when blood glucose levels fall, the stored glycogen is broken down and glucose is added to the blood.

## GLYCOGENESIS

Refers to the synthesis of glycogen from glucose.

***Site:*** Liver and muscle are the major organs.

***Subcellular site***: Cytosol.

### Reaction Sequence

*Step 1*: Glucose is phosphorylated to glucose-6-phosphate (as in glycolysis).

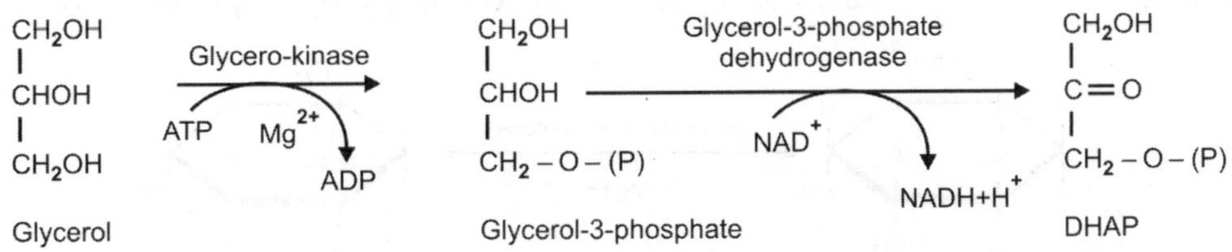

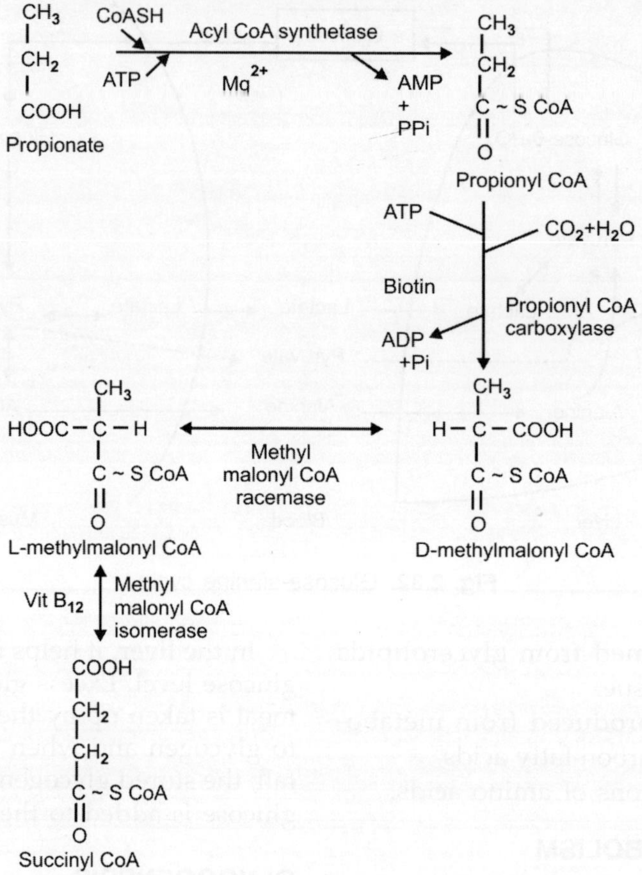


**Fig. 2.33:** Conversion of propionate to succinyl CoA

*Enzyme*: *Hexokinase* in muscle.

*Glucokinase* in liver.

ATP and $Mg^{++}$ are required.

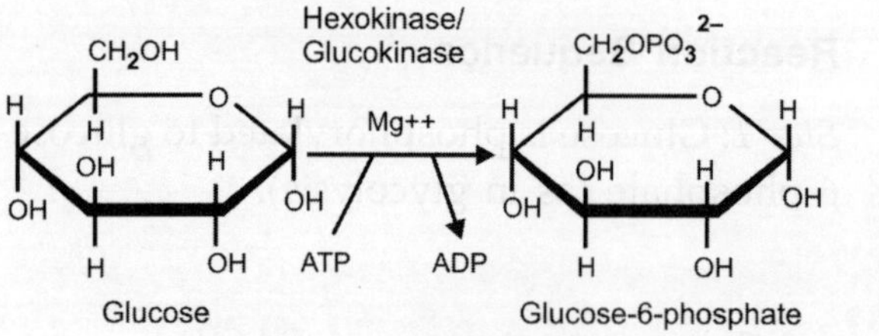


*Step 2*: Conversion of glucose-6-phosphate to glucose-1-phosphate.

*Enzyme: Phosphoglucomutase*

$Mg^{++}$ is required

*Step 3*: Formation of UDP glucose.

Glucose-1-phosphate reacts with uridine triphosphate (UTP) to form the active nucleotide, uridine diphosphate glucose (UDP-Glc) liberating pyrophosphate (PPi).

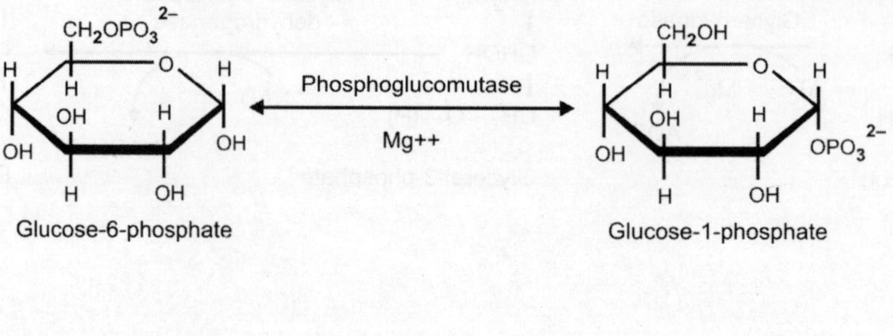

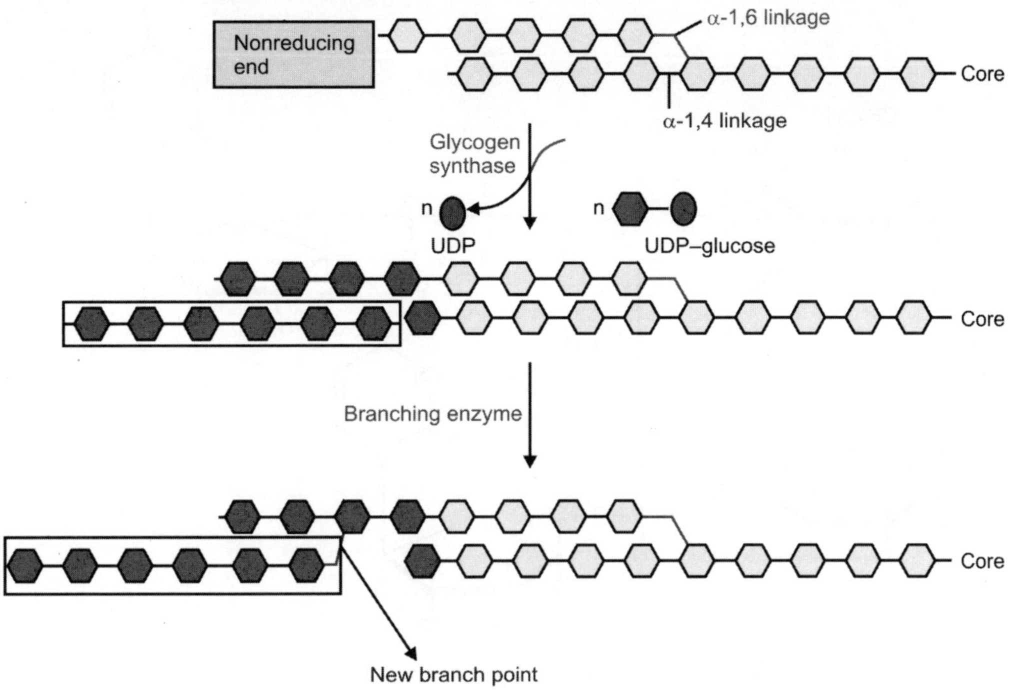


**Fig. 2.34:** Glycogen synthesis *(For colour version see Plate 1)*

*Enzyme*: *UDP-Glc pyrophosphorylase.*

The energy released during the hydrolysis of pyrophosphate to phosphate drives the reaction towards formation of UDP-Glc.

UDP-Glc is a highly reactive complex of glucose and serves as a donor of glucose for incorporation into oligosaccharides and gangliosides.

*Step 4: Transfer of glucose from UDP-Glc to oligosaccharide chain.*

*Enzyme: Glycogen synthase (Fig. 2.34).*

- Glycogen synthase can transfer glucose from UDP- Glc to the nonreducing end of a pre-existing oligosaccharide (glycogen primer), but not to another free glucose molecule.
- While initiating glycogenesis, the glucose unit from UDP-Glc is transferred to the hydroxyl group of tyrosine residue of a protein called *glycogenin*, which acts as an initiating primer.
- The next glucose is bound to the $C_4$ of the glycogenin bound glucose, followed by successive additions of glucose residues from UDP-Glc to the nonreducing end of the growing oligosaccharide held by glycogenin to form α-1, 4-glycosidic bonds.

$$\text{UDP-Glc} + (\text{glucose})_n \text{ glycogen} \xrightarrow{\textit{Glycogen synthase}} \text{UDP} + (\text{glucose})_{n+1} \text{ glycogen}$$

- UDP may be rephosphorylated to UTP utilising a phosphate group of ATP by the enzyme nucleoside diphosphokinase.

$$\text{UDP} + \text{ATP} \xrightarrow[\textit{Nucleoside diphosphokinase}]{Mg^{++}} \text{UTP} + \text{ADP}$$

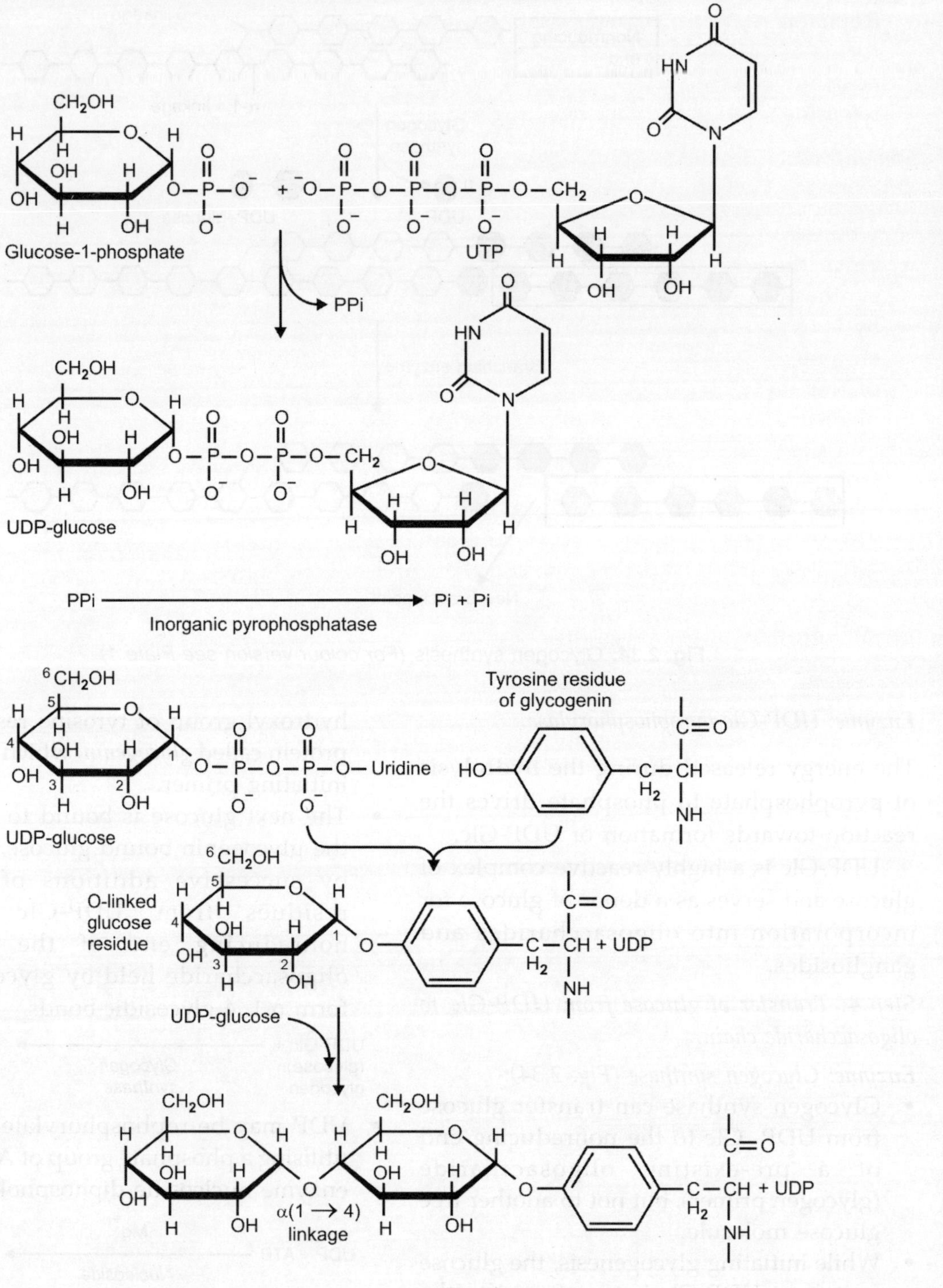

Glucose-1-phosphate
UTP
PPi
UDP-glucose
PPi → Pi + Pi
Inorganic pyrophosphatase
Uridine
UDP-glucose
Tyrosine residue of glycogenin
O-linked glucose residue
+ UDP
UDP-glucose
α(1 → 4) linkage
+ UDP

- Repeated action of glycogen synthase forms unbranched polyglucose chains adding one glucose molecule and releasing UDP molecule at each step.

*Step 5*: Branching of the chain.

*Enzyme*: Branching enzyme or *amylo [1,4] → [1,6] transglucosidase.*

- Once the repeated action of glycogen synthase has extended the chain length by 11 or more glucose residues, branching enzyme splits an α-1,4 glycosidic bond of the chain and transfers the released oligosaccharide of 6 or more glucose units to a site on the chain which is separated from a pre-existing branch point by at least 4 glucose residues.
- The transferred oligosaccharide is linked by an α-1,6 glycosidic bond to the C6 of the glucose residue at the new branching site, producing a new 1, 6 branching.
- The two new branches created are elongated at the nonreducing ends by the action of glycogen synthase. Elongation and branching are repeated several times to form large branched glycogen molecule.

### Energy Expenditure

| | |
|---|---|
| Glucose to glucose-6-phosphate | 1 ATP |
| Formation of UDP-Glc | 2 ATP |
| | 3 ATP |

Three high energy bonds are spent in incorporating each molecule of glucose into glycogen.

## GLYCOGENOLYSIS

Glycogenolysis is the breakdown of glycogen to glucose (Fig. 2.35).

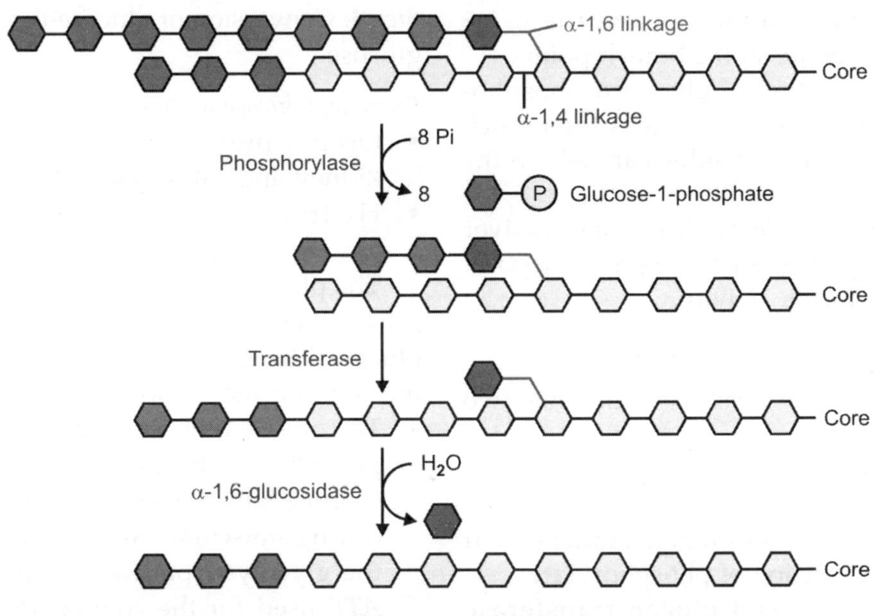


**Fig. 2.35:** Glycogenolysis *(For colour version see Plate 2)*

## Site

*Liver*: Occurs during interdigestive period; glycogen is converted to glucose.

*Muscle*: Occurs during exercise. Glycogen is converted to glucose-6-phosphate and by anaerobic glycolysis is converted to lactic acid.

***Subcellular site***: All the steps except the last occur in the cytosol. The last step occurs in the endoplasmic reticulum.

## Reactions

*Step 1*: Release of glucose units as glucose-1-phosphate.

*Enzyme*: *Glycogen phosphorylase*.

- The enzyme catalyzes the phosphorolysis of α-1,4 glycosidic bonds in the stored glycogen with the help of inorganic phosphate ($HPO_4^{2-}$) starting from the non-reducing end of the outer branches of the molecule and releases glucose as glucose-1-phosphate molecule, one at a time.
- This enzyme cannot cleave the α-1,6 glycosidic bonds at the branch points.
- Successive action of glycogen phosphorylase progressively shortens the branch until four glucose residues are left on the branch.
- *Muscle phosphorylase* is different from liver phosphorylase in that it requires *pyridoxal phosphate* for its activity.

*Step 2*: Removal of the branches.

*Enzyme*: *Debranching enzyme* which has two enzyme activities, *viz.,*

- *α-[1,4] →α-[1,4] glucan transferase* and
- *amylo 1,6 glucosidase.*
  - — Acts on the shortened branches left after the action of phosphorylase.
  - — The α-1,4 →1,4 glucan transferase transfers a trisaccharide unit from one branch to the other exposing α-1,6 branch points and joins the trisaccharide to the latter by a new α-1,4 glycosidic bond.
- The amylo-1,6 glucosidase hydrolyzes the exposed 1,6 glycosidic bond at the branch points liberating glucose residue as free glucose.
- About 90% of glucose residues of glycogen are released as glucose-1-phosphate by glycogen phosphorylase, while the other 10% are released as glucose by the debranching enzyme.
- Once the branches are removed, further action by phosphorylase proceeds.

*Step 3*: Isomerization of glucose-1-phosphate to glucose-6-phosphate.

*Enzyme: Phosphoglucomutase.*

$Mg^{++}$ is required.

*Step 4*: Conversion of glucose-6-phosphate to glucose.

*Enzyme: Glucose-6-phosphatase*

- Occurs only in microsomes of liver, kidney and intestinal mucosa.
- Hydrolyzes glucose-6-phosphate to glucose and Pi.

The plasma membrane is impermeable to glucose-6-phosphate but permeable to glucose. The glucose formed passes out from the cell into the blood.

- The muscle cells *do not* contain glucose-6-phosphatase. In the muscle, the glucose-6-phosphate formed by glycogenolysis during muscular exercise directly enters the glycolytic pathway, thus *saving one ATP* used for the conversion of glucose to glucose-6-phosphate.

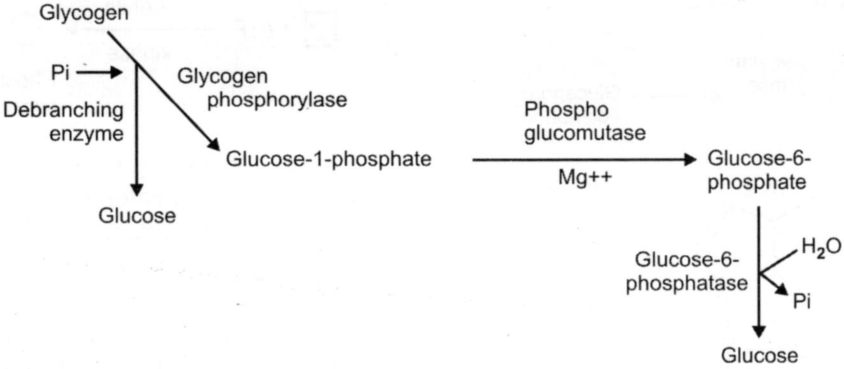


### Lysosomal Degradation of Glycogen

Some amount of glycogen is continuously degraded by a lysosomal enzyme, α-1,4 glucosidase (acid maltase). The purpose of this is unknown.

## REGULATION OF GLYCOGEN METABOLISM

In the liver, glycogen synthesis occurs in the well fed state, when the excess glucose is converted to glycogen and glycogenolysis occurs during periods of fasting, to add glucose to the blood, thus helping to keep the blood glucose within the reference range.

In the muscle, glycogenolysis occurs during exercise to provide energy to the contracting muscle and glycogenesis occurs during resting stage.

So, these two processes occur under different or opposing circumstances and are reciprocally regulated.

The regulatory enzymes are:
- *Glycogen synthase* for glycogenesis and
- *Glycogen phosphorylase* for glycogenolysis.

These two enzymes are controlled by
a. Allosteric modulation and
b. Reversible covalent modification.

### Allosteric Modulation

*Glycogen synthase:*
*Activator:* Glucose-6-phosphate

*Glycogen phosphorylase:*

*Activators:* AMP and in muscle also $Ca^{++}$

*Inhibitors*: Glucose, glucose-6-phosphate, ATP.

### Reversible Covalent Modification

The regulation of these enzymes by covalent modification is mediated through the hormones glucagon, epinephrine and insulin via cAMP, by a cascade system.

#### *i. Formation of cAMP (Fig. 2.36)*

- Glucagon is secreted by the alpha cells of the islets of Langerhans in pancreas during starvation and epinephrine by adrenal medulla in response to stress.
- When these hormones are released in response to stimuli, they bind to a receptor on the target cell.
- This activates an enzyme called the adenylate cyclase present on the inner surface of the cell membrane.
- Adenylate cyclase brings about the conversion of ATP to 3' 5' cyclic adenylic acid (cAMP), which acts as the second messenger.

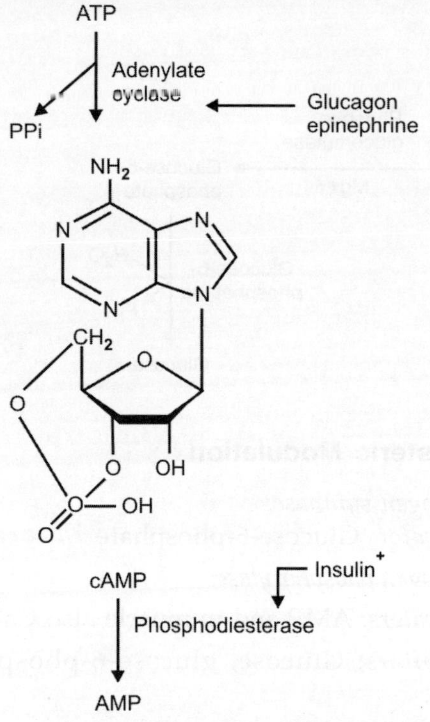


**Fig. 2.36:** Structure, formation and breakdown of cAMP

- cAMP may be destroyed by phosphodiesterase which is stimulated by insulin.

*ii. Activation of cAMP Dependent Protein Kinase*

- The cAMP dependant protein kinase in its inactive form consists of four subunits; two identical catalytic and two identical regulatory subunits.
- The cAMP binds to the regulatory subunits and brings about a conformational change in these subunits.
- This releases the catalytic subunits. The catalytic subunits are active only when they are dissociated from the regulatory subunit.
- The active protein kinase then brings about the phosphorylation of specific serine residues of the different enzymes using ATP.

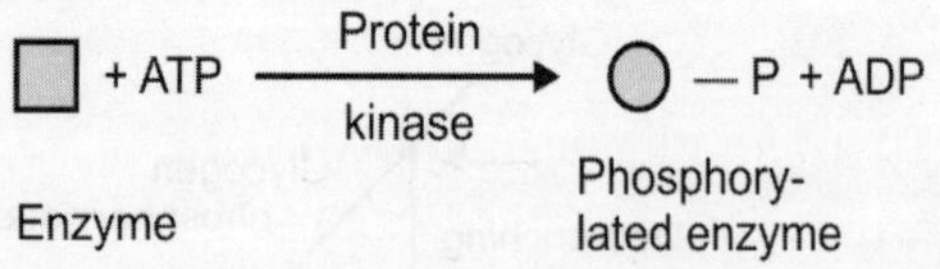


The phosphorylation of enzymes subject to regulation by covalent modification causes a change in conformation, which affects the active site. The change in conformation may greatly increase or greatly decrease the activity of the enzyme, depending on the enzyme involved. Only a few of the known enzymes are subject to this type of regulation called as *covalent modification*. Regardless of whether phosphorylation or dephosphorylation activates the enzyme, the active form of the enzyme is always called the *'a'* form and the inactive, the *'b'* form.

Likewise, regardless of the effect of phosphorylation on catalytic activity, the action of cAMP dependant protein kinase is always opposed by a phosphoprotein phosphatase which brings about dephosphorylation.

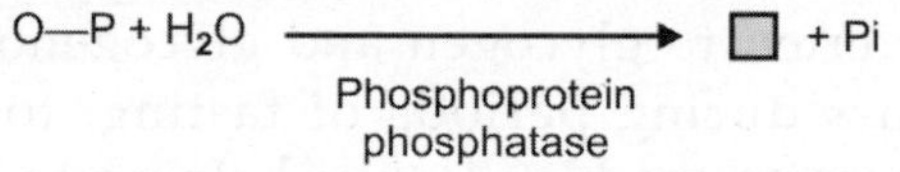


So, the relative activity of an enzyme is dependant on the activity of cAMP dependant protein kinase and phosphoprotein phosphatase, which in turn is dependant upon the relative amounts of the hormones present.

*Regulation of Glycogen Phosphorylase*

Glycogen phosphorylase exists in two forms: Active glycogen phosphorylase a, which is the phosphorylated form and inactive glycogen phosphorylase b, which is the dephosphorylated form.

Phosphorylase a is phosphorylated and active in the presence or absence of AMP.

Phosphorylase b is dephosphorylated and active only in the presence of AMP, its allosteric activator; so that, the enzyme is active during exercise when AMP level rises.

Phosphorylase is activated by epinephrine released during stress or glucagon released as a result of hypoglycaemia, by a cascade mechanism (Fig. 2.37).

- When these hormones bind to their receptors on target tissues, adenylate cyclase is activated.
- Adenylate cyclase converts ATP to cAMP.
- cAMP binds to the regulatory subunits of cAMP dependant protein kinase, releasing the catalytic subunits.
- The catalytic subunits of protein kinase phosphorylate phosphorylase kinase b to active phosphorylase kinase a.
- Phosphoprylase kinase a phosphorylates inactive phosphorylase b to phosphorylase a.

### *Role of Calcium*

Calcium also acts as a second messenger. Glycogenolysis increases several hundred fold immediately after the onset of contraction. Nervous excitation of muscle activity is mediated via changes in intracellular concentration of $Ca^{2+}$.

Nerve impulses cause membrane depolarization which in turn causes release of $Ca^{2+}$ from sarcoplasmic reticulum into the sarcoplasm of the muscle cells.

The phosphorylase kinase has four subunits:

- The alpha and beta subunits are phosphorylated by the cAMP dependant protein kinase.
- The delta subunit is identical to a regulatory protein called calmodulin, i.e., calcium modulating protein, which is present as a component of many proteins. In the presence of micromolar quantities of $Ca^{2+}$, calmodulin undergoes a conformational change, which is transmitted cooperatively to the catalytic subunit of specific enzyme kinases.

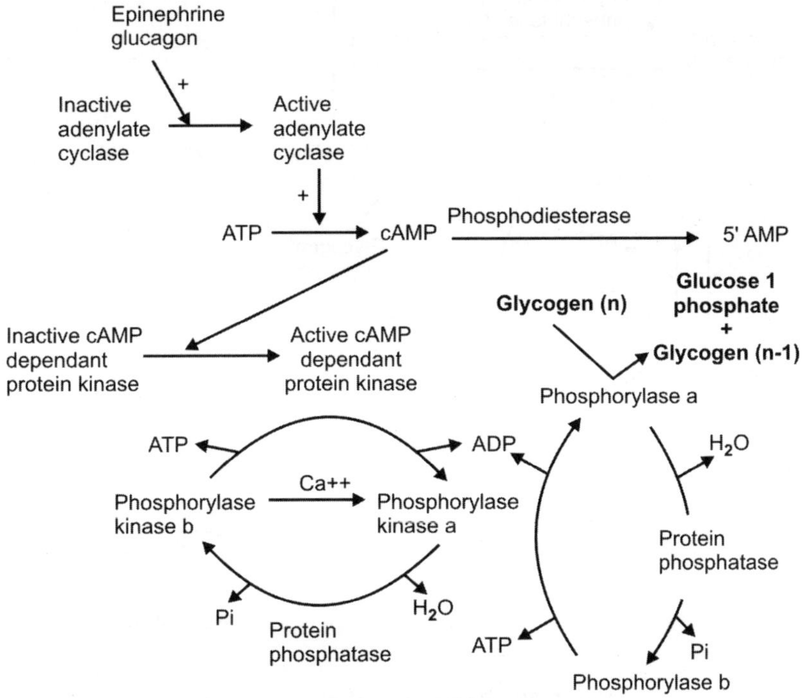


**Fig. 2.37:** Regulation of glycogenolysis

- The gamma subunit is the catalytic subunit, whose activity is regulated both by cAMP and $Ca^{2+}$.

Phosphorylase kinase activates glycogen phosphorylase which converts glycogen to glucose-6-phosphate. Glucose-6-phosphate enters the glycolytic pathway, producing ATP, to meet the demands of muscle contraction.

### *Regulation of Glycogen Synthase (Fig. 2.38)*

- Glycogen synthase also exists in two forms. But, unlike phosphorylase, active form in this case is dephosphorylated glycogen synthase *'a'* while phosphorylated glycogen synthase *'b'* is inactive.
- Glycogen synthase *a* may be inactivated to glycogen synthase *b* by phosphorylation on seven different serine residues by five different protein kinases. The enzyme has four identical subunits and each subunit has seven phosphorylation sites.
- Phosphorylation is catalyzed by cAMP dependent protein kinase, phosphorylase kinase and glycogen synthase kinase. As more sites are phosphorylated the $K_m$ of the enzyme for UDP glucose rises progressively and activity of the enzyme decreases progressively; requiring increasingly higher levels of glucose-6-phosphate for its allosteric activation.
- When phosphorylase is activated by a rise in concentration of cAMP, glycogen synthase is simultaneously converted to

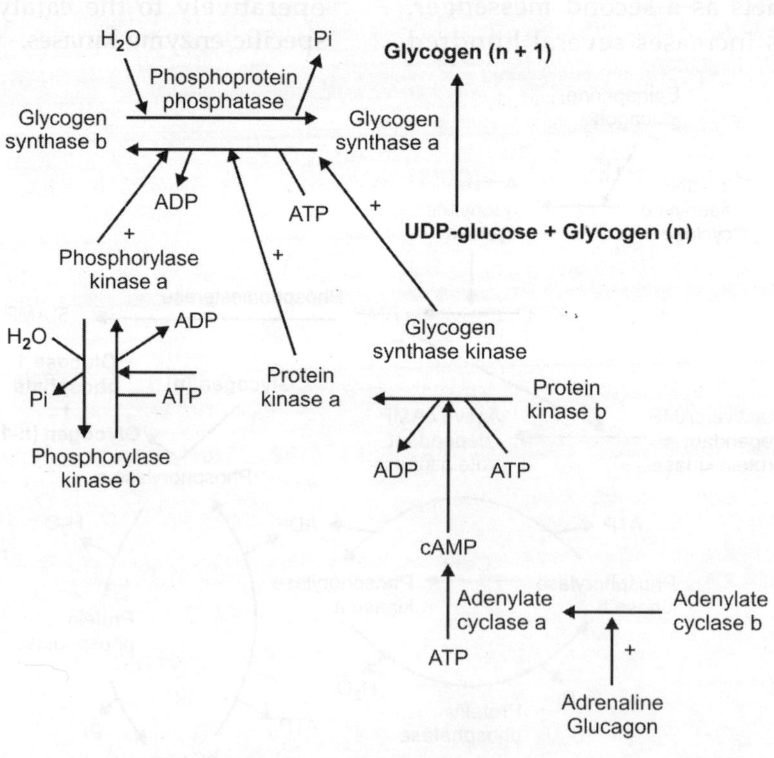


**Fig. 2.38:** Regulation of glycogenesis

the inactive form. Thus inhibition of glycogenesis enhances glycogenolysis and vice versa (Fig. 2.39). Both processes do not occur simultaneously since glycogen phosphorylase, and synthase are reciprocally regulated.

### Glycogen Storage Diseases (Glycogenoses)

- Are a group of inherited disorders involving glycogen metabolism, each characterised by deficiency of an enzyme (Table 2.9).
- They are characterised by deficient mobilisation and abnormal deposition of normal or abnormal forms of glycogen in muscle or liver.
- Characteristic feature of the defect in liver is fasting hypoglycaemia, along with hyperlipidemia and ketosis.
- Defect in a muscle enzyme leads to muscular weakness or even death.
- Diseases are designated by Roman numerals.
- All are inherited as autosomal recessive traits except Type VIII which is sex linked recessive trait.

## REGULATION OF BLOOD GLUCOSE LEVEL (TABLE 2.10)

Glucose is the principal sugar of mammalian blood. Normally, most of the constituents of the blood and other body fluids remain within fairly narrow limits of concentration.

When blood glucose level goes above normal range, the individual is said to have impaired glucose tolerance.

Condition where the values are above the normal range is termed *"hyperglycaemia"* and values are below the normal range is termed *"hypoglycaemia."*

Blood glucose level is maintained by a balance between two sets of factors (Fig. 2.40).

a. Rate of supply of glucose to the blood.
b. Rate of removal of glucose from the blood stream.

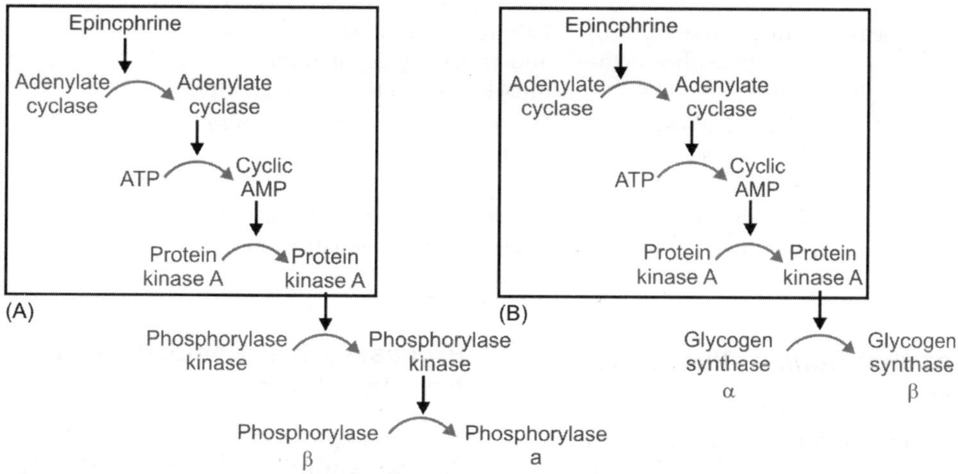


**Fig. 2.39:** Reciprocal regulation of glycogen phosphorylase and glycogen synthase

**Table 2.9:** Glycogen storage diseases

| *Type and name* | *Deficient enzyme* | *Symptoms* | *Treatment* |
|---|---|---|---|
| I Von Gierke's disease | Glucose-6-phosphatase | Glycogen deposition in the liver and kidney, hepatomegaly, fasting hypoglycaemia, hyperlipidemia | Frequent feeding in small amounts |
| II Pompe's disease | Lysosomal α-glucosidase (acid maltase) | Accumulation of glycogen in all tissues, hypoglycaemia, massive cardiac enlargement. Respiratory failure causes death in infancy | No effective treatment |
| III Cori's disease or Forbe's disease or Limit dextrinosis | Amylo 1,6 glucosidase or debranching enzyme | Accumulation of highly branched glycogen (limit dextrin), hepatomegaly, hypoglycaemia, ketosis | |
| IV Anderson's disease or Amylopectinosis | Amylo 1,4 →1,6 transglucosidase or branching enzyme | Accumulation of amylopectin with long, unbranched chains in liver, muscle, spleen. Death due to liver cirrhosis or cardiac failure in the 1st year of life | No effective treatment |
| V McArdle's disease | Muscle glycogen phosphorylase | Increased muscle glycogen, inability to utilise glycogen during exercise. So, inability to perform heavy exercise, low blood lactate level, severe muscle cramps | Avoid heavy exercise; otherwise healthy |
| VI Her's disease | Hepatic glycogen phosphorylase | Failure of hepatic glycogenolysis, high liver glycogen, hypoglycaemia, ketosis, hepatomegaly | |
| VII Tarui's disease | Muscle and erythrocyte phosphofructokinase | Failure of glycolysis in muscle and erythrocytes, increased muscle glycogen, muscle cramps | |
| VIII | Hepatic phosphorylase kinase | Decreased glycogenolysis, mild hypoglycaemia and mild hepatomegaly | |
| IX | Hepatic glycogen synthase | Failure of glycogenesis, fasting hypoglycaemia, seizures, mental impairment | |

### Processes that Supply Glucose to the Blood

1. Absorption of dietary carbohydrates
2. Glycogenolysis
3. Gluconeogenesis, from noncarbohydrate sources.

### Processes that Remove Glucose from the Blood

1. Glycolysis, both aerobic and anaerobic
2. Oxidation by HMP shunt pathway
3. Uronic acid pathway
4. Glycogenesis
5. Lipogenesis.

**Table 2.10:** Blood glucose level under various conditions

| | *Blood glucose level (mg/100 ml)* | | |
|---|---|---|---|
| | *Normal tolerance* | *Impaired glucose tolerance* | *Diabetes mellitus* |
| Fasting (12-14 hr after the last meal) | 70-110 | 110-126 | > 126 |
| Postprandial (2 hr after meal) | < 140 | 140-199 | > 200 |

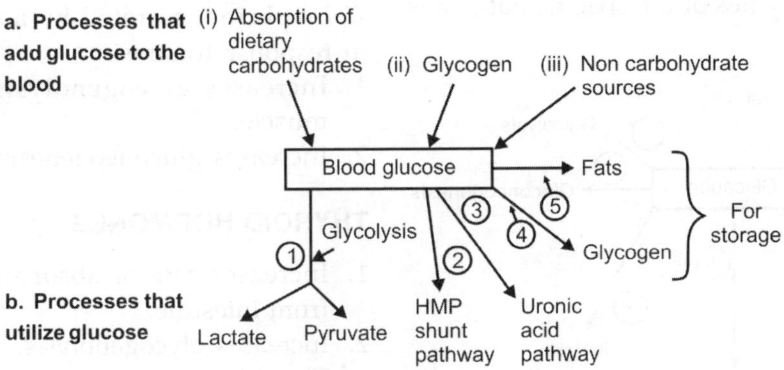


**Fig. 2.40:** Blood glucose homeostasis

Several hormones function to maintain the blood glucose levels within normal range.

Hormones which increase blood glucose level are called *hyperglycaemic hormones*. These are:

- Glucagon
- Epinephrine
- Thyroxine
- Glucocorticoids
- Adrenocorticotropic hormone (ACTH).

The action of these hormones is opposed by the *hypoglycaemic hormone*, which decreases the blood glucose level, i.e.,

- Insulin, secreted by the β-cells of the islets of Langerhans' of pancreas.

### Actions of Insulin (Fig. 2.41)

Hyperglycaemia induces the secretion of insulin, the action of which results in hypoglycaemia.

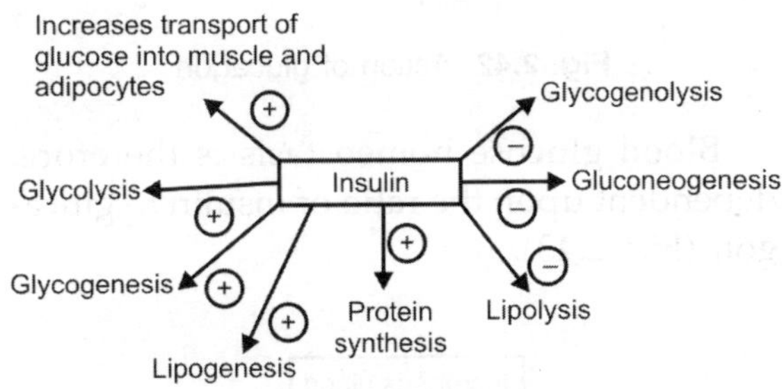


**Fig. 2.41:** Action of Insulin

1. Insulin enhances the uptake of glucose by tissues except the liver.
2. Increases glycolysis
3. Increases glycogenesis
4. Increases lipogenesis
5. Increases protein synthesis
6. Decreases glycogenolysis
7. Decreases gluconeogenesis
8. Decreases lipolysis.

In effect, insulin increases utilisation of glucose and decreases the formation of glucose.

Hypoglycaemia, on the other hand, causes release of glucagon and mobilisation of glucose from the liver.

All the actions of insulin and glucagon are brought about by increase or decrease in the regulatory enzymes of the specific pathways (Fig. 2.42).

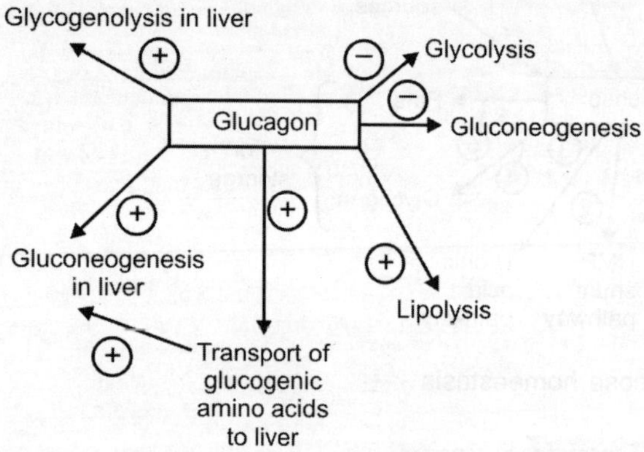


**Fig. 2.42:** Action of glucagon

Blood glucose homeostasis is therefore, dependent upon the ratio of insulin to glucagon (Fig. 2.43).

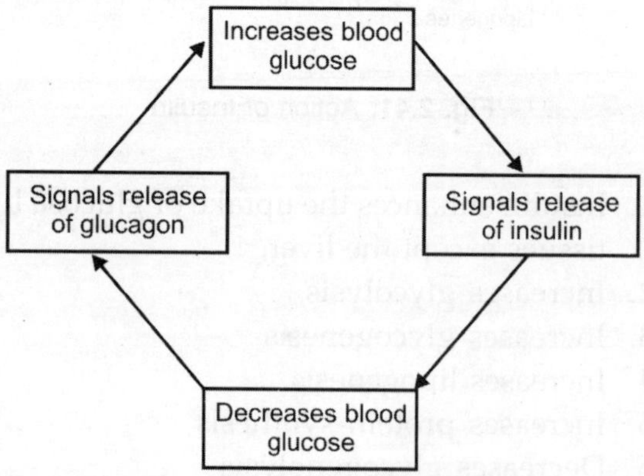


**Fig. 2.43:** Homeostatic control of glucose by the two key hormones

**Glucocorticoids** are released by adrenal cortex; actions are antagonistic to insulin.

1. Increases protein catabolism.
2. Increases hepatic uptake of amino acids.
3. Increases activity of transaminases.
4. Increases gluconeogenesis.
5. Decreases utilisation of glucose by extrahepatic tissues.

**Epinephrine,** secreted by adrenal medulla in response to stress.

1. Increases glycogenolysis in liver and muscle.
2. Increases gluconeogenesis.

## THYROID HORMONES

1. Increases rate of absorption of glucose from intestine.
2. Increases glycogenolysis.

Thus, the above hormones play an important role in maintaining the blood glucose level.

## RENAL THRESHOLD FOR GLUCOSE

Glucose that is filtered at the glomerulus is almost completely reabsorbed by the kidney tubules. The kidney tubules can reabsorb glucose, when the glucose that is filtered at the glomerulus is less than 350 mg/min.

If this increases, some glucose is excreted in the urine to cause glycosuria. This occurs when the blood glucose level rises above 160 mg/dl. This is termed the *renal threshold* for glucose.

## DIABETES MELLITUS

Diabetes mellitus is a chronic disease characterised by derangements in carbohydrate, fat and protein metabolism.

*Cause:* Due to a failure of either secretion or action of the hormone insulin.

### Effects

1. The transport of glucose into adipocytes and muscle cells is greatly reduced. Due to the inability to clear blood of glucose, the blood glucose level is elevated.
2. Since blood glucose level is increased, the renal threshold is often exceeded and glucose spills into the urine, causing *glycosuria.*
3. Since glucose is osmotically active, it is accompanied by increased excretion of urine—*polyuria.*
4. Since there is polyuria, there is increased loss of water resulting in excessive thirst – *polydypsia.*
5. Though blood glucose level is elevated, this glucose cannot be used by the tissues because of the absence of insulin, resulting in hunger—*polyphagia.*
6. The increased blood glucose leads to non-enzymic glucosylation of various proteins like the lens proteins — leading to the opacity of lens, basement membrane of the kidney leading to nephropathy, plasma albumin, haemoglobin, poor healing of wounds and neuropathy.

### Metabolic Consequences

Since the uptake of glucose to adipose tissue is decreased, there is decreased glycolysis leading to decreased formation of glycerol-3-phosphate required for synthesis of triacylglycerols. In the absence of insulin, there is no inhibition of hormone sensitive lipase causing lipolysis. All these result in increased level of free fatty acid (FFA) in the blood which gets accumulated in the liver causing *fatty liver*.

The absence of insulin favours glycogenolysis and gluconeogenesis, thereby the hyperglycaemic state is exaggerated.

Since the diabetic patient is not able to use glucose for energy purposes, the body uses FFAs for energy purposes. Fatty acids are mobilised from the adipose tissue and oxidized to acetyl CoA. All the acetyl CoA formed cannot be oxidized to $CO_2$ and $H_2O$ because of inadequate supply of oxaloacetate (oxaloacetate is a substrate for gluconeogenesis) and decreased activity of TCA cycle. So, the excess acetyl CoA forms ketone bodies, which accumulate and cause a condition called *ketoacidosis*.

All the fatty acids taken up by the liver cannot be oxidized, so, results in *hypertriglyceridemia*.

### Types

There are two types of diabetes mellitus:

1. *Type I* diabetes mellitus or *juvenile onset diabetes mellitus*.
   - Results from failure of β cells to secrete insulin, due to destruction of β cells (due to drugs, viruses or autoimmunity).
   - Usually starts in the young, abruptly, in a severe form.
   - Patient is usually thin and underweight.
   - Plasma insulin level decreases.
   - Susceptible to hypoglycaemia, in case of insulin overdosage.
2. *Type II* or *adult onset* diabetes mellitus.
   - Relatively mild and starts at an advanced age.
   - Arises from peripheral insensitivity to insulin.
   - Decreases binding of insulin to its receptor, which are also reduced in number.
   - Patient is usually obese and overweight.
   - Not prone to ketosis.
   - Plasma insulin level not decreased.
   - Not easily susceptible to insulin-overdose hypoglycaemia.

**Diagnosis**

1. By estimation of blood glucose—fasting and postprandial.
2. Testing urine for presence of sugar by Benedicts test.
3. Estimation of glycated haemoglobin ($HbA_{1C}$).

Normally 4-7% of the total haemoglobin is glycosylated. But in diabetes mellitus, this increases by 2-3 folds. Since the half-life of haemoglobin is about 60 days, the $HbA_{1C}$ level reflects the blood glucose level of the patient over the previous two months.

# 3 Lipids

Lipids include fats, oils, steroids, waxes and related compounds, which are heterogenous organic molecules, insoluble in water and soluble in organic solvents. Fats are important fuels, which can be stored in adipose tissue. Some of the lipids are major structural components and their ability to associate (so as to exclude water and other polar compounds) makes complex organisms possible through the formation of membranes. Membranes separate cells within the tissues and organelles within the cells, creating compartments, so as to permit distinct organisation and regulation. A brief outline of lipid chemistry and metabolism is given in this chapter.

## BIOMEDICAL IMPORTANCE

A knowledge of lipid biochemistry is essential for understanding the basis of major clinical problems such as atherosclerosis and obesity. Current biomedical areas of research include lipid peroxidation, antioxidants and role of polyunsaturated fatty acids in nutrition.

## CHEMISTRY

### Definition

Lipids are heterogenous group of compounds relatively insoluble in water (**hydrophobic**) and soluble in organic solvents (**nonpolar**) such as ether, benzene and chloroform.

### Functions of Lipids

1. Lipids are important dietary constituents, not only because of high energy value (9 kcal/gm) but also because of the fat soluble vitamins and the essential fatty acids contained in the fat of natural foods, e.g., milk, butter, meat.
2. Fat serves as an efficient source of energy both directly and potentially when stored in adipose tissue, e.g., triglycerides or triacylglycerols (TG or TAG).
3. Fat acts as thermal insulator in the subcutaneous tissues and around certain organs.
4. Combination of lipid and protein (lipoproteins) help in transporting lipids in the blood, e.g., chylomicrons, VLDL, LDL, HDL.
5. Some lipids are constituents of membranes and regulate the membrane permeability, e.g., phospholipids. Lipid content of nerve tissue is particularly high and they serve as electrical insulator allowing rapid propagation of depolarisation along the myelinated nerves, e.g. sphingomyelin.
6. Some lipids are important as cellular and metabolic regulators responsible for body homeostasis, e.g. prostaglandins and steroid hormones.

### Classification of Lipids (Modified from Bloor)

Broadly classified into:
- Simple lipids
- Compound lipids
- Derived lipids and precursor lipids

#### *Simple Lipids*

These are esters of *fatty acids* (FA) with different alcohols.

Simple lipids are further classified into fats and waxes.

**Fats:** These are esters of FA and glycerol, which is a trihydric alcohol, e.g., triglycerides.

**Waxes:** These are solid esters of long chain FA with higher molecular weight monohydric alcohols (containing 26-34 carbon atoms). For example, cetyl palmitate is a major component of bees wax.

### Compound Lipids

These are also esters of fatty acids (FA) with different alcohols, but in addition carry other nonlipid substances such as—

- Phosphate
- Carbohydrate
- Protein
- Nitrogenous base

Compound lipids are further divided into the following:

*Phospholipids*: Contain phosphoric acid and frequently a nitrogenous base in addition to FA and glycerol. Phospholipids are classified into two types.

i. Glycerophospholipids or phosphoglycerides—in which the alcohol is glycerol.
ii. Sphingophospholipids—in which the alcohol is sphingosine (amino alcohol).

*Glycolipids*: Also known as glycosphingolipids. These are fatty acid esters of sphingosine carrying a carbohydrate. They do not contain phosphoric acid, e.g., cerebrosides, gangliosides and sulphatides.

*Lipoproteins*: These are macromolecular complexes of lipids and proteins. The lipid part may be fats, phospholipids, cholesterol and its ester, e.g., chylomicrons, VLDL, LDL, HDL.

### Derived Lipids and Precursor Lipids

These are obtained from hydrolysis of simple or compound lipids. Examples are:

- Fatty acids
- Steroids and sterols
- Alcohols such as glycerol, sphingosine, cetyl alcohol
- Hydrocarbons, fat soluble vitamins
- Polyisoprenoids such as dolichol and ubiquinone
- Eicosanoids

*Note: Glycerides, cholesterol and cholesteryl esters being uncharged are termed neutral lipids.*

We have seen that, fatty acids are integral components of lipids. Now let us know some of the important aspects and nomenclature of fatty acids.

*Fatty acids:* These are water insoluble long chain hydrocarbons with one COOH group at the end of the chain, which may be saturated or unsaturated. They have the general structural formula, R–COOH where R represents an alkyl chain composed of carbon and hydrogen atoms. COOH or carboxylic group represents the functional group.

- Fatty acids occur mainly as esters of glycerol in natural fat and oils.
- In unesterified form they occur as free fatty acids (FFA). In plasma they are transported in the unesterified form.
- Fatty acids predominantly occurring in natural fats are straight chain derivatives containing even number of carbon atoms, e.g., palmitic acid 16C, stearic acid 18C. A few odd chain fatty acids occur, among which propionic acid is most important.
- The alkyl chain may be saturated or unsaturated.
- Plants and fish fats contain more polyunsaturated fatty acids than animal fats.

*Saturated Fatty Acids*: Contain no double bonds. Some of the important fatty acids and their sources are given in Table 3.1.

*Unsaturated Fatty Acids:* Contain one or more double bonds and are subdivided as follows.

**Table 3.1:** Commonly occurring saturated fatty acids

| *Common names* | *Number of C atoms* | *Sources* |
|---|---|---|
| Acetic acid | 2 | End product of carbohydrate digestion by ruminants |
| Propionic acid | 3 | Same as above |
| Butyric acid | 4 | Present in butter |
| Lauric acid | 12 | Coconut oils, cinnamon, palm kernel |
| Myrisitic acid | 14 | Nutmeg, palm kernel, coconut oils |
| Palmitic acid | 16 | Common in all animal and plant fats |
| Stearic acid | 18 | |
| Arachidic acid | 20 | Peanut oil |
| Lignoceric acid | 24 | Cerebrosides, peanut oil |

**Table 3.2:** Commonly occurring unsaturated fatty acids

| *Common name* | *Number of* | | *Sources* |
|---|---|---|---|
| | *Carbon atoms* | *Double bonds* | |
| Palmitoleic acid | 16 | 1 | Present in all fats |
| Oleic acid | 18 | 1 | Most common fatty acid in natural fats |
| Linoleic acid | 18 | 2 | Corn, peanut, cotton seed and many plant oils |
| Linolenic acid | 18 | 3 | Linseed oil |
| Arachidonic acid | 20 | 4 | Animal fats and peanut oil. Important component of phospholipids in animals |
| Timnodonic acid | 20 | 5 | Important component of fish oils |

- Monounsaturated →has one double bond
- Polyunsaturated → has two or more double bonds

Some of the important unsaturated fatty acids and their sources are given in Table 3.2.

## Nomenclature

Systematic naming of a fatty acid is based on the number of carbons, with *–oic* being substituted for final – *e*. Thus saturated fatty acids end in *"anoic"* and unsaturated acids end in *"enoic* acid", e.g.,

Octadecane ($C_{18}$) – octadec*anoic acid*

Octadecane ($C_{18}$ : 1) – octadec*enoic acid*

- Carbon atoms are numbered from the carboxyl carbon (Carbon No. 1).

The carbon atoms adjacent to the carboxyl carbon i.e., 2, 3, 4 are also known as α, β and γ carbons respectively. Terminal methyl carbon is known as ω or n-carbon.

$$\underset{\omega\,(n)}{\overset{6}{CH_3}} - \overset{5}{CH_2} - \underset{\gamma}{\overset{4}{CH_2}} - \underset{\beta}{\overset{3}{CH_2}} - \underset{\alpha}{\overset{2}{CH_2}} - \overset{1}{COOH}$$

- Various conventions are used for indicating the number and position of the double bonds. The short hand notations for representing the fatty acids are generally used (Fig. 3.1). The general rule is that the total number of carbon atoms are written first, followed by the number

| | | |
|---|---|---|
| Palmitic acid | 16:0 | $CH_2 - (CH_2)_{14} - COOH$ |
| Oleic acid | 18:1;9 | 9<br>$CH_3 - (CH_2)_7\ CH = CH - (CH_2)_7 - COOH$ |
| Linolenic acid | 18:3;9,12,15 | 15 12 9<br>$CH_3 - CH_2 - CH = CH - CH_2 - CH = CH - CH_2 - CH = CH - (CH_2)_7 - COOH$ |

**Fig. 3.1:** Structures of some unsaturated fatty acids

of double bonds and finally position of the double bonds, starting from the carboxyl end.

*Note:*

9 – *Indicates position of double bond relative to the carboxyl end. Hence double bond is present between 9 and 10 and not 8 and 9.*

12 – *Indicates position of double bond between 12 and 13.*

- The ω (n) numbering system is based on the numbering of carbon atom starting from the methyl terminal, e.g.,

$$\underset{\omega}{CH_3} - \underset{\omega 1}{CH_2} - \underset{\omega 2}{CH_2} - \underset{\omega 3}{CH_2} - CH_2 \text{------} COOH$$

ω3 indicates double bond on the third carbon counting from the ω carbon.

In the ω-numbering system palmitoleic acid (16:1) is referred to as ω-7 or n-7, indicating that unsaturated bond is located seven carbon atoms away from ω-carbon atom.

- In animals additional double bonds are introduced only between the existing double bond and the carboxyl carbon, leading to four series of fatty acids known as ω9, ω7, ω6 and ω3 families (Table 3.3). Each class is made up of a family of fatty acids and all members of that family can be synthesised biologically from the parent fatty acid. For example, arachidonic acid is synthesized from the parent fatty acid of the ω6 class linoleic acid. A fatty acid of one class cannot be converted biologically to another class.

**Table 3.3:** Various families of unsaturated fatty acids

| Class | Parent fatty acid | Structure |
|---|---|---|
| ω-7 | Palmitoleic acid | 9<br>$CH_3 - (CH_2)_5 - CH = CH - (CH_2)_7 - COOH$ |
| ω-9 | Oleic acid | 9<br>$CH_3 - (CH_2)_7 - CH = CH - (CH_2)_7 - COOH$ |
| ω-6 | Linoleic acid | 12<br>$CH_3 - (CH_2)_4 - CH = CH - CH_2 - CH = CH - (CH_2)_7 - COOH$ |
| | Arachidonic | 14<br>$CH_3 - (CH_2)_4 - CH = CH - CH_2 - CH = CH - CH_2 - CH$<br>8 5<br>$= CH{-}CH_2 - CH = CH - (CH_2)_3 - COOH$ |
| ω-3 | Linolenic acid | 15 12 9<br>$CH_3 - CH_2 - CH = CH - CH_2 - CH = CH - CH_2 - CH = CH - (CH_2)_7 - COOH$ |

*Eicosanoids:* These are compounds derived from $C_{20}$ (Eicosa) polyenoic fatty acids. These comprise the prostanoids, leukotrienes (LT) and lipoxins (LXs).

### Isomerism in Naturally Occurring Fatty Acids (Fig. 3.2)

A type of geometric isomerism occurs in unsaturated fatty acids depending on the orientation of the atoms or groups around axis of the double bonds, which do not allow rotation. If the acyl chains are on the same side of the bond it is *cis* configuration, e.g., oleic acid. If the acyl chains are on the opposite sides, it is *trans* configuration, e.g., *elaidic acid.*

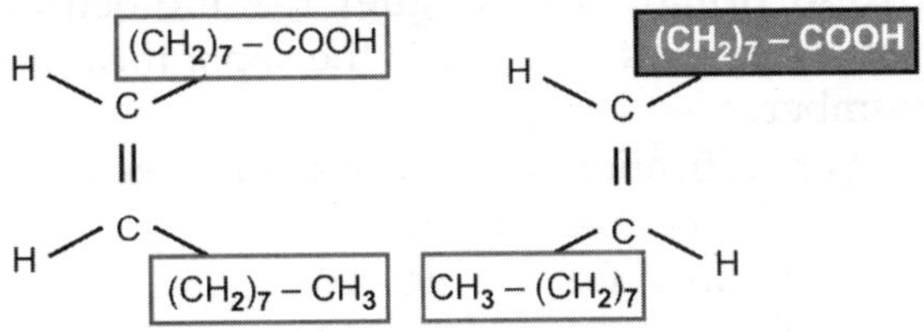

Oleic acid — *Cis form*

Elaidic acid — *trans form*

**Fig. 3.2:** Isomerism in fatty acids

- Naturally occurring unsaturated long chain fatty acids are of the ***cis***—configuration, the molecules being bent 120° at the double bonds.
- If there is more than one double bond they are spaced at three carbon intervals.
- Increase in the number of cis double bonds leads to a variety of possible spatial configurations of the molecule. This influences the fluidity and has profound significance on the *molecular packing in membrane normal*.

### Essential Fatty Acids (EFA)

Some of the PUFA cannot be synthesised by the body and have to be obtained in the diet and are called *essential fatty acids.* (linoleic 18:2; 9,12 and linolenic 18:3; 9,12,15). There are no human enzyme systems that can introduce a double bond beyond the 9th carbon atom (9-10 position) of a fatty acid chain.

#### *Importance*

- EFA are components of phospholipids and biomembranes.
- Required for synthesis of eicosanoids
- Required for the esterification and excretion of cholesterol and therefore are hypocholesterolemic and have antiatherogenic effect.
- Prevents fatty liver, by formation of VLDL, which helps in mobilisation of fats from the liver.

*Source:* Vegetable oils like ground nut oil, gingelly oil.

Deficiency of EFA causes **phrynoderma** (toad like skin), which is characterised by horny eruptions over back and buttocks, loss of hair and poor wound healing.

#### *Clinical Importance*

*Dietary PUFA:* Ordinary dietary oils rich in PUFA contain predominantly linoleic acid belonging to ω6 series.

- Fish oils such as menhaden oil and salmon oil are rich in ω3 PUFA especially eicosapentaenoic acid (EPA; 20:5). It is effective in reducing plasma TAG concentration and also appears to protect against thrombosis and certain inflammatory diseases.
- There is a growing interest in marine oils for therapeutic diets instead of plant oils that contain ω6 PUFA.

### Triglycerides or Triacylglycerols (TAG)

Are esters of glycerol and three fatty acids.

Glycerol ⎡ FA<br>
⎢ FA<br>
⎣ FA

These triglycerides are also termed as *"neutral fats"* and are the main storage form of fatty acids.

- Monoacyl and diacylglycerol, where in one and two fatty acids are esterified with glycerol respectively are found in the tissues.These are formed as intermediates in the synthesis and hydrolysis of triacyl glycerols.
- As per the IUB—International Union of Biochemistry nomenclature, the carbon atoms of glycerol are designated as α, β and α[1] or 1, 2 and 3 as shown in Figure 3.3.

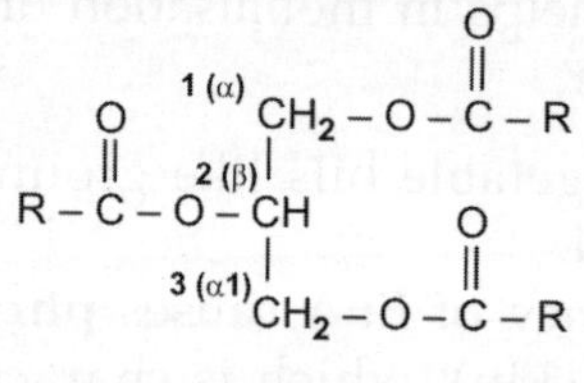


**Fig. 3.3:** Structure of triglyceride

- If all three hydroxyl (OH) groups of glycerol are esterified with the same fatty acid, it is termed as *simple triglyceride*, e.g., Tripalmitin.
- When different fatty acids are esterified to the hydroxyl group of glycerol, it is termed as *mixed triglyceride*, e.g., 1, palmitoyl, 2,3 distearin.
- When a PUFA is present, it is commonly esterified to the 2nd or β carbon.
- Naturally occurring fats and oils are mixed triglycerides. Triglycerides which are liquid at room temperature are referred to as oils. Their melting point decreases with the increase in content of unsaturated fatty acids.

### *Characteristics of fats and oils*

- Fats are rich in long chain saturated fatty acid and remain solid at room temperature, whereas oils contain unsaturated short chain fatty acids and are liquids at room temperature.
- Fats have higher melting point whereas oils have lower melting point.
- Fats containing long chain saturated fatty acids are termed hard fat (e.g., lard), while those containing medium chain unsaturated fatty acids are termed soft fat (e.g., butter).

## Indices to assess the quality of fat

***Saponification number***—Defined as number of mg of potassium hydroxide (KOH) required to saponify 1 gm of fat. It is an index of molecular weight of the fat and is inversely proportional to it. Higher the molecular weight of the fat, lower is the saponification number.

Saponification number of some important fats / oils are given below:

- Human fat 94 - 198
- Butter 10 - 230
- Coconut oil 53 - 262

***Iodine number***: It is defined as the number of grams of iodine, taken up by 100 grams of fat. It is an index of degree of unsaturation and directly proportional to the content of unsaturated fatty acids. Higher the iodine number greater is the degree of unsaturation.

Iodine number of some important oils / fats are given below:

- Coconut oil — 7-10
- Butter — 25-28
- Groundnut oil — 85-100
- Sunflower oil — 125-135

Thus sunflower oil has more unsaturated fatty acids compared to the other oils.

## Rancidity

The term *'rancidity'* refers to the offensive odour and taste of fats and oils on exposure to air, moisture, warm temperature and some metals.

Rancidity is of two types—

i. Hydrolytic rancidity—due to partial hydrolysis of ester bonds in the fat by traces of hydrolytic enzymes present in naturally occurring fats and oils.
ii. Oxidative rancidity—results from partial oxidation of unsaturated fatty acid with formation of epoxides and peroxides.

## Waxes

- Waxes are water insoluble, solid esters of higher fatty acids with long chain monohydric alcohols (50–100 carbon atom) or with sterols. They are widespread in nature and form secretions of insects, leaves and fruits of plants. For example,
  - — Cetyl palmitate is a major component of bees wax.
  - — Lanolin (wool fat), is a mixture of fatty acid esters of the sterols, lanosterol and agnosterol.
- They can take up water without dissolving and are used as a base for ointments and cosmetics.
- They are also used for the preparation of polishes, lubricants and candles.

## Phospholipids

Are the major lipid constituents of cellular membrane and occur in high concentration in the lipids of glandular organs, plasma, egg yolk and in the seeds of legumes.

- ***Glycerophospholipids or phosphoglycerides*** contain glycerol as alcohol.
- ***Sphingophospholipids*** contain sphingosine as alcohol.

The various glycerophospholipids (phosphoglycerides) are given below. All of these are *phosphoacylglycerols*. It is to be noted that most phospholipids have a saturated acyl group in the 1 position and an unsaturated acyl group in the 2 position of glycerol (Figs 3.4A to I).

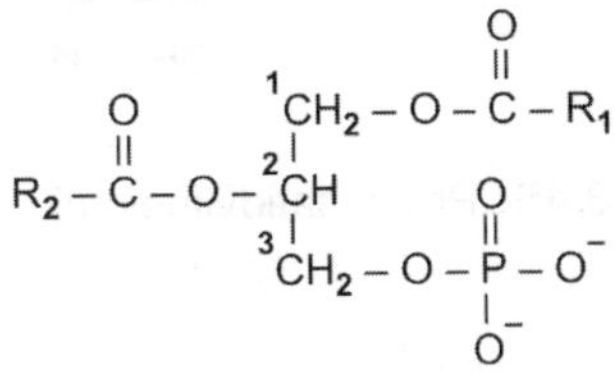

**Fig. 3.4A:** Phosphatidic acid

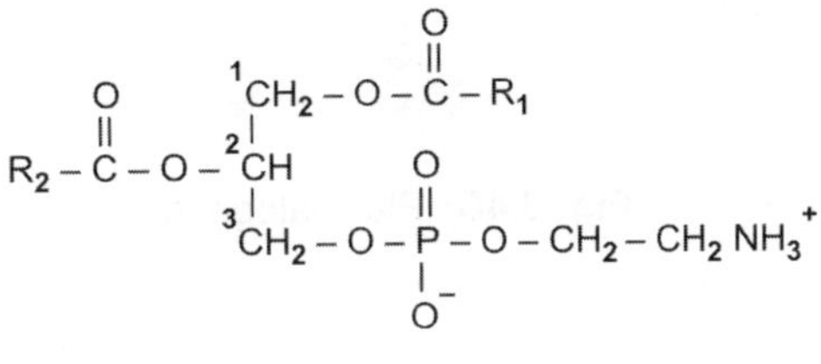

**Fig. 3.4B:** Phosphatidyl ethanolamine

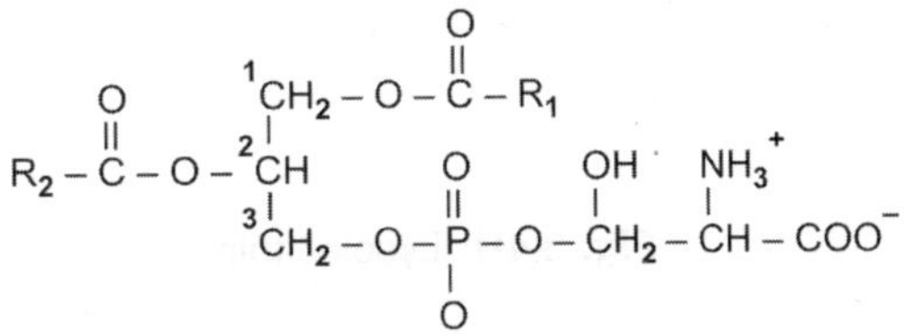

**Fig. 3.4C:** Phosphatidyl serine

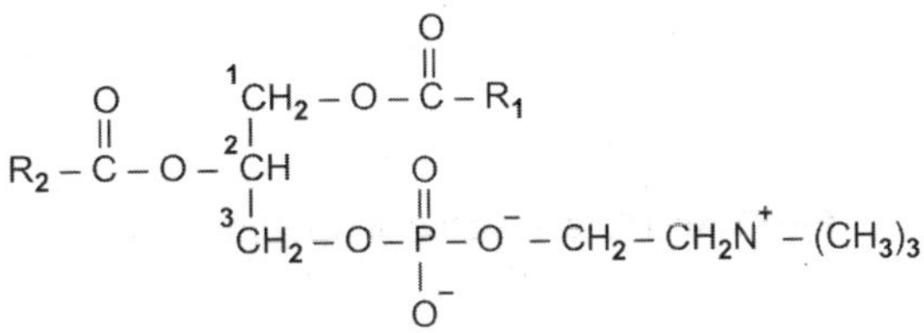

**Fig. 3.4D:** Phosphatidylcholine

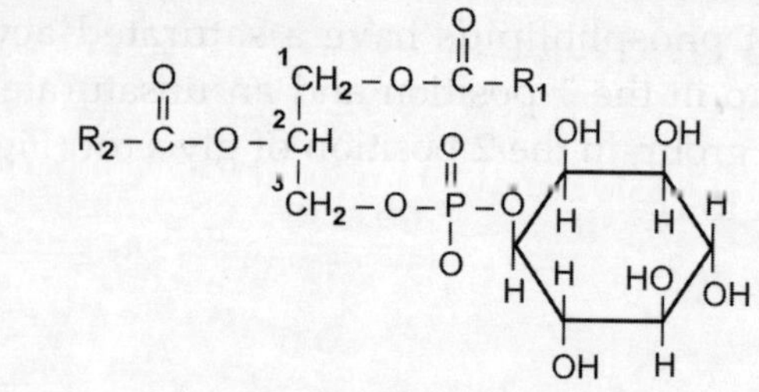

Fig. 3.4E: Phosphatidylinositol

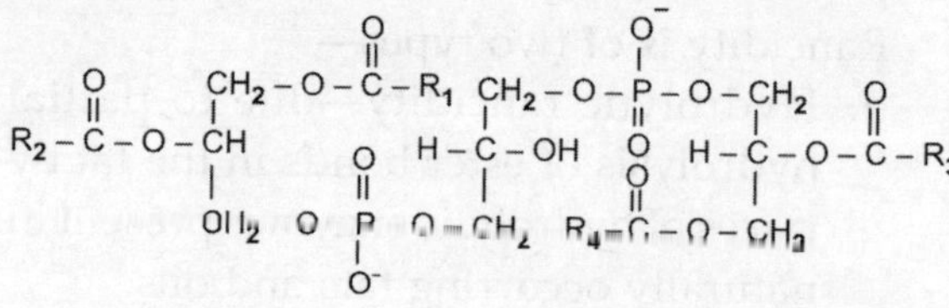

Fig. 3.4F: Diphosphatidylglycerol

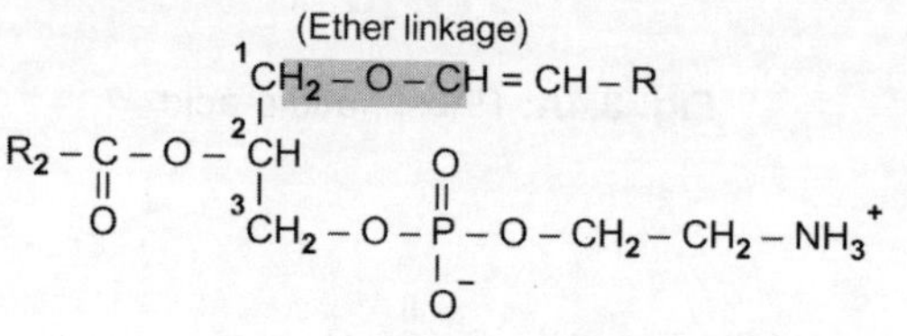

Fig. 3.4G: Plasmalogens

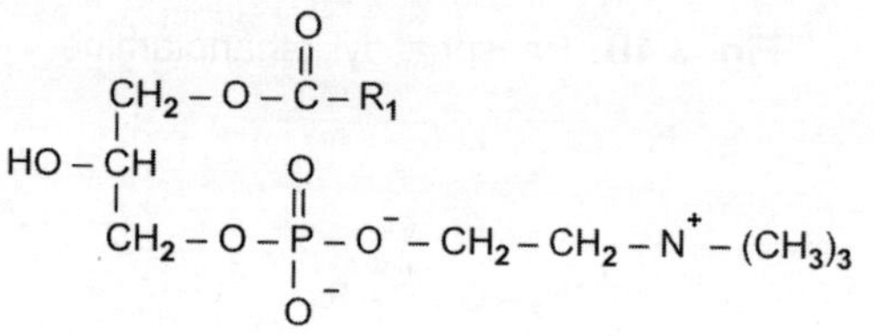

Fig. 3.4H: Lysolecithin

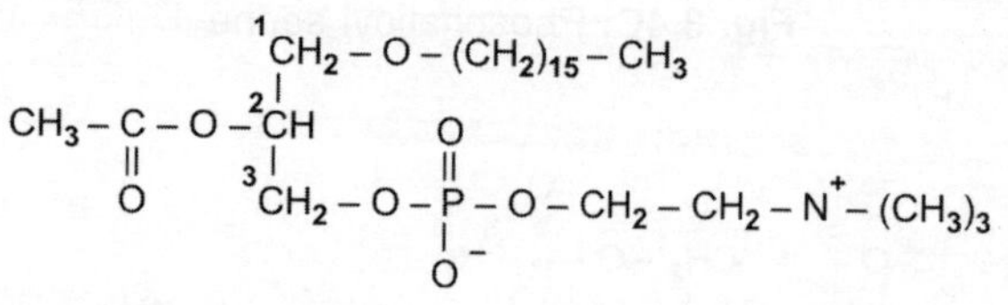

Fig. 3.4I: Platelet activating factor (PAF)

1. *Phosphatidic acid*: is the simplest of the phospholipids. Contains a molecule of phosphoric acid esterified to diacyl glycerol. The various glycerophospholipids may be regarded as derivatives of phosphatidic acid.

*Function*

- It is important as an intermediate in the synthesis of triacylglycerol as well as phosphoglycerides, but is not found in great quantity in any tissue.

*2. Phosphatidylethanolamine (Cephalin)*

Contains the nitrogenous base ethanolamine which is esterified to phosphatidic acid. It is found in most tissues.

*3. Phosphatidylserine*

Contains the amino acid serine which is esterified to phosphatidic acid. Like cephalin, it is also found in most of the tissues.

*4. Phosphatidylcholine (Lecithin)*

- The base choline is replace ethanolamine.
- Phosphoacyl glycerol containing choline are the most abundant phospholipids of cell membrane and represent a large proportion of the body's store of choline.

- Dipalmitoyl lecithin is a very effective surfactive agent and is a major constituent of lung surfactant lining the alveoli. It prevents adherence, due to surface tension, of the inner surface of the lungs. Its deficiency in lungs of premature infants causes *respiratory distress syndrome.*
- Choline is important in nerve transmission as acetyl choline.
- It is also a store of labile methyl groups.

### 5. *Phosphatidylinositol*

Has the carbohydrate myo-inositol which is esterified to phosphoric acid.

- It is further phosphorylated to phosphatidylinositol 4,5-bisphosphate which is an important constituent of cell membrane and has a role in signal transduction. On stimulation by suitable hormone agonists, it is cleaved into diacylglycerol (DAG) and inositol triphosphate ($IP_3$), both of which act as internal signals or "second messengers." Thus, phosphatidylinositol bisphosphate is a precursor of second messengers.

### 6. *Diphosphatidylglycerol or Cardiolipin*

Has two phosphatidic acid molecules attached to glycerol.

- Cardiolipin is a major lipid present in inner mitochondrial membrane.
- Only phospholipid which has antigenic properties.

### 7. *Plasmalogens (Phosphatidyl Ethanolamine)*

Structurally, plasmalogens resemble phosphatidyl ethanolamine, but possess an ether linkage on the 1st carbon, instead of ester linkage found in acyl glycerols. Typically the alkyl radical at position 1 is an unsaturated alcohol. Ethanolamine may often be replaced by choline, serine or inositol.

### 8. *Lysophospholipids*

Are those phospholipids in which one of the two acyl groups is removed. Most common is lysophosphatidyl choline (lysolecithin). Two forms are present. One contains the acyl group in sn-1 position and the other in sn-2 position. They have strong detergent action.

*Note: One of the poisonous principles in snake venom is the enzyme phospholipase which produces lysolecithin and is responsible for the lysis of RBC.*

### 9. *Platelet Activating Factor (PAF)*

Is an alkyl phospholipid. It has the following structural features.

1. Presence of a saturated alkyl chain containing 16 carbon atoms at 1 position.
2. Presence of acetic acid at second position. Therefore, PAF is 1-o-alkyl-2 acetyl glycerylphosphorylcholine.

Naturally occurring phosphatidal choline derivative has several potent physiologic actions.

- Activates the blood platelets causing them to aggregate.
- Reduces blood pressure.
- Mediates inflammation.

## Sphingophospholipids

Contain the complex amino alcohol **sphingosine** unlike glycerol. The combination of sphingosine and a fatty acid is known as ceramide.

Sphingosine is an unsaturated amine alcohol containing 18 carbon atoms and 2 hydroxyl groups. The fatty acid is attached to the amine group. The terminal hydroxyl group is esterified with the phosphoric acid and contains the nitrogenous base choline.

An important example of a sphingophospholipid is *sphingomyelin* (Fig. 3.5).

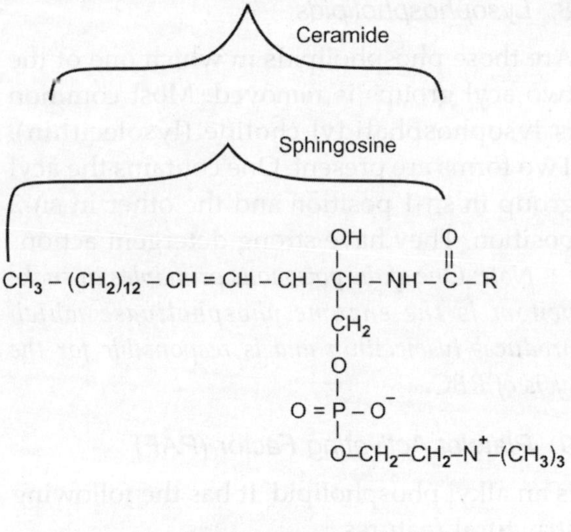


**Fig. 3.5:** Structure of sphingomyelin (R = Fatty acid)

- ***Sphingomyelin:*** They are found in large quantity in the brain and nerve tissue. On hydrolysis the sphingomyelins yield a fatty acid, phosphoric acid, choline and the complex amino alcohol sphingosine.

*Glycolipids (Glycosphingolipids)*: Belong to the class of sphingolipids. They contain ceramide and one or more sugars but no phosphoric acid. Simple glycolipids contain sphingosine, galactose or glucose and a high molecular weight fatty acid.

- Widely distributed in the brain and the cell membrane.
- They occur particularly in the outer leaflet of the cell membrane contributing to cell surface carbohydrates.

Four classes of glycolipids have been distinguished.

*Cerebrosides:* They are ceramide monohexosides. Most important of them being galactocerebroside (galactosyl ceramide) and glucocerebroside.

*Galactosyl* ceramide is the major glycolipid of the brain and other nervous tissue (Fig. 3.6).

*Glucosyl ceramide* is a predominant glycolipid of the extra neural tissues.

Cerebrosides are differentiated by the type of fatty acid in the molecule. They contain characteristic $C_{24}$—fatty acids belonging to four main types.

i. Kerasins: Lignoceric acid
ii. Cerebrons: Cerebronic acid (α-hydroxy lignoceric acid)
iii. Nervons: Nervonic acid (unsaturated homolog of lignoceric acid)
iv. Oxy nervons: Hydroxynervonic acid

*Sulfatides or Sulfolipids* — These are cerebrosides containing a sulfate group (Fig. 3.7) which is substituted at one of the OH group of the galactose moiety. For example, β

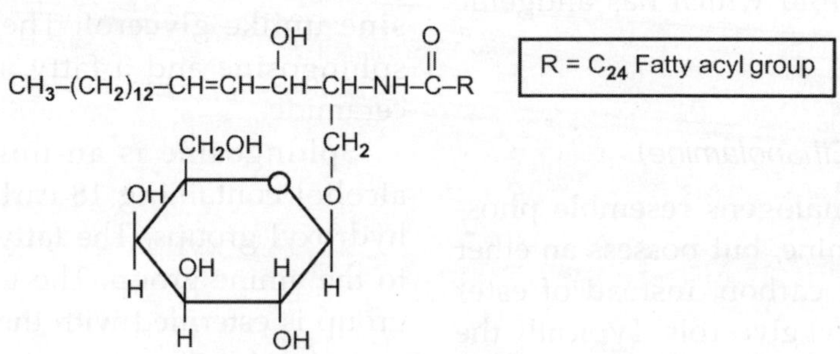


**Fig. 3.6:** Structure of galactosyl ceramide

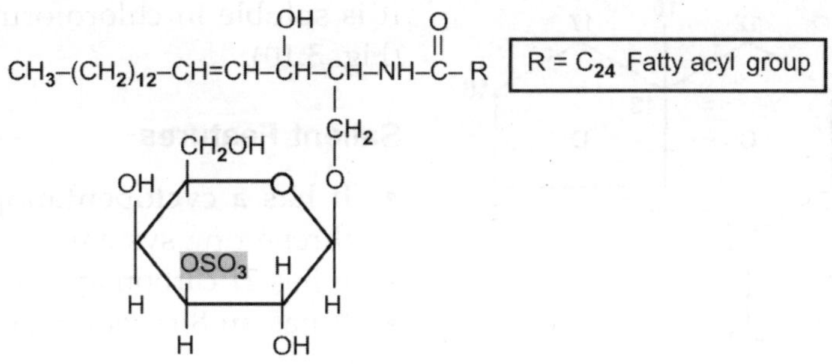


**Fig. 3.7:** Structure of sulphatide

sulfogalacto-cerebroside is a major sulfolipid in the white matter of CNS.

*Globosides*—These are ceramide oligosaccharides. They contain two or more sugar molecules (most often galactose, glucose or N-acetylgalactosamine). Basic structure is represented in the abbreviated form as shown below.

Gal NAc——Gal——Gal——Glc——**Cer**

An important example for globoside is *Lactosyl ceramide* formed during the breakdown of sphingolipids

### *Gangliosides (Fig. 3.8)*

- Are ceramide oligosaccharides that have sialic acid. For example, N-acetyl-neuraminic acid (NANA) and 3 molecules of

$GM_3$ : GAL – GLC – CER
|
NANA

$GM_2$ : GALNAC – GAL – GLC – CER
|
NANA

$GM_1$ : GAL – GALNAC – GAL – GLC – CER
|
NANA

**Fig. 3.8:** Important gangliosides

hexose. Hexosamine is a common constituent of naturally occurring gangliosides.

- NANA content varies between 1-5 molecules
- They are designated as $GM_1$, $GM_2$, $GM_3$ etc depending on the complexity of the oligosaccharides present.
- The simplest ganglioside found in most tissues is $GM_3$.
- They are found in higher concentration in ganglion cells of the CNS and lower concentration in the membrane of most cells.
- They act as receptors.

*Note: G represents ganglioside; M represents monosialo containing species; the subscript 1, 2, 3 etc, is a number assigned on the basis of chromatographic separations.*

### Steroids (Fig. 3.9)

***Chemistry:*** These are derivatives of cyclopentanoperhydrophenanthrene ring.

- The ring A, B and C represent the perhydro (completely saturated) phenanthrene ring to which is attached a cyclopentane ring 'D'.
- It is to be noted that, in structural formula of steroids, simple, hexagonal ring denotes, completely saturated 6-carbon ring.

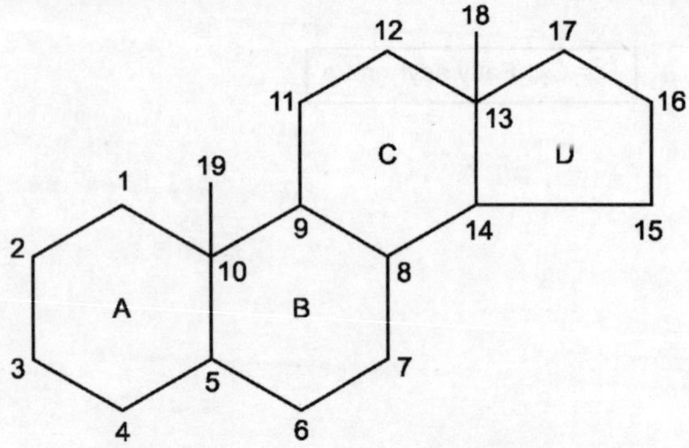


**Fig. 3.9:** Structure of the steroid ring

- The methyl ($CH_3$) side chain – $C_{18}$ and $C_{19}$ are shown as single bonds that are extending out from $C_{10}$ and $C_{13}$.

## Sterols

These are class of steroids that contain the hydroxyl (OH) group at $C_3$ and an aliphatic chain of at least 8-carbon atoms attached to it. An important example is cholesterol.

### *Cholesterol*

Is the main sterol in the human tissue. The word cholesterol is derived from the greek word, *chole*-bile, *steros*-solid, *ol*-alcohol. Cholesterol is a light yellow crystalline solid. It is soluble in chloroform and fat solvents (Fig. 3.10).

## Salient Features

- It has a cyclopentanoperhydrophenanthrene ring system
- It has 27 carbon atoms
- It has an 8 carbon side chain attached to $C_{17}$.
- The hydroxyl group (OH) in cholesterol is at $C_3$.
- A double bond is present in the B ring between carbon atoms 5 and 6
- Presence of methyl groups at positions 10, 13, 20 and 25
- Esterification of cholesterol takes place by the attachment of fatty acid residue to the OH group.

### *Function*

- Cholesterol is widely distributed in the human body. It occurs only in animals.
- It is a major constituent of plasma membrane.
- It is the precursor from which the following biochemically important products are synthesised.

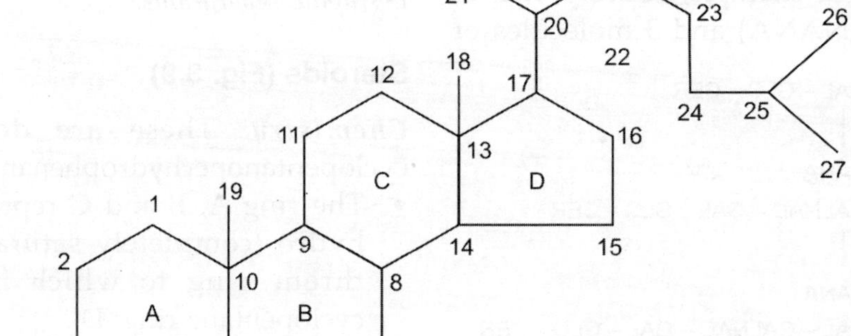


**Fig. 3.10:** Structure of cholesterol

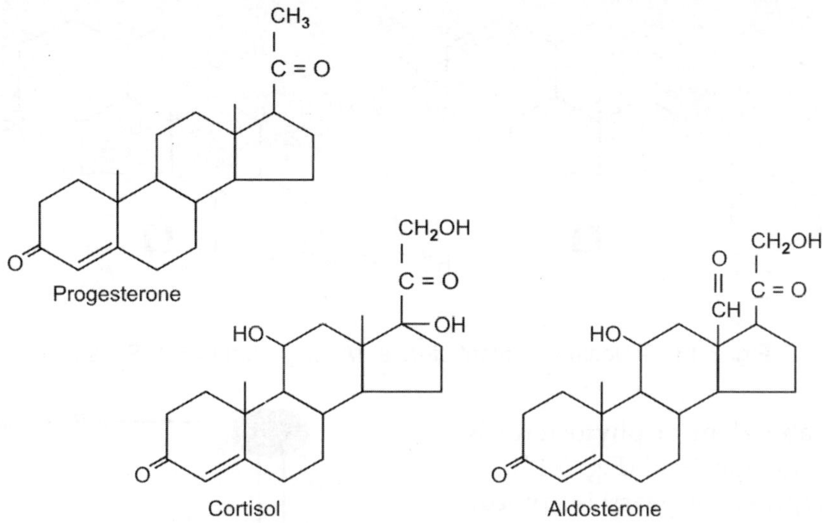


**Fig. 3.11:** Structures of some steroid hormones

— Steroid hormones
— Bile acids
— Vitamin D.

- The various steroid hormones synthesised from cholesterol are given below (Fig. 3.11).

| | |
|---|---|
| — Cortisol, | Adrenal cortex |
| — Cortisone | Adrenal cortex |
| — Aldosterone | Adrenal cortex |
| — Testosterone | — Testis |
| — Estradiol | — Ovary |
| — Progesterone | — Placenta |

- Bile acids derived from cholesterol are cholic, chenodeoxycholic, and deoxycholic acids. Bile acids play an important role in the digestion and absorption of lipids (Fig. 3.12).
- The fat soluble vitamin $D_3$ (cholecalciferol) is also derived from cholesterol.

### *Plant Sterols*

- Plants do not contain cholesterol but they have other sterols that are known as phytosterols.

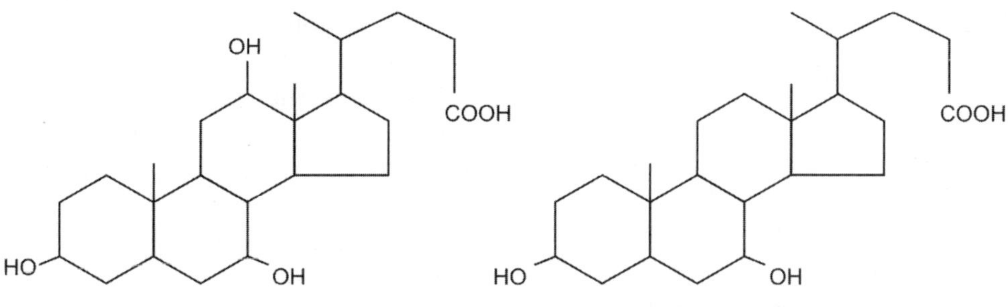


Cholic acid

Chenodeoxycholic acid

**Fig. 3.12:** Structures of bile acids

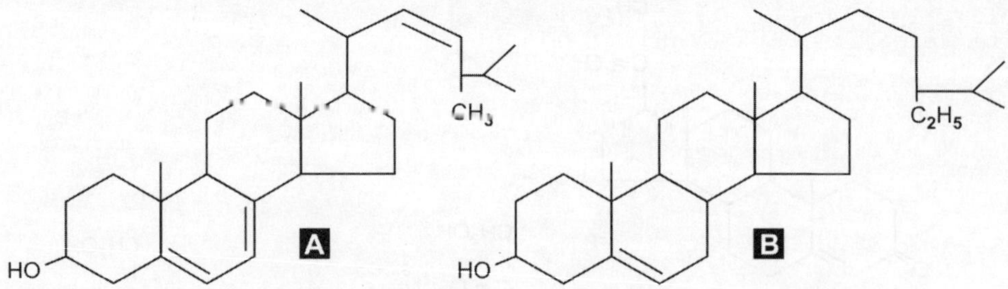


**Fig. 3.13:** Structures of plant sterols: (A) Ergosterol (B) β-Sitosterol

- The most abundant of phytosterol is β-sitosterol and ergosterol (Fig. 3.13).
- Ergosterol (present in yeast) is a precursor of Vit D.
- Unlike plant and animals, microbes do not contain sterols.

*β-sitosterol:* The ring structure of β-sitosterol is identical to that of cholesterol. The side chain attached to C-17 contains 10 carbon atoms (10C) rather than 8 carbon atoms (8C). In addition an ethyl group is attached to $C_{24}$ of the chain.

*Ergosterol*: is the sterol of yeast and contains a methyl group at $C_{24}$, a double bond in the side chain and a second double bond in the B ring. It is a precursor of ergocalciferol (vitamin $D_2$).

*Polyisoprenoids:* Although not steroids, these compounds are related because they are synthesised like cholesterol from 5-carbon units (isoprene).

$$-CH=\underset{}{\overset{CH_3\atop|}{C}}-CH=CH-$$

They include the following

i. Ubiquinone – member of respiratory chain in mitochondria.
ii. Dolichol – long chain alcohol which participates in glycoprotein synthesis.

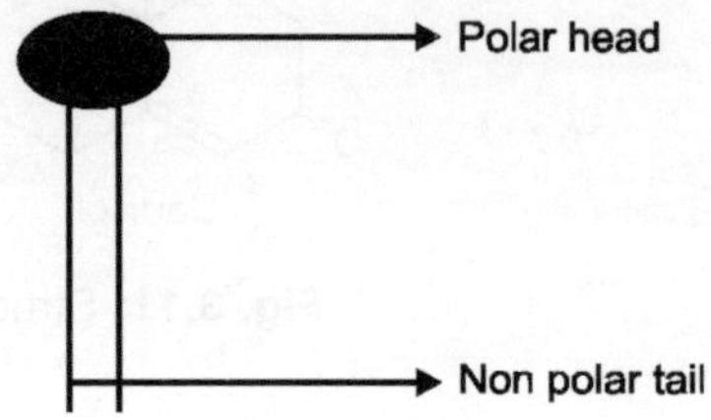


**Fig. 3.14:** Amphipathic lipid

## Amphipathic Lipids (Fig. 3.14)

- Lipids are in general, insoluble in water, due to the presence of hydrocarbon group (non polar).
- However, some lipids do contain polar groups. For example, fatty acids, phospholipids, sphingolipids, bile salts and cholesterol (to a lesser extent).
- Therefore, only part of the molecule is hydrophobic or water insoluble. Such molecules are described as ***amphipathic,*** e.g., phospholipid.
- An important feature of the amphipathic lipids is that they become oriented at oil-water interface with the polar group towards the water phase and nonpolar group towards the oil phase forming micelles, emulsions, lipid bilayer and liposomes.
- *Micelles*: When a critical concentration of amphipathic lipids, is present in an aqueous medium they form *'micelles.'* Aggregation of bile salts into the micelles,

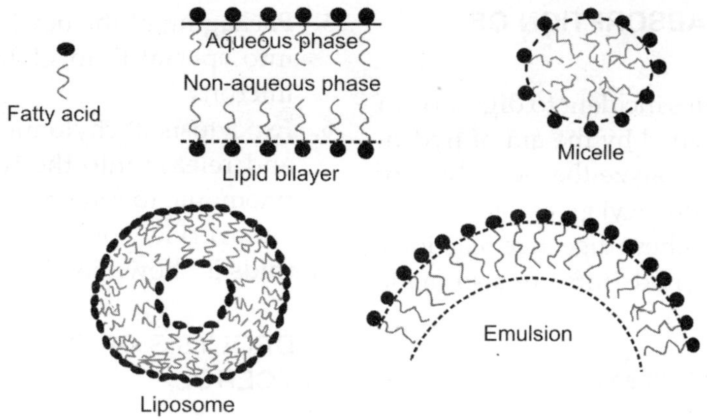


**Fig. 3.15:** Formation of lipid bilayer, micelle, liposome and emulsion from amphipathic lipids

along with the products of fat digestion are important in facilitating the absorption of lipids from the intestine (Fig. 3.15).

### Emulsions

- These are much larger particles formed by nonpolar lipids in an aqueous medium. E.g., An emulsion of oil and water. Emulsions may be stabilised by the addition of amphipathic lipids such as lecithin. The process is called *'emulsification.'* It increases the surface of the substance that is emulsified (Fig. 3.15).

  ***Lipid Bilayer***: A bilayer of amphipathic lipids is regarded as the basic structure in biologic membranes. This arrangement is mainly adapted by amphipathic molecules with two nonpolar groups such as phospholipids (Fig. 3.15).

  ***Liposome:*** These are formed when an amphipathic lipid in an aqueous medium is subjected to *'sonication'*. These consist of spheres of lipid bilayers, that enclose part of the aqueous medium (Fig. 3.15). They are of potential clinical use, particularly when combined with tissue specific antibodies, such as Phospholipids (Fig. 3.15)

- Carrier of drugs in circulation, targeted to specific organs, e.g., cancer therapy.
- In gene transfer into vascular cell.
- For topical and transdermal delivery of drugs and cosmetics.

### Lipid Peroxidation

Peroxidation (auto-oxidation) of lipids exposed to $O_2$ is responsible not only for deterioration of foods but also for damage to tissues *in vivo* leading to cancer, inflammatory diseases, atherosclerosis and aging.

- The deleterious effects are considered to be caused by *free radicals.*
- The free radicals (ROO, RO and OH) are produced during peroxide formation from PUFA.
- Lipid peroxidation is a chain reaction leading to formation of *free radicals* that initiate further peroxidation.
- The whole process takes place in three steps—initiation, propagation and termination.
- Other products of auto-oxidation of physiologic significance include *oxysterols* and *isoprostanes* (formed from cholesterol and prostanoids respectively).

## DIGESTION AND ABSORPTION OF LIPIDS

The chemical changes incident to digestion of lipids are accomplished by the aid of hydrolytic enzymes that catalyze the hydrolysis of triacylglycerol to monoacyl glycerol and fatty acid. In the course of these digestive reactions, the vitamins in the foodstuffs are also made assimilable.

## BIOMEDICAL IMPORTANCE

Lipid malabsorption syndromes (steatorrhoea) often lead to nutritional deficiencies. For example, Vit D malabsorption leads to rickets and osteomalacia.

An average adult human ingests 50-150 gm of lipids per day, of which > 90% is normally triacylglycerol (TAG). The rest is made up of cholesterol, cholesteryl esters (CE), phospholipids (PL) and free fatty acids (FFA). In addition, cholesterol (1-2 gm) and lecithin (7-22 gm) are secreted into the small intestinal lumen as constituents of bile.

Poor water solubility of lipids present problems for digestion because the substrates are not easily accessible to the digestive enzymes in the aqueous phase and also the hydrolyzed products may aggregate to form large complexes that make poor contact with cell-surface. These problems are overcome by:

- increasing the interfacial area between the aqueous and lipid phase.
- solubilisation of lipids with detergents.

The steps involved in the process of digestion and absorption are as follows:

1. **H**ydrolysis of TAG to FFA and glycerol.
2. **S**olubilisation of FFA and monoglycerides by bile acids.
3. **T**ransport of lipids from the intestinal lumen toward the cell surface.
4. **U**ptake of FFA and MAG into the cell and resynthesis of TAG.
5. **P**ackaging of the newly synthesized TAG into special lipid globules called chylomicrons.
6. **E**xocytosis of chylomicrons from the cells and release into the lymph. {**HSTUPE** - pnemonic to remember}.

An overview of digestion and absorption of lipids is shown in Figure 3.16.

## HYDROLYSIS OF TAG TO FFA AND GLYCEROL

### Gastric Digestion

Digestion of lipids is initiated in the stomach, by the lingual lipase (acid stable) and the gastric lipases that attack the sn-3 ester bond forming 1,2 diacyl glycerol and fatty acid.

- Lingual lipase from the mouth, enters the stomach along with food.
- Lingual and gastric lipase degrade the fats containing short chain and medium chain FA at neutral pH *(e.g. FA present in milk.)* This is of particular importance only in infants. This is because of neutral pH of stomach contents in infants (Fig. 3.17).
- Food is retained in the stomach for 2-4 hr. The heat in the stomach is important in liquidising the dietary lipids and emulsification takes place by peristaltic movements.
- Gastric digestion accounts for *about 30%* of the total TAG hydrolysed.

### 2. Emulsification of Dietary Lipids in the Small Intestine

- The ingested TAG, form the separate lipid phase with limited water—lipid interface. The lipases adsorb to that interface and convert TAG into DAG and FA (Fig. 3.15).
- The importance of initial hydrolysis is that some of the water immiscible TAG are

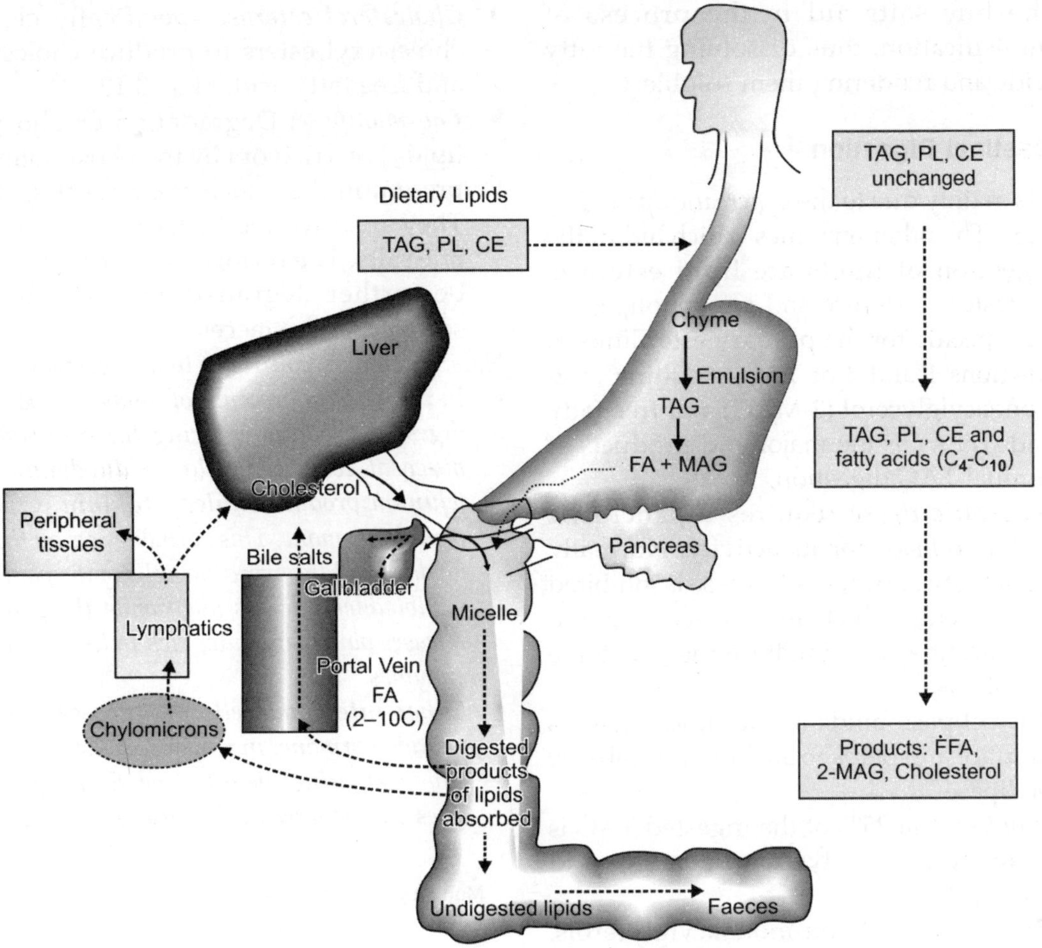


**Fig. 3.16:** Overview of lipid digestion and absorption

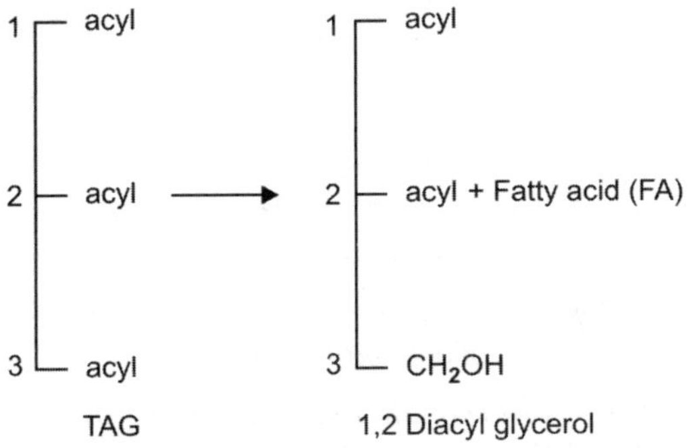


**Fig. 3.17:** Action of lingual and gastric lipase

converted to products, that possess polar and nonpolar groups. Such surfactive products spontaneously adsorb to water-lipid interface and confer hydrophilic surface to lipid droplets, thereby providing the stable interface with aqueous environment.

- Increase in interfacial area, results in dispersion of lipid phase into smaller droplets (emulsification) and provide more sites for adsorption of lipase molecules.

- The bile salts aid in the process of emulsification, thus dissolving the fatty acids and rendering them soluble.

### 3. Intestinal Digestion

- It is mainly due to the *pancreatic lipase (Fig. 3.18)*. The other enzymes which aid in the digestion of lipids are lipid esterase, cholesterol esterase and phospholipases.
- It is specific for the primary ester links in positions 1 and 3 of TAG, resulting in 2-monoacylglycerol (2-MAG) and free fatty acids (FFA) as the major end products of luminal TAG digestion.
- *Pancreatic lipase* requires a pancreatic protein *colipase* for its activity. Normally the activity of pancreatic lipase is inhibited by bile acids which are also secreted into the intestine. This inhibition is overcome by colipase.
- This colipase binds at the lipid aqueous interface and helps to anchor and stabilize the lipase.
- *Only less than 25%* of the ingested TAG is completely hydrolysed to glycerol and fatty acids.
- *Lipid esterase* acts on monoacylglycerols, cholesteryl esters, vitamins etc, to liberate free fatty acids. The presence of bile acids is essential for the activity of lipid esterase.

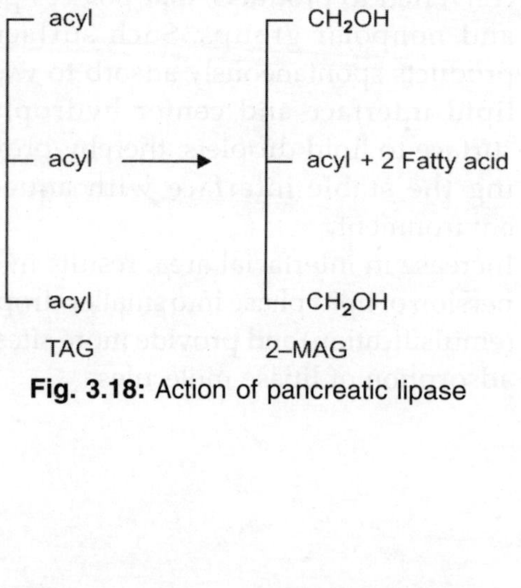


TAG 2–MAG

**Fig. 3.18:** Action of pancreatic lipase

- *Cholesterol esterase* specifically cleaves cholesteryl esters to produce cholesterol and free fatty acids (Fig. 3.19).
- *Phospholipase*: Degradation of phospholipids is carried out by the phospholipases present in the pancreatic juice (Fig. 3.20). They are activated by trypsin and require bile salts. The nitrogenous base (NB) may be further degraded and absorbed or secreted in the faeces.

*Note: Hormonal control of lipid digestion.*

- *In response to presence of lipids and partially digested proteins entering the intestine the mucosal cells in the lower duodenum and jejunum produce* ***cholecystokinin*** *(CCK), a peptide hormone. This hormone acts on the gall bladder and the exorcrine cells of the pancreas. Gallbladder releases bile, while the pancreas releases pancreatic juice rich in lipid-digestive enzymes.*
- *Other intestinal cells produce* ***secretin,*** *a peptide hormone, in response to low pH of the chyme entering the intestine. Secretin stimulates the pancreas to release watery solution rich in bicarbonate that helps in bringing the pH of the intestinal contents to neutrality which is essential for enzymic digestion.*

$$\text{Monoacyl glycerol} \xrightarrow[\text{H}_2\text{O, Bile acids}]{\text{Lipid esterase}} \text{Glycerol + FA}$$

$$\text{Cholesteryl ester} \xrightarrow[\text{H}_2\text{O, Bile acids}]{\text{Lipid esterase}} \text{Cholesterol + FA}$$

$$\underset{\text{(Retinyl ester)}}{\text{Vitamin A}} \xrightarrow[\text{H}_2\text{O, Bile acids}]{\text{Lipid esterase}} \text{Retinol + FA}$$

$$\text{Cholesterol – FA} \xrightarrow[\text{H}_2\text{O, Bile acids}]{\text{Cholesterol esterase}} \text{Cholesterol + FA}$$

**Fig. 3.19:** Action of lipid esterase and cholesterol esterase

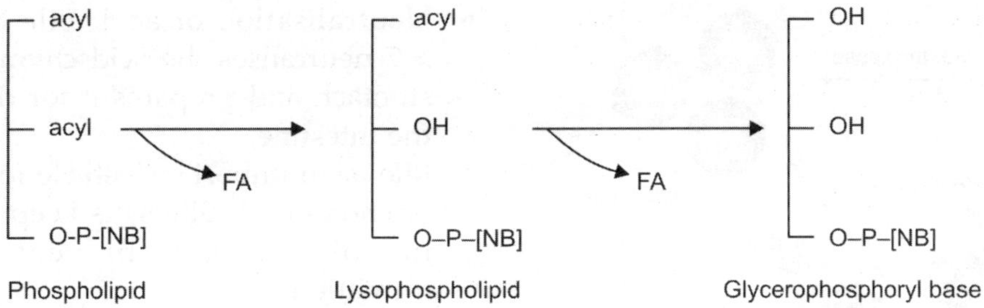


**Fig. 3.20:** Action of phospholipase

### Solubilisation of Free-fatty Acids and Monoglycerides by Bile Acids

- Bile acids are biological detergents, synthesized by liver and secreted as conjugates of glycine or taurine with bile into the duodenum. They are, (i) cholic acid and chenodeoxycholic acid—primary bile acids, (ii) deoxycholic acid and lithocholic acid—secondary bile acids. (For structure of bile acids refer chemistry of lipids)
- Bile salts are essential for solubilising the lipids during the process of digestion. The terms bile salts and bile acids are often used interchangeably.
- The bile salts interact with both the lipid particles and the aqueous duodenal contents, thereby stabilising the particles as they become smaller and preventing them from coalescing. The bile salts form *micelles* above the critical micellar concentration of 2-5 mM.
- The minimum concentration necessary for micelle formation is called the *critical micellar concentration.*
- Micelles are aggregates formed in aqueous solution by an amphipathic lipid. The non polar components orient themselves inside the aggregate whereas the polar groups are on the outside, where they interact with the surrounding water molecules
- The micelles in contrast to emulsified lipids have well defined sizes and are much smaller than emulsion droplets. They typically range between 4-60 nm depending on the concentration of bile acids in lipids.
- Bile acid micelles can solubilise other lipids such as fatty acids, phospholipids etc, and are called as *mixed micelles*. They are made up of more than one amphipathic compound. For example, lipid constituents of bile, phosphatidyl choline, cholesterol and conjugated bile acids exist together in aggregates that are mixed micelles.
- These mixed micelles have disc like shapes, whereby, phospholipid and fatty acids form a bilayer and bile acids occupy the edge position rendering the edge of the disc hydrophilic.
- Within the mixed micelle other water insoluble lipids such as cholesterol and fat soluble vitamins can be accommodated and therefore stabilised. Thus the products of triacyl glycerol hydrolysis are continuously transferred from the emulsion droplets to the micelles (Fig. 3.21).

### Bile

Liver plays an important role in digestion of lipids by producing bile and the gallbladder stores it. The composition of hepatic bile (Table 3.4) differs from that of gallbladder bile

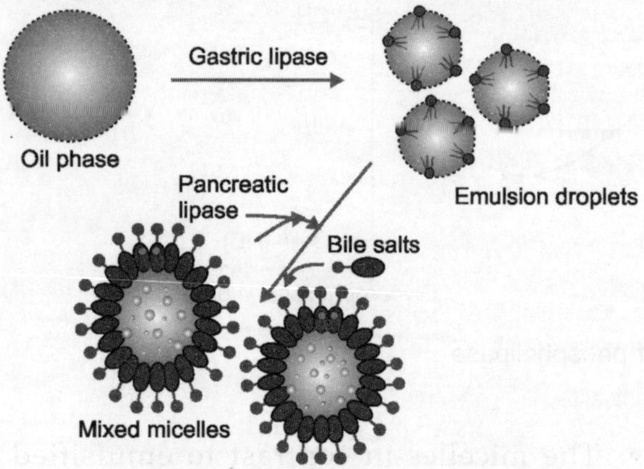


**Fig. 3.21:** Formation of micelles

**Table 3.4:** Composition of bile

| | *Hepatic bile (as secreted)* | *Bladder bile* |
|---|---|---|
| | *Percent of total bile* | *Percent of total bile* |
| Water | 97.00 | 85.92 |
| Solids | 2.52 | 14.08 |
| Bile acids | 1.93 | 9.14 |
| Mucin and pigments | 0.53 | 2.98 |
| Cholesterol | 0.06 | 0.26 |
| Fatty acids | 0.14 | 0.32 |
| Inorganic salts | 0.84 | 0.65 |
| Specific gravity | 1.01 | 1.04 |
| pH | 7.1 – 7.3 | 6.9 – 7.7 |

During digestion, the gallbladder contracts and supplies bile rapidly to duodenum by way of common bile duct. The pancreatic secretions mix with bile before its entry into duodenum.

### Functions of Bile

- Emulsification—bile salts have the ability to lower the surface tension and have detergent like action. Thus it enables to emulsify the fats in the intestine. Presence of bile in intestine is an important adjunct to accomplish the digestion and absorption of fats as well as fat soluble vitamins A, D, E and K.
- Neutralisation of acid: Bile having pH > 7 neutralises the acid chyme from the stomach and prepares it for digestion in the intestine.
- Bile is an important vehicle for excretion of cholesterol. Bile salts keep cholesterol in solution and prevent gallstone formation.
- It also removes many drugs, toxins and bile pigments. Various inorganic substances excreted by bile are copper, zinc and mercury.

**Absorption of lipids:** Is mediated by the transport of lipids from the intestinal lumen towards the cell surface (Fig. 3.22).

### Transport of Lipids Towards the Cell Surface

It is mediated by the micelles. The fluid layer next to the mucosal surface is poorly mixed (*unstirred layer*). The major transport mechanism for solute flux across the *unstirred layer* is diffusion, down the concentration gradient.

Obviously the *unstirred fluid layer* presents problems for sparingly soluble or insoluble nutrients. Bile acid micelles overcome this problem for lipids by increasing the effective concentration in unstirred layer. Thus micelles mediate the transport of lipid digest through the aqueous environment of the gut lumen to the brush border of the enterocyte and then to intestinal epithelial cell.

*Note: Short and medium chain FA do not require the assistance of micelles for absorption by the intestinal mucosa. This is important in dietary therapy for individuals with malabsorption of lipids.*

### Uptake of FFA, MAG and Synthesis of TAG

Absorption is virtually complete for FA and 2 MAG which are slightly water soluble. It is less efficient for water insoluble lipids, e.g., only 30-40% of dietary cholesterol is absorbed.

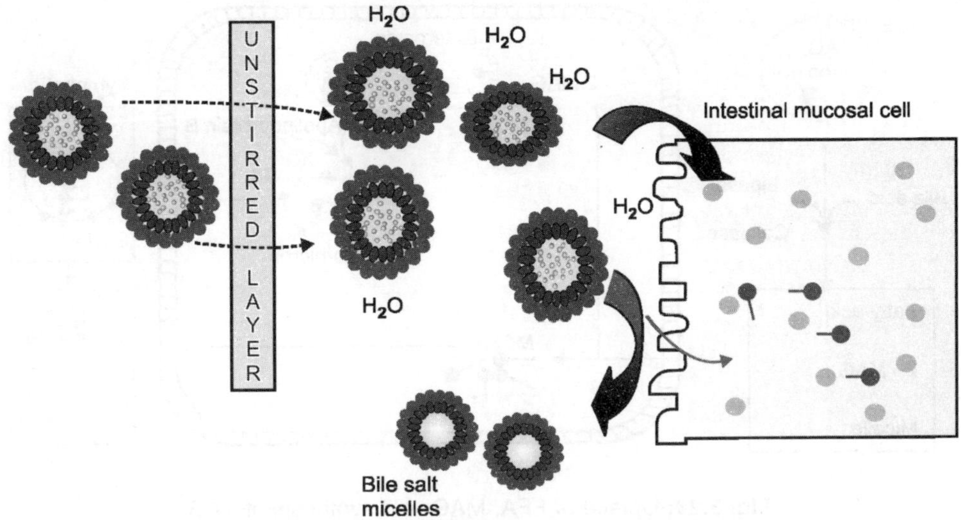


**Fig. 3.22:** Absorption of lipids contained in the micelles

Within the intestinal cell, fate of the absorbed FA depends on chain length. Medium and short chain FA (<10 C-atoms) pass through the cell into the portal blood, without modification. Long chain FA (> 12 C-atoms) are bound to fatty acid binding protein and transferred to rough endoplasmic reticulum (RER) for resynthesis of TAG.

The glycerol for this process is provided by the following three pathways.

- Absorbed 2 MAG (Monoacyl glycerol path way)
- Hydrolysis of 1 MAG produces free glycerol.
- Via glycerol-3-$PO_4$ produced from DHAP, an intermediate of glycolysis.

### Packaging of newly synthesised TAG into special lipid globules called chylomicrons

Resynthesised TAG form lipid globules along with phospholipids and special proteins termed apolipoproteins. These are called **chylomicrons**.

The lipid globules (chylomicrons) migrate within membrane bound vesicles through the golgi to the contraluminal plasma membrane (Fig. 3.23).

### Exocytosis of chylomicrons from the cells and release into the lymph

- Chylomicrons are finally released into the intracellular spaces by exocytosis and these travel through the lacteals to the thoracic duct and then conveyed to the left subclavian vein and enter blood.
- Protein component, Apo B-48, synthesized in the intestinal cells is essential for the final release of chylomicrons from the enterocyte.
- Dietary short chain FA and medium chain FA (MCFA) reach the liver directly via the portal blood. The long chain FA (LCFA) reach the adipose tissue and muscle via the systemic circulation, before reaching the liver.
- This bypass of liver has been evolved to protect the liver from lipid overload.

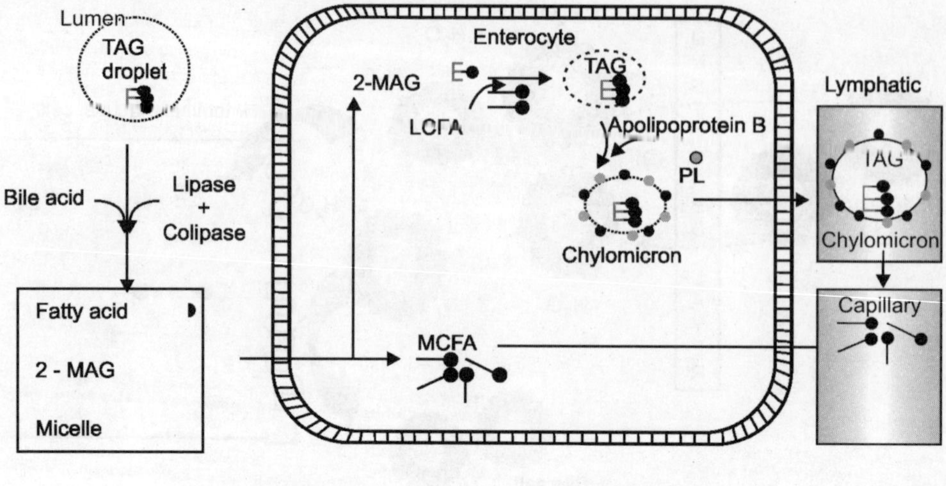


**Fig. 3.23:** Uptake of FFA, MAG and synthesis of TAG

*Note: The fat and muscle cells take up large amount of dietary lipid for storage or metabolism.*

## Clinical Importance

*Chyluria*: Urine appears milky due to the presence of lipid droplets. There is abnormal connection between the urinary tract and lymphatic drainage system of the intestine.

*Chylothorax*: Chyle may leak into the pleural cavity from the thoracic duct or its tributaries as a result of trauma (stab wounds, crush injuries, surgical injury) or obstruction by tumour. It is commonly mistaken for pus. Chyle contains fat globules.

*Cholelithiasis (Gallstones)*: In the absence of bile salts, cholesterol may get precipitated producing cholesterol stones (gallstones) and the condition is called cholelithiasis.

*Steatorrhoea (Fig. 3.24):* Characterised by foul smelling fatty stools due to excessive loss of lipids (> 6 gm/day). This may be due to defective digestion (defect in the secretion of bile or pancreatic juice into the intestine) or impaired absorption, resulting in loss of lipids (including vitamins A, D, E and K and essential fatty acids) in the faeces. It is commonly seen in disorders associated with pancreas, biliary obstruction, severe liver dysfunction etc.

Other disorders caused due to defective digestion and absorption are given in Table 3.5.

## METABOLISM

### Oxidation of Fatty Acids

Fatty acids (FA) are major fuels for humans both as primary supply from the diet and also as secondary supply synthesized from other dietary components. They are stored within the cells as fats (triglycerides) and later released through the blood stream to meet the demands of many tissues, especially muscles. The complete combustion of FA to $CO_2$ and $H_2O$ occurs in the mitochondria, generating ATP.

- The combustion occurs in two stages,
  - i. The FA is first oxidized to convert all of its carbons to acetyl CoA.
  - ii. The acetyl CoA is then oxidized by the reactions of the citric acid cycle.

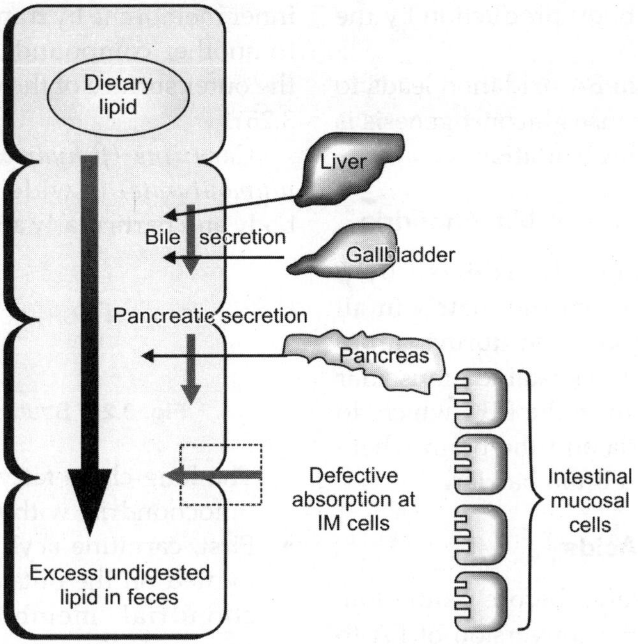


**Fig. 3.24:** Steatorrhoea

**Table 3.5:** Disorders associated with lipid digestion and absorption

| *Defective digestion* | | *Defective absorption* |
|---|---|---|
| Chronic diseases of pancreas | $\downarrow$ pancreatic lipase | Diseases of intestine—enteritis |
| Surgical removal of pancreas | | Coeliac disease—(non tropical sprue). There is permanent intolerance to *gluten*—a wheat protein leading to villous atrophy and malabsorption [incidence more in UK]. Responds to a gluten free diet |
| Defective synthesis of bile acids due to liver damage | $\downarrow$ bile | Tropical sprue—characterized by villous atrophy and malabsorption of fats and vitamins (mainly seen in the Middle East). Does not respond to gluten free diet. Can be treated with folate and vit $B_{12}$ alongwith an antibiotic |
| Gallstones | | |

- Approximately 40% of the energy produced in this process is conserved by the cell in the form of ATP. The remainder of the energy is released as heat.

## BIOMEDICAL IMPORTANCE

- Increased FA oxidation is characteristic of *starvation and diabetes mellitus* leading to increased ketone body production by the liver *(ketosis).*

- Any impairment in FA oxidation leads to *hypoglycaemia,* because gluconeogenesis is dependent upon FA oxidation.

### Oxidation of FA Occurs in Mitochondria

Fatty acids are aerobically oxidized by *β oxidation* in the mitochondrial matrix in all tissues for energy production during carbohydrate deprivation and sustained muscular work. It does not occur in the RBC which do not have mitochondria and the brain where FA do not cross blood brain barrier.

### Activation of Fatty Acids

Fatty acids are activated before oxidiation. Activation involves the conversion of FA to 'active FA' or *acyl CoA* in the presence of coenzyme A (CoA) and is accompanied by the expenditure of high energy phosphate.

- The acyl CoA formed is a highly polar thiol ester which is easily soluble in the aqueous phase of the cell.
- The presence of inorganic pyrophosphatase ensures complete activation by hydrolysing the pyrophosphate which is a high energy compound.

### Transport of Acyl CoA into Mitochondria as Acylcarnitine

The mitochondrial inner membrane is impermeable to coenzyme A and its derivatives. Instead, the acyl groups are moved across the inner membrane by transferring the acyl CoA to another compound, **carnitine,** present at the outer surface of the inner membrane (Fig 3.25).

*Carnitine (β-hydroxy γ-trimethylammoniumbutyrate)* is widely distributed in the body and particularly abundant in the muscle.

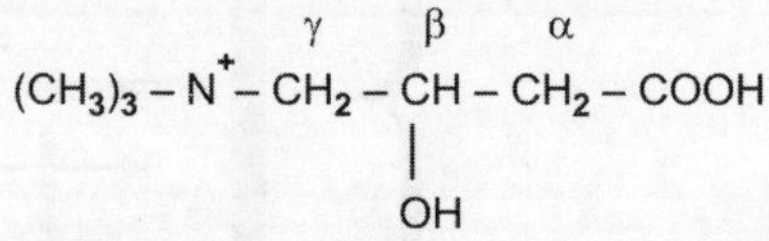


**Fig. 3.25:** Structure of carnitine

- The long chain fatty acyl CoA enters the mitochondria, with the help of carnitine.
- First, carnitine acyl transferase I (CAT-I) located on the outside of the inner mitochondrial membrane catalyzes the transacylation reaction.

$$\text{Acyl CoA + Carnitine} \xrightleftharpoons{\text{CAT-I}} \text{Acylcarnitine + CoA} - \text{SH}$$

- A translocase (carnitine-acylcarnitine translocase) a membrane protein transports the acylcarnitine across the mitochondrial membrane into the mitochondrial matrix.
- The entry of acylcarnitine is coupled with the transport of one molecule of carnitine to the outer side.
- In the mitochondrial matrix acylcarnitine reacts with CoA and releases the acyl CoA and free carnitine. This reaction is catalyzed by carnitine acyl transferase II (CAT – II).

Acyl CoA synthetase (Thiokinase)

$$H_3C{-}(CH_2)n{-}COO^- + ATP \xrightarrow{\text{CoASH}} H_3C{-}(CH_2)n{-}CO-S-CoA + AMP + PPi$$

$$PPi \xrightarrow{\text{Pyrophosphatase}} 2Pi$$

CAT - II

Acylcarnitine + CoASH → AcylCoA + Carnitine

- Carnitine fatty acyl carrier system depends on presence of CoA on both sides of inner mitochondrial membrane. This transport process involving carnitine is also known as *carnitine shuttle (Fig. 3.26).*

## β-oxidation

- The process of β-oxidation involves, successive cleavage of two carbon atoms at a time, starting at the carboxyl end.
- The β carbon (C -3) of the fatty acid is first oxidized and then split between α and β carbon (C-2 and C- 3). Hence, the name β-oxidation.
- The two carbon units cleaved are released as acetyl CoA and further oxidized by citric acid cycle.
- The process of β-oxidation occurs in a sequence of 4 reactions (Fig. 3.27) viz,

i. **Oxidation**
ii. **Hydration**
iii. **Oxidation**
iv. **Cleavage.**

### *i. Oxidation by FAD linked dehydrogenase*

- The first step is removal of two hydrogen atoms from α and β-carbon atoms, catalyzed by acyl CoA dehydrogenase, requiring FAD. This results in the formation of an unsaturated derivative, $\Delta^2$-trans enoyl- CoA and $FADH_2$.

### *ii. Hydration*

- Water is added to saturate the double bond forming β-hydroxy acyl CoA and this reaction is catalyzed by $\Delta^2$ - enoyl- CoA hydratase.

### *iii. Oxidation by NAD dependent dehydrogenase*

- β hydroxy acyl CoA (3C) undergoes further dehydrogenation on the β-carbon (3rd carbon) to form the corresponding β

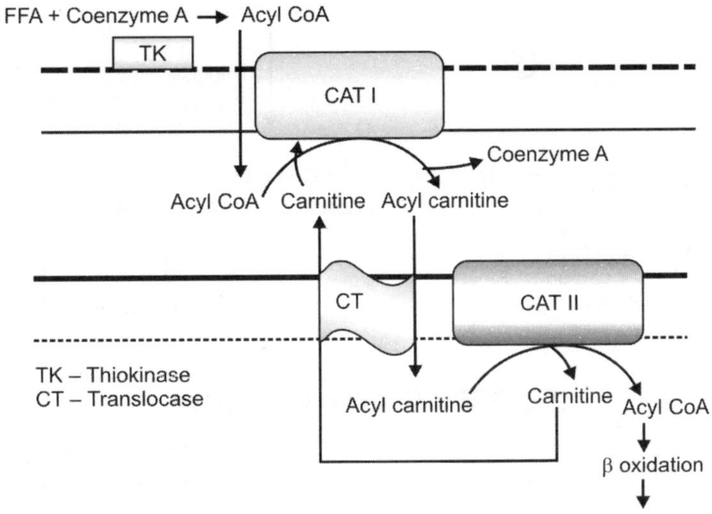


**Fig. 3.26:** Carnitine Shuttle

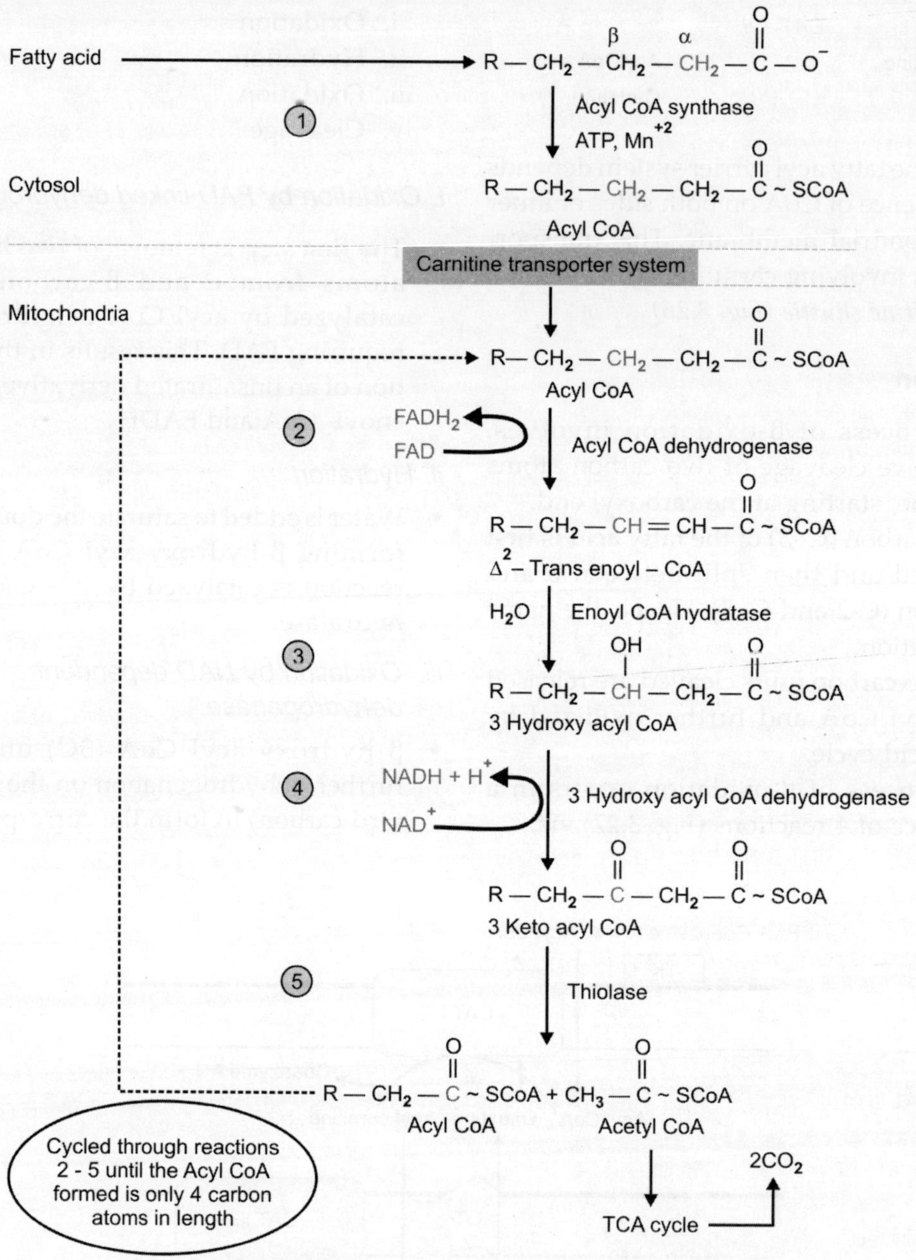


**Fig. 3.27:** β Oxidation of fatty acids

***Energetics***

| | | |
|---|---|---|
| 1 cycle of β - oxidation →5 ATP | | |
| (1NADH ~ 3 ATP) | | |
| ($1FADH_2$ ~ 2ATP) | | |
| 7 cycles of β - oxidation →5 x 7 | | = 35 ATP |
| 1 acetyl CoA by citric acid cycle →12 ATP | | |
| 8 acetyl CoA by citric acid cycle →12 x 8 | | = 96 ATP |
| | Total | 131 ATP |
| Energy utilised for activation of acyl CoA | | – 2 ATP |
| | Net gain | = **129 ATP** |

***Efficiency***

| | |
|---|---|
| Standard free energy of palmitate | = 2340 Kcal |
| Energy conserved by formation of 129 ATP (129 x 7.3) | = 940 Kcal |
| Efficiency of energy conservation by FA oxidation | $= 940 \times \frac{100}{2340}$ |
| | = **40%** |

ketoacyl CoA. The reaction is catalyzed by 3 ketohydroxyacyl CoA dehydrogenase. In this case, coenzyme involved is $NAD^+$.

### *iv. Cleavage*

- β ketoacyl CoA undergoes thiolytic cleavage at the α β positions (C-2, 3) by thiolase (3 ketoacyl CoA thiolase) forming acetyl CoA and the new acyl CoA (2 carbon atoms shorter than the acyl CoA molecule).
- The acyl CoA formed in the cleavage reaction reenters the oxidative pathway. In this way a long chain FA may be degraded completely to acetyl CoA.
- Thus, even chain FA like palmitoyl CoA (16C) forms 8 acetyl CoA molecules by *7 cycles* of β-oxidation and stearoyl CoA (18C) forms 9 acetyl CoA molecules by *8 cycles* of β-oxidation.
- The acetyl CoA produced is finally oxidized to $CO_2$ and $H_2O$ by citric acid cycle and the complete oxidation of FA is thus achieved.

It is often said that *'fats burn in the flame of carbohydrates'*—how can this be explained?

It may be recalled that oxidation of acetyl CoA via TCA cycle requires the presence of *oxaloacetate* which is obtained mainly from glucose. It means that oxidation of fatty acids requires *glucose.* Thus we may rightly say that *"fats burn in the flame of carbohydrates."* This saying is very true and applicable to the burning of an oil lamp which requires the wick. (The wick is made up of cellulose which is a carbohydrate).

### *Regulation of Fatty Acid Oxidation*

At times of high energy charge in the cells FA oxidation is inhibited by the increased concentration of malonyl CoA because FA synthesis is stimulated. This is mediated by the inhibition of *carnitine acyl transferase-I,* by malonyl CoA.

When the energy charge is low, there is an accompanying decrease in concentration of malonyl CoA, NADH and acetyl CoA, leading to stimulation of FA degradation (Fig. 3.28). The result is that FA synthesis is inhibited when FA oxidation is stimulated and vice versa.

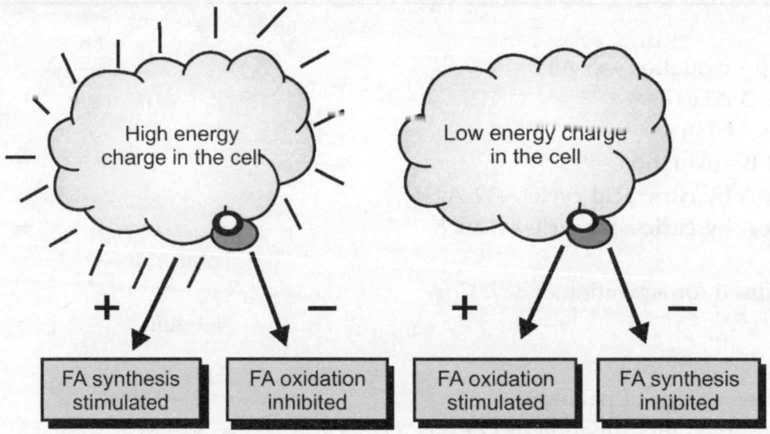


**Fig. 3.28:** Regulation of fatty acids oxidation

### Oxidation of FA with an Odd Number of Carbon Atoms

These FA are also oxidized by the pathway of β-oxidation producing acetyl CoA until a 3-carbon residue (propionyl CoA) remains. This compound is finally converted to succinyl CoA, a constituent of citric acid cyc1e which can be completely oxidized to $CO_2$ and $H_2O$ or to glucose by gluconeogenesis. Hence, propionyl residue from an odd chain FA, is the only part of FA which is *glucogenic.* The propionyl CoA is then converted to succinyl CoA by gluconeogenesis (Ref Chapter 2).

### Oxidation of Unsaturated FA

The unsaturated FA are also oxidised in a similar manner as saturated FA. However, two additional enzymes—an *isomerase* and an *epimerase* are needed to degrade a wide range of unsaturated FA.

### α and ω-Oxidation of Fatty Acids

Animal tissues contain minor pathways that involve oxidation of FA at the α and ω-carbons. The products of α and ω-oxidation can enter β-oxidation.

*α-oxidation*

- α-oxidation occurs in the endoplasmic reticulum (ER).
- It is of importance in the metabolism of dietary FA that are methylated.
  *Example*: Phytanic acid, a metabolic product of chlorophyll. It is also a significant constituent of milk lipids and animal fats.
- Normally phytanic acid is metabolised by initial α-oxidation followed by dehydrogenation and decarboxylation (Fig. 3.29).
- β-oxidation cannot occur initially because of the presence of three methyl groups but can proceed after decarboxylation.
- α-oxidation is an important process in brain. It does not require activation, nor yield energy and is a method of generating hydroxy fatty acids which are components of glycolipids.

*ω-oxidation*

- It is a minor pathway taking place in microsomes with the help of hydroxylase enzymes involving NADPH and cyt-P450.
- The terminal carbon (ω-carbon) is oxidized to COOH group to produce a dicarboxylic acid.

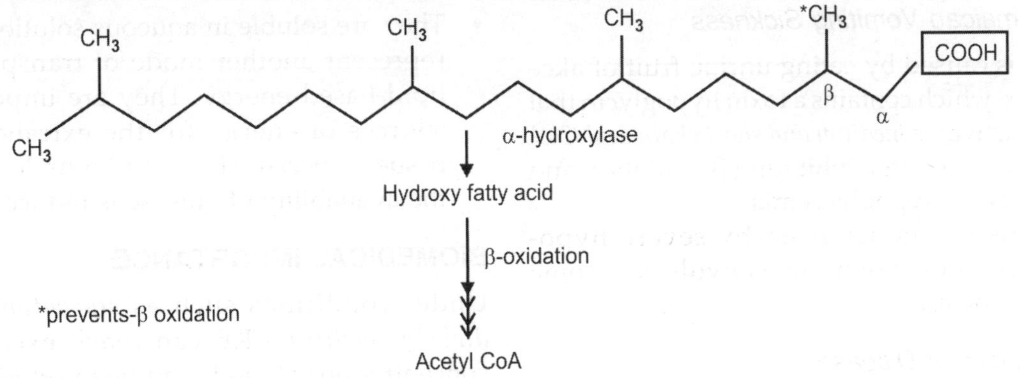


**Fig. 3.29:** Structure of phytanic acids

- This undergoes β-oxidation to adipic acid (C6) and suberic acid (C8) which is excreted in urine.

*Peroxisomal Oxidation*

- Modified form of β-oxidation found in the peroxisomes and leads to formation of acetyl CoA and $H_2O_2$.
- $H_2O_2$ formed is reduced to water by catalase.
- This dehydrogenation in peroxisomes is not linked directly to phosphorylation and the generation of ATP. This system facilitates the oxidation of very long chain FA containing $C_{26} - C_{38}$.

### *Clinical Correlation*

*1. Carnitine Deficiency*

Congenital absence of carnitine acyl transferases in skeletal muscle or low concentration of carnitine (due to defective synthesis) results in inability to use long chain FA as a metabolic fuel. Important features are hypoglycaemia and cramps.

Hypoglycaemia—because gluconeogenesis is inhibited and is usually precipitated by periods of fasting.

Cramps—caused by strenuous exertion damage the muscles.

- Destruction of muscle cells is manifested by myoglobinuria with elevated levels of creatine kinase.
- Congenital absence of carnitine leads to more serious impairment of muscle function than congenital absence of carnitine acyl transferase enzyme.
  — *The treatment for all inherited disorders of carnitine transport system is avoidance of starvation—and a diet devoid of long chain FA. Supplementation of dietary fat containing medium chain FA which are oxidised by carnitine independent process proves beneficial.*

Inherited defects in enzymes of β-oxidation lead to nonketotic hypoglycaemic coma and fatty liver. Defects are usually found in medium and short chain acyl CoA dehydrogenases.

*2. Sudden Infant Death Syndrome (SIDS)*

The sudden unexpected death of healthy infants usually overnight is due to deficiency of *medium chain acyl CoA dehydrogenase.*

- It is more prevalent than phenylketonuria.
- Metabolic complications of medium chain acyl CoA dehydrogenase deficiency are seen during carbohydrate deprivation.

### 3. *Jamaican Vomiting Sickness*

- It is caused by eating unripe fruit of akee tree which contains a toxin hypoglycin that inactivates *medium and short chain acyl CoA dehydrogenases,* inhibiting β-oxidation and causing hypoglycaemia.
- This is characterised by severe hypoglycaemia, vomiting, convulsions, coma and death.

### 4. *Refsum's Disease*

- It is a rare neurological disorder, due to defect in the enzyme *phytanate α-hydroxylase* resulting in accumulation of large quantities of phytanic acid in tissues and plasma.
- The restriction of dietary diary products and meat result in lowering of plasma phytanic acid and regression of neurological symptoms.

### 5. *Zellweger's (cerebro hepato renal) syndrome*

- Occurs in individuals with a rare inherited absence of peroxisomes in all the tissues.
- Characterized by poor oxidation of very long chain FA, accumulation of polyenoic fatty acids ($C_{26}$-$C_{38}$) in the brain and impaired synthesis of bile acids and plasmalogens.

## KETONE BODIES—METABOLISM

Ketone bodies (KB) are an alternate fuel for cells. Excess acetyl CoA derived from FA oxidation is diverted to synthesize KB in the liver. The compounds categorized as KB are **acetoacetate, β-hydroxybutyrate** and **acetone.** They are also called as acetone bodies. However, the term *'ketone'* is incorrect since, β-hydroxy butyrate is not a ketone and there are other ketones in the blood that are not ketone bodies, e.g., pyruvate and fructose.

- They are soluble in aqueous solution and represent another mode of transport of lipid-based energy. They are important sources of energy for the extrahepatic tissues especially brain and muscle, when the availability of glucose is reduced.

## BIOMEDICAL IMPORTANCE

Under conditions such as *starvation* and *diabetes mellitus*, KB can reach excessive concentration in blood, leading to ketosis and acidosis *(diabetic ketoacidosis).*

### *Synthesis of ketone bodies (Ketogenesis)*

- Synthesis of KB occurs primarily in the liver mitochondria (Fig. 3.30). The starting material for ketogenesis is acetoacetyl - CoA which arises by either of the two processes.
  - i. Directly during the course of β-oxidation.
  - ii. Condensation of two acetyl CoA molecules.
- Pathway involves condensation of another molecule of acetyl CoA with aceto acetyl CoA to produce 3-hydroxy, 3-methyl glutaryl CoA (HMG CoA) in the presence of *HMG CoA synthase*. This step is the rate limiting step in the synthesis of KB. The HMG CoA synthase is present in significant quantities in the liver.
- HMG CoA is then cleaved to produce acetoacetate and acetyl CoA in the presence of *HMG CoA lyase.* The carbon atoms split off as acetyl CoA are derived from the original acetoacetyl molecule.
- Both enzymes *HMG CoA synthase* and *lyase* must be present in mitochondria for ketogenesis to take place.
- Acetoacetate produced is in equilibrium with 3-hydroxybutyrate (β-hydroxy butyrate) catalyzed by *3-hydroxybutyrate*

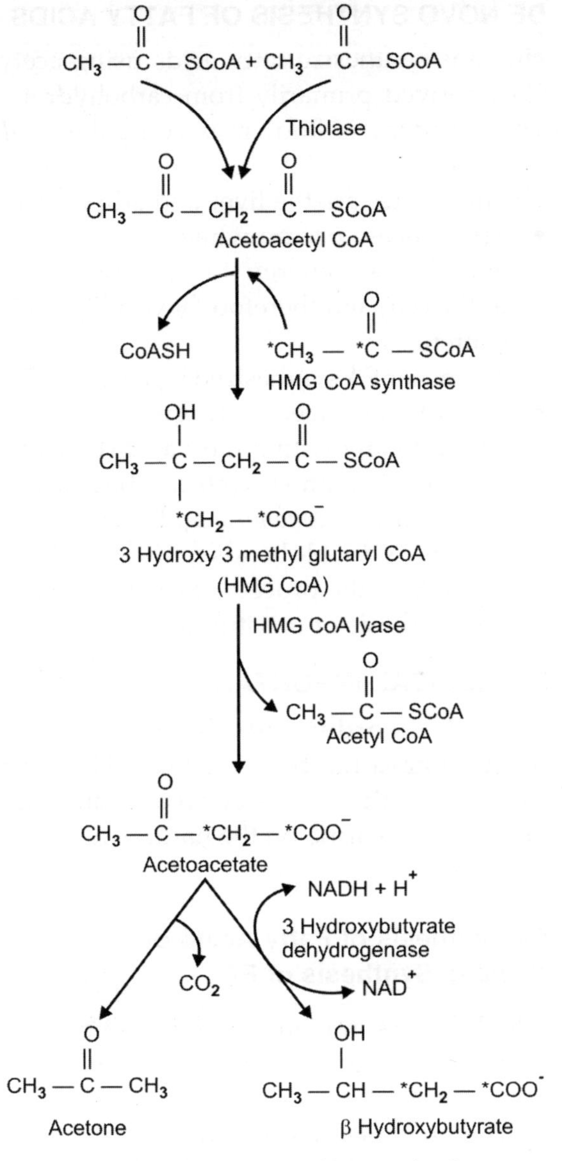

**Fig. 3.30:** Formation of ketone bodies

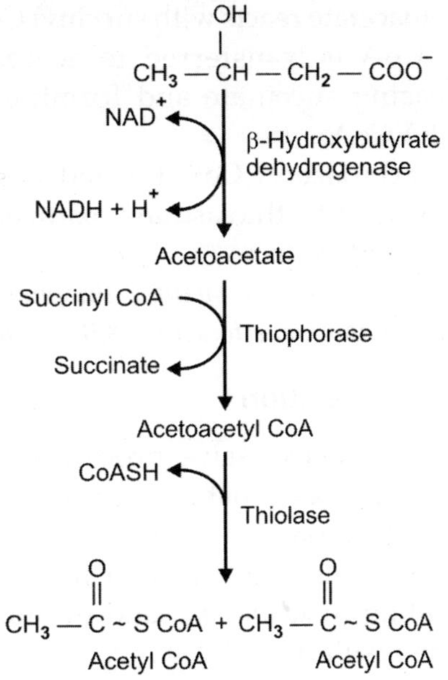

**Fig. 3.31:** Utilisation of ketone bodies

*dehydrogenase* or can be *spontaneously decarboxylated* to form acetone.

- β-hydroxybutyrate (3-hydroxybutyrate) is quantitatively the predominant KB present in the blood and urine in ketosis. The concentration of total KB in the blood of well fed mammals, does not normally exceed 3 mg/dl. Loss via the urine is usually less than 1 mg/24 hrs.

### *Regulation of KB synthesis*

Ketogenesis is regulated at three crucial steps.

i. Availability of FFA – lipolysis.
ii. Entry of FA into the mitochondria for oxidation pathway by carnitine palmitoyl transferase – I (CPT-I)
iii. Partition of acetyl CoA between ketogenic pathway and the pathway of oxidation to $CO_2$ via citric acid cycle.

### *Utilisation of Ketone Bodies*

KB synthesized in the liver are transported in the blood, primarily to the peripheral tissues (extrahepatic tissues).

- The main pathway of utilisation of KB, includes the activation of acetoacetate to acetoacetyl CoA, (Fig. 3.31) which involves, succinyl CoA and the enzyme *succinyl CoA–acetoacetate CoA transferase (thiophorase).*

- Acetoacetate reacts with succinyl CoA and the CoA is transferred to acetoacetate releasing succinate and forming aceto-acetyl CoA.
- The acetoacetyl CoA formed is split to acetyl CoA by thiolase and oxidized in the citric acid cycle
- *The liver lacks the enzyme thiophorase and therefore is unable to utilise KB as a fuel.*

### Clinical Correlation

***Ketosis*** – It is excessive production of KB when the rate of formation is greater than the rate of utilisation.

- Their levels begin to rise in the blood resulting in *ketonaemia* and eventually in urine resulting in *ketonuria.*
- These two conditions are most often seen in case of *starvation* and *severe diabetes mellitus (uncontrolled DM)* due to decreased utilisation of carbohydrates. As a result there is increased lipolysis and fatty acid oxidation.
- In diabetic individuals with severe ketosis, blood levels may reach 90 mg/dl; urinary excretion may be as high as 5000 mg/24 hr.
- An elevation of KB concentration in blood results in acidemia, and also excretion of glucose in urine resulting in dehydration of the body. Acetone odour is often detectable in the breath of these individuals.
- The triad of ketonaemia, ketonuria and acetone odour of breath is known as *ketosis.*

*Note : The carboxyl group of ketone body has a pKa around 4. Thus each ketone body loses a proton ($H^+$) as it circulates in the blood (resulting in increased $H^+$ ions and decreased pH). When the concentration of ketone body increases, pH of the blood is lowered due to the increased $H^+$ resulting in acidaemia.*

## DE NOVO SYNTHESIS OF FATTY ACIDS

Humans synthesize fatty acids using acetyl CoA derived primarily from carbohydrates. This pathway is known as complete or ***de novo synthesis*** and it occurs in the *cytoplasm.* The main sites are the liver and adipocytes.

- The process occurs when the cells have abundance of energy (high energy charge in the cell) and therefore have a high ATP content.
- The major FA synthesized is *palmitate ($C_{16}$).*
- In addition, acetate fragments can be added to an existing fatty acid (FA), by a process of chain *elongation* which occurs in both the *mitochondria* and *microsomes.*
- Finally, *cis* double bonds can be introduced into FA by the process of *desaturation,* for the synthesis of unsaturated fatty acids.

## BIOMEDICAL IMPORTANCE

Critical disorders of the biosynthetic pathway have not been reported. However, variations in the activity between individuals may have a bearing on the nature and extent of obesity.

### Biosynthesis of Fatty Acids or *de novo* Synthesis of FA

*Site:* It occurs in many tissues such as liver, adipose tissue, mammary gland, brain, kidney and lungs.

The enzymes that synthesize fatty acids are localised in the cytosol and are completely different from the mitochondrial enzymes that catalyze fatty acid degradation.

#### *Requirements*

a. Acetyl CoA
b. Fatty acid synthase complex
c. Cofactors—NADPH, ATP, $Mn^{++}$, biotin and bicarbonate.

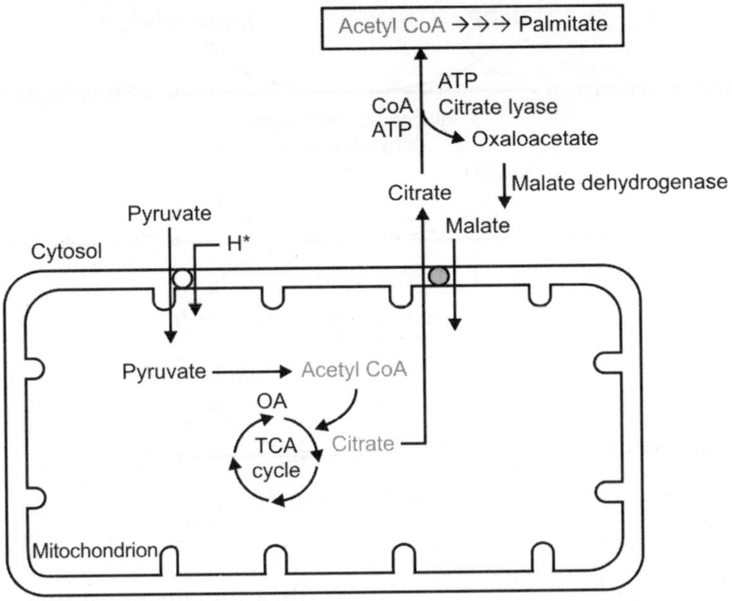


**Fig. 3.32:** Transport of acetyl CoA
o - Tricarboxylate transporter
● - Pyruvate transporter

The FA synthesis may be learnt under three headings:

i. Production of acetyl CoA and NADPH.
ii. Conversion of acetyl CoA to malonyl CoA.
iii. Reactions of FA synthase complex.

### *I. Production of acetyl CoA and NADPH*

Acetyl CoA is the principal building block of FA. It is formed within the mitochondria by the oxidation of pyruvate. However, acetyl CoA does not diffuse readily into the cytosol, the principal site of FA synthesis.

- The transfer of acetyl CoA to cytosol is made by condensation of acetyl CoA with oxaloacetate to form citrate (Fig. 3.32).
- This is followed by translocation of citrate into the cytosol via the tricarboxylate transporter which requires malate, to exchange with citrate.
- In the cytosol, the citrate undergoes cleavage to acetyl CoA and oxaloacetate by ATP citrate lyase.
- 8 Acetyl CoA is thus made available for FA synthesis.

*NADPH:* The main sources of NADPH are:

i. The HMP shunt pathway and
ii. Malic enzyme.

i. HMP shunt pathway is the main source of NADPH for the *de novo* synthesis of FA (Fig. 3.33). Tissues having active lipogenesis (liver, adipose tissue and lactating mammary gland) also possess active HMP shunt pathway. The reactions producing NADPH are given below.

Since both the processes take place in cytosol, the transfer of NADPH from one pathway to the other is made easy unlike the transfer of acetyl CoA.

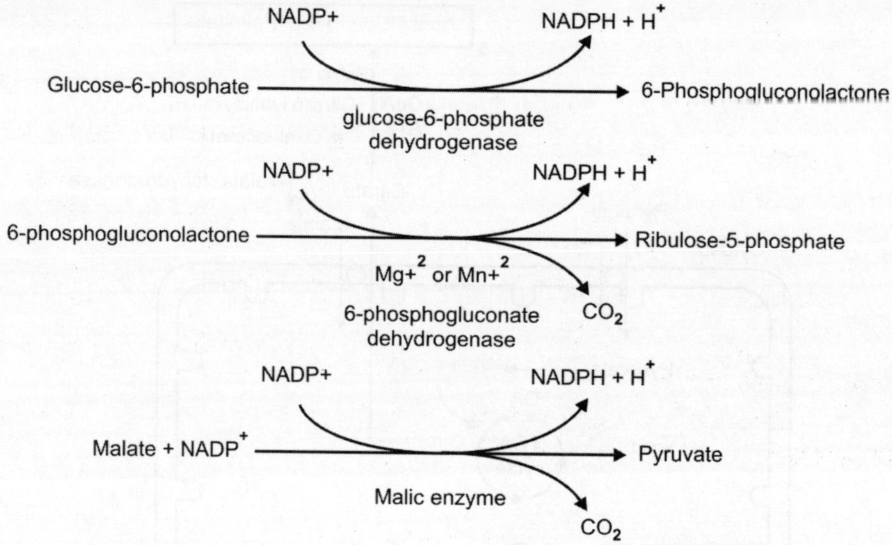


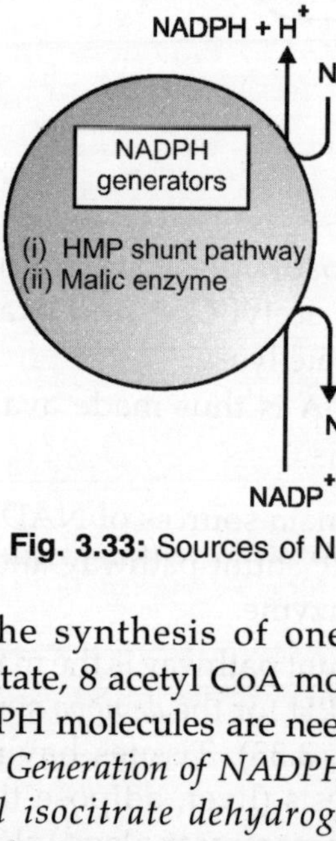


**Fig. 3.33:** Sources of NADPH

- For the synthesis of one molecule of palmitate, 8 acetyl CoA molecules and 14 NADPH molecules are needed.

**Note:** *Generation of NADPH via extramitochondrial isocitrate dehydrogenase is not a substantial source in humans. However, it is more important in ruminants, because of the deficiency of malic enzyme.*

### *II. Conversion of acetyl CoA to malonyl CoA*

It is the initial and controlling step in FA synthesis. The reaction is as follows:

- Bicarbonate as a source of $CO_2$ is required in the initial reaction for the carboxylation of acetyl CoA to malonyl CoA by acetyl CoA carboxylase and this requires ATP and biotin.

$$\underset{\text{Acetyl CoA}}{CH_3{-}CO{\sim}SCoA} + HCO_3^- + ATP \xrightarrow{\text{Acetyl CoA carboxylase}} \underset{\text{Malonyl CoA}}{^-OOC{-}CH_2{-}CO{\sim}SCoA} + ADP + Pi$$

- Acetyl CoA carboxylase contains variable number of identical subunits, and is therefore a multienzyme protein and the chief regulatory enzyme.

### *III. Reactions of FA synthase complex*

The remaining reactions of FA synthesis are catalyzed by multifunctional enzyme complex known as *FA synthase complex.*

In eukaryotic cells, FA synthase exists as dimer with two identical subunits. Each monomer possesses, activities of *seven* different enzymes and an acyl carrier protein *(ACP)* bound to 4' phosphopantetheine—SH group (Fig. 3.34). The seven enzymes forming

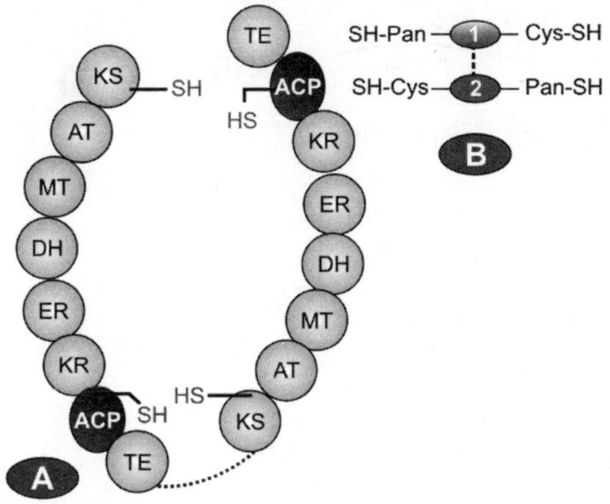


**Fig. 3.34:** (A) Fatty acids synthase multi enzyme complex (B) Short hand notation of fatty acid synthase

the fatty acid synthase complex are given below.

1. *Acetyl transacylase (AT)*
2. *Malonyl transacylase (MT)*
3. *Ketoacyl synthase (KS)*
4. *Ketoacyl reductase (KR)*
5. *Dehydratase (DH)*
6. *Enoyl reductase (ER)*
7. *Thioesterase (TE)*

- The two monomers lie in a head to tail configuration.
- In close proximity is another thiol ester of cysteine residue of 3 ketoacyl synthase of the other monomer.
- In this system, ACP takes over the role of CoA.
- Since both thiols participate in the synthase activity, only the dimer is active.

The aggregation of all the enzymes of one multifunctional enzyme unit offers *great efficiency* and *freedom* from interference by competing reactions. Effect of compartmentalisation of the process within the cell is thus achieved.

- Another advantage is that, synthesis of all the enzymes in the complex are *coordinated,* since it is encoded by a single gene.
- The sequence of reaction for biosynthesis of palmitate (C-16) is described below (Fig. 3.35).

Fatty acid synthesis involves four steps viz,

i. **C**ondensation
ii. **R**eduction
iii. **D**ehydration
iv. **R**eduction

- Initially, a priming molecule of acetyl CoA combines with the cysteine-**SH** of one of the monomers. The reaction is catalyzed by *acetyl transacylase.*
- Next, malonyl CoA is added, on to the adjacent-**SH** group on 4'-phosphopantotheine of the ACP of the other monomer. This reaction is catalysed by *malonyl transacylase.*
- The acetyl group attacks the methylene group of the malonyl residue catalyzed by *3-ketoacyl synthase.* This reaction liberates $CO_2$, forming ketoacyl enzyme. Decarboxylation allows the reaction to go to completion and acts as a pulling force for the whole sequence of reactions.
- The ketoacyl group is then reduced, in the presence of *3-ketoacyl reductase* forming 3-hydroxyacyl enzyme.
- This is followed by removal of water by *dehydratase* forming 2,3 – unsaturated acyl enzyme.
- The final step involves the reduction of the unsaturated acyl enzyme by *enoyl reductase* to form saturated butyryl enzyme.
- The saturated acyl residue which is attached to the SH group of 4'- phosphopantotheine of ACP of the second monomer is displaced to the free cysteine-SH group of the first monomer.

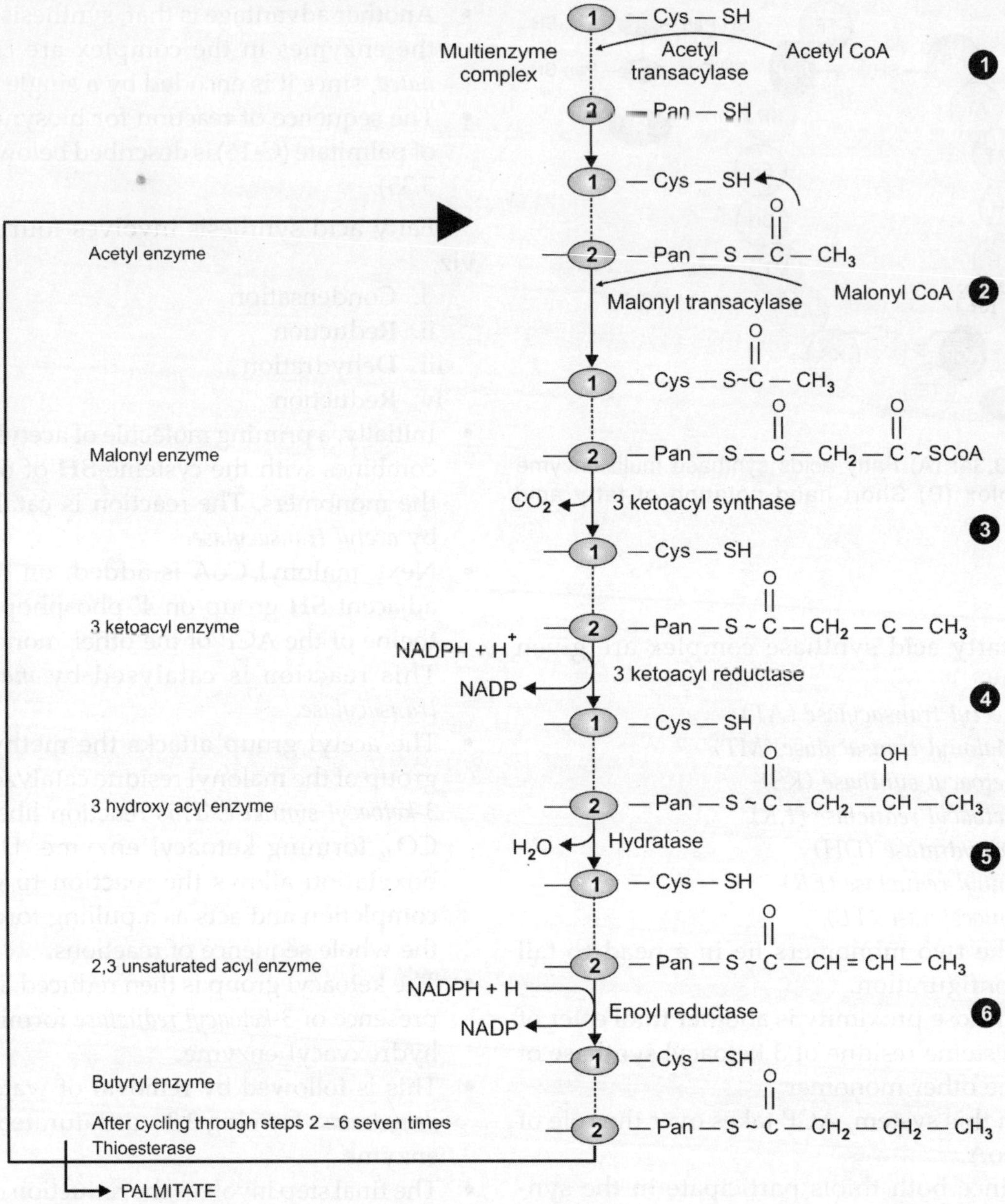


**Fig. 3.35:** Reactions of the FA synthase complex leading to synthesis of palmitate

- New malonyl CoA is added on to the SH group of 4' phosphopantotheine.
- This sequence of reactions mentioned above is repeated six more times. Thus two

carbon atoms are incorporated during each sequence, using malonyl residues until a saturated 16 carbon acyl group (palmityl) has been assembled.

- Palmityl ($C_{16}$) group is liberated from the enzyme complex by the activity of *thioesterase (deacylase)* of the other monomer in the complex.
- During synthesis, the growing FA chain migrates back and forth between the cysteine—SH of one monomer and the Pan-SH of ACP of the other monomer in the complex.
- Thus, there are *two* centres of activity in one dimer, that function simultaneously to form two molecules of palmitate.

Equation for overall synthesis of FA is as given below:

$$CH_3CO \sim S\text{-}CoA + 7\ HOOC.\ CH_2CO.\ S\text{-}CoA + 14NADPH + 14H^+ \longrightarrow CH_3(CH_2)_{14}COOH + 7CO_2 + 6H_2O + 8\text{-}CoASH + 14\ NADP^+$$

### *Regulation of FA Synthesis*

- Nutritional state is the main factor regulating the rate of lipogenesis. Thus the rate is higher in well fed state, when the diet contains high proportion of carbohydrate. It is depressed:
  - — Under conditions of fasting, intake of high fat diet and when there is a deficiency of insulin as in diabetes mellitus.
  - — All these conditions are associated with increased concentration of plasma free fatty acids.
- The most important enzyme regulating the *de novo* synthesis is the acetyl CoA carboxylase, regulated both allosterically and by covalent modification. It is activated by citrate. Activation of the enzyme involves, conversion of monomeric form of the enzyme into polymeric form (Fig. 3.36). Short term regulation is mediated by allosteric and covalent modification of the enzyme acetyl CoA carboxylase.
  - — Malonyl CoA and long chain acyl CoA molecules formed during lipolysis, inhibit the synthesis.
  - — The activity of acetyl CoA carboxylase is controlled by cAMP mediated phosphorylation/dephosphorylation. The phosphorylated form is inactive and the dephosphorylated form is active.
  - — Insulin activates the acetyl CoA carboxylase by promoting the dephosphorylation and thus stimulates *de novo* synthesis (lipogenesis).
  - — Acetyl CoA carboxylase is also an inducible enzyme. Its synthesis is induced by insulin, whose level is high in the fed state.
  - — Glucagon, epinephrine and norepinephrine inactivate the enzyme by promoting the phosphorylation and thus inhibit the synthesis.
  - — Long term regulation is mediated by increase in the synthesis of FA via increased synthesis of acetyl CoA carboxylase and FA synthase.

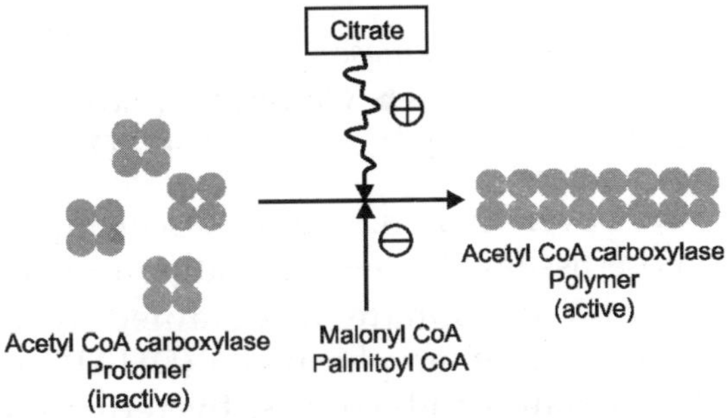


**Fig. 3.36:** Allosteric regulation of acetyl CoA carboxylase

The palmitic acid synthesized can further be elongated and / or desaturated to produce other fatty acids found in various tissues. These are then used for the synthesis of TAG, compound lipids and cholesterol esterification.

**Synthesis of Triacylglycerols:** Synthesis occurs mainly in the liver, adipose tissue, and intestinal mucosa and to a lesser extent in other tissues. Synthesis requires *fatty acids* and *glycerol.* The two processes involved are as follows:

i. Activation of (a) fatty acids and (b) glycerol.
ii. Incorporation of acyl CoA into glycerol 3 phosphate.

*Activation*

a. Fatty acids are activated to acyl CoA by the enzyme thiokinase.

$$\text{Fatty acid + ATP + CoASH} \xrightarrow[Mg^{+2}]{\textit{Thiokinase}} \text{Acyl CoA + AMP + PPi}$$

b. Glycerol is activated to glycerol -3-phosphate.
In tissues such as liver, kidney and lactating mammary gland, glycerol-3- phosphate is formed by the action of glycerokinase.

$$\text{Glycerol + ATP} \xrightarrow[Mg^{+2}]{\textit{Glycerokinase}} \text{Glycerol-3-phosphate + ADP + Pi}$$

However this enzyme is *absent* in the adipose tissue. Hence in the adipose tissue gycerol-3 phosphate is obtained from dihydroxy acetone phosphate (DHAP) an intermediate of glycolysis, by reduction with NADH. The reaction is catalyzed by the enzyme glycerol-3-phosphate dehydrogenase.

*II. Incorporation of acyl CoA into glycerol-3-phosphate*

- Two fatty acyl groups are added at position 1 and 2 respectively of glycerol-3-phosphate to yield lysophosphatidic acid and phosphatidic acid.
- Phosphate group from phosphatidic acid is hydrolysed to yield 1,2 diacyl glycerol (DAG ).
- 1,2 DAG is subsequentially acylated with another fatty acyl CoA at the 3 position to yield TAG. This is known as the *'Diacyl glycerol (DAG ) pathway'* of TAG synthesis (Fig. 3.37).

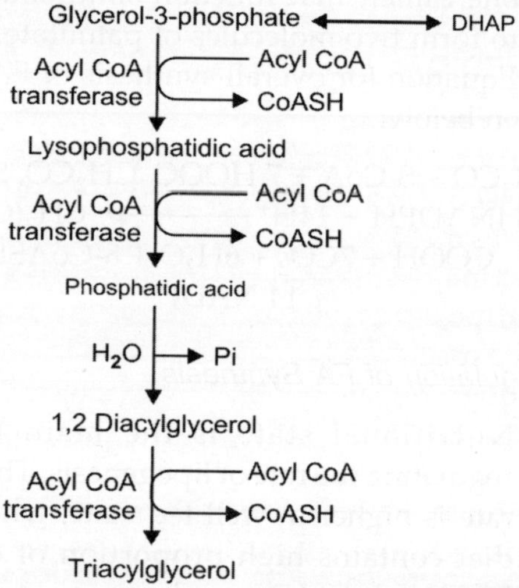


**Fig. 3.37:** Synthesis of TAG by DAG pathway

*Note: An alternate pathway that uses DHAP as the substrate involves initial acylation at the 1 position followed by reduction at the 2 position to yield lysophosphatidic acid.*

*Note: MAG Pathway of TAG synthesis: Intestinal mucosa synthesizes TAG primarily from 2 MAG and fatty acids released by the digestion of dietary lipids.*

- *2 MAG is acylated at 1 position to form 1,2 DAG.*
- *1,2 DAG is further acylated at 3 position to form TAG.*

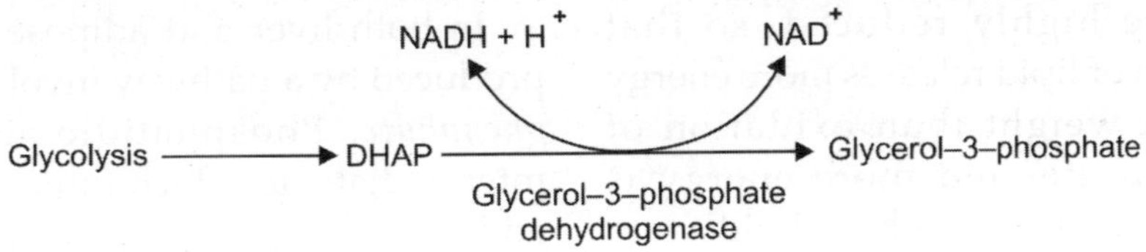


- *In this pathway phosphatidate is not involved as an intermediate.*

## ADIPOSE TISSUE

Fats in the form of triacylglycerol (TAG) is a major storage form of energy in humans. It is mainly stored in the adipose tissue. Capacity to store fat in the adipose tissue is large and expandable.

## BIOMEDICAL IMPORTANCE

Excess fat deposit causes obesity. Abdominal obesity, in particular is a risk factor for increased mortality, hypertension, Type II diabetes mellitus, hyperlipidemia and various endocrine dysfunctions.

Adipose tissue is a type of connective tissue, specialised for the synthesis, storage and degradation of triacylglycerols. Most adipose tissue is found in subcutaneous and intramuscular deposits surrounding the abdominal cavity and the mammary gland.

- It is second in importance to the liver with respect to distribution and maintenance of fuel levels in blood.
- Adipose tissue is currently gaining importance for its endocrine functions (*for the synthesis of the hormone leptin*)
- Adipose tissue is of two types.
  i. White adipose tissue.
  ii. Brown adipose tissue.

### *i. White Adipose Tissue*

Is mainly concerned with energy storage. It is made up of specialised cells known as *adipocytes* or fat cells. The cytoplasm of the adipocytes contain a large central lipid droplet which occupies 90% of the cell volume. These cells contain a large quantity of stored TAG. As a result, the nucleus, mitochondria and all the other organelles are pressed into a very small volume along the periphery of cell (Fig. 3.38). This cell is entirely devoted to fat storage and its structure is adapted to this purpose.

The TAG in adipose tissue is the body's major deposit of stored energy. TAG have two properties that equip them for long-term energy storage—

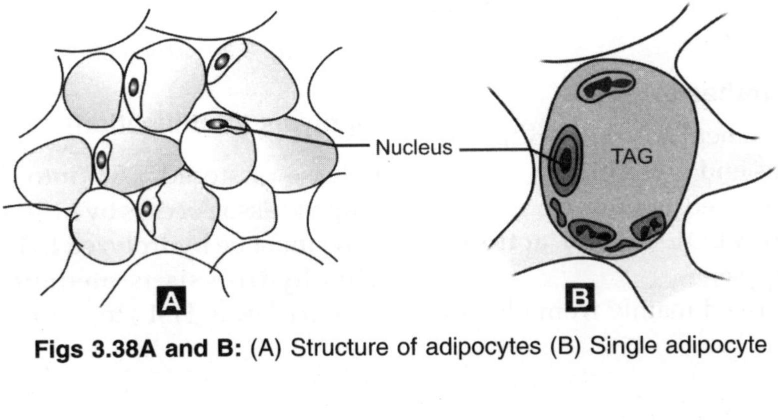


**Figs 3.38A and B:** (A) Structure of adipocytes (B) Single adipocyte

i. They are highly reduced, so that oxidation of lipid releases more energy per unit weight than oxidation of carbohydrate, and more energy is stored in a given weight of lipid.

ii. TAG are relatively nonreactive, so they can be stored for long periods under normal conditions without undergoing unwanted chemical alterations.

*Note: In some animals TAG stored under the skin serves not only as an energy store but also as insulation against very low temperature. For example, seals, walruses, penguins are amply padded with triacylglycerol.*

### Metabolism of Adipose Tissue

TAG stores in the adipose tissue are continually undergoing lipolysis and re-esterification. They are entirely different pathways involving different reactants and enzymes. This allows many of the nutritional, metabolic and hormonal factors that regulate the metabolism of adipose tissue to act either upon the process of lipogenesis (esterification) or on the process of lipolysis.

- The net result of these two processes determines the magnitude of the FFA pool in the adipose tissue which in turn is the source and determinant of the level of FFA, circulating in the plasma.
- The level of FFA in plasma has most profound effects upon the metabolism of other tissues, particularly the liver and muscles.

### Lipogenesis (Synthesis)

The prime requirement for the synthesis of TAG is fatty acids and glycerol.

- The FA are derived mainly from chylomicrons and VLDL by the action of lipoprotein lipase.
- Glycerol is derived mainly from glucose.

In both liver and adipose tissue, TAG is produced by a pathway involving *glycerol-3-phosphate*. Phosphatidic acid acts as an intermediate, to which three acyl groups are added resulting in the formation of TAG.

However, the source of glycerol-3-phosphate is different in the two tissues. In the liver, the source is glycerol via the action of glycerokinase; while in the adipose tissue, source is glucose (which is formed from dihydroxy acetone phosphate (DHAP) formed from glucose during glycolysis) since the adipose tissue lacks the enzyme glycerokinase.

The storage of fatty acids in the adipose tissue can occur only in the fed state when the concentration of glucose and insulin are high. Insulin is essential for the uptake of glucose, since it increases the number of GLUT-4 transporters which is involved in the transport of glucose into this tissue.

*Lipoprotein lipase (LPL)* hydrolyses the triglycerides contained in the chylomicrons and VLDL to yield monoacylglycerol, fatty acids and glycerol (Fig. 3.39). It resides on the capillary walls of most tissues, predominantly the adipose tissue and cardiac and skeletal muscle. It is activated by apo C-II (a protein present in the lipoproteins).

$$\underset{\text{Chylomicrons}}{\text{TAG}} \xrightarrow{\text{LPL}} \text{MAG} + 2\ \text{FA}$$

**Fig. 3.39:** Action of Lipoprotein lipase

### Lipolysis (Mobilisation)

Release of stored TAG into the blood from adipose tissue occurs by a different route. The TAG must be hydrolyzed to FA and glycerol. This hydrolysis is mediated by *hormone sensitive lipase* (HSL) to form *FFA and glycerol.*

This enzyme is distinct from LPL that catalyses lipoprotein TAG hydrolysis, before its uptake into the extrahepatic tissues (Fig. 3.40).

***Hormone sensitive lipase*** (HSL): Release of TAG from the adipose tissue in the fasting state is accomplished by hormone sensitive lipase present within the adipocyte (Fig. 3.41). The hormone sensitive lipase is normally in an inactive form but is activated by phosphorylation requiring a cAMP dependent protein kinase. TAG hydrolysis is thus influenced by the level of cAMP which is generated from ATP by the enzyme adenylate cyclase.

Since glycerol cannot be utilised (absence of glycerol kinase) in this tissue, it diffuses out into the plasma, and it is utilised by tissues such as liver and kidney, which possess an active glycerol kinase to form glycerol-3-phosphate which is re-esterified with fatty acids to form TAG.

**Thus, there is continuous cycle of lipolysis and re-esterification within the tissues.**

However, when the rate of esterification is not sufficient to match the rate of lipolysis, FFA accumulate and diffuse into the plasma, where they bind to albumin and raise the plasma-*FFA* which is an important source of fuel for many tissues.

- In the *well fed* (post prandial) state the concentration of glucose and insulin are high. Hence *lipogenesis* is stimulated.
- In *starvation* due to the decreased availability of glucose, *lipolysis* is stimulated. Also the lipoprotein lipase activity of adipose tissue is low. Hence circulating TAG of lipoproteins is not available for lipogenesis.

***Hormonal regulation of fat mobilisation (Fig. 3.42):*** The rate of release of free fatty acids from adipose tissue is affected by many hormones (Table 3.6). These hormones influence either the rate of esterification or the rate of lipolysis.

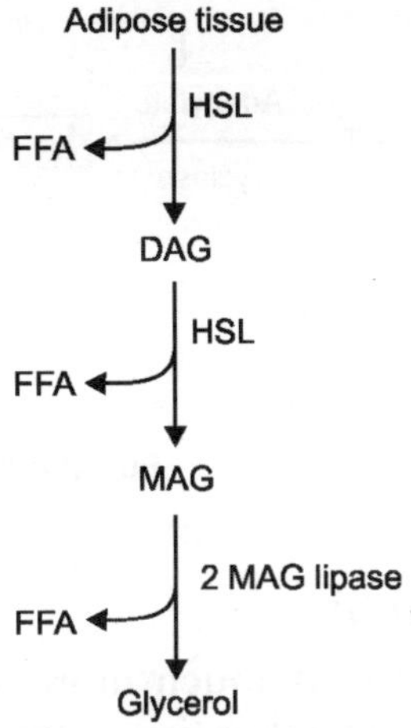


**Fig. 3.40:** Lipolysis mediated by HSL

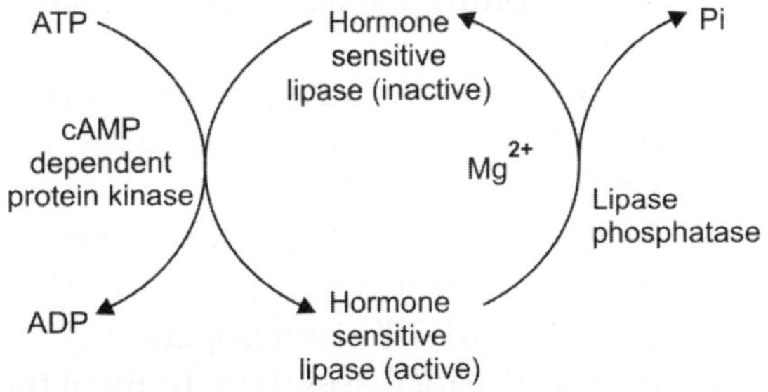


**Fig. 3.41:** Activation of hormone sensitive lipase

**Table 3.6:** Hormones regulating fat metabolism in adipose tissue

| *Lipogenesis* | *Lipolysis* |
|---|---|
| Insulin | Catecholamines<br>Glucagon<br>Growth hormone (GH)<br>Glucocorticoids<br>Adrenocorticotropic hormone (ACTH)<br>Thyroid stimulating hormone (TSH)<br>Thyroxine |

- TAG hydrolysis is influenced by the level of cyclic AMP which is generated from ATP by the enzyme adenylate cyclase. The hormonal regulation is mainly brought about by increasing or decreasing the levels of cyclic AMP.

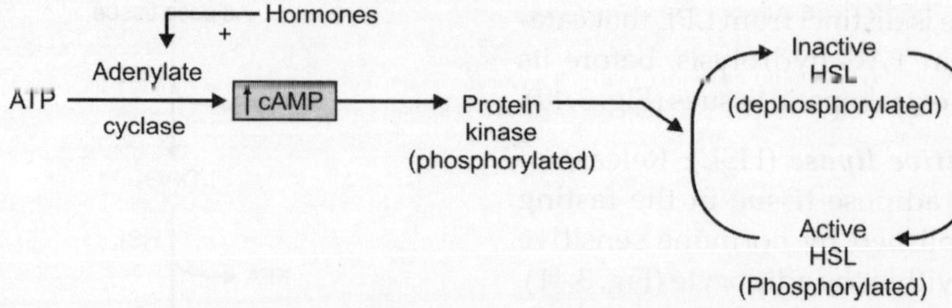


**Fig. 3.42:** Hormonal regulation of fat mobilisation

### *Action of Insulin*

Adipose tissue is much more sensitive to insulin than any other tissue. Insulin inhibits the release of free fatty acids from the adipose tissue, by inhibiting the activity of hormone sensitive lipase.

i. It enhances the transport of glucose into the tissue and increases the synthesis of acylglycerol.
ii. Increases the oxidation of glucose via the pentose phosphate pathway. Thus provides *NADPH* for lipogenesis.
iii. Insulin also increases the activity of the enzymes of fatty acid and TAG synthesis.
iv. It inhibits lipolysis by converting the hormone sensitive lipase to the dephosphorylated form which is inactive.

### *Action of Other Lipolytic Hormones*

Most of the lipolytic hormones mentioned above stimulate lipolysis by activating the hormone sensitive lipase through cAMP by activating adenylate cyclase.

### Catecholamines

Act rapidly in promoting lipolysis by stimulating the activity of adenylate cyclase. Catecholamines are liberated in the adipose tissue by the sympathetic nervous system and plays a central role in the mobilisation of free fatty acids by exerting a tonic influence even in the absence of augmented nervous activity.

*Note: Thus, stress results in an increase in the FFA due to the release of catecholamines.*

### Action of Other Hormones

- The remaining hormones do not increase lipolysis markedly but act in a facilitatory or permissive capacity with respect to other endocrine factors.
- For an optimal effect, the presence of glucocorticoids and thyroid hormones are required. The lipolytic stimulus is switched off by removal of the stimulating hormone.

*Note: Drinking of coffee causes elevation of plasma FFA in humans. This is due to the inhibition of phosphodiesterase, the enzyme which destroys cAMP, by caffeine (a methyl xanthine).*

### *Role of Leptin*

- Leptin is a small molecular weight protein produced by white adipose tissue and has numerous metabolic effects.
- Interaction of leptin with its receptors lead to suppression of food intake through the release of corticotrophin releasing (CRH) hormone and suppression of neuropeptide 'Y'.

**(NPY) is released from the nerves during periods of stress. It inhibits the cells of the immune system that normally destory the pathogens in the body.*

### Brown Adipose Tissue

- Is involved in metabolism particularly at times when heat generation is necessary.
- The tissue is extremely active in some species during arousal from hibernation, in animals exposed to cold and during heat production in the new born animals.
- Brown adipose tissue cells are polygonal with abundant cytoplasm and small lipid droplets. Brown colour is due to the presence of numerous mitochondria which have high content of cytochromes.
- In normal individuals, brown adipose tissue appears to be responsible for diet induced thermogenesis which may account for how some persons can eat well and not get fat.
- Brown adipose tissue is reduced or absent in obese persons. Norepinephrine released from the sympathetic nerve endings in cold conditions promote lipolysis in this tissue. Fatty acids are oxidized and the energy produced is not used for ATP synthesis but released as heat.

## LIPOPROTEINS

Lipids being insoluble in water, cannot be readily transported in the aqueous environment of the plasma. Hence lipids are associated with specialised proteins to mediate their transport between the various tissues and organs for utilisation and storage. These complexes of lipids and proteins are called *"lipoproteins."* It can be compared to a boat carrying someone who cannot swim across a river.

## BIOMEDICAL IMPORTANCE

Abnormalities of lipoprotein metabolism occur at the sites of production or utilisation of lipoproteins causing various hypo or hyperlipoproteinaemias. Lipoproteins play an important role in the development of *atherosclerosis,* a phenomenon that underlies a range of diseases of the cardiovascular system, such as *coronary heart disease (CHD)* and *stroke.*

### Lipid Transport

Lipids must be carried through the aqueous phase of plasma from one tissue to the other. An overview of this process, is shown in the Figure 3.43.

It involves the following:

1. Transport of TAG from the intestine to other tissues in the body.
2. Transport of TAG formed in the liver to various tissues for oxidation or to adipose tissue for storage.
3. Transport of TAG from the adipose tissue to other tissues when they require energy.
4. Transport of cholesterol from one place to another in the body.

### Lipoprotein Structure

The plasma lipoproteins are large complexes made up of physically combined lipids and proteins (Fig. 3.44).

- They have spherical structure with nonpolar lipids contained in the hydrophobic core that is surrounded by amphipathic lipids and proteins.
- Triglycerides and cholesterol esters are located in the central hydrophobic core of the macromolecular complex.
- The amphipathic lipids surrounding the hydrophobic core are mainly phospholipids and unesterified cholesterol.

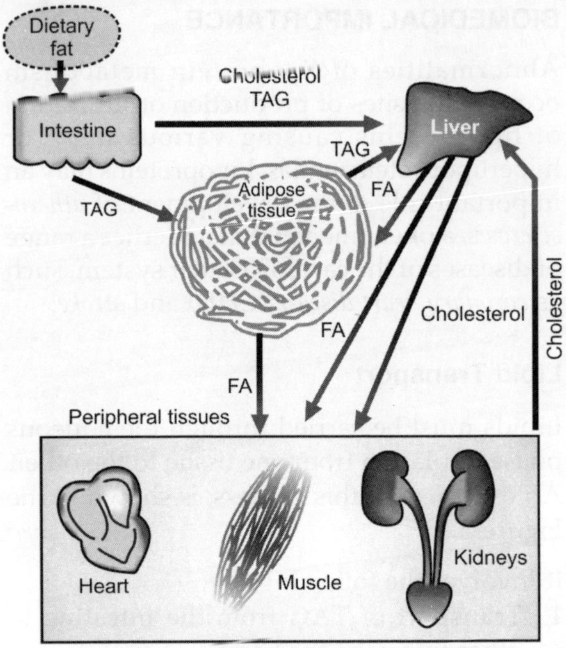


**Fig. 3.43:** Overview of lipid transport

- Proteins present are of one or more types and are commonly referred to as *apolipoproteins* or *apoproteins.*
- Some apoprotein, (e.g., apo-B) are deeply embedded into the particles, and are known as *integral proteins.* While others (e.g., apo-C) are located on the surface and can be easily exchanged between different particles and are known as *peripheral proteins.*
- The polar groups of amphipathic lipids interact with water and form ionic constituent of plasma, thereby stabilizing the macromolecular complex in the aqueous environment.

## Classification

Two systems of classifying lipoproteins are in use.

They are based on (i) density and (ii) electrophoretic mobility.

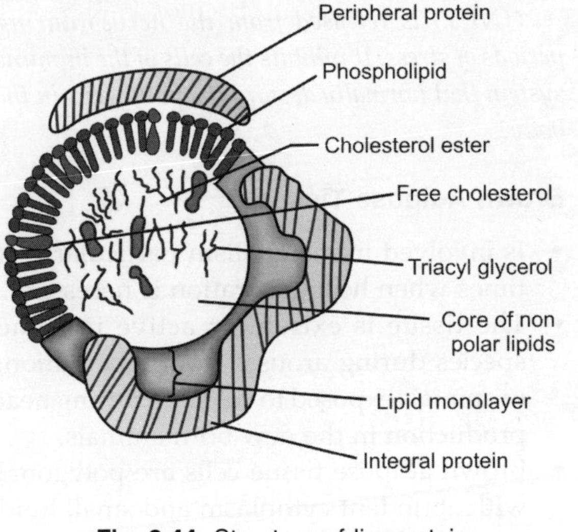


**Fig. 3.44:** Structure of lipoprotein

### *(i) Density*

This system of classification is based on the density (d) of lipoprotein particles as determined by floatation rate on centrifugation with sodium chloride solution (1.063 density) Lipoprotein with high lipid content have a low density and so float on centrifugation. Those with a high protein content have a high density and sediment easily (Fig. 3.45). Depending on the floatation constant (sf), four fractions are distinguished and their chara-cteristics are given below.

### *Electrophoretic Mobility*

This system of classification is based on the separation of the lipoproteins when subjected to an electric field. Four major bands can be visualised following plasma electrophoresis (Fig. 3.46).

- One band remains at the origin and contains *chylomicrons.*
- Second band called *β-lipoprotein* migrates with the *β*-globulin and contain the low density lipoprotein *(LDL).*

| Lipoprotein | Density |
|---|---|
| 1. Chylomicrons | < 0.94 |
| 2. Very low density lipoprotein (VLDL) | 0.94–1.006 |
| 3. Low density lipoprotein (LDL) | 1.019–1.063 |
| 4. High density lipoprotein (HDL) | 1.063–1.21 |

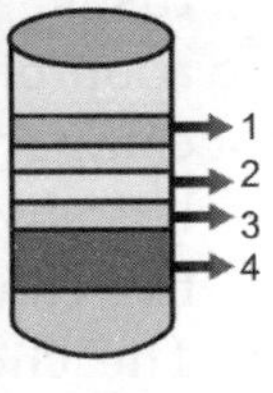


**Fig. 3.45:** Lipoprotein fractions obtained by centrifugation

- A third band migrates in front of the *β*-region and is called *pre-β-lipoprotein.* This fraction contains very low density lipoprotein *(VLDL).*
- The fourth band migrates with the *α-globulin* and contains high density lipoprotein *(HDL).* It has a higher mobility and moves ahead of other lipoproteins.

## Apoproteins

Protein component of lipoproteins are called *"apoproteins."* Letters are used to designate the apoproteins, e.g. apo A, apo B etc. To distinguish the various proteins within a class the roman numerals are used, e.g., apo A-I, A-II, etc. Each type of lipoprotein contain a characteristic set of apoproteins.

Apoproteins interact with cellular receptors and therefore determine the metabolic fate of lipoprotein particles. They also serve as activators and inhibitors of enzymes involved in lipoprotein metabolism.

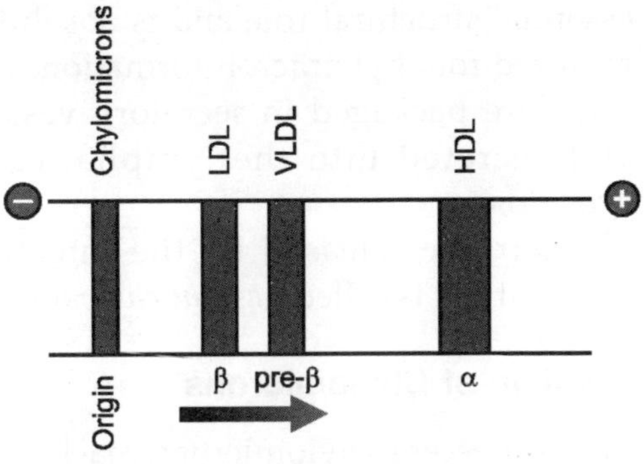


**Fig. 3.46:** Lipoprotein electrophoresis

The most important ones are apo-A, apo B, apo-C and apo-E.

The important features of the various lipoproteins are listed in Table 3.7.

## Metabolism of Chylomicrons

***Function:*** *Delivery of dietary TAG to peripheral tissues (mainly adipose tissue) and cholesterol to the liver.*

- Chylomicrons are synthesized in the intestinal mucosal cells from the absorbed dietary lipids. Inside the cell the TAG are resynthesised. *(See digestion and absorption).* Addition of cholesterol, apoproteins and phospholipids to the resynthesised TAG takes place. The resulting complex of lipids and proteins formed is termed *chylomicrons.* The main apoprotein present in chylomicron is *apo-B–48.* It plays an

**Table 3.7:** Characteristic features of lipoproteins

| *Lipoprotein* | *Sources* | *Density* | *Apoprotein* | *% Composition Total* | |
|---|---|---|---|---|---|
| | | | | *Lipid* | *Protein* |
| Chylomicrons | Intestine | < 0.95 | apo-A, B – 48,C, E | 98 | 2 |
| VLDL | Liver, Intestine | 0.95-1.006 | apo-B – 100, C,E | 90 | 10 |
| LDL | VLDL | 1.019-1.063 | apo-B – 100 | 79 | 21 |
| HDL | Liver, intestine, VLDL, Chylomicrons | > 1.063 | apo-C, A, E | 68 | 32 |

essential structural role and is absolutely required for chylomicron formation.

- They are packaged in secretory vesicles and secreted into the lymphatics by exocytosis.
- The particle released by the intestinal mucosal cell is called *nascent chylomicrons.*

### Catabolism of Chylomicrons

When the nascent chylomicrons reaches the plasma, it is rapidly modified by receiving apo-E and apo-C (apo–CII) from HDL.

- Catabolism takes place by rapid lipolysis of TAG via *lipoprotein lipase* (LPL) which is present in the endothelial cells of the capillaries of most tissues, but predominantly in the adipose tissue and cardiac and skeletal muscle. *(Apo C-II is necessary for the activation of lipoprotein lipase.)*
- Lipoprotein lipase hydrolyses TAG, contained in the particle, to yield MAG, FA and glycerol (Fig. 3.47).

$$\text{TAG} \xrightarrow[\text{Apo C-II}]{\text{Lipoprotein lipase}} \text{MAG + FA + glycerol}$$

**Fig. 3.47:** Action of lipoprotein lipase

As the chylomicron circulates with the triglyceride in its core, it is degraded by lipoprotein lipase. Consequently, the size of the particles decreases and density increases (Fig. 3.48).

- The apo-C is returned to the HDL. The remaining particle is now called *chylomicron remnant.* In humans, these chylomicron remnants are removed from the circulation by the liver. Hepatocyte membranes contain lipoprotein receptors that recognize the combination of apo B-48 and apo-E.
- Chylomicron remnants bind to these receptors and are taken into the cells by endocytosis. The endocytosed vesicle then fuses with the lysosome and the apoproteins, cholesteryl esters and other components of the remnant are hydrolytically degraded releasing amino acid, free cholesterol and fatty acids.
- The cholesterol released from chylomicrons regulate the rate of *de novo* cholesterol synthesis in the liver.

### Metabolism of VLDL

***Function:*** *Delivery of endogenous TAG from the liver to the peripheral tissues.*

VLDL are primarily synthesised in the liver via a process that is similar to chylomicron synthesis.

- The nascent VLDL particle released from the liver is predominantly composed of TAG and apo-B-100. They obtain apo C-II and apo E from circulating HDL.
- As VLDL pass through the circulation, their structure is altered. TAG is removed by lipoprotein lipase, causing the VLDL to decrease in size and become more dense and is called as *"VLDL remnant"* (IDL) (Fig. 3.49). Apo C-II which was originally donated by HDL is now transferred back to HDL. IDL undergoes partial modification by losing triacylglycerol and becoming richer in cholesterol to form LDL.

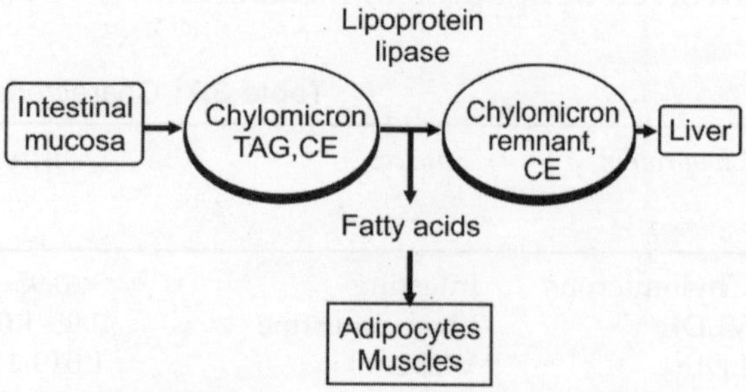


**Fig. 3.48:** Metabolism of chylomicrons. TAG, triglycerides, CE, cholesteryl esters

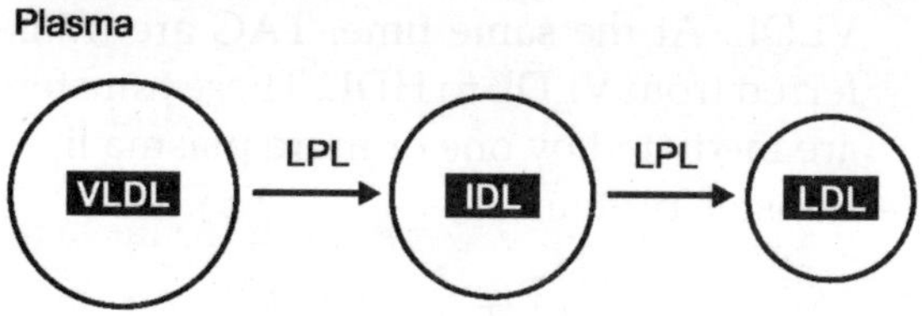


**Fig. 3.49:** Catabolism of VLDL

*Note: The VLDL remnant is also referred to as intermediate density lipoprotein (IDL).*

### Metabolism of LDL

***Function:*** *Delivery of cholesterol from the liver to the extrahepatic tissues.*

LDL is synthesized in the plasma from VLDL. It can be considered as the *"terminal VLDL remnant"* in which 80-90% of original TAG component is removed leaving the cholesterol enriched core, apo B-100 as the sole apoprotein.

- Most of the cholesterol contained in the plasma of a normal human being after an overnight fast is present in LDL. About 75% of cholesterol in LDL is in the form of cholesterol esters.
- About half of circulating LDL is taken up by extrahepatic tissues to satisfy their requirement of cholesterol and the remaining is taken up by the liver.

### Uptake of Cholesterol (LDL Clearance)

- Approximately 70% of LDL clearance is mediated by 'LDL receptors' which are present mainly in the extrahepatic tissues as well as the liver. The LDL receptor binds apo-B-100 of LDL by endocytosis. The pathway of cholesterol uptake is known as LDL receptor pathway or receptor mediated endocytosis (Fig. 3.50). The remainder is cleaved by scavenger receptors or by a non receptor mediated pathway.

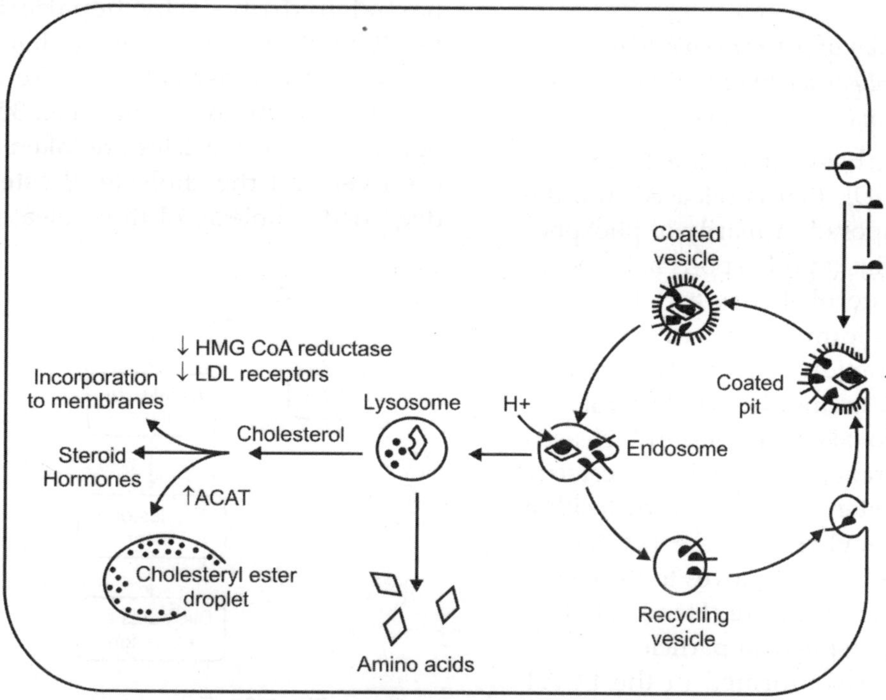


**Fig. 3.50:** Mechanism of receptor mediated endocytosis

*LDL Receptor Pathway*
*(Receptor mediated endocytosis)*

Although, all tissues can synthesize cholesterol, LDL cholesterol is preferentially used. The activity of LDL receptor pathway is regulated by cellular cholesterol needs.

Internalisation and degradation of LDL by the cells cause increased intracellular cholesterol which results in the following:

i. Suppression of HMG CoA reductase activity.
ii. Stimulation of acyl CoA cholesterol acyl transferase (ACAT) activity.
iii. Suppression of LDL receptor synthesis.

### Metabolism of HDL

***Function:***

i. *Transport of excess cholesterol from the extrahepatic tissues to the liver for excretion. This is known as reverse cholesterol transport.*
ii. *Esterification of plasma cholesterol.*
iii. *Act as catalysts facilitating the catabolism of VLDL and chylomicrons.*

HDL is synthesised in the liver and intestine. The HDL that is released into the plasma are composed primarily of phospholipids, apoproteins (apo-E, apo-A and apo-C) and free cholesterol. It has a flat discoid structure. In this form they are called *nascent HDLs.* They are rapidly converted to spherical particles as they accumulate cholesterol.

- The HDL also take up cholesterol released from the extrahepatic tissues and esterify it. Free cholesterol is esterified by lecithin cholesterol acyl transferase (LCAT) to cholesteryl ester, that moves to the centre of the discoid structure. Hence HDL is converted to a spherical particle.
- Cholesterol ester formed in the LCAT reaction (Fig. 3.51) are transferred to VLDL. At the same time, TAG are transferred from VLDL to HDL. These transfers are mediated by one or more plasma lipid transfer proteins.

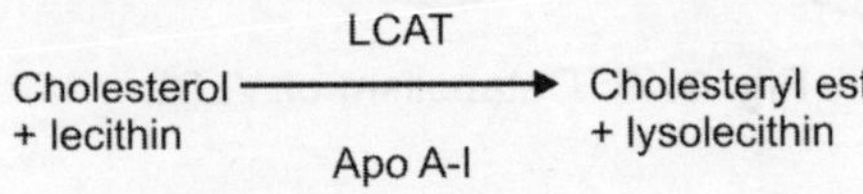


**Fig. 3.51:** Action of LCAT

- The resulting HDL is enriched with triglyceride and is larger. It is also called $HDL_2$ (Fig. 3.52).

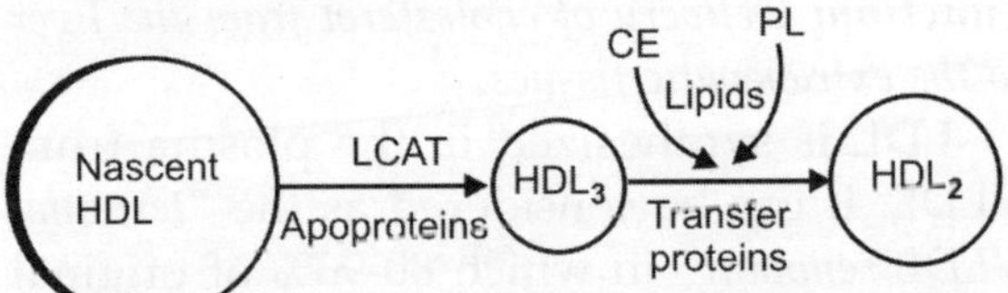


**Fig. 3.52:** Metabolism of HDL

- Two pathways are open to $HDL_2$—one is partial hydrolysis by hepatic lipase resulting in the conversion of $HDL_2$ back to $HDL_3$. The other is the uptake by the liver by binding to receptor (Fig. 3.53).
- Spherical HDL particles are taken up by the liver and the cholesteryl esters are degraded. Cholesterol thus released can

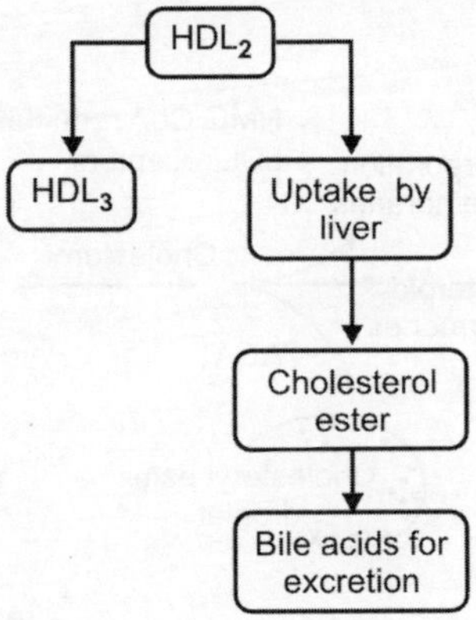


**Fig. 3.53:** Fate of $HDL_2$

either be repackaged, converted into bile acids or secreted into the bile for removal from the body.

### Free Fatty Acids (FFA)

About 95% of the fatty acids present in the plasma are in the form of fatty acid esters (i.e., TAG, phospholipid, and cholesteryl esters). About 2-5% of the fatty acid is present in the unesterified form and is bound to plasma albumin. These unesterified fatty acids are referred to as *free fatty acids (FFA).*

$$\text{FFA - albumin} \rightleftharpoons \text{FFA + albumin}$$

Thus the complexes continuously disassociate and reassociate as mentioned above. This allows unbound FFA to be made available to the tissues. Albumin is considered as a lipid transport protein although it is unlike the other lipoproteins which we have discussed so far.

*Source:* Plasma FFA are derived from the following two ways:

1. Intracellular hydrolysis of TAG in hepatic or adipose cells.
2. The action of LPL on circulating lipoproteins.

FFA circulate in complexes with albumin and are rapidly cleared from the plasma and have a half life of approximately two minutes.

*FFA and fuel metabolism:* Metabolism of FFA is closely linked to fuel metabolism through dual effect of insulin on TAG hydrolysis and synthesis. The two major fuels being (i) glucose and (ii) fatty acids.

*Decreased utilisation:* FFA utilisation as fuel is low in the postprandial state. In this state when there is adequate amount of glucose, the activity of hormone sensitive lipase is inhibited and TAG synthesis is stimulated, by insulin.

*Increased utilisation:* FFA are preferred fuels in the following conditions.

#### *Fasting*

- Plasma glucose levels and insulin secretion drop to a low level.
- Decreased insulin stimulates hormone sensitive lipase in adipocytes due to decreased availability of glycerol-3-phosphate for esterification of fatty acids.
- Uptake of glucose by the adipose tissue is also decreased resulting in increased lipolysis. Hence free fatty acids are released into the circulation.

*Diabetes mellitus:* Low levels of insulin associated with this condition stimulates hormone sensitive lipase which liberates FFA. Subsequently FFA utilisation for fuel increases.

### Dyslipidemias

The defects in various components of lipoprotein metabolism lead to changes in plasma lipoprotein levels and the disorders are known as *"dyslipidemias"*. These can be *"hyper or hypolipoproteinemias."* The hyperlipoproteinemias (hyperlipidemias) can be primary, (due to the enzymatic defect) or secondary (due to other causes, e.g. diabetes mellitus, nephrotic syndrome and obstructive jaundice). The classification is mainly based on the type of lipoprotein increased or deneased (Tables 3.8 and 3.9).

## CHOLESTEROL AND ITS IMPORTANCE

Cholesterol is the major sterol in the human body. It is a structural component of cell membranes. It is the most important animal sterol from which important compounds such

**Table 3.8:** Hyperlipoproteinaemias

| Name of the disorder | Metabolic defect | Lipoprotein fraction increased | Features | Management |
|---|---|---|---|---|
| 1. ***Type – I*** *Familial lipoprotein lipase deficiency | Lipoprotein lipase deficiency | *chylomicrons* | Abdominal pain, Hepatomegaly, Pancreatitis | Reducing fat content and increasing complex carbohydrates in diet |
| 2. ***Type – IIA*** Familial hypercholestrolemia | Defective LDL receptor | *LDL* | Atherosclerosis, CAD, Tuberous xanthoma, Corneal arcus | Low cholesterol diet decreased intake of saturated fats, supplementation with PUFA and administration of bile acid resins |
| ***Type –II-B*** | Defective Apo-B | *LDL and VLDL* | | |
| 3. ***Type-III*** * Familial dysbeta lipoproteinemia *Broad β-disease | Abnormal apo-E, which does not react with apo-E receptors | *VLDL remnant and chylomicrons* | Palmar xanthoma higher incidence of vascular disease | Weight reduction and restriction of fat and cholesterol. |
| 4. ***Type – IV*** *Familial hyperglyceridemia | Over production or decreased catabolism of VLDL | *VLDL* | Obesity, features of hyper insulinism, glucose intolerance. | ,, |
| 5. ***Type – V*** *Familial mixed hyperlipoproteinemia | Secondary to other causes | *VLDL chylomicrons* | Hyper triglyceridemia, ↑ coronary artery disease (CAD) hypercholesterolemia, pancreatitis | High PUFA diet hypolipidemic drugs-clofibrate and gemfibrozil. |

**Table 3.9:** Hypolipoproteinemias

| | | | | |
|---|---|---|---|---|
| 1. Abeta-lipoproteinemia | Defect in TAG transfer protein which prevents the loading of apo-B with the lipid | ↓*LDL* or absent | Malabsorption, mental and physical retardation, acanthocytosis | Administration of large doses of vitamin E |
| 2. Familial hypo α-lipoproteinemia (Tangier's disease) | Defect in HDL biogenesis or metabolism | ↓*HDL* or absence of HDL | Orange tonsils or adenoids, corneal opacities, hepato splenomegaly, neuropathy | |

as hormones, bile salts and vitamin D are synthesized. Widely distributed in animal tissues and absent in the plants.

- Most tissues synthesize cholesterol but the bulk of endogenous cholesterol is synthesized in the liver.

## BIOMEDICAL IMPORTANCE

Any abnormality in cholesterol metabolism leads to the development of *atherosclerosis* resulting in *myocardial infarction, stroke* or *aneurysms.*

*Gallstones* occur in 20% of the population in Western countries and is characterised by cholesterol rich stones in the gallbladder.

### *Dietary Sources*

Food derived from animal products are rich in cholesterol. For example, eggs, butter, cheese, cream and most meats. Some of the cholesterol in these products is in the form of cholesteryl esters. An ordinary diet contains a mixture of cholesterol and cholesteryl esters.

Dietary cholesterol is absorbed by the intestinal cells (normally about 200-300 mg) and transported by chylomicrons (For details refer digestion and absorption). When the dietary intake is relatively small the absorption is efficient. But when the intake exceeds approximately 500 mg, cholesterol absorption becomes less efficient.

*Plant Sterols:* Unlike cholesterol, plant sterols are poorly absorbed by humans. Feeding large quantities of plant sterols such as β-sitosterol actually inhibits cholesterol absorption.

## Cholesterol Biosynthesis

In addition to dietary intake, cholesterol can be synthesized by the human tissues.

*Site*: Liver and intestine are the main sites of cholesterol synthesis. It is also synthesised in the glands that produce steroid hormones e.g., adrenal cortex, testis and ovaries.

*Subcellular site:* Cytosol.

Cholesterol synthesis occur by a series of condensation reactions that build the polyunsaturated hydrocarbon squalene. The squalene formed is then converted to cholesterol (Fig. 3.54). The biosynthesis of cholesterol can be studied in three stages—

I. Stage—Conversion of acetyl CoA to HMG CoA and mevalonate.
II. Stage—Conversion of mevalonate to squalene.
III. Stage—Cyclization of squalene to cholesterol.

### *Regulation of Cholesterol Biosynthesis*

Cholesterol biosynthesis is regulated by the following factors:

1. Dietary cholesterol intake.
2. Caloric intake

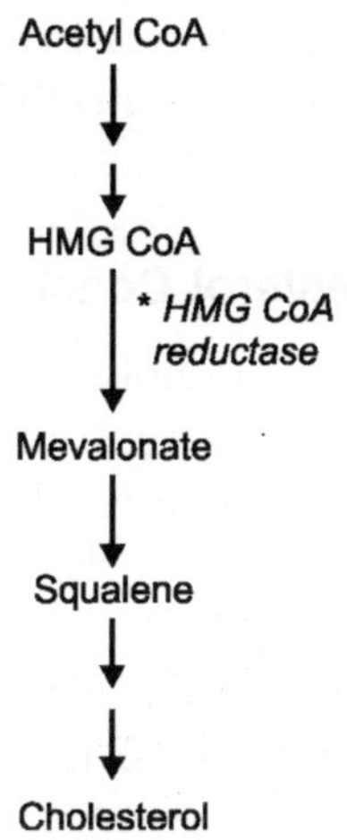


**Fig. 3.54:** Biosynthesis of cholesterol

*HMG CoA reductase is the rate limiting step in cholesterol biosynthesis.

3. Hormones
4. Bile acids

1. *Dietary cholesterol intake:* Dietary cholesterol has a strong feedback inhibitory effect on the cholesterol synthesis in the liver by inhibiting HMG CoA reductase.
2. *Caloric intake*: Excess caloric intake leads to synthesis of cholesterol due to availability of acetyl CoA.
3. *Hormones:* The hormones such as insulin, thyroxine, glucagon and cortisol have regulatory actions on cholesterol synthesis in the liver. It is mediated through HMG CoA reductase.
   - Insulin and triiodothyronine ($T_3$) increase HMG CoA reductase activity.
   - Glucagon and cortisol decrease the enzyme activity.
4. *Bile acids:* As the synthesis of bile acid increases, the *de novo* synthesis of cholesterol decreases.

### Transport and Cellular Uptake of Cholesterol

Cholesterol is transported in the plasma in the form of lipoproteins (For cellular uptake refer lipoproteins).

### Plasma Cholesterol Concentration

- Normal plasma cholesterol values range between 150–250 mg/dl.
- About 65% of cholesterol in the plasma of normal subjects in the fasting state is esterified. The lipoprotein fraction richest in cholesterol is *LDL.*
- The normal level of LDL cholesterol is 60–160 mg/dl.
- The normal level of HDL cholesterol is 30–60 mg/dl.
- These measurements are usually after an overnight fast of 12–14 hours.

Increase in plasma cholesterol is associated with increased incidence of atherosclerosis and coronary heart disease. Increase in LDL cholesterol and decrease in HDL cholesterol are considered as risk factors.

*Note: The plasma of a normal person will have no chylomicrons after a period of fasting. Males have relatively low HDL than premenopausal females.*

**Hypercholesterolemia:** Plasma cholesterol levels above the normal range is referred to as *"hypercholesterolemia."* It is observed in the following disease conditions:
- Hypothyroidism
- Nephrotic syndrome
- Diabetes mellitus
- Obstructive jaundice
- Familial hypercholesterolemia.

**Hypocholesterolemia:** Plasma cholesterol levels below the normal range is referred to as hypocholesterolemia. It is mainly observed in the following conditions:
- Hyperthyroidism
- Starvation

### Fate of Cholesterol

Cholesterol obtained from the diet or synthesized in the liver is used for the following purpose:

i. Incorporation into membranes.
ii. Conversion to bile acids.
iii. Synthesis of steroid hormones.
iv. Synthesis of vitamin $D_3$.

### Synthesis of Bile Acids

- Major metabolic fate of cholesterol is its conversion to bile acids.
- Bile acids are synthesized in the liver, released into the intestine and stored in the gallbladder.

- The liver converts cholesterol into primary bile acids. The two major primary bile acids are *cholic acid* and *chenodeoxycholic acid.*
- The secondary bile acids are *deoxycholic acid* and *lithocholic acid.*
- Bile acids production is the most important catabolic pathway for cholesterol. Continuous conversion of cholesterol to bile acid prevents the body from being overloaded with cholesterol.
- Excessive accumulation of cholesterol in the tissues is harmful.
- Like other metabolites cholesterol cannot be destroyed by oxidation to $CO_2$ and $H_2O$, because mammalian tissues do not have the enzymes capable of catabolising the steroid nucleus.
- Therefore, the only way to rid the body of excess cholesterol is to excrete it either as intact sterols or by conversion to bile acids.

*Note: Mammalian cells are unable to degrade cholesterol ring structures, therefore, bile acids are the end products of cholesterol metabolism.*

### *Synthesis of Steroid Hormones*

Cholesterol is the precursor for the synthesis of all the *"five classes"* of steroid hormones.

a. Glucocorticoids, e.g., cortisol.
b. Mineralocorticoids, e.g., aldosterone.
c. Progestins, e.g., progesterone.
d. Androgens, e.g., testosterone.
e. Estrogen, e.g., estradiol.

A brief outline is given in the (Fig. 3.55).

### *Synthesis of Vitamin D*

Vitamin D (Cholecalciferol), a fat soluble vitamin is synthesized (Fig. 3.56) from 7-dehydrocholesterol, an intermediate in the synthesis of cholesterol. *[Refer Chapter 7-fat soluble vitamins].*

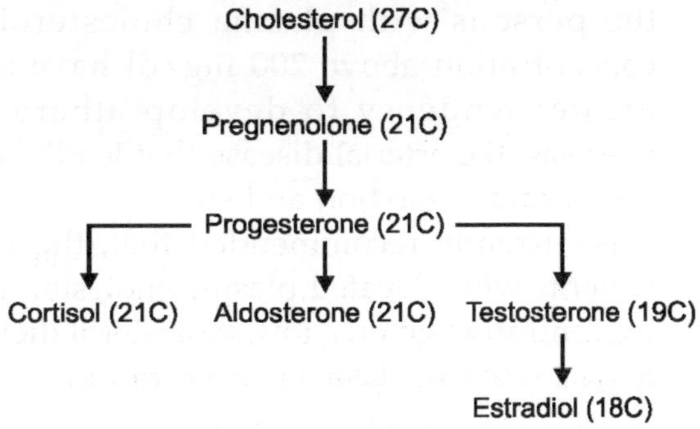


**Fig. 3.55:** Outline of steroid hormone synthesis

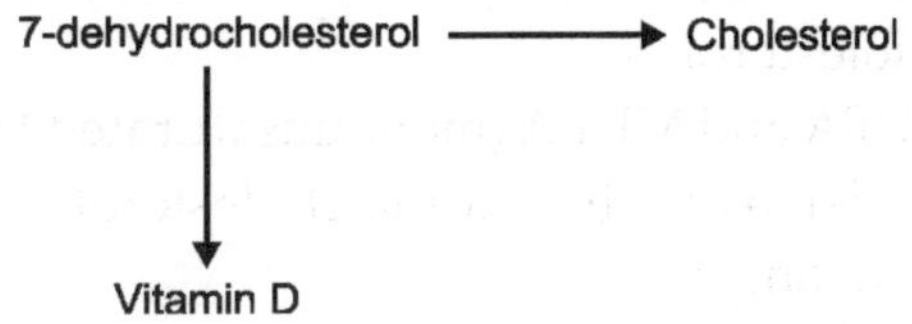


**Fig. 3.56:** Outline of vitamin D synthesis

## Excretion of Cholesterol

The total daily excretion of cholesterol is 1 g. Approximately half of this is excreted as bile acids. The remaining amount is secreted into the intestine via the bile and acted upon by the intestinal bacterial enzymes. Unabsorbed dietary cholesterol is also acted upon by the intestinal flora. The principal product, *coprostanol* and *cholestanol* (neutral sterols) formed are excreted in feces.

## Clinical Correlation

### *Atherosclerosis and Coronary Heart Disease (CHD)*

Plasma cholesterol has a direct correlation with atherosclerosis and CHD.

- Atherosclerosis is characterised by the deposition of cholesterol and cholesteryl esters in the connective tissues of arterial walls. Epidemiologic studies indicate that the persons with plasma cholesterol concentration above 200 mg/dl have a greater tendency to develop atherosclerosis, the arterial disease that leads to myocardial infarction and stroke.

- It is currently recommended that, those patients with elevated plasma cholesterol concentration go on a low cholesterol diet to decrease the plasma concentration.

### Factors and Lifestyle Affecting Plasma Cholesterol

- Diet: Saturated fatty acids raise the plasma cholesterol.
  PUFA and MUFA (monounsaturated fatty acids) lower the plasma cholesterol.
- Smoking
- High blood pressure
- Obesity
- Lack of exercise
- Drinking of soft water
- Emotional stress
- Excessive coffee drinking.

All the above factors tend to raise the plasma cholesterol levels.

### Lifestyle Modifications

- Regular exercise is beneficial.
- PUFA rich diet—inclusion of PUFA rich oils such as sunflower oil, soya bean oil and olive oil in the diet (substitution of PUFA and MUFA for saturated fatty acids is most beneficial).
- Avoidance of beef fat, butter fat and palm oil containing high proportion of SFA.
- Moderate consumption of sucrose and fructose rich food stuffs.

### Note

*i. The PUFA become incorporated into plasma lipoproteins changing their structures so that they incorporate more cholesterol. This leads to an enhanced transport of cholesterol from the tissues.*

*ii. Increased quantity of PUFA entering the liver lead to the synthesis of phospholipids containing a much higher proportion of these FA. These phospholipids are capable of transporting much more cholesterol and bile salts in the bile than phospholipids containing saturated FA. As a consequence the excretion of cholesterol is significantly increased.*

*These hypotheses are in accordance with the observation that steroid excretion in the feces is increased, when the intake of PUFA is also increased.*

### Hypolipidemic Drugs

If the dietary modifications fail to reduce the plasma cholesterol and TAG levels then hypolipidemic drugs are used.

Significant reduction of plasma cholesterol can be achieved by the use of following drugs (conservative treatment).

i. Cholestipol: Blocks the reabsorption of bile acids.
ii. Sitosterol: Blocks the absorption of cholesterol from the GI tract.
iii. Mevastatin and lovastatin: Inhibit HMG CoA reductase and reduces the cholesterol synthesis.
iv. Clofibrate and gemfibrozil: Diverts the hepatic inflow of free fatty acids from the pathway of esterification into those of oxidation.
v. Probucol: Increases the LDL catabolism.

# 4 Proteins

Proteins are dynamic molecules whose functional diversity ranges from being structural components of hair, muscle, bones and teeth to catalysis and regulation of metabolic reactions in the living organisms. The synthesis of proteins is dependant on the genetic material present in the cell. Proteins are synthesized in the cell at the ribosomes, by assembling amino acids (which are the building blocks of proteins) as dictated by the genetic code. This chapter gives an overview of the structure and properties of amino acids and a brief insight into the general reactions of amino acid metabolism for a better understanding of the structure and functions of proteins.

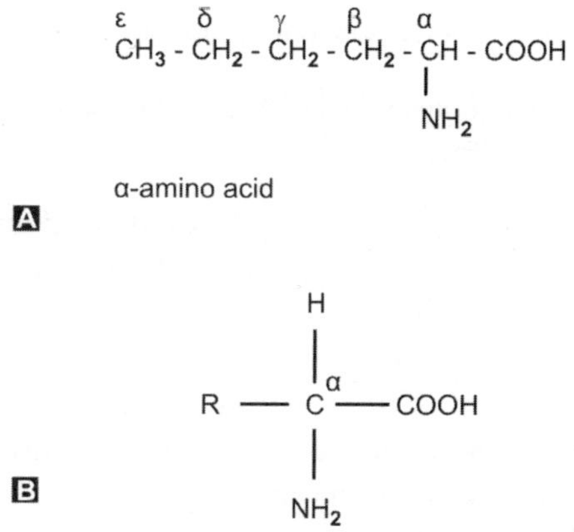


**Figs 4.1A and B:** (A) Amine substituent group at the α position of an organic acid—an α-amino acid (B) The 4 substituent groups on the α carbon atom of an α-amino acid H, COOH, $NH_2$ and R group (side chain)

## BIOMEDICAL IMPORTANCE

- Adequate amounts of amino acids are required to support infant growth and to maintain health in adults.
- Genetic defects in metabolism of some amino acids can result in severe illness like phenylketonuria, maple syrup disease, etc.
- Derivatives of amino acids participate in diverse functions like nerve transmission, regulation of cell growth, biosynthesis of small molecules like porphyrins, purines and pyrimidines.
- Proteins also have diverse functions and mutations in the genetic code for a protein can result in major disease like sickle cell anaemia.

*Amino acids* are small, organic biomolecules. They are organic acids with amine functional group at the α position. (Fig. 4.1A).

Carbon has a valency of 4 i.e., it is capable of forming four covalent bonds with four groups which might be similar or different. In an amino acid, the α carbon atom forms 4 covalent bonds (Fig. 4.1B) with:

i. H atom
ii. Carboxyl group –COOH
iii. Amino group –$NH_2$
iv. Side chain—R group (side chain), containing C, H and in some amino acids O, S or additional N.

Amino acids (AA) are molecules having two functional groups at the α carbon atom

**A**

$\overset{\beta}{NH_2 - CH_2} - \overset{\alpha}{CH_2} - COOH$

**B**

$NH_2 - \overset{\gamma}{CH_2} - \overset{\beta}{CH_2} - \overset{\alpha}{CH_2} - COOH$

**Fig. 4.2:** Structures of some of the non α-amino acids
(A) β-alanine – Component of vitamin pantothenic acid
(B) γ-amino butyric acid (GABA) – Inhibitory neuro transmitter

i.e., carboxyl (acid) group –COOH and amine group $-NH_2$ (Fig. 4.1B). Hence they are named as *α-amino acids*.

Amino acids with amino substituent groups at positions other than α are also known to be of biological significance and are called as *non α-amino acids* (Fig. 4.2).

Though these amino acids are biologically important they are not found in proteins.

There are as many as 300 amino acids in nature. Of these only 19 α-amino acids and one imino acid are found in proteins of different forms of life from bacteria to vertebrates. These 20 amino acids differ from one another in the R group (Table 4.1).

## Classification of Amino Acids

The amino acids are classified in several ways based on

I. Structure
II. Location
III. Nature of the R group
IV. Nutritional requirement
V. Metabolic fate.

### *I. Classification of amino acids based on structure:*

i. *Aliphatic amino acids*
   Simple amino acids—Gly, Ala
   Branched chain amino acids—Val, Leu, Ile
ii. *Hydroxy amino acids*—Ser, Thr
iii. *Sulphur containing amino acids*—Cys, Met
iv. *Amide group containing amino acids*—Asn, Gln
   These are monoamino, monocarboxylic amino acids—*Neutral amino acids*
v. *Acidic amino acids*—Asp, Glu
   Monoamino, dicarboxylic amino acids
vi. *Basic amino acids*—Lys, Arg,
   Diamino, monocarboxylic amino acids
vii. *Aromatic amino acids*—Phe, Tyr, Trp, His
viii. *Imino acids*—Pro

*Note: His is both basic as well as heterocyclic*

### *II. Classification based on location*

a. *Standard amino acids*—All the amino acids mentioned above are present in proteins and have a definite genetic code. For example, aliphatic, hydroxy, sulphur containing, acidic, amides, basic, aromatic, heterocyclic amino acids and imino acids.
b. *Derived amino acids*—The amino acids that are present in proteins and have no genetic code (Fig. 4.3).
   For example, hydroxy proline, hydroxy lysine—in collagen.
   Gamma carboxy glutamic acid residues in clotting factors.
c. *Non Standard amino acids*—These are those which are not present in proteins but seen free in the cells (Fig. 4.4).
   For example, ornithine, citrulline, homoserine, homocysteine.

### *III. Classification based on nature of the R group (Fig. 4.5)*

In this classification, the amino acids having the capacity to attract water (hydrophilic) or to repel water (hydrophobic) are grouped together. The hydrophilic amino acids are further classified based on their charge. The

**Table 4.1:** Amino acids—R group (side chain), structures, abbreviations (abbr) and alphabet codes

| *Sl. No.* | *Name of AA* | *Structure* | *Abbr* | *Alphabet code* |
|---|---|---|---|---|
| 1. | Glycine | $\mathbf{H}-CH-COOH$<br>\|<br>$NH_2$ | Gly | **G** |
| 2. | Alanine | $\mathbf{H_3C}-CH-COOH$<br>\|<br>$NH_2$ | Ala | **A** |
| 3. | Valine | $\mathbf{H_3C}$ \ <br>$\mathbf{CH}-CH-COOH$<br>$\mathbf{H_3C}$ / \|<br>$NH_2$ | Val | **V** |
| 4. | Leucine | $\mathbf{H_3C}$ \ <br>$\mathbf{CH-CH_2}-CH-COOH$<br>$\mathbf{H_3C}$ / \|<br>$NH_2$ | Leu | **L** |
| 5. | Isoleucine | $\mathbf{H_3C-CH_2-CH}-CH-COOH$<br>\| \|<br>$\mathbf{CH_3}$ $NH_2$ | Ile | **I** |
| 6. | Serine | $\mathbf{HOCH_2}-CH-COOH$<br>\|<br>$NH_2$ | Ser | **S** |
| 7. | Threonine | $\mathbf{H_3C-CH}-CH-COOH$<br>\| \|<br>$\mathbf{OH}$ $NH_2$ | Thr | **T** |
| 8. | Cysteine | $\mathbf{HS-CH_2}-CH-COOH$<br>\|<br>$NH_2$ | Cys | **C** |
| 9. | Methionine | $\mathbf{H_3C-S-CH_2-CH_2}-CH-COOH$<br>\|<br>$NH_2$ | Met | **M** |
| 10. | Phenylalanine | 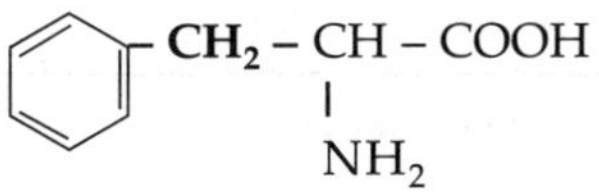 | Phe | **F** |
| 11. | Tyrosine | 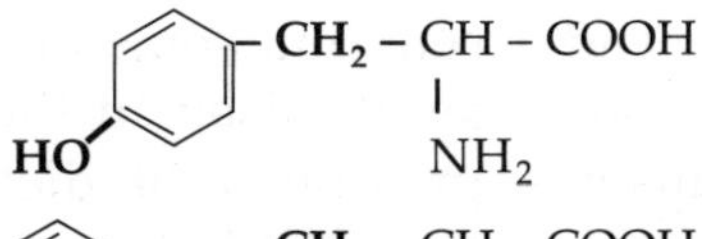 | Tyr | **Y** |
| 12. | Tryptophan | 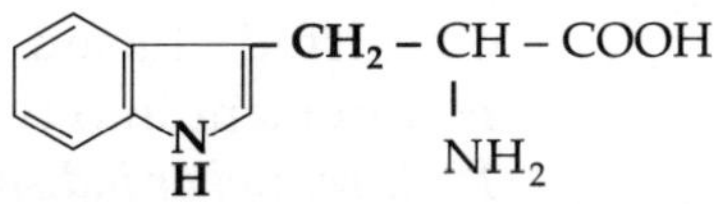 | Trp | **W** |

*Contd...*

*Table contd...*

| Sl. No. | Name of AA | Structure | Abbr | Alphabet code |
|---|---|---|---|---|
| 13. | Histidine | 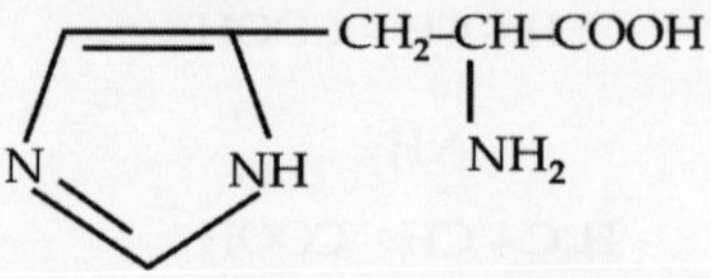 | His | **H** |
| 14. | Asparatic acid | $\mathbf{HOOC-CH_2-}\underset{NH_2}{\underset{\vert}{CH}}-COOH$ | Asp | **D** |
| 15. | Glutamic acid | $\mathbf{HOOC-CH_2-CH_2-}\underset{NH_2}{\underset{\vert}{CH}}-COOH$ | Glu | **E** |
| 16. | Asparagine | $\mathbf{H_2N\,OC-CH_2-}\underset{NH_2}{\underset{\vert}{CH}}-COOH$ | Asn | **N** |
| 17. | Glutamine | $\mathbf{H_2N\,OC-CH_2-CH_2-}\underset{NH_2}{\underset{\vert}{CH}}-COOH$ | Gln | **Q** |
| 18. | Lysine | $\mathbf{NH_2-CH_2-CH_2-CH_2-CH_2-}\underset{NH_2}{\underset{\vert}{CH}}-COOH$ | Lys | **K** |
| 19. | Arginine | $\mathbf{(H_2N)(HN{=})C-NH-CH_2-CH_2-CH_2-}\underset{NH_2}{\underset{\vert}{CH}}-COOH$ | Arg | **R** |
| 20. | Proline | 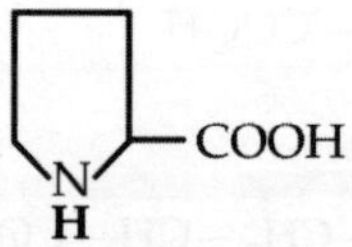 | Pro | **P** |
| 21. | Selenocysteine | $\mathbf{HSe-CH_2-}\underset{NH_2}{\underset{\vert}{CH}}-COOH$ | Secys | **Sec** |

The 21st amino acid to be discovered is selenocysteine (Secys/Sec).

charged polar are divided into positively and negatively charged. The nonpolar amino acids are further classified based on the groups in the side chains (Fig. 4.5).

### *Classification based on nutritional requirement (Fig. 4.6)*

Synthesis of proteins in the body requires all the twenty amino acids to be present in the cell. But, some of these amino acids can be synthesised in the body and need not be present in the diet. Hence amino acids are classified based on their nutritional requirement into two groups:

1. *Essential or indispensable amino acids*: These cannot be synthesized in the body and have to be supplied through the diet. There are ten essential amino acids: **Met, Arg,**

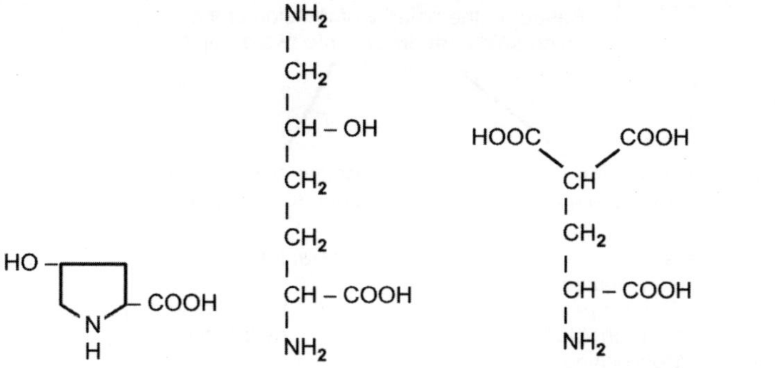

Hydroxy proline　　Hydroxy lysine　　Gamma carboxy glutamic acid

**Fig. 4.3:** Derived amino acids—which are present in proteins but have no genetic code

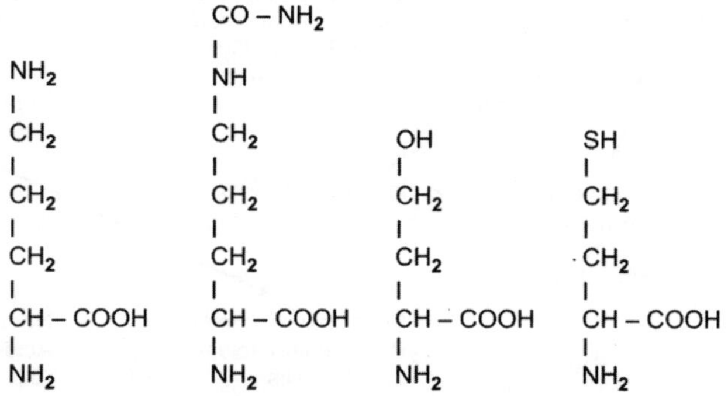

Ornithine　　Citrulline　　Homoserine　　Homocysteine

**Fig. 4.4:** Non Standard amino acids—which are not present in proteins but seen free in the cells

**T**yr, **T**rp, **V**al, **I**le, **L**eu, **P**he, **H**is and **L**ys. A code for the essential amino acids is **"MATT VIL PHLy"** (first alphabet of each amino acid). Two of the essential amino acids can be partly synthesized by adult humans but the quantity synthesized does not meet the requirement of the body. Hence these are called as *semi essential amino acids*. They are Arg and His. The remaining eight are called as *absolutely essential amino acids*.

2. ***Non-essential / Dispensable amino acids:*** The remaining ten amino acids can be synthesized in the body in sufficient quantities to meet the requirement and need not be present in the diet. For example, Gly, Ala, Cys, Glu, Asp, Asn, Gln, Tyr, Pro.

### *IV. Classification of amino acids based on metabolic fate of the carbon skeleton*

Catabolism of amino acids involves the removal of the amino group as ammonia. The structure of the amino acid after removal of the amino group is called as the *carbon skeleton*. Degradation of the carbon skeleton may give rise to compounds which lead to formation of either glucose (glucogenic amino

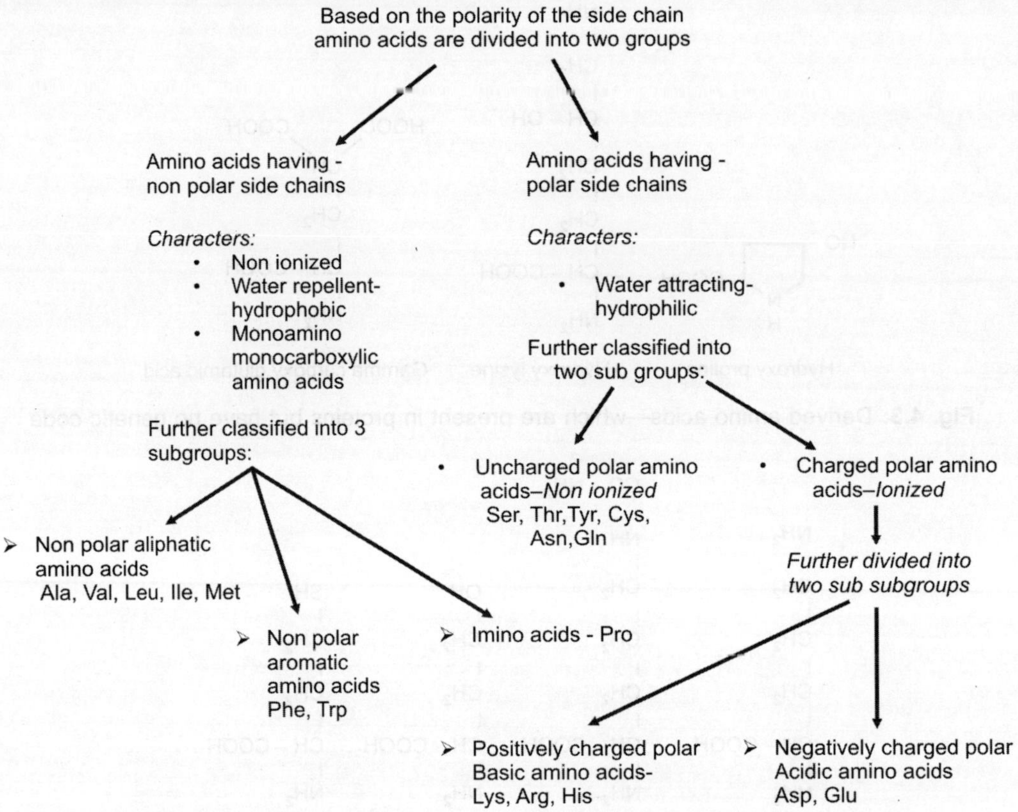


**Fig. 4.5:** Classification based on polarity of the R group (side chain)

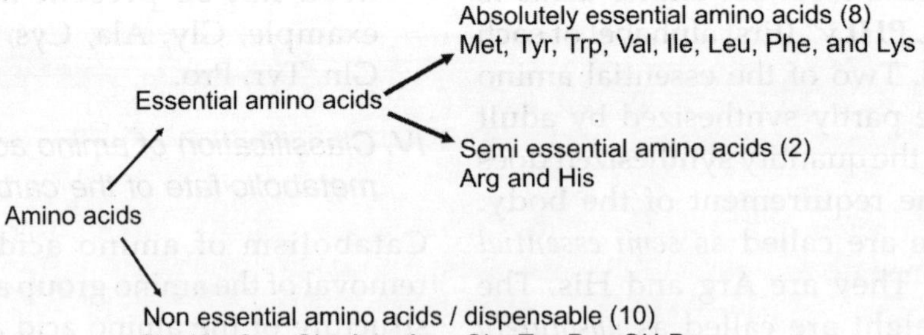


**Fig. 4.6:** Classification based on nutritional requirement of amino acids

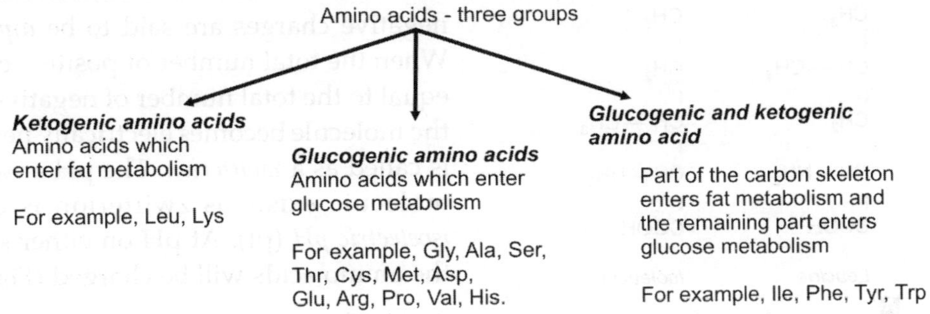


**Fig. 4.7:** Classification of amino acids based on metabolic fate

acids) or acetyl CoA (ketogenic amino acids) or to both (glucogenic and ketogenic amino acids).

Based on the metabolic destination of the carbon skeleton of an amino acid they are classified into three groups (Fig. 4.7 ).

## General Properties of Amino Acids

### *Stereoisomerism*

Amino acids exhibit DL isomerism, optical isomerism and positional isomerism.

### *DL isomerism of amino acids*

This type of stereoisomerism is *based on the orientation of the* $NH_2$ *group* and the *H atom around the* $\alpha$ *carbon atom Fig. 4.8.*

If the $NH_2$ group is oriented on the right side it is D-amino acid (Fig. 4.8A) and if it is on the left side it is L-amino acid (Fig. 4.8 B).

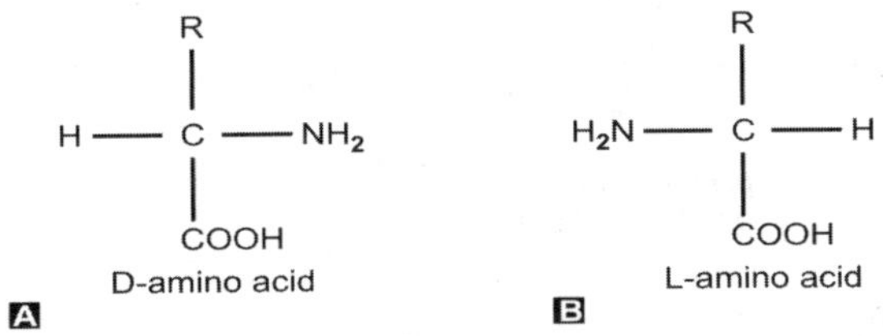


**Fig. 4.8:** DL stereoisomerism of amino acids

*Biological significance*: Natural proteins are made up of L-amino acids. D-amino acids occur in the cell wall of bacteria.

### *Optical isomerism*

All amino acids except glycine have one or more asymmetric carbon atoms. Hence they exhibit optical isomerism i.e., rotate the plane of polarized light passing through it. Consequently all amino acids except glycine show optical activity.

### *Positional isomerism*

Two amino acids having *same molecular formula* but *different structural formula* are called as *positional isomers*. Leucine and isoleucine are positional isomers (Fig. 4.9A). Another example is alanine and β-alanine (Fig. 4.9B).

### *Acid base properties of amino acids*

The COOH group of amino acids has *acidic* properties since it can donate a proton and acquire a negative charge.

$$R{-}\underset{\large NH_2}{CH}{-}COOH \longrightarrow R{-}\underset{\large NH_2}{CH}{-}COO^- + H^+$$

Anion Proton

The $NH_2$ group of amino acids is *basic* in nature since it can accept a proton and acquire a positive charge.

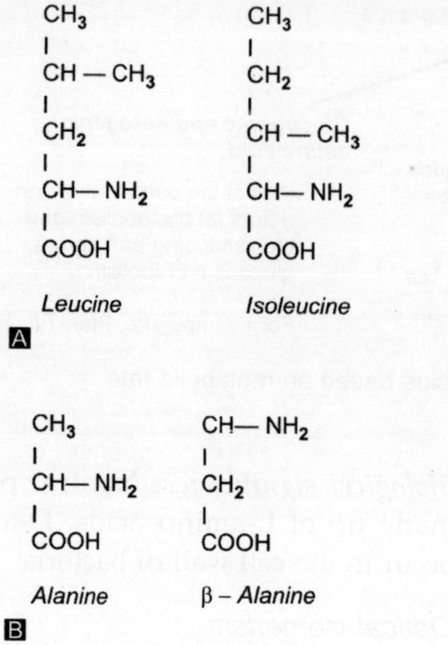


**Figs 4.9A and B:** (A) Positional isomerism in leucine and isoleucine. (B) Alanine and β-alanine

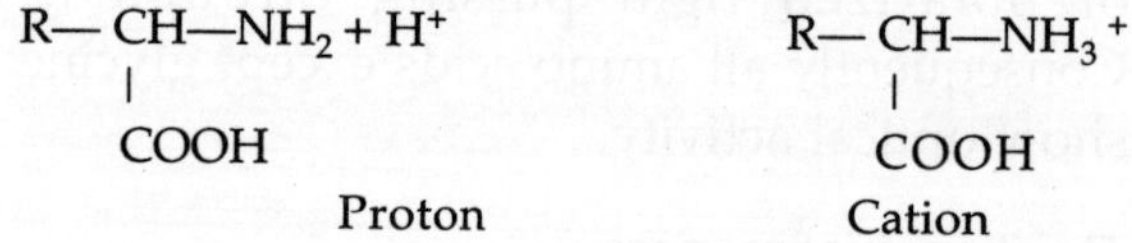


Presence of both these functional groups in the amino acid makes it an *ampholyte*.

*Ampholyte* is a substance which can act both as an acid as well as a base. At physiological pH of 7.4, the carboxyl group is dissociated and amino group is protonated. Such molecules, which have both positive and negative charges are said to be *dipolar ions*. When the total number of positive charges is equal to the total number of negative charges the molecule becomes electrically neutral and is called as a *zwitterion*. The pH at which the molecule exists as zwitterion is called as *isoelectric pH* (pI). At pH on either side of pI the amino acids will be charged (Fig. 4.10).

At pI: The protein
- Will be electrically neutral.
- Will not move in an electric field.
- Has minimum solubility. So it will precipitate. This is called as *isoelectric precipitation.*

### *Formation of a peptide bond:*

$NH_2$ and –COOH groups on two different amino acids react to form a peptide bond (Fig. 4.11A).

- *Characteristics of the peptide bond*: (Fig. 4.11B).
  - Peptide bond is an *amide linkage (C–N)*.
  - *It has four atoms*—carbon and oxygen of the carboxyl group, nitrogen and hydrogen of the amino group.
  - All the four atoms are *planar*—in the same plane.
  - The peptide bond has *partial double bond* character.
  - It is *rigid* and does not rotate freely.

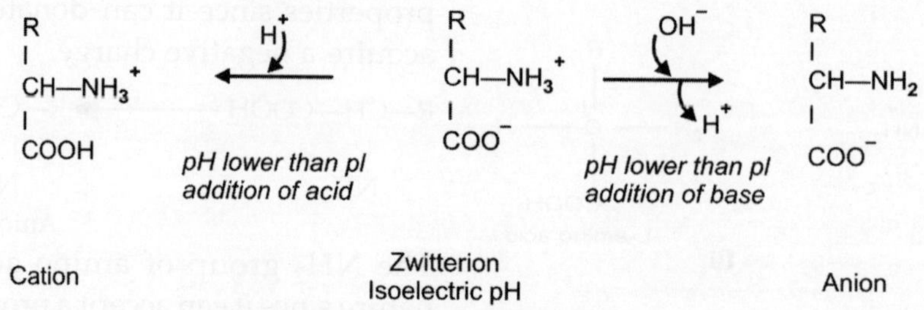


**Fig. 4.10:** Ionization of an amino acid

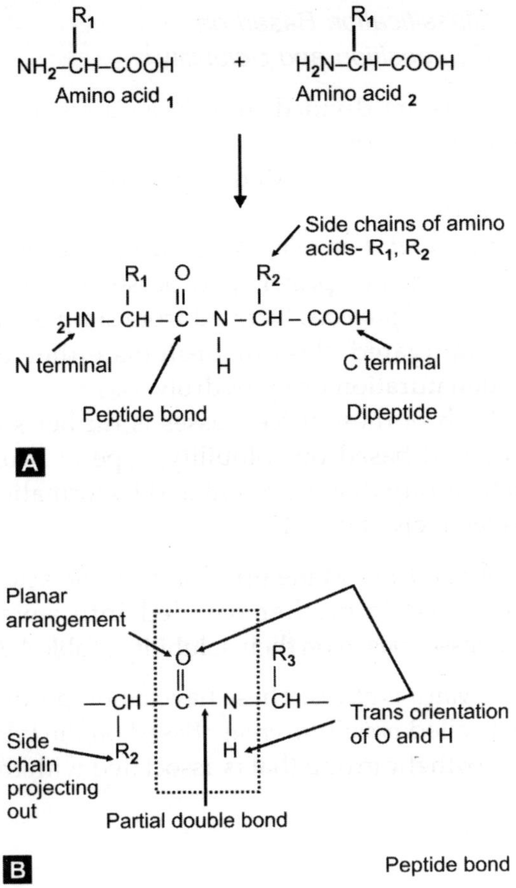


**Figs 4.11A and B:** (A) Formation of a peptide bond (B) Structure of peptide bond

- The *'O'* of the carbonyl group and *'H'* of the amide group are oriented in *"trans fashion* (opposite sides of the C-N linkage).

- Amino acids in a polypeptide are named by changing the suffix 'ine' to 'yl'. For example glutathione which has glutam*ic acid*, cyste*ine* and glyc*ine* is referred as γ-glutam*yl*- cystein*yl*-glycine. (Fig. 4.12)

*Importance:* Two amino acids are linked by a peptide bond to form a *dipeptide.*

Three are linked by two peptide bonds to form a *tripeptide.*

In general, *'n'* number of amino acids will be linked by *'n-1'* number of peptide bonds. Polymerisation of a few amino acids will form an *oligopeptide* and large numbers of amino acids polymerise to form *proteins*.

### *Dipeptides of significance*

- Carnosine – β alanyl histidine
- Anserine – β alanyl, methyl histidine
  *Both present in muscles.*
- Aspartame – L-aspartyl phenylalanyl methyl ester – *artificial sweetener.*

### *Tripeptides*

- Glutathione – GSH, γ-glutamyl cysteinyl glycine.
  *Importance*: Has many functions in the cell. One of them is destruction of free radicals generated during metabolism.
- Thyrotropin releasing hormone (TRH) pyro Glu-His-Pro-$NH_2$.
  *Importance:* Hypothalamic hormone stimulating release of TSH.

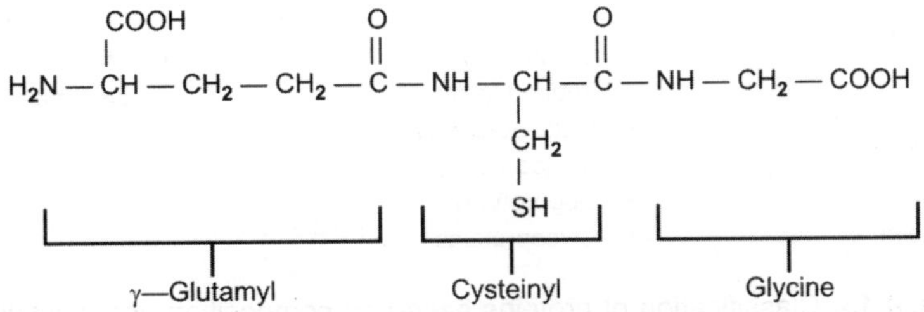


**Fig. 4.12:** Structure of glutathione

*Other peptides*

- *Pentapeptides*: Methionine enkephalin
- *Octapeptide*: Angiotensin II
- *Nonapeptide*: Posterior pituitary hormones, Bradykinin
- *Decapeptide*: GnRH, substance P, Gramicidin S (antibiotic)
- *14 AAs*: Somatostatin
- *29 AAs*: Glucagon
- *51 AAs*: Insulin

## PROTEINS

Proteins are macromolecules varying in amino acid composition, structure, shape and properties like solubility. Proteins can be classified in several ways on the following basis:

1. Composition and solubility
2. Shape
3. Function
4. Nutritive value
5. Number of subunits.

### 1. Classification Based on Composition and Solubility

Proteins are divided into *three classes* based on composition.

- *Simple proteins:* Made up of only amino acids
- *Conjugated proteins:* Made up of a protein part and non protein part-prosthetic group
- *Derived proteins:* Derived from simple and conjugated class of proteins either by denaturation or by hydrolysis.

Each of these three classes is further sub classified based on solubility, type of prosthetic group that is present and its formation respectively (Fig. 4.13).

*Simple proteins*: Made up of only amino acids. This class is further divided into seven subclasses based on their solubility (Table 4.2).

*Conjugated proteins*: These have a non protein part called *prosthetic group*. Based on the type of prosthetic group that is associated with the

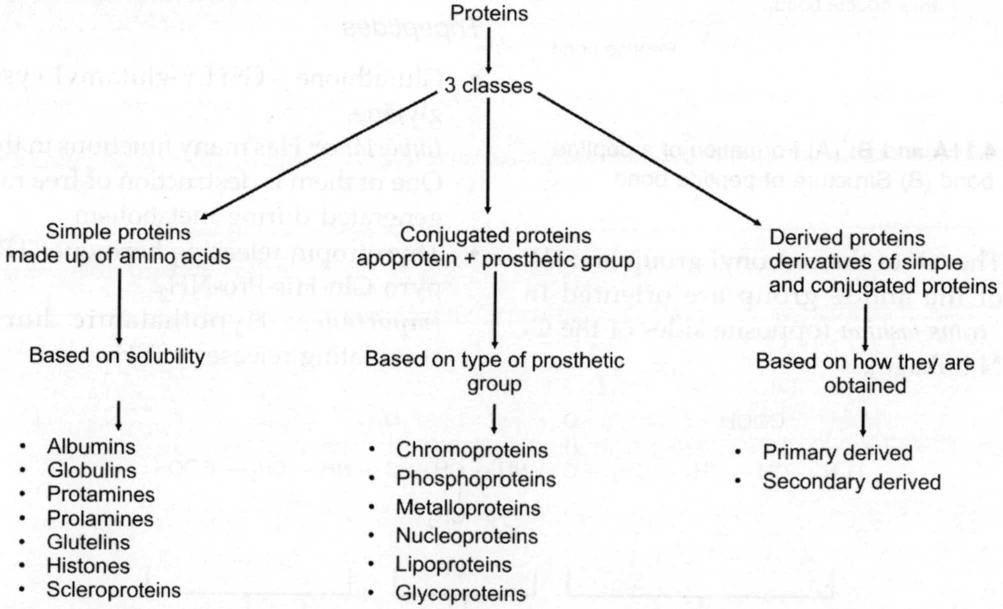


**Fig. 4.13:** Classification of proteins based on composition and solubility

**Table 4.2:** Subclassification of simple proteins based on solubility

| *Sl. No.* | *Subclass* | *Solubility* | *Examples* |
|---|---|---|---|
| 1. | Albumin | Soluble in water | Serum albumin, lactalbumin ovalbumin |
| 2. | Globulins | Soluble in dilute salt solutions | Serum globulins, globulins of milk and egg. |
| 3. | Protamines | Soluble in water, dilute acids and alkali | Associated with nucleic acids |
| 4. | Prolamines | Soluble in 70% alcohol | Zein of corn, gliadin of wheat |
| 5. | Glutelins | Soluble in dilute acids and alkali | Glutelin of wheat, oryzenin of rice |
| 6. | Histones | Soluble in water, dilute acids and alkalis but insoluble in ammonia. | Associated with nucleic acids |
| 7. | Scleroproteins | Insoluble in water, salt solutions, organic solvents, but soluble in hot strong acids | Collagen of bone, keratin of hair |

**Table 4.3:** Subclassification of conjugated proteins based on type of prosthetic group

| *Sl. No.* | *Subclass* | *Prosthetic group* | *Examples* |
|---|---|---|---|
| 1. | Chromoproteins | Colored prosthetic group | Heme in haemoglobin, riboflavin in flavoproteins |
| 2. | Phosphoproteins | Phosphate group | Casein of milk, ovovitellin of egg |
| 3. | Metalloprotein | Metal ions | $Zn^{++}$ in carboxypeptidase, carbonic anhydrase, $Fe^{++}$ in haemoglobin, cytochrome oxidase |
| 4. | Nucleoprotein | Nucleic acids - DNA | Protamines, histones |
| 5. | Lipoproteins | Lipids | Chylomicrons<br>VLDL<br>LDL<br>HDL |
| 6. | Mucoproteins<br>Proteoglycans | Carbohydrates | Immunoglobulins<br>Receptors<br>Hormones |

proteins, conjugate class of proteins is further classified into six subclasses (Table 4.3).

*Derived proteins*: These are derivatives of simple or conjugated class of proteins. They are obtained either by denaturation or by hydrolysis. They are further classified depending on how they are obtained (Table 4.4).

### *II. Classification based on function of proteins*

Proteins are classified into nine classes based on the function they perform (Table 4.5).

**Table 4.4:** Subclassification of derived proteins

| Subclass | Obtained by | Features |
|---|---|---|
| 1. Primary derived protein | Protein denaturation | • Primary structure remains intact<br>• Loss of higher structures<br>• No change in molecular weight in the primary derived protein.<br>• Examples:<br>Proteans<br>Metaproteins<br>Coagulated proteins |
| 2. Secondary derived protein | Cleavage of peptide bond – enzymatic or nonenzymatic hydrolysis | • All levels of structure are lost<br>• Decrease in molecular weight from native protein to secondary derived protein<br>*Examples:*<br>Native proteins (MW)<br>↓<br>Proteoses (MW decrease ↓)<br>↓<br>Peptones<br>↓<br>Peptides (least MW) |

**Table 4.5:** Functional classification of proteins

| Sl. No. | Class of protein | Examples |
|---|---|---|
| 1. | Structural proteins | Keratin of hair, collagen of connective tissue |
| 2. | Catalytic proteins | Enzymes–hexokinase, trypsin, carbonic anhydrase |
| 3. | Contractile proteins | Actin and myosin in muscle |
| 4. | Regulatory proteins | Hormones–insulin, glucagon |
| 5. | Transport proteins | Hemoglobin–respiratory gases, transferrin–$Fe^{+++}$ |
| 6. | Storage proteins | Ferritin–$Fe^{+++}$, myoglobin-$O_2$ |
| 7. | Genetic proteins | Histones and protamines |
| 8. | Protective/defense proteins | Immunoglobulins, clotting factors, mediators of inflammation like cytokines |
| 9. | Receptor proteins | Receptors of hormones like insulin, viruses |

### *III. Classification of the proteins based on nutritional value*

Proteins are made of amino acids and amino acids may be essential, semi-essential or non essential (see Fig. 4.6). This classification of proteins depends on the essential amino acid content of the protein. Based on the essential amino acids content proteins can be classified into two classes (Table 4.6).

### *IV. Classification of proteins based on the number of subunits*

Some proteins are made up of a single polypeptide chain or a single unit called as a

**Table 4.6:** Classification of proteins based on nutritional requirement

| *Sl. No.* | *Class* | *Characters* | *Examples (limiting amino acid)* |
|---|---|---|---|
| 1. | Complete/nutritionally rich/first class proteins | They contain all the ten essential amino acids in required proportions to promote growth and maintain the body. | Casein of milk, albumin of egg |
| 2. | Incomplete proteins | These proteins lack one or more essential amino acids. Hence they are not able to promote growth in young age but are able to sustain body weight in adults. The missing essential amino acid is called as *limiting amino acid* | Wheat protein (Lys)<br>Rice protein (Thr)<br>Pulse protein (Met)<br>Cereal protein (Lys) |

**Table 4.7:** Classification of proteins based on number of subunits

| *Sl. No.* | *Class* | *Feature* | *Examples* |
|---|---|---|---|
| 1. | Monomeric proteins | Made up of a single polypeptide chain | Ribonuclease<br>Hexokinase<br>Myoglobin |
| 2. | Oligomeric proteins— | Made up of more than one polypeptide chains (subunits/monomer units) which maybe of the same type or different type | Creatine phosphokinase<br>LDH, haemoglobin |
| | sub classified: Based on number of subunits-- | 2 subunits–dimeric<br>4 subunits–tetrameric | |

monomer/subunit. Certain others are made of two or more number of polypeptide chains or subunits. The subunits may be of same type or of different types. Based on the number of subunits proteins are classified into two classes (Table 4.7).

### *V. Classification based on shape of the protein*

Axial ratio of a protein is the ratio of the length of the protein to its breadth. When the axial ratio of a protein is less, it is oval, compact in shape, and when the axial ratio is high the protein is elongated, stretched out. Hence based on the axial ratio proteins are classified into two classes (Table 4.8).

## Structure of Proteins

Proteins are polymers of L-α amino acids in which the amino acids are linked by peptide bonds. The order of amino acids in a protein is responsible for its three dimensional structure as well as its biological function.

There are *four levels* in the structural organization of protein.

- Primary structure
- Secondary structure
- Tertiary structure
- Quaternary structure.

The secondary, tertiary and quaternary structures are referred to as the *higher levels of structural organisation* of proteins.

**Table 4.8:** Classification of proteins based on shape

| *Sl. No.* | *Class* | *Feature* | *Examples* |
|---|---|---|---|
| 1. | Globular proteins | *Axial ratio:* less than 10–length is almost equal to breadth<br>*Shape*: spherical, oval, compact<br>*Solubility*: easily soluble in water<br>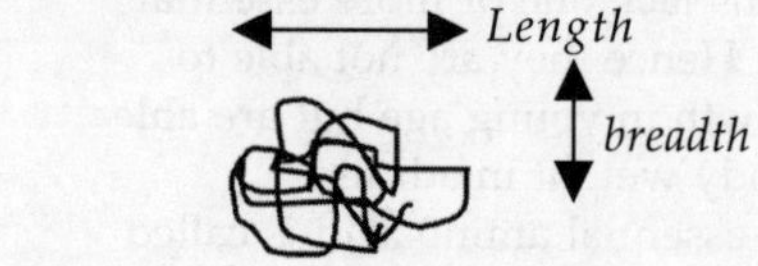<br> | Plasma proteins – albumin, globulins |
| 2. | Fibrous protein | *Axial ratio:* more than 10–length is much more than the breadth.<br>*Shape*: elongated, stretched, lengthy<br>*Solubility*: is minimum (insoluble)<br>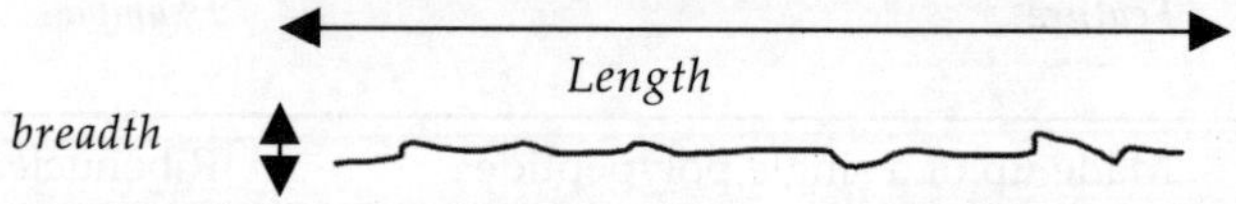<br> | Collagen, elastin, keratin |

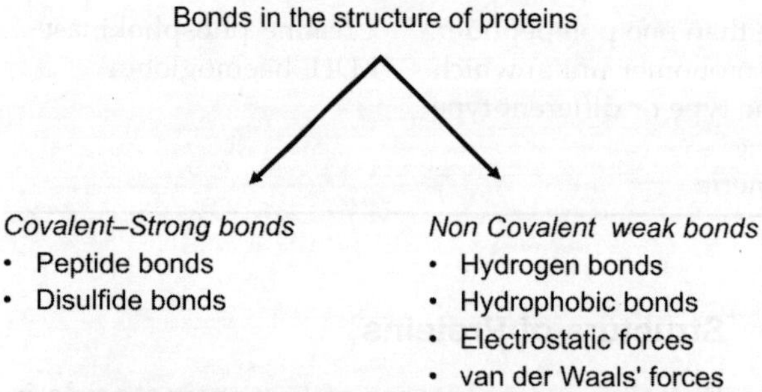


**Fig. 4.14:** Bonds in the structure of proteins

The structure of proteins has *two types of bonds* (Fig. 4.14).

- Covalent bonds:
- Non covalent bonds:

### *Formation of disulfide bonds/bridges*

Disulfide bonds are formed by dehydrogenation between –SH groups in the side chains of two Cys residues (Fig. 4.15).

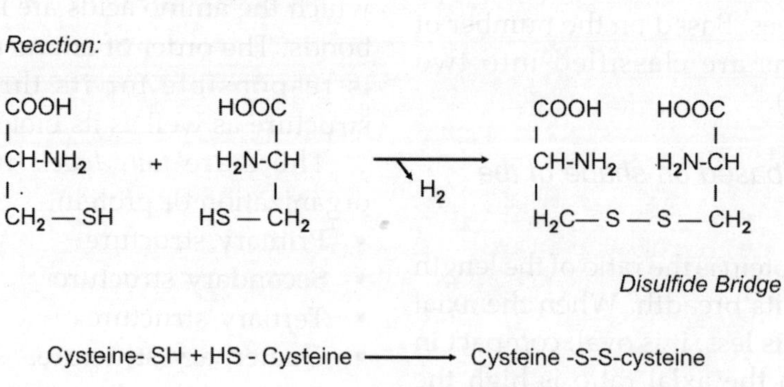


**Fig. 4.15:** Formation of cystine and disulfide bridge

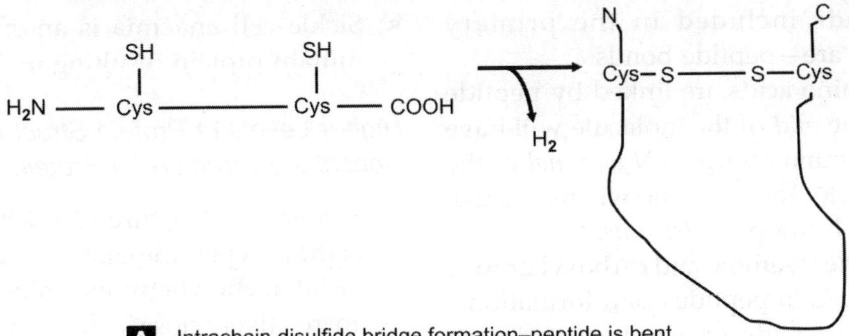


**A** Intrachain disulfide bridge formation–peptide is bent.

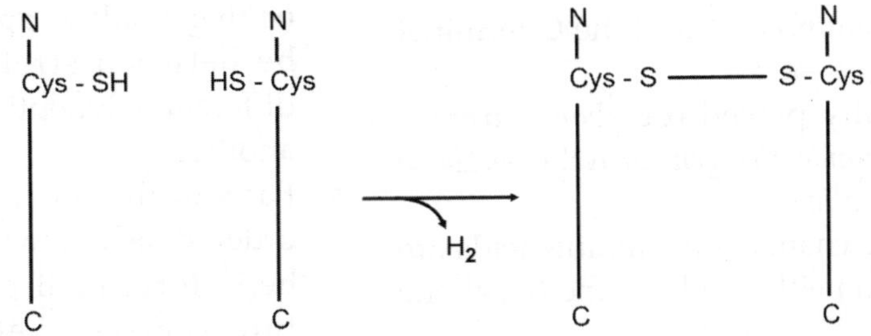


**B** Interchain disulfide bridge formation–linking of two polypeptide chains

**Figs 4.16A and B:** Disulfide bridge formation (A) intrachain (B) interchain

*Importance*: Intrachain disulfide bridge formation results in folding of the molecule (Fig. 4.16A). Interchain disulfide bridge formation holds different polypeptides together (Fig. 4.16B).

Disulfide bonds though covalent are involved in stabilizing tertiary and quaternary structure of proteins.

## *Primary Structure of Proteins (Fig. 4.17)*

- The linear sequence of amino acids in a protein is the *'primary structure of proteins'* (1°).
- It gives the following data: Total number of amino acids, quantity of each type of amino acid, and the N and C terminal amino acid residues of a protein.

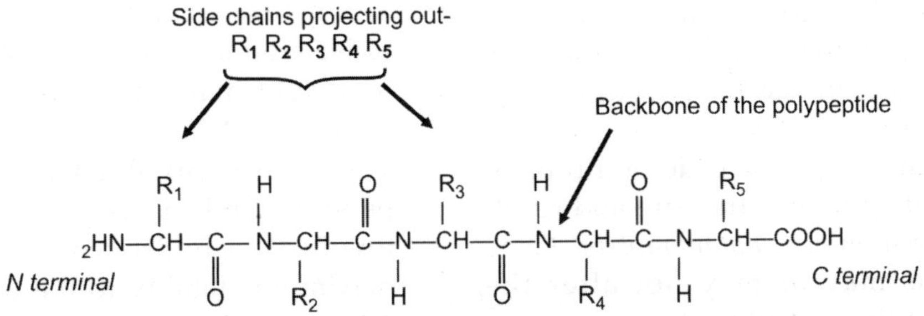


**Fig. 4.17:** Primary structure of a polypeptide

- The bonds included in the primary structure are—peptide bonds
- When amino acids are linked by peptide bonds, one end of the molecule will have a free α-amino group — *N terminal* of the polypeptide. The other end will have a free α-carboxyl group—*C terminal.*
- All the other α-amino and carboxyl groups are involved in peptide bond formation.
- The amino acids in a polypeptide are serially numbered from the N terminal to the C terminal end. Hence the N terminal is the first amino acid and the C terminal is the last (N →C).
- The centrally placed *α-carbon, carboxyl carbon and amide nitrogen* form the *backbone of the polypeptide.*
- 'R', the side chains of the amino acids are projecting on either side of the backbone (Fig. 4.17).

### *Biological Importance of Primary Structure*

- The primary structure of a polypeptide/ protein is dictated by the DNA sequence in the gene that codes for the specific polypeptide/protein.
- The higher levels of structural organization i.e., secondary, tertiary and quaternary structures, give the protein a three dimensional conformation. All of them are dependant on the primary structure.
- The function of the protein is dependant on its three dimensional structure.
- Any change in the primary structure due to alterations in the DNA sequence is called as '*gene mutation*'.
- Changes in the genetic code of a protein will alter its primary structure and result in the formation of a '*mutant protein*'. Gene mutations may or may not alter the function of the mutant protein.
- Sickle cell anaemia is an example for a mutant protein resulting in disease.

### *Higher Levels of Protein Structure: Secondary Structure of Proteins (2°)*

- Secondary structure *(2°)* denotes folding of the polypeptide chain along its long axis by interaction between amino acids which are relatively close in the primary structure of the protein.
- The interacting amino acids may be four or five residues apart or interaction may be between stretches of the same or different polypeptides lying parallel to one another.
- Parts of the polypeptide is folded in an ordered fashion which may be either in α-helix form or β-pleated sheet fashion. Interspersed between the ordered structures lies the disordered parts i.e., random coil and β-bends.
- 2° structure is stabilized and preserved by noncovalent bonds such as hydrogen bonds between NH and C=O groups (NH - - - O = C) of different peptide bonds.

### *α-helix (Figs 4.18A to C)*

- Pauling and Corey described the α-helix in 1951.
- It is seen in proteins such as haemoglobin and myoglobin.
- This type of secondary structure is tightly coiled around the long axis of the molecule in a right handed manner and appears spiral (Fig. 4.18B).
- The groups involved in formation of peptide bond in the polypeptide are all involved in hydrogen bonding, giving maximum stability to the molecule (Fig. 4.18A).

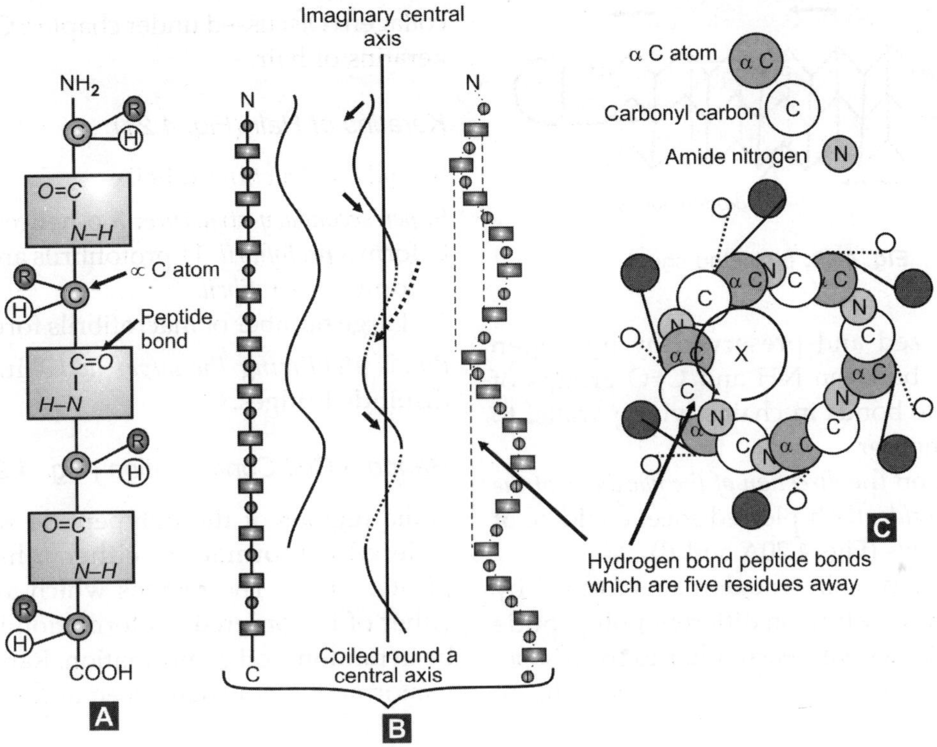


**Figs 4.18A to C:** (A) Polypeptide (B) Formation of α-helix and hydrogen bond (C) Top view of the α-helix *(For colour version see Plate 2)*

- The intrachain hydrogen bonding is between 'O' of the C=O of one peptide bond and 'H' of the N-H group of another peptide bond which is four amino acid residues away. The hydrogen bonds are parallel to the long axis of the molecule.
- All the side chain groups will project out from the molecule (Fig. 4.18C).
- The amino acids which destabilize the α-helix are hydroxy Pro, Gly, Glu, Asp, Arg, Lys, Ile, Thr and Ser. Pro residues break the helix.
- A single protein can have more than one α-helical domain.

*β-pleated Sheet (Fig. 4.19)*

- This is the second structure to be discovered, hence the name β.
- It is seen in *keratins of reptilian claws, fish scales, spider web and silk fibres.*
- Structure of *β-pleated sheet* is *non helical, fully extended, slightly zig zag* and *sheet like.*
- It occurs in sections of the same polypeptide which turn back in direction a couple of times and lie parallel to one another or between 2-5 different polypeptides lying adjacent to one another.

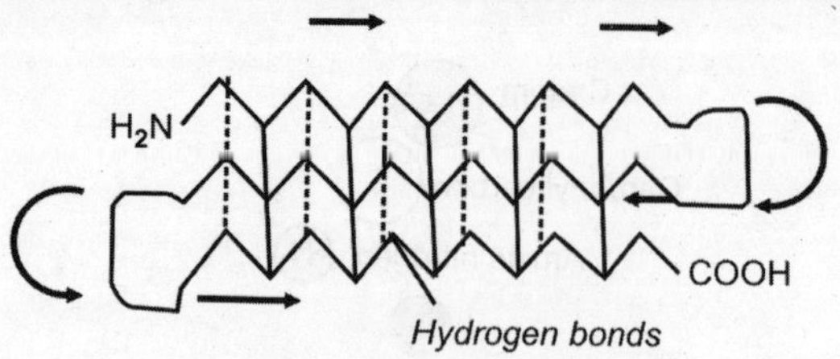


**Fig. 4.19:** β-pleated sheet

- Stabilized and preserved by hydrogen bonds between NH and C=O groups of peptide bonds in chains lying parallel to one another.
- Based on the *direction of the backbone of the polypeptide, the* β-pleated sheet could be of two types (Figs 4.20A and B)

In some proteins the domains formed by secondary structure on different polypeptide chains interact with each other to form *super secondary structure. Example:* Triple helix of collagen (Discussed under chapter: Collagen), keratins of hair.

### Keratins of Hair (Fig. 4.21)

*Secondary structure:* α-helix

*Super secondary structure:* 3 α-helices twisted to form a *protofibril.* 11 protofibrils are twisted to form a *macrofibril.*

Large number of macrofibrils forms a *hair.*

*Bonds stabilizing the super helix:* Interchain disulfide bridges.

### Random Coil Conformation (Fig. 4.22)

Some regions of the polypeptide will be in ordered conformation, either α-helix or β pleated sheet. The regions which are not in either of the ordered conformation is said to be in random coil conformation. Random coil conformation is usually seen at the terminal

β-pleated sheet

*Parallel*

- The adjacent polypeptide chains run in the same direction
  N ⟶ C
  N ⟶ C
- The consecutive hydrogen bonds are at an angle to the long axis of the molecule (Fig. 4.20A)
- For example: Flavodoxin
  Carbonic anhydrase.

*Anti parallel*

- The adjacent polypeptide chains run in the opposite direction
  N ⟶ C
  C ⟵ N
- The consecutive hydrogen bonds are perpendicular to the long axis of the molecule and parallel to one another (Fig. 4.20B)
- For example, Silk fibroin
  Amyloid protein in Alzheimer's disease

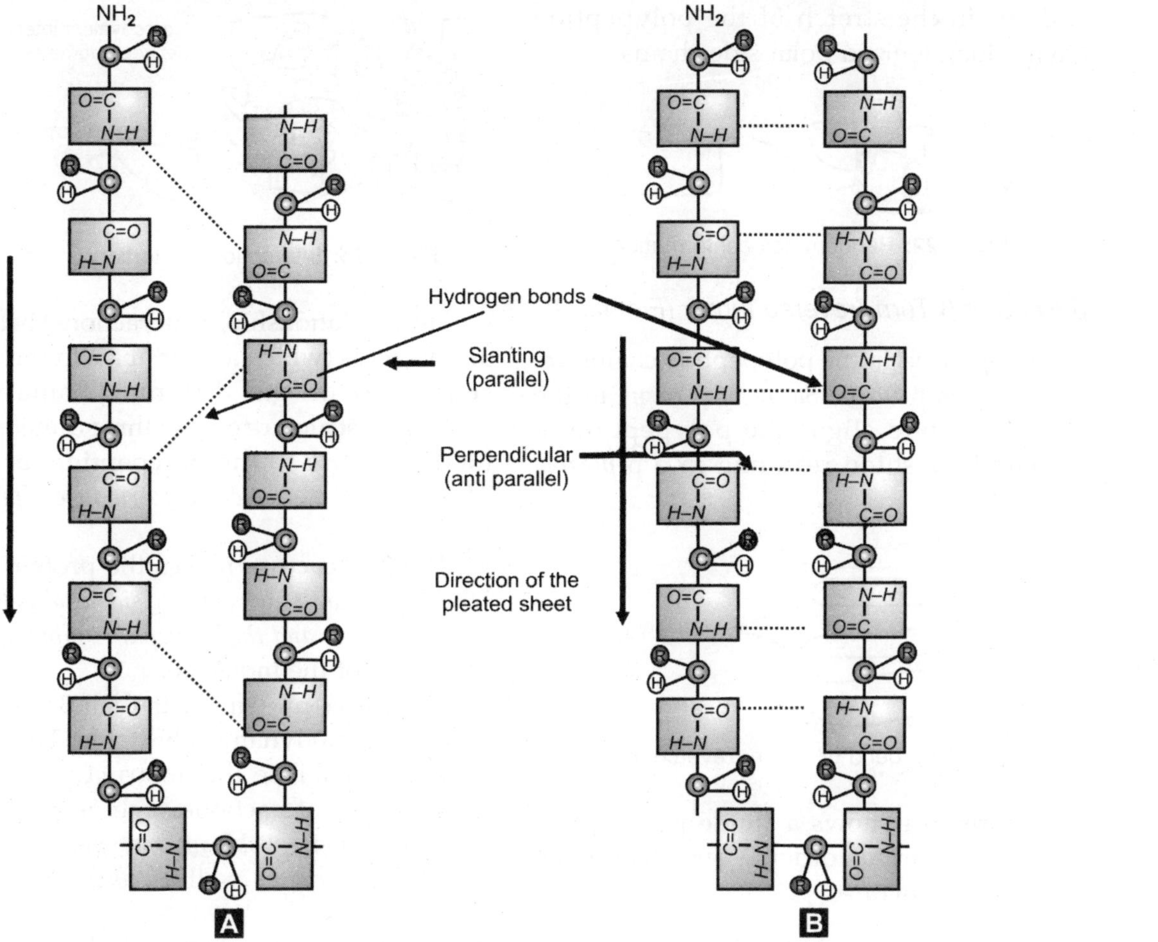


**Figs 4.20A and B:** (A) Parallel β pleated sheet (B) Antiparallel β pleated sheet
*(For colour version see Plate 3)*

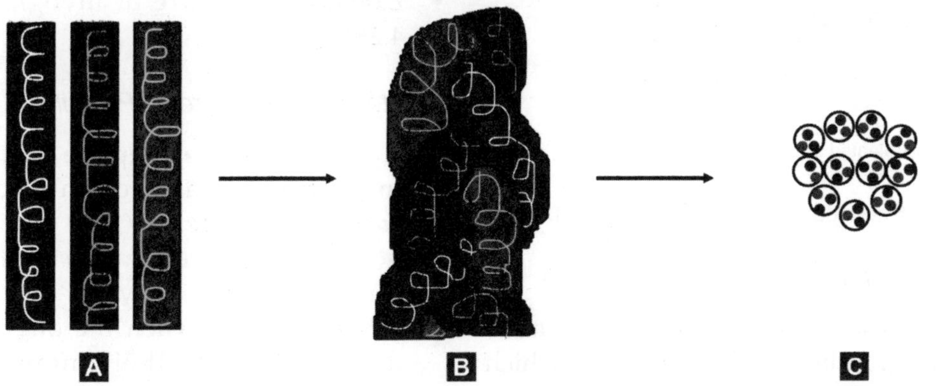


*Three α- helices* | *Coiled into a protofibril* | *Eleven protofibrils coiled into a macrofibril*

**Figs 4.21 A to C:** Super secondary structure of keratin of hair
*(For colour version see Plate 3)*

ends or in the stretch of the polypeptide having long ionized polar side chains.

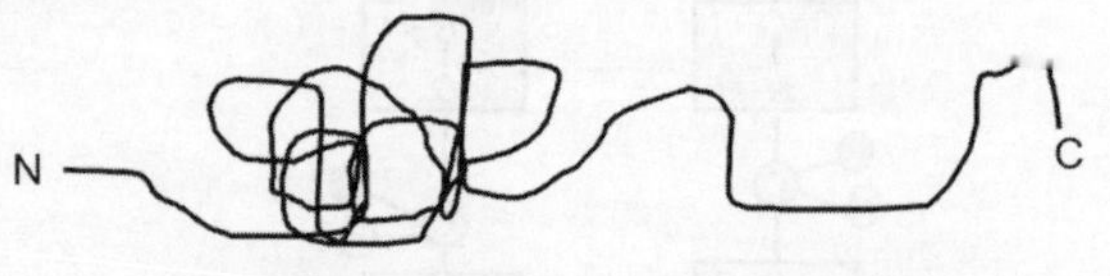


**Fig. 4.22:** Random coil conformation

### *β Bend or β Turn (reverse turns) (Fig. 4.23)*

- *Abrupt turns* in the polypeptide chains are called as β *turns*. *Series of* β *turns* in the polypeptide aligns the polypeptide in parallel fashion resulting in a β-*pleated sheet*.

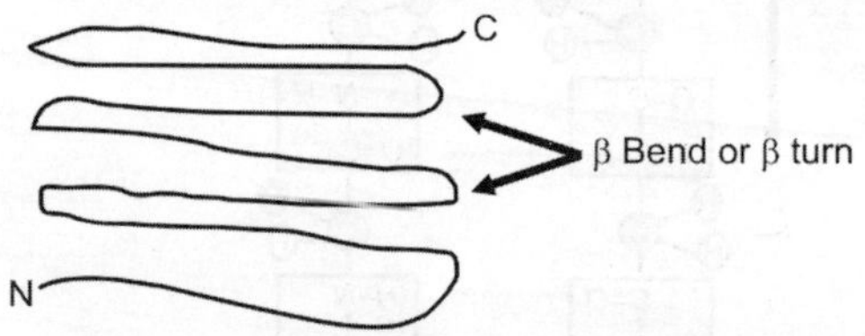


**Fig. 4.23:** β bend or β turn (reverse turns)

Figure 4.24 shows a single polypeptide having the various ordered and disordered secondary structures

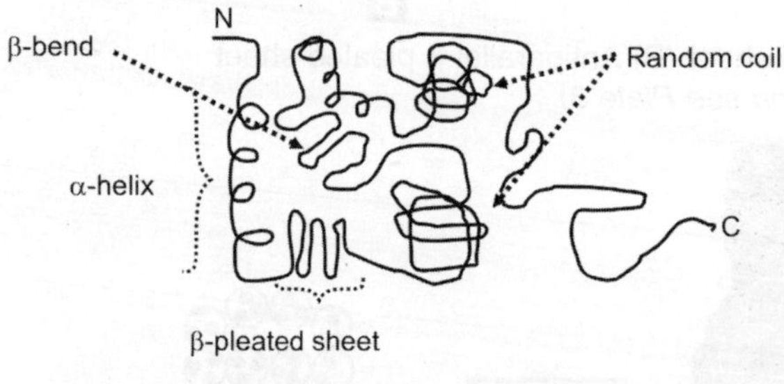


**Fig. 4.24:** Secondary structure in a polypeptide

### *Tertiary Structure of Proteins (Fig. 4.25)*

As already discussed, secondary structure is the relationship between amino acids which are 3-5 residues apart. So it denotes organisation among the amino acid residues which are close to one another.

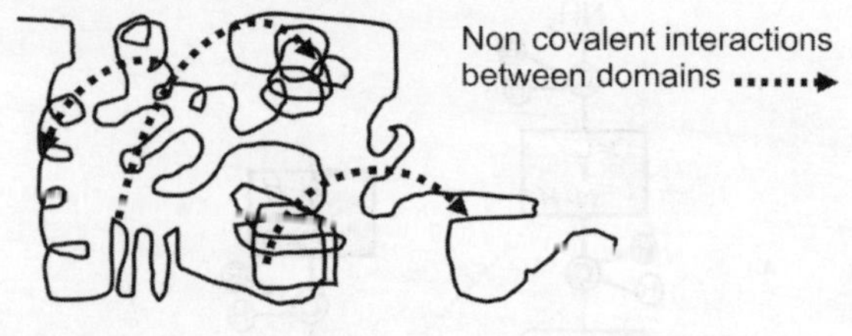


**Fig. 4.25:** Interactions in tertiary structure

Steric relationships (interactions) between amino acids (which are far apart in the linear sequence) of the different domains of secondary structure gives a three dimensional (3D) structure to the polypeptide/protein. This is called as the *tertiary structure of proteins* ($3^o$).

- The $3^o$ structure makes the protein more *compact with hydrophobic side chains held in the interior and the hydrophilic groups on the surface* of the molecule.
- The native proteins with their $3^o$ structure are thermodynamically stable. The bonds which stabilise the $3^o$ structure are non covalent interactions such as *ionic interactions, hydrophobic interactions and van der Waals' forces* as well as *covalent disulfide linkages*.
- *Domain* is a compact globular, independent, functional unit of a protein.
- *Example*: Structure of myoglobin (Fig. 4.26).

### *Quaternary Structure of Proteins*

- Aggregates of polypeptide chains, forming a functional protein is called as *quaternary structure* ($4^o$).
- $4^o$ structure is seen in *oligomeric proteins*. The monomers may be of *similar or dissimilar* type. When the oligomers are dissociated into their monomers the protein loses its biological function.
- $4^o$ structure is stabilised by *non covalent interactions* as well as *covalent disulfide*

A. **1° structure.** Single polypeptide of 153 amino acids.

B. **2° structure.** The polypeptide has 8 right handed helical segments, of 7-20 amino acids length. They are numbered from A to H. The helical domains are separated by 7 interhelical spaces AB, BC, CD, DE, EF, FG and GH respectively.

C. **3° structure.** Non covalent interactions between the helices gives a 3D globular structure to the protein part of myoglobin with which the heme will interact.

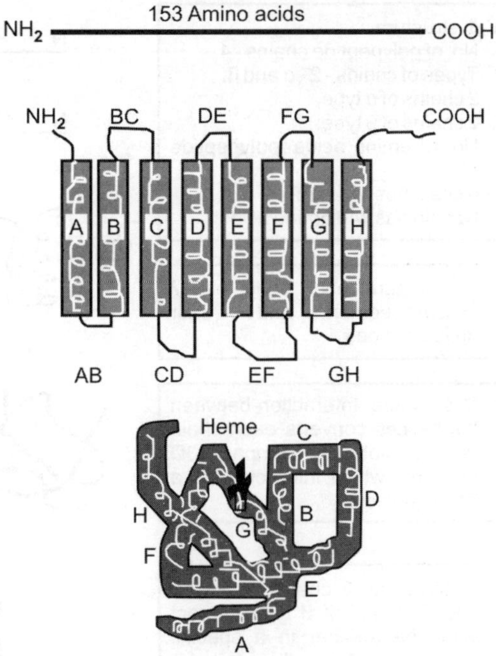


**Fig. 4.26:** Different levels of structural organisation of myoglobin (A) 1° structure (B) 2° structure (C) 3° structure

*linkages*. For example, antibodies, insulin receptors, haemoglobin, lactate dehydrogenase, aspartate transcarbamoylase.

Structural levels of haemoglobin and myoglobin are shown in Figure 4.27.

### Denaturation of Proteins

Functional proteins occurring in animal and plant tissues are called as *'native proteins.'* The higher levels of structure of proteins can be destroyed with certain physicochemical agents. The loss of higher structural organisation of proteins results in alterations of physical, chemical and biological properties of the molecule. This is called *denaturation of proteins*. For an oligomeric protein, dissociation into its monomers may result in loss of biological activity. The physicochemical agents which bring about denaturation are called as *denaturing agents*. Denaturing agents can be of different types (Fig. 4.28).

### Alterations that Occur during Denaturation

- The higher levels of protein structure is lost but the 1° structure remains unchanged (peptide bonds are not broken).
- So biological property of the protein is lost but molecular weight remains unchanged.
- There is a decrease in solubility and increase in precipitability of protein.
- Denatured protein is easily digested by proteolytic enzymes compared to native proteins due to increased exposure of peptide bonds.

A. 1° structure:
No. of polypeptide chains - 4
Types of chains - 2 - α and β.
2 chains of α type
2 chains of β type.
No. of amino acids /polypeptide chain:
α chain has 141 and
β chain has 146 amino acids

B. 2° structure: each chain has 7 helical segments and 6 inter helical regions

C. 3° structure: Interaction between the helices converts each chain into a globular, compact 3D molecule which interacts with a heme molecule

D. 4° structure: 2 chains of α type and 2 chains of β type interact with one another in a specific manner to form the globular tetrameric hemoglobin. There is no αα and ββ interactions only αβ interactions

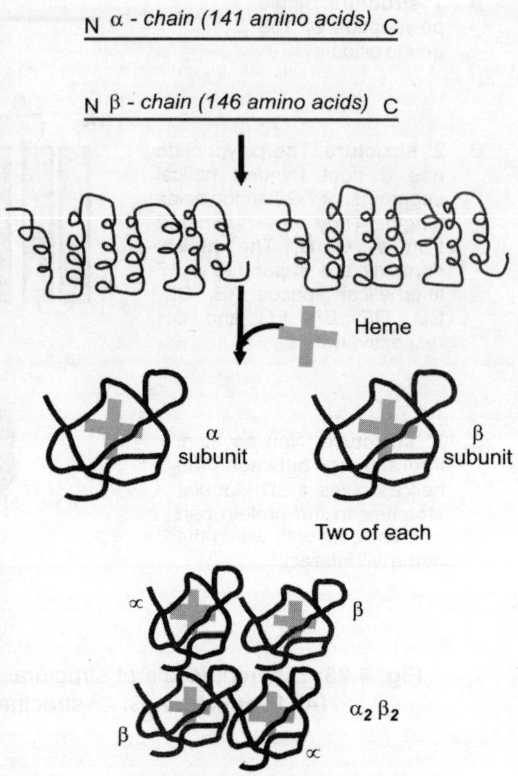


**Fig. 4.27:** Different levels of structural organisation of haemoglobin (A) 1° structure (B) 2° structure (C) 3° structure (D) 4° structure
*(For colour version see Plate 4)*

Native proteins ⟶ Denatured proteins

Denaturing agents

Physical agents
- Heat
- Violent shaking
- UV rays
- X-rays
- High pressure
- Ultrasound

Chemical agents
- Acids
- Alkalis
- Salts of heavy metal ions
- Urea
- Detergents like sodium dodecyl sulphate (SDS)
- Alcohol

**Fig. 4.28:** Types of denaturing agents

- *Example*: Denaturation of enzyme ribonuclease by urea is reversible. When urea is removed from the enzyme by dialysis the enzyme goes back to its active native conformation.
- Denaturation can be of two types (Fig. 4.29).
- Some proteins, when heated, though denatured are still soluble.
- When the pH of such a solution is adjusted to its pI by using acids such as acetic acid the denatured protein becomes less soluble and forms a thick floating semisolid precipitate. This is called as *Coagulum.*
- *Significance of coagulation reaction*: This principle is used for *detection of albumin in urine* by heat and acetic acid test.

## DIGESTION OF PROTEINS

### Sources of Proteins

a. Exogenous or dietary protein: 70–100 g/day (1 g/kg body weight).
b. Endogenous protein : Enzymes and other cellular proteins (35–200 g/day)

*Proteins lost from the body*: Amount is same as that of intake. It is lost in urine, feces and sweat.

Enzymes that digest proteins are called as *"proteolytic enzymes"*.

### Synthesis of Proteolytic Enzyme

*Site*: Endoplasmic reticulum.

*Modified*: Golgi apparatus.

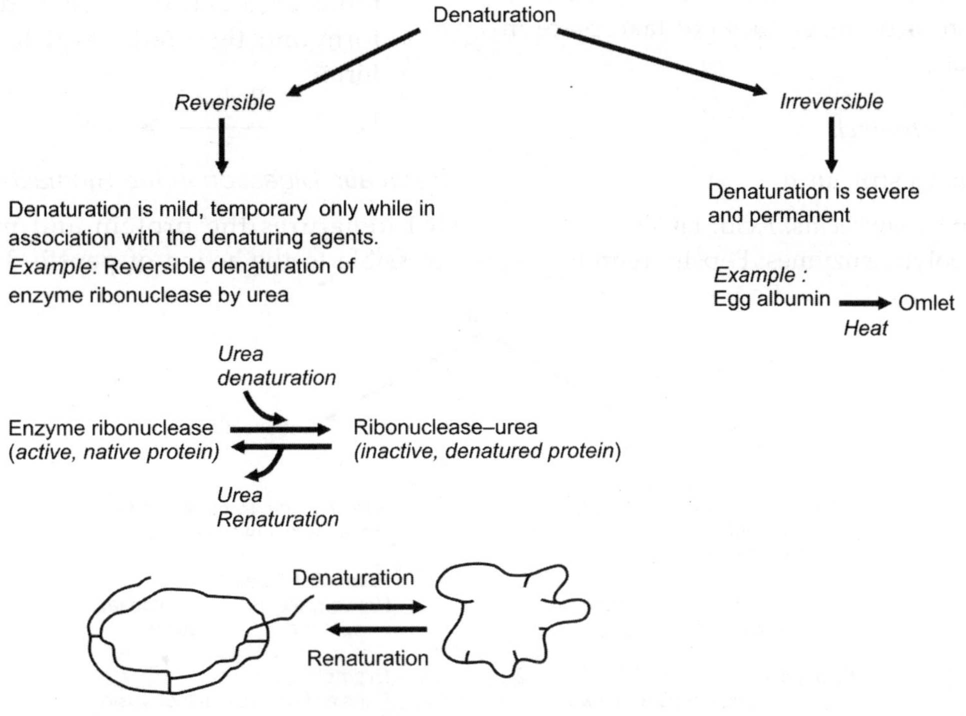


**Fig. 4.29:** Denaturation of proteins

*Stored*: As membrane bound vesicles in inactive proenzyme form.

*Secretion of inactive enzymes*: By exocytosis.

*Stimulus for secretion*: Hormones, secretogogues, entry of food chyme into the digestive tract.

*Activation of proenzymes*: Initial (a) HCl (for pepsinogen) (b) Enterokinase (for trypsinogen) (c) Trypsin (for most of the other proteolytic enzymes)

Later by autocatalysis.

### Classification of Proteolytic Enzymes

*Based on site of action* (Fig. 4.30).

### Digestion

**a. *In the mouth***

*Secretion*: Saliva

Saliva has no proteolytic enzymes. So digestion of proteins does not take place in the mouth.

**b. *In the stomach***

*Secretion*: Gastric juice

*Proteolytic components*: Acid: HCl

Proteolytic enzymes: Pepsin, rennin

### Functions of HCl

i. Denaturation of native proteins to acid metaproteins (for efficient action of pepsin).
   Native proteins → acid metaproteins
ii. HCl brings the pH of gastric juice to 2.0 for activation of pepsinogen and optimum activity of pepsin.
iii. Maintains the optimum pH for absorption of vitamin $B_{12}$.
   Intrinsic factor + Extrinsic factor (Vitamin $B_{12}$) → Complex formation
iv. Hydrolysis of sucrose: Sucrose → Glucose + Fructose. These products are easily absorbed across the wall of the stomach directly into the blood.
v. Destruction of micro organisms-*antiseptic / bacteriolytic* effect.
vi. For absorption of iron in the ferrous form: Releases iron from bound organic form and then reduces it to ferrous form.

$$Fe^{+++} \xrightarrow{Reduction} Fe^{++}$$

### Enzymatic Digestion in the Stomach

HCl denatures the protein and makes it accessible to the action of pepsin. HCl also

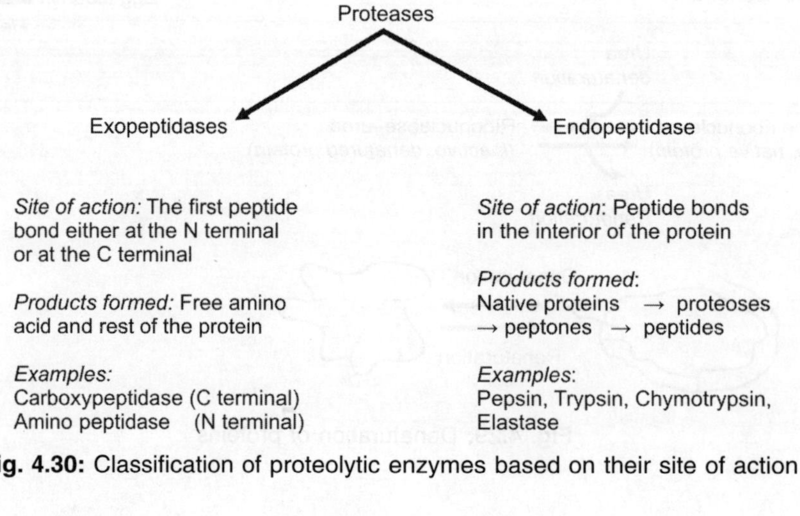


**Fig. 4.30:** Classification of proteolytic enzymes based on their site of action

activates inactive pepsinogen to pepsin. Small amounts of pepsin formed is capable of activating more pepsinogen, this is called as *autocatalysis*.

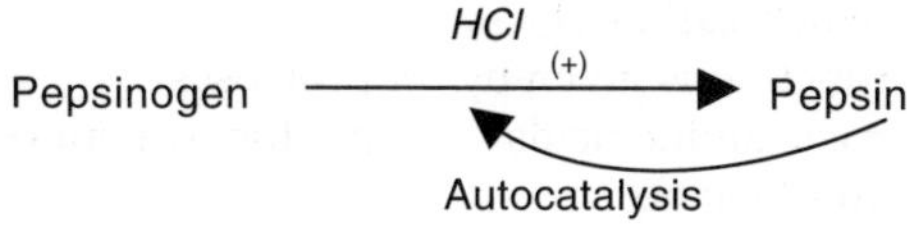


Both pepsin and rennin are endopeptidases (Table 4.9).

*Digestion of proteins in the small intestine:* It is aided by secretion of pancreatic juice along with bile.

*Proteolytic enzymes in the pancreatic secretion:* Trypsin, chymotrypsin, carboxypeptidase and elastase.

*Trypsin:* It is secreted in the inactive form trypsinogen. It is activated by enterokinase of the intestinal juice and later by autocatalysis as shown below:

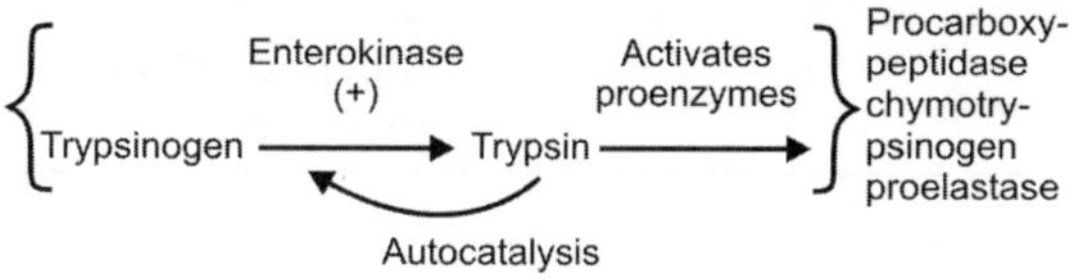


### Functions of Trypsin

- It hydrolyses dietary proteins into proteoses, peptones, peptides.
- It activates the other pancreatic enzymes: proteolytic enzymes: procarboxypeptidase, chymotrypsinogen and proelastase Lipid hydrolyzing enzymes: prophospholipase $A_2$.

Table 4.10 shows the action of the pancreatic proteolytic enzymes.

### Intestinal Secretion

The oligopeptides and dipeptides formed by the action of pancreatic enzymes are completely digested to amino acids by the enzymes present in the *succus* (intestinal juice) *entericus* (Table 4.11).

These are enzymes of surface digestion. The enzymes are attached to the microvillous surface of the intestinal mucosal cells by connector peptide molecules (Fig. 4.31).

### Intestinal Absorption of Amino Acids

*Site:* Small intestine
*Type:* Secondary active transport.

**Table 4.9:** Proteolytic enzymes of gastric juice

| *Enzyme* | *Proenzyme* | *Activation by* | *Group specificity* | *Products formed* |
|---|---|---|---|---|
| Pepsin | Pepsinogen | HCl followed by autocatalysis | Peptide bonds involving aromatic amino acids, methionine, leucine | Proteins<br>↓<br>Proteoses, peptones. peptides |
| Renin | Prorenin | HCl | Peptide bonds in which carbonyl function is given by aromatic amino acids | Casein (milk protein)<br>↓ *curdling action*<br>Whey protein + paracasein<br>↓ $Ca^{++}$<br>Calcium paracaseinate |

**Table 4.10:** Action of pancreatic proteolytic enzymes in the intestine

| *Enzyme* | *Proenzyme* | *Activation by* | *Group specificity* | *Products formed* |
|---|---|---|---|---|
| ***Endopeptidases*** | | | | |
| Trypsin | Trypsinogen | Enterokinase/ autocatalysis | Peptide bonds in which carbonyl function is given by basic amino acids and leucine | Proteins ↓ proteoses, peptones, peptides |
| Chymotrypsin | Chymotrypsinogen | Trypsin/ autocatalysis | Peptide bonds in which carbonyl function is given by aromatic amino acids, methionine, leucine | Proteins ↓ proteoses, peptones, peptides |
| Elastase | Proelastase | Trypsin | Peptide bonds in which carbonyl function is given by Gly, Ser and Ala amino acids | Proteins ↓ proteoses, peptones, peptides |
| ***Exopeptidases:*** | | | | |
| Carboxy-peptidase | Procarboxy-peptidase | Trypsin | Terminal peptide bonds adjacent to the C terminal amino acid residue | Proteins ↓ free amino acid + rest of the peptide ↓ dipeptide (final product) |

**Table 4.11:** Intestinal digestion of proteins

| *Enzyme* | *Site of action* | *Product formed* |
|---|---|---|
| Enterokinase/Enteropeptidase | Activates trypsinogen | Active trypsin |
| Amino peptidase | Exopeptidase | Oligopeptides →one amino acid at a time from N terminal end →end product dipeptide |
| Dipeptidyl peptidase | Endopeptidase | $\text{Oligopeptides}_n$ →dipeptide + $\text{Oligopeptides}_{n-2}$ number of amino acids |
| Tripeptidyl peptidase | | $\text{Oligopeptides}_n$ →tripeptide + $\text{oligopeptides}_{n-3}$ number of amino acids |
| Dipeptidase | | Dipeptide →2 amino acids |
| Tripeptidase | | Tripeptide→free amino acid + dipeptide |

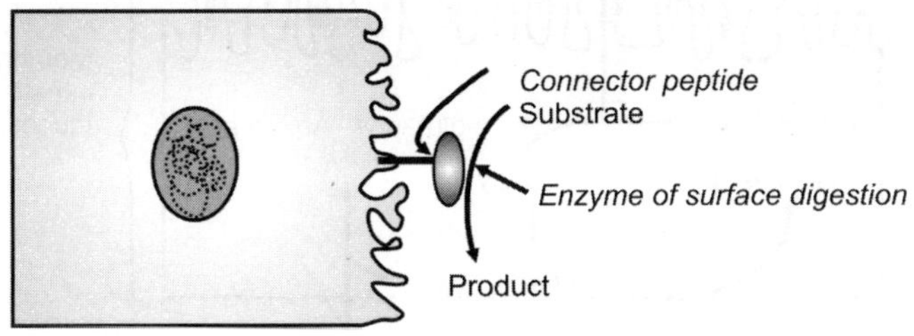


**Fig. 4.31:** Enzymes of surface digestion

### *Features*

- Carrier mediated
- Energy dependant
- Against concentration gradient
- $Na^+$ dependant.

*Transport systems*: There are *five transport systems* for the different classes of amino acids. They are transport system for:

1. Neutral amino acids – Ala, Val, Leu, Ile, Met, Phe, Tyr.
2. Basic amino acids – Arg, Lys, His.
3. Acidic amino acids – Asp, Glu.
4. Imino acids – Pro, hydroxy pro.
5. For β-amino acids.

Mechanism of amino acid absorption are shown in Figure 4.32.

### *Absorption of Amino Acids by γ-glutamyl cycle (Miester cycle): (Fig. 4.33)*

*Site*: Intestinal mucosal cells, kidney tubular cells and brain.

*Type of amino acids absorbed*: neutral amino acids.

*Requirement:* Glutathione (tripeptide: γ-Glutamyl-cysteinyl-glycine).

### *Absorption of Oligopeptides*

In infants, colostrum, the principal diet contains immunoglobulins, which are absorbed by endocytosis, enters the blood and contribute to the immunity of the infant.

### *Disorders of Amino Acid Absorption*

1. *Hartnup's disease*: It is a genetic defect of the carrier molecules of aromatic and hydrophobic amino acids in the intestine as well as in the kidney tubular cells. Hence there is decreased absorption of

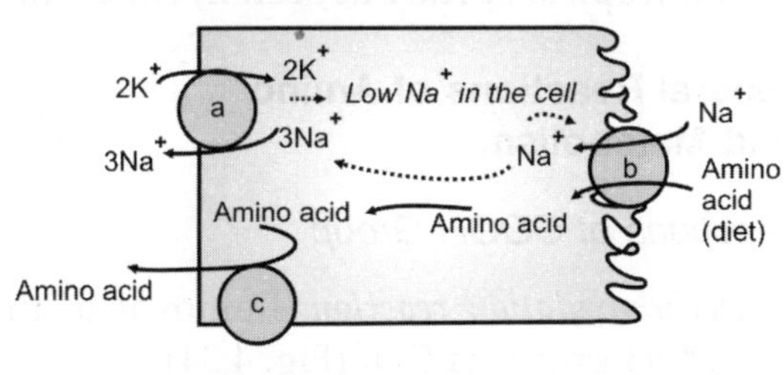


**Fig. 4.32:** Intestinal absorption of amino acids—"Secondary active transport"

a. $Na^+$ - $K^+$ ATPase—maintains low Na concentration in the cells.
b. $Na^+$ dependant amino acid co-transport—facilitated diffusion. The carrier binds to $Na^+$ from the lumen to form a binary complex which then binds to amino acids from the lumen to form a ternary complex. This ternary complex crosses the membrane and releases both the $Na^+$ and amino acid into the intestinal mucosal cell.
c. $Na^+$ independent amino acid transport-transports amino acid from cell into blood down concentration gradient.

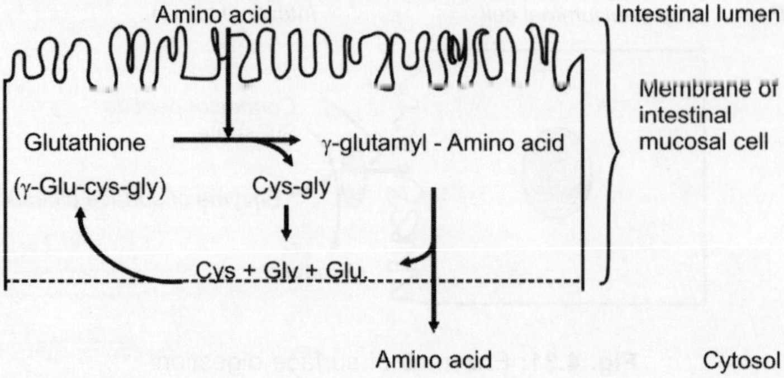


**Fig. 4.33:** γ-glutamyl cycle (Miester cycle)—absorption of neutral amino acids

dietary amino acids in the intestine and decreased reabsorption of the amino acids from the glomerular filtrate in the kidney. The amino acids are excreted in the urine —*amino aciduria*.

2. *Non-tropical sprue*: It is due to inflammation and atrophy of the intestinal mucosal cells in reaction to a toxic wheat protein, *gluten.*
3. *Celiac disease*: It is a condition similar to non-tropical sprue but seen in children.

## General Reactions of Amino Acid Metabolism

### *Reactions of COOH Group*

- *Decarboxylation reaction*: Removal of the COOH group as $CO_2$ (Fig. 4.34).

  Product formed: An amine (biogenic amines).

*General reaction:*

*Examples:*

Histidine →Histamine (mediator of allergic reactions)

5-Hydroxy tryptophan →5-hydroxy tryptamine (serotonin) – neurotransmitter

Glutamic acid →GABA (neurotransmitter)

Dihydroxyphenylalanine (DOPA) → Dopamine (neurotransmitter)

*Biological importance of biogenic amines*: Biogenic amines have a wide range of functions. As shown above GABA, dopamine and serotonin are neurotransmitters. Putresceine (from ornithine) is a growth factor. Taurine (from Cys) is required for formation of conjugated bile acid. Epinephrine and nor epinephrine are adrenal medullary hormones.

### *Reactions of $NH_2$ Group*

i. *Transamination*

- *Transamination*: Exchange of amino and keto groups at the α positions of an α-amino acid and an α-keto acid (Fig. 4.35). *Product formed*: A new amino acid corresponding to the original keto acid and a

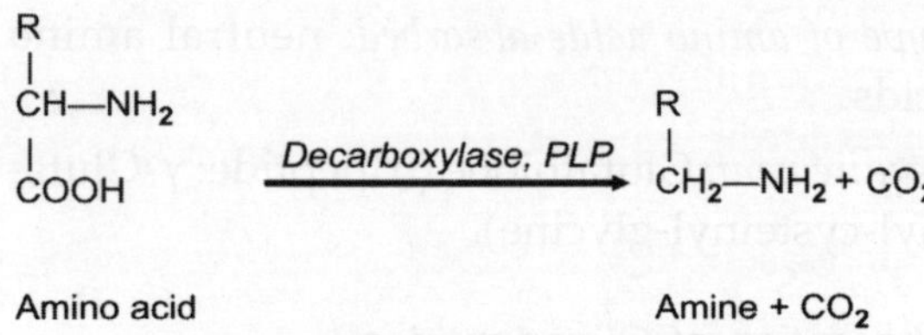


**Fig. 4.34:** Formation of an amine by decarboxylation reaction

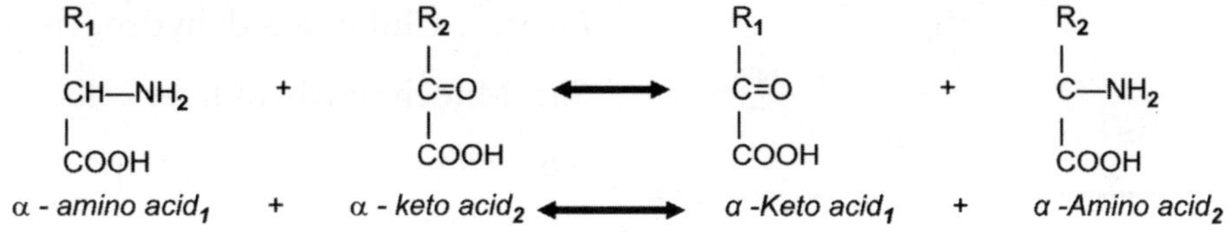


**Fig. 4.35:** General reaction of transamination

new keto acid corresponding to the original amino acid.

The enzymes that catalyze transamination reaction are *transaminases* or *amino transferases,* which belong to the transferase class of enzymes. These enzymes are present both in the mitochondria as well as the cytoplasm fractions of the cell. They are pyridoxal phosphate (PLP) dependant enzymes.

### *Metabolic Significance of Transamination*

- Synthesis of non-essential amino acids from their α-keto acids. For example, alanine from pyruvate, aspartic acid from oxaloacetate and glutamic acid from α-ketoglutarate.
- After transfer of its amino group, the carbon skeleton of the amino acid enters other metabolic pathways. For example, oxaloacetate and α-ketoglutarate enter TCA cycle for generation of energy.
- Amino acids can be interconverted by transamination when their concentrations are disproportionate.
- In all tissues, amine groups from other amino acids can be concentrated in glutamic acid (which is the transport form of ammonia to the liver). In the liver, it undergoes oxidative deamination to release it as ammonia for urea formation.

*Clinical significance:* Two transaminases of clinical significance are –

*Serum glutamate pyruvate transaminase (SGPT):*

*Synonyms:* Alanine aminotransferase (ALT)

*Reaction catalyzed:*

Alanine + α-ketoglutarate ⟷ (*ALT, PLP*) Pyruvate + Glutamic acid

The normal level of this enzyme in serum is 4-20 U/ L. The serum concentration of this enzyme is found to increase in liver diseases like hepatitis.

*Serum glutamate oxaloacetate transaminase (SGOT)*

*Synonym*: Aspartate aminotransferase (AST)

*Reaction catalyzed*:

Aspartate + α-ketoglutarate ⟷ (*AST, PLP*) Oxaloacetate + Glutamic acid

The concentration of this enzyme is very high in the cardiac muscle. The normal level of this enzyme in serum is 3-20 U/L. The serum concentration of this enzyme is found to elevate abnormally in cardiac diseases like myocardial infarction and also in hepatic disorders.

The α-keto acid involved in the transamination reaction is usually α-ketoglutarate (an intermediate of TCA cycle). The L-glutamic acid formed can subsequently undergo deamination. Hence the two reactions are coupled and referred to as *transdeamination.*

### *Deamination*

Deamination is removal of the amine group as ammonia (Fig. 4.36).

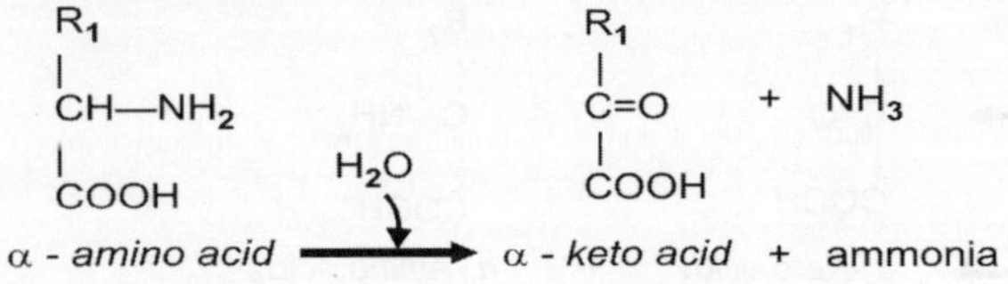


**Fig. 4.36:** General reaction of deamination.

*Product formed:* α-keto acid and ammonia
*Enzyme/Coenzyme:* Deaminase, PLP

*Importance*: Deamination reaction produces ammonia in the liver for urea cycle.

Deamination reactions maybe of two types:

i. Oxidative deamination.
ii. Non-oxidative deamination.

### *Oxidative Deamination*

*Mechanism*: Oxidative deamination involves two steps (Fig. 4.37A)

- In the first step the amino acid undergoes dehydrogenation. Two atoms of hydrogen are removed to form *α-imino acid*. These hydrogens are accepted by a hydrogen acceptor ($NAD^+$/$NADP^+$/FMN/FAD) to form reduced acceptor.
- In the second step, the imino group reacts with water to form ammonia and *α-keto acid*.

  For example, enzymes—glutamate dehydrogenase and amino acid oxidases.

*Enzyme:* Glutamate dehydrogenase.

*Site:* Mitochondria of liver cells.

*Importance:* The amine groups of all the other amino acids is funnelled into glutamate which can undergo oxidative deamination by glutamate dehydrogenase to release ammonia in the liver which is the site of urea synthesis.

*Transdeamination* is a coupled reaction of transamination followed by deamination. For example, L-glutamate dehydrogenase (Fig. 4.37B).

a. Amino acid oxidases: There are two amino acid oxidases which are flavoproteins.
   - *L-amino acid oxidase* which is FMN dependant, oxidizes L isomer of amino acids. Hydrogen peroxide formed is destroyed by the catalase enzyme in the peroxisomes.
   - *D-amino acid oxidase* which is FAD dependant oxidizes D isomer of amino acid.

*Mechanism*: It is same as that of the general reaction of oxidative deamination.

### *Non-oxidative Deamination (Fig. 4.38)*

The amino acids which undergo non-oxidative deamination are hydroxy amino acids (ser), sulphur containing amino acids (cysteine) and histidine. These amino acids undergo dehydration/desulfuration followed

**General reaction:**

**Fig. 4.37A:** General reaction of oxidative deamination

***Reaction:***

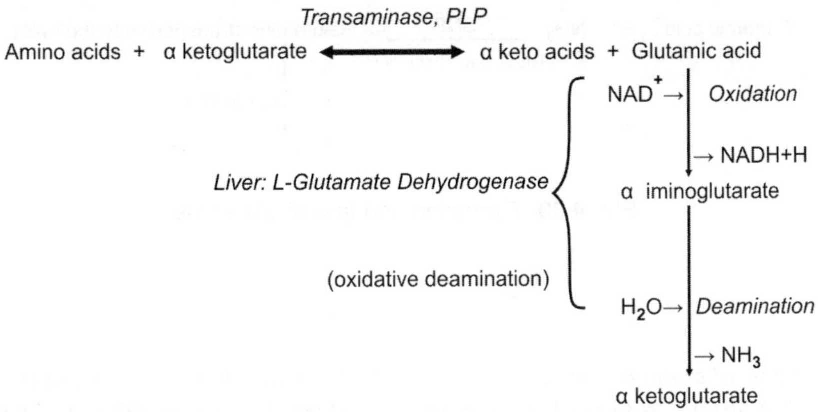


**Fig. 4.37B:** Oxidative deamination by glutamate dehydrogenase

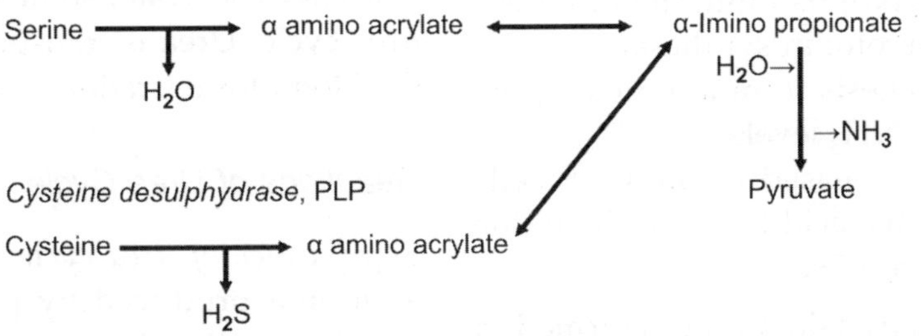


**Fig. 4.38:** Non-oxidative deamination by serine dehydratase and cysteine desulphydrase

by deamination. The enzymes are called as *dehydratases* (serine dehydratase and threonine dehydratase) and *desulphydrases* (cysteine desulphydrase).

PLP is the prosthetic group of these enzymes.

Histidine undergoes deamination by Histidase enzyme to give urocanic acid.

$$\text{Histidine} \xrightarrow{\textit{Histidase}} \text{Urocanic acid} + NH_3$$

### *Metabolism of Ammonia*

Ammonia is the product of deamination reactions and bacterial putrefaction of the unabsorbed amino acids in the intestine. It is produced by most of the cells in the body. Ammonia is highly toxic especially to the CNS. The symptoms of ammonia toxicity are:

- Flapping tremors
- Blurring of vision
- Slurring of speech
- Ultimately coma and death.

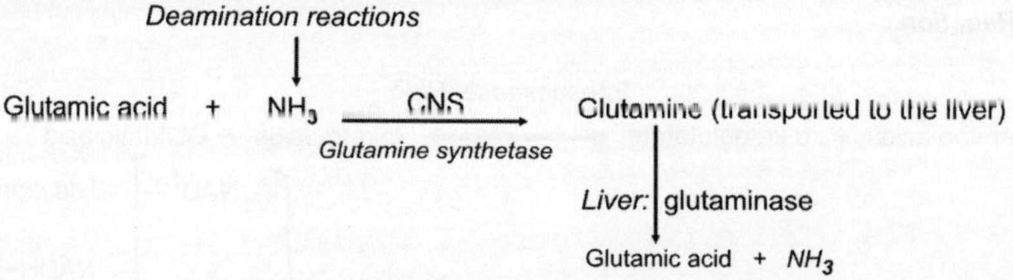


**Fig. 4.39:** Formation and fate of glutamine

*Causes for symptoms of ammonia toxicity:*

- ↓ Rate of TCA cycle: α-ketoglutarate is diverted glutamic acid synthesis.
- ↑ Synthesis of glutamine from glutamic acid in the CNS, so ↓ Glu for GABA synthesis.
- ↑ Transport of glutamine from the brain, so entry of tryptophan into the brain is ↓ and hence ↓ serotonin synthesis.
- ↑ Gluconeogenesis from amino acids so there is rise in $NH_3$ levels.

The ammonia released in the CNS cells reacts with glutamic acid to form glutamine as shown in Figure 4.39.

*Transport of ammonia*: Glutamine is formed in the CNS (Fig. 4.39). In other tissues, ammonia from other amino acids is concentrated in glutamic acid (from α-ketoglutarate) and alanine (from pyruvate in muscle) by transamination. Hence glutamine and alanine are the major transport forms of ammonia in blood.

### Regeneration of Ammonia

In the liver, ammonia is regenerated by the following reactions:

- Action of glutaminase enzyme on glutamine (Fig. 4.39).
- Deamination of glutamate by glutamate dehydrogenase (Fig. 4.37B).

### Fate of Ammonia

The toxic ammonia is brought to the liver from all the tissues for detoxification. It is converted to a less toxic, water soluble form i.e., urea by urea cycle. Urea is transported to kidney by the blood for excretion in the urine.

### Reactions of Urea Cycle

*Significance of urea cycle:* Detoxification of ammonia produced by protein and amino acid metabolism is the major function of this cycle.

*Site*: Liver.

*Subcellular site*: Mitochondrial matrix and cytosol.

*Starting materials*: $CO_2$ and $NH_3$.

*Sources of carbon dioxide*: TCA cycle.

*Sources of ammonia*: Deamination, L-glutamate dehydrogenase.

### *Reactions of Urea Synthesis*

There are *five reactions* as shown below:

| | |
|---|---|
| I. Synthesis of carbamoyl phosphate | Mitochondrial matrix |
| II. Synthesis of citrulline | |
| III. Synthesis of argininosuccinate | |
| IV. Synthesis of arginine | Cytosol |
| V. Synthesis of urea and regeneration of ornithine. | |

I. Synthesis of carbamoyl phosphate:

*Site: mitochondrial matrix*

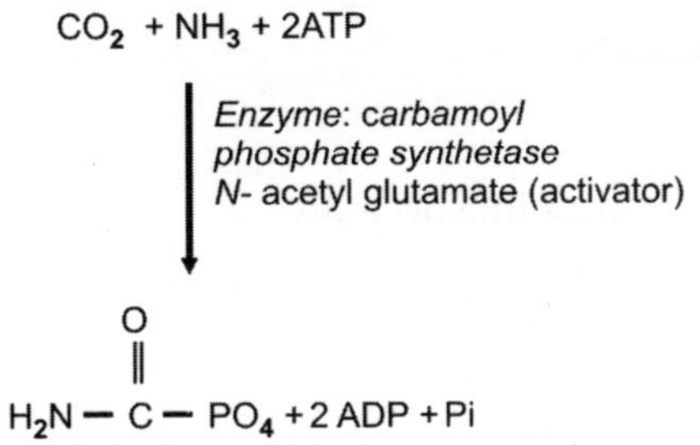


*carbamoyl phosphate*

*Energy:* two high energy bonds are required for the synthesis of carbamoyl phosphate

II. Synthesis of Citrulline:

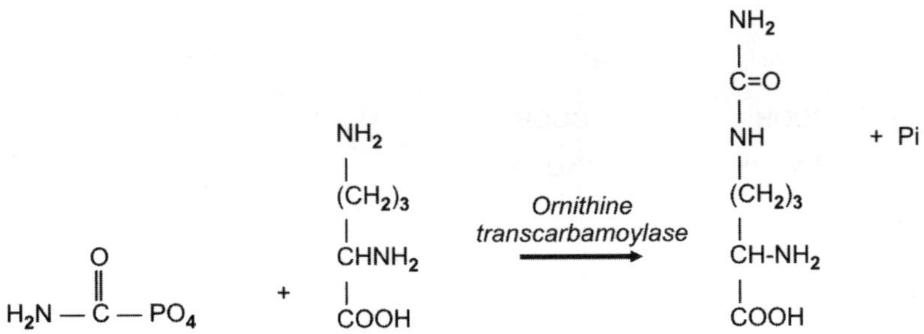


*Carbamoyl Phosphate* *Ornithine* *Citrulline*

Citrulline is transported from the mitochondrial matrix into the cytosol

III. Synthesis of argininosuccinate.

*Site: cytosol.*

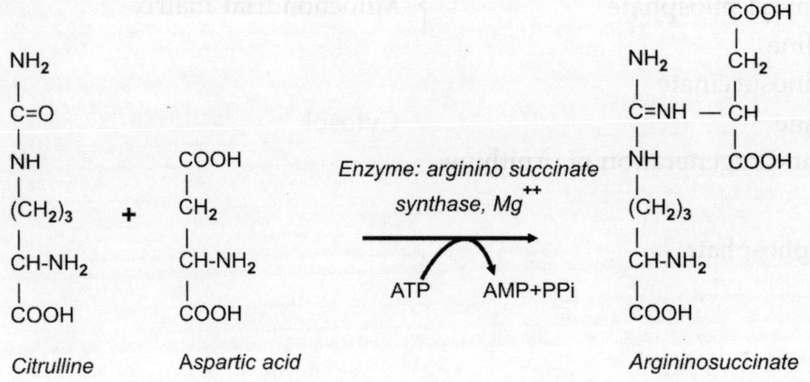

*Energy:* Two high energy bonds are required for the synthesis of argininosuccinate

IV. Synthesis of arginine:

The carbon skeleton of aspartic acid is released as fumarate.

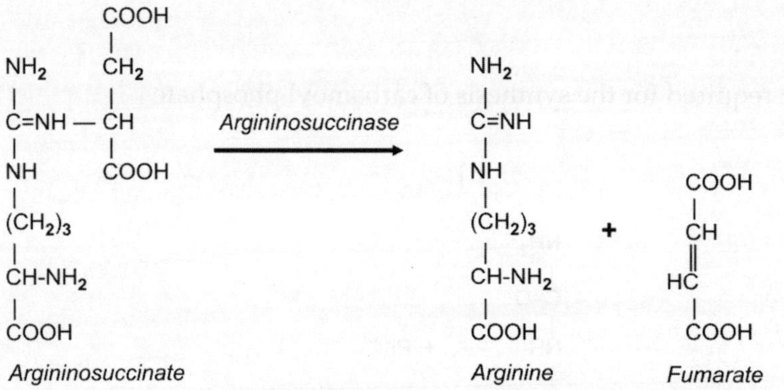

V. Formation of urea from arginine:

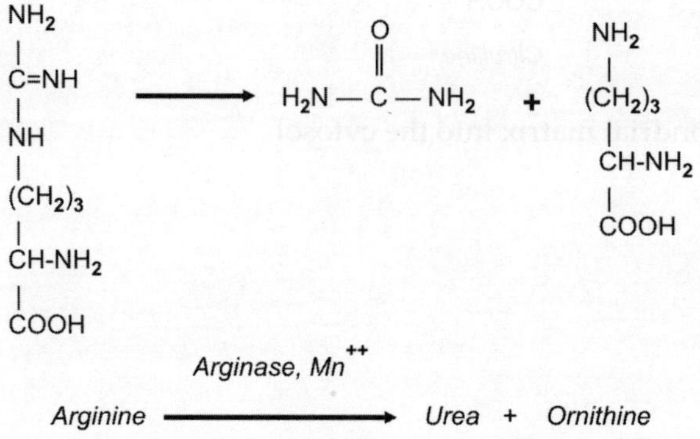

$$\text{Arginine} \xrightarrow{\text{Arginase, } Mn^{++}} \text{Urea} + \text{Ornithine}$$

### Fate of Fumarate:

Aspartic acid is regenerated from fumarate by the following reactions:

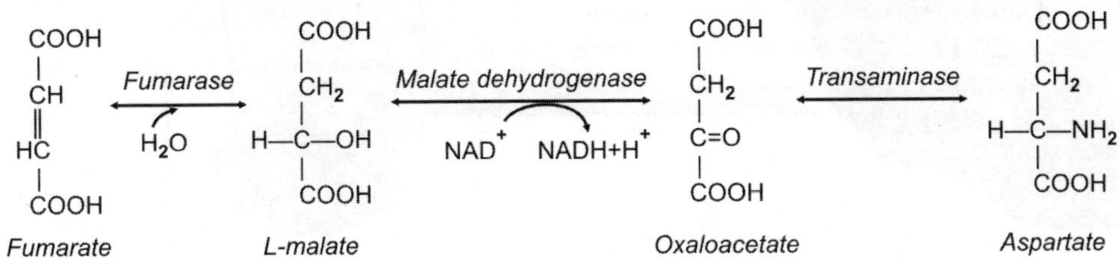


Malate and oxaloacetate are the intermediates. Aspartic acid which is used for urea cycle is regenerated by these reactions.

### Energy Expenditure

Four high energy bonds are needed for the synthesis of one molecule of urea—two for synthesis of carbamoyl phosphate and two for synthesis of argininosuccinate.

### Energy Gain

*Three molecules of ATP are synthesized* from the NADH+$H^+$ that is generated during regeneration of aspartic acid from fumarate.

Reactions of urea cycle in the hepatic cell are shown in Figure 4.40.

### Metabolism of Urea (Fig. 4.41)

The urea synthesised in the liver is transported to the kidney by blood. The normal blood urea level is 12-36 mg/100 ml. Urea gets filtered in the kidney and excreted in urine. About 20-30 g of urea is excreted in urine everyday. Small amounts of urea are excreted in sweat.

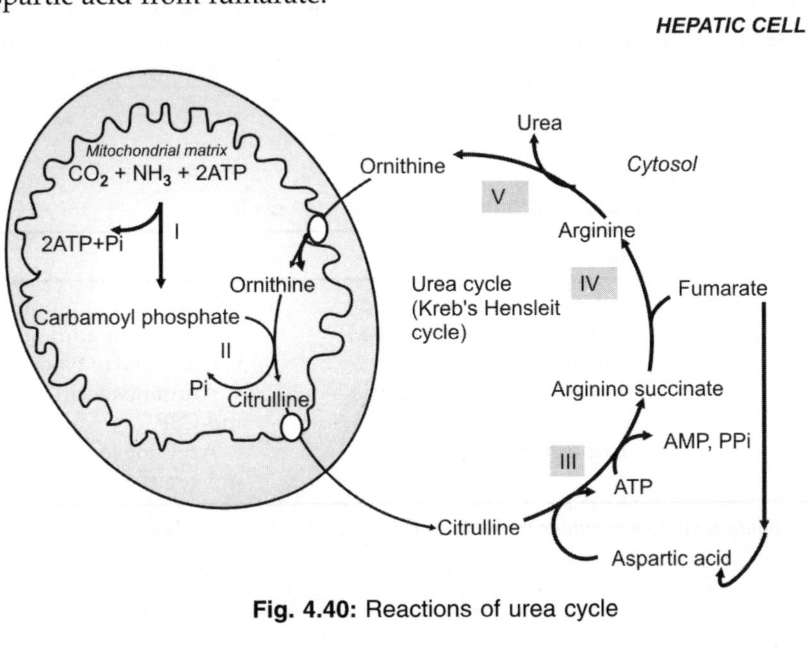


**Fig. 4.40:** Reactions of urea cycle

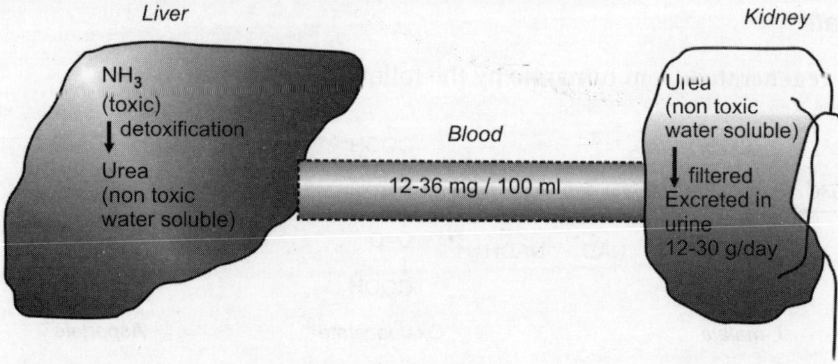


**Fig. 4.41:** Metabolism of urea

### *Clinical Significance of Blood Urea Levels*

Increased blood urea levels is called as *uraemia*, seen in kidney diseases.

### *Causes*

*Pre-renal* – haemoconcentration as in vomiting, diarrhoea and diabetes mellitus.

*Renal* – kidney diseases-acute glomerulonephritis, hydronephrosis.

*Post renal* – obstruction to the flow of urine.

Decreased level of urea in blood is called as *hypouremia*, seen in advanced liver disorders.

### *Regulation of Urea Cycle*

*Rate limiting enzyme*: Carbamoyl phosphate synthase I.

*Allosteric activator:* N-acetyl glutamate.

### *Inborn Errors of Urea Cycle*

Inborn errors of all the enzymes of urea cycle are known but are rare. The substrate of the defective enzyme will increase both in blood and urine and bring about feedback inhibition of the previous enzyme of the pathway (Table 4.12).

**Table 4.12:** Inborn errors of urea cycle

| *Defective enzyme* | *Inborn error* | *Features* |
|---|---|---|
| Carbamoyl phosphate synthase I | Hyperammonaemia type I | Symptoms of ammonia toxicity |
| Ornithine transcarbamoylase | Hyperammonaemia type II | Symptoms of ammonia toxicity |
| Argininosuccinate synthase | Citrullinaemia | ↑ Citrulline in blood and urine |
| Argininosuccinase | Argininosuccinic aciduria | ↑ Argininosuccinate in blood, urine and CSF |
| Arginase | Argininaemia | ↑ Arginine levels in blood, CSF and brain |

*Note: Symptoms of ammonia toxicity are milder in the later three than the first two disorders.*

## Specialized compounds that are generated from amino acids:

Amino acids themselves or part of their structures are converted to compounds of biochemical interest. The functions of these amino acid derivatives are discussed under functions of amino acids. The derivatives of some of the amino acids from each group are listed below.

- Neurotransmitters – GABA, serotonin, DOPA, DOPamine
- Skin pigment melanin – from Tyr
- Vitamin niacin – from Trp
- Creatine in muscle – from Arg, Gly, Met
- Bile salts for fat digestion and absorption – from Gly, taurine
- Heme of haemoglobin and hemoproteins – from Gly

***Aliphatic amino acids:***

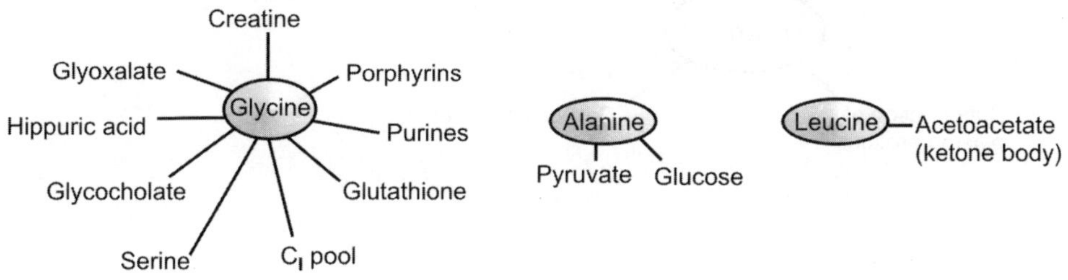


***Acidic amino acids:***

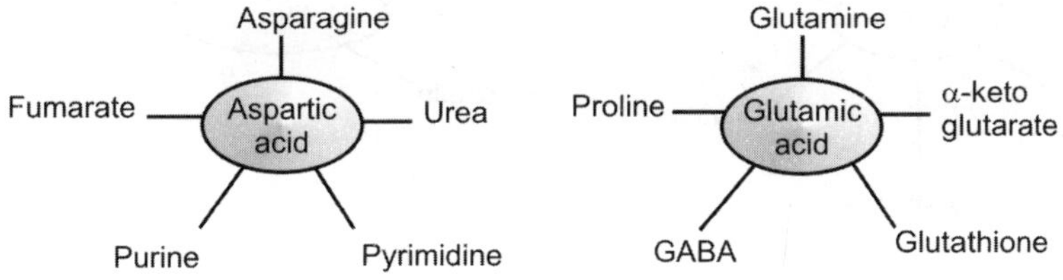


### *Functions of Amino Acids*

- Constituent of proteins
- One carbon pool – Gly, Ser, Met, His
- Hormones – catecholamines, thyroid hormones
- Biogenic amines – histamine, GABA, tryptamine, tyramine, DOPA, DOPamine
- Purines and pyrimidines of nucleic acids – from Gly, Asp, Gln
- Glutathione – from Glu, Cys, Gly
- Detoxification of toxic compounds – with Gly, Cys, Gln
- Sphingosine of sphingophospholipids (sphingomyelin) – from Ser

***Basic amino acids:***

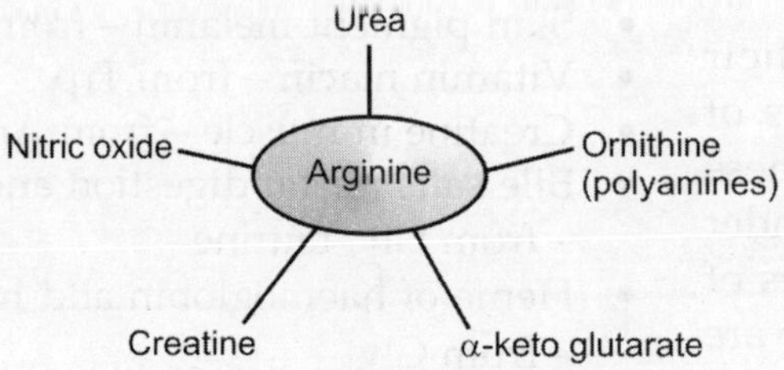


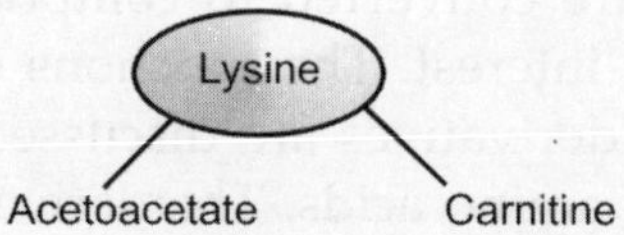


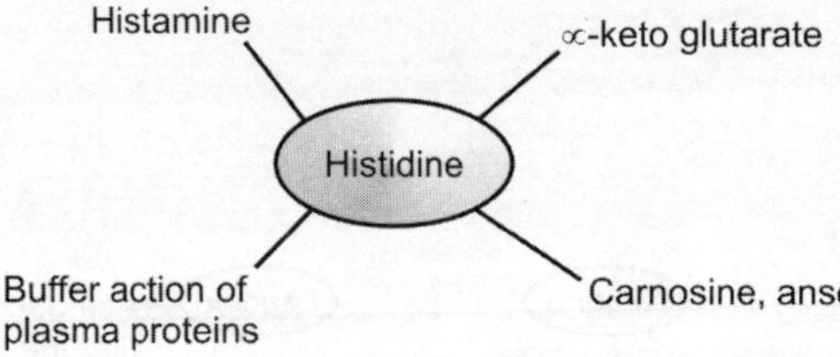


***Sulphur containing amino acids:***

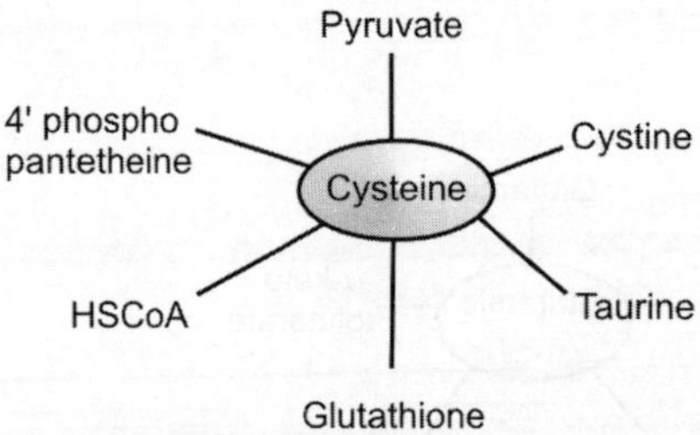


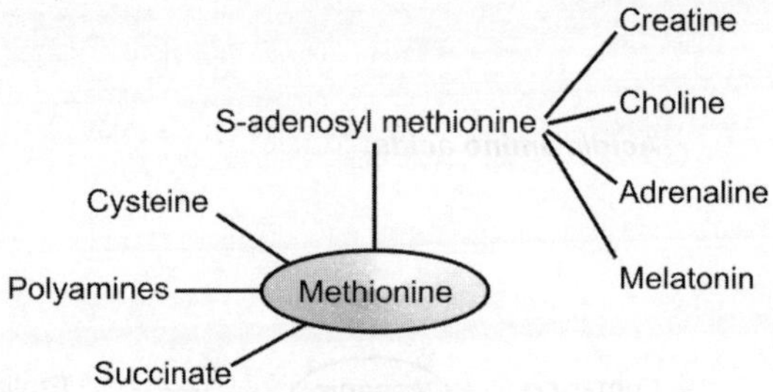


***Hydroxy amino acids:***

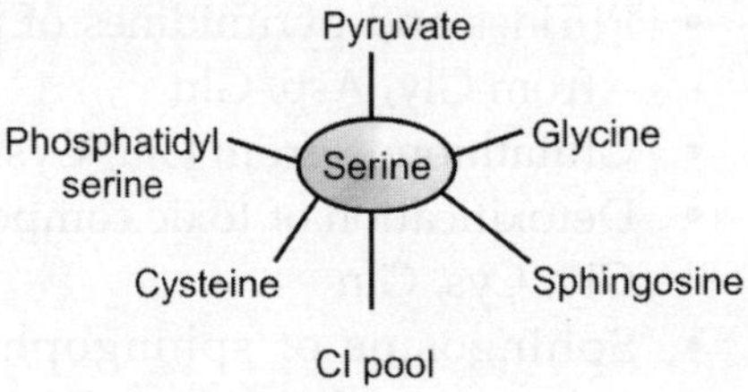

***Aromatic amino acids:***

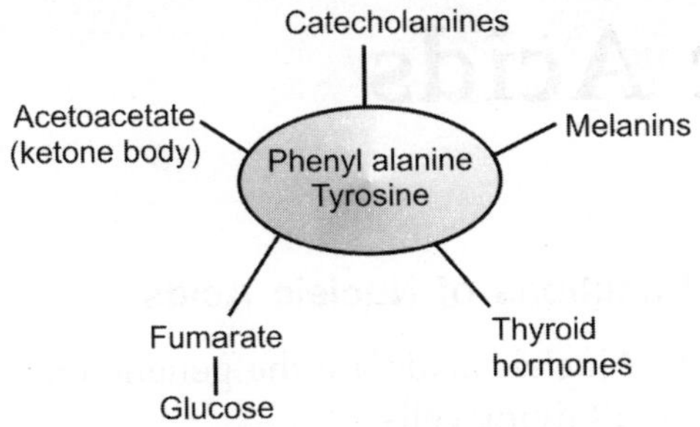


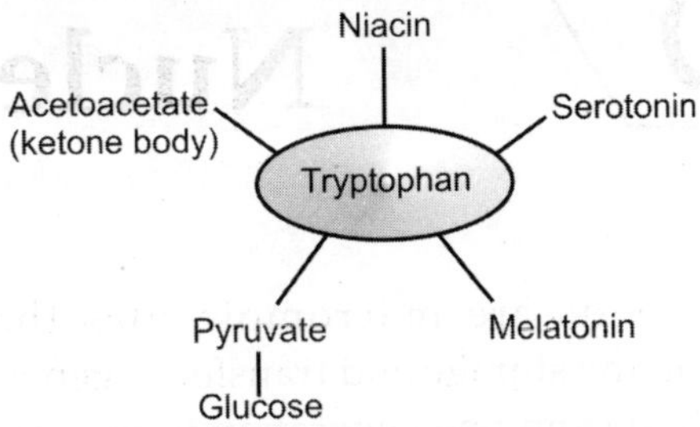


- Carnitine for transport of fatty acids into mitochondria – from Lys, Met
- Polyamines – spermine and spermidine – from Arg, Met, Ornithine.

# 5 Nucleic Acids

Nucleic acids are macromolecules that function in the storage and transfer of genetic information from one generation to the next. Nucleic acids were so named because DNA was first isolated from the nuclei even though they occur in other parts of the cell. Each type of nucleic acid is distinguished by the sequence of the characteristic nitrogenous bases in their nucleotide monomers. In this chapter, we discuss the structure of the bases, nucleosides and nucleotides which serve not only as building blocks of nucleic acids but have other important functions in the cell as well. The structure of DNA and different types of RNA for a better understanding of their functions is discussed next. A brief insight into the central dogma of molecular biology is also discussed to understand the processes of replication, transcription and translation.

## BIOMEDICAL IMPORTANCE

- DNA the *storage of genetic information* is responsible for life and heredity.
- The role of nucleotides in synthesis of DNA is made use of by using structural analogs of the nucleotides in *therapy of diseases such as cancer and AIDS*.
- The structure of DNA, the central dogma of molecular genetics and techniques like PCR is the basis of *recombinant technology/genetic engineering*.
- *Gout* is a disease associated with increased catabolism of purine bases.

## Functions of Nucleic Acids

- Nucleic acids are the genetic material for all living cells.
- It is involved in the storage and transfer of genetic material from one generation to the next.
- The genetic material of all the cells in a living organism is the same and has all the information required for making an identical organism.
- Nucleic acids determine the phenotype of an organism.
- Other functions of nucleic acids include catalysis (enzyme like action, e.g., ribozyme) and coenzyme action (RNA acts as coenzyme for the enzyme telomerase).

## Classification of Nucleic Acids

As the name indicates nucleic acids are acidic in nature. Nucleic acids can be classified into two classes as shown in Figure 5.1.

## Chemistry of Nucleotides

Nucleotides are the building blocks of nucleic acids. Each nucleotide is made of three components. They are:

- Nitrogenous base
- Pentose sugar
- Phosphate.

### *Nitrogenous Bases*

Nitrogenous bases are nitrogen containing, heterocyclic compounds. There are two types of nitrogenous bases. They are:

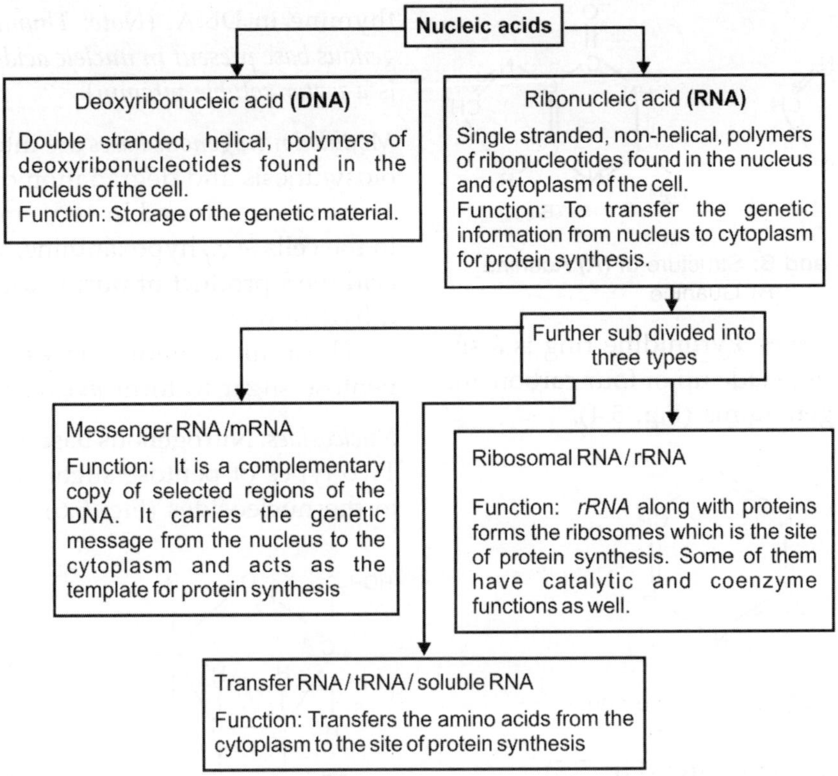


**Fig. 5.1:** Classification and functions of nucleic acids

- Purines
- Pyrimidines
- *Purine bases*: Purine ring is a nine member ring made up of five carbon and four nitrogen atoms. It has a pyrimidine ring fused with an imidazole ring. (Fig. 5.2).

There are two types of purine bases (Fig. 5.3):

- Adenine: 6-amino purine.
- Guanine: 2-amino, 6-oxo purine.

*Importance:* Adenine and guanine are present in both the nucleic acids—DNA and RNA.

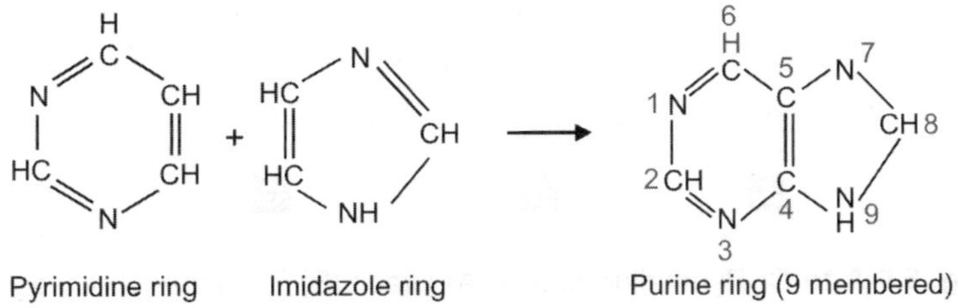


**Fig. 5.2:** Purine ring

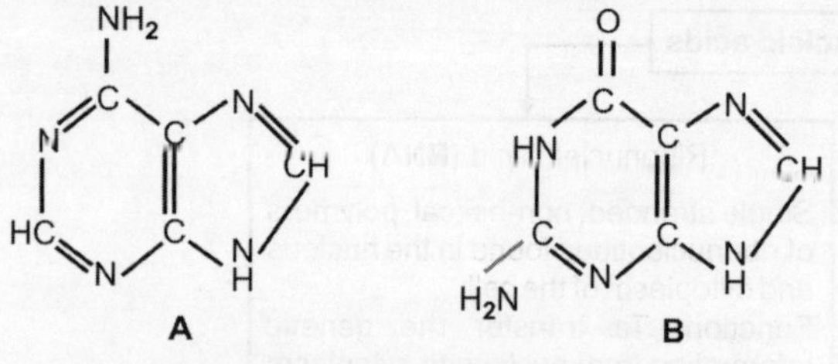

**Figs. 5.3A and B:** Structure of (A) Adenine (B) Guanine

- *Pyrimidine bases*: Pyrimidine ring is a six-member ring made up of four carbon and two nitrogen atoms (Fig. 5.4).

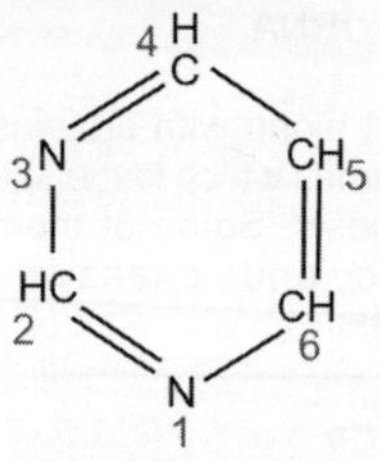

**Fig. 5.4:** Pyrimidine ring

The pyrimidine bases are: (Fig. 5.5)
- Cytosine: 2-oxo, 4-amino pyrimidine
- Thymine: 2, 4-dioxo 5-methyl pyrimidine
- Uracil: 2, 4-dioxo pyrimidine.

*Importance:* Cytosine is present in both DNA and RNA. Uracil is present in RNA and thymine in DNA. *(Note: Thymine is a nitrogenous base present in nucleic acids and thiamine is a water-soluble vitamin).*

*Modified nitrogenous bases* are intermediates in biosynthesis and degradation of these bases. These are also found in very small quantities in the cells, *e.g.*, hypoxanthine, xanthine, uric acid (end product of purine catabolism) and dihydrouracil.

These nitrogenous bases are linked to pentose sugar to form *nucleosides.*

*Nucleosides:* Nitrogenous base + Pentose sugar

Two types of pentose sugars can be present in the nucleosides (Fig. 5.6).

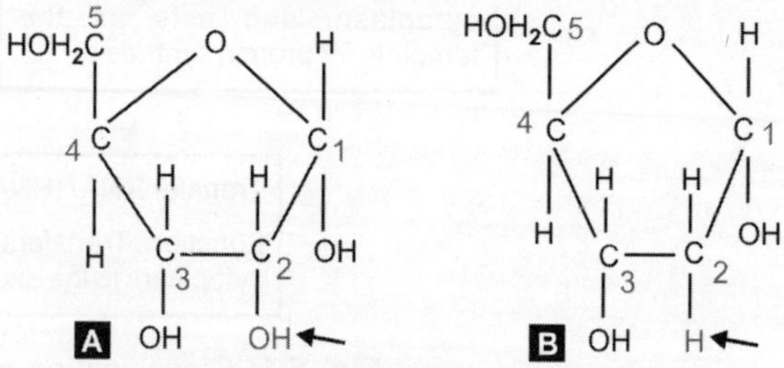

**Fig. 5.6:** Structure of (A) D-ribose (B) 2-D-deoxyribose

Based on the type of sugar that is present in the nucleoside, they are divided into two classes (Fig. 5.7).

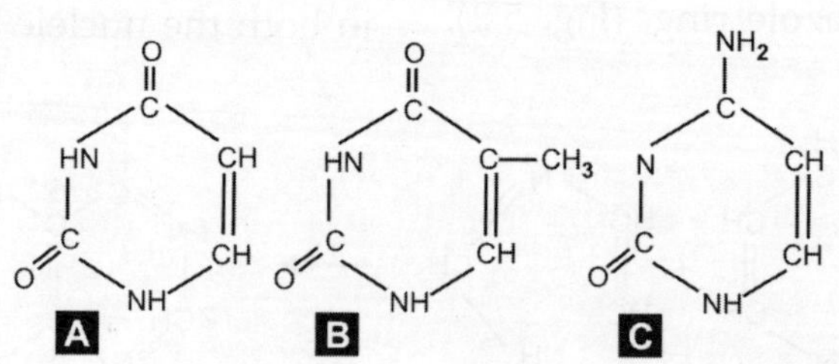

**Fig. 5.5 A to C:** Pyrimidine bases (A) Uracil (B) Thymine (C) Cytosine

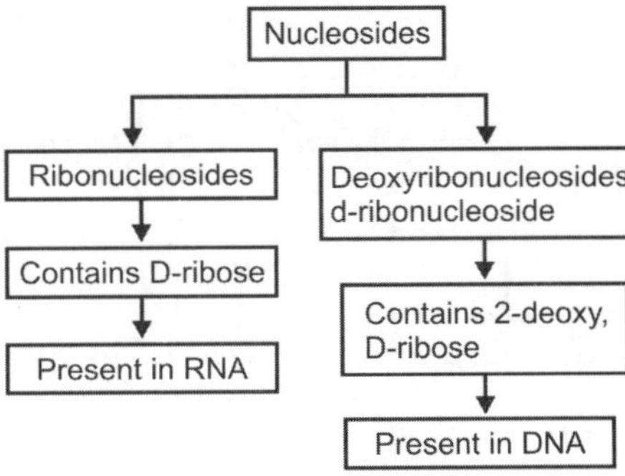


**Fig. 5.7:** Types of nucleosides

The pentose sugar is linked to the nitrogenous base by a *β-N-glycosidic linkage* as shown in Figure 5.8.

The N-glycosidic linkage in the nucleoside is formed by dehydration between the OH-group either on the $C_1$ carbon of the pentose sugar and hydrogen either on the $N_9$ of the purine base (Fig. 5.8) or on $N_1$ of pyrimidine base (Fig. 5.9).

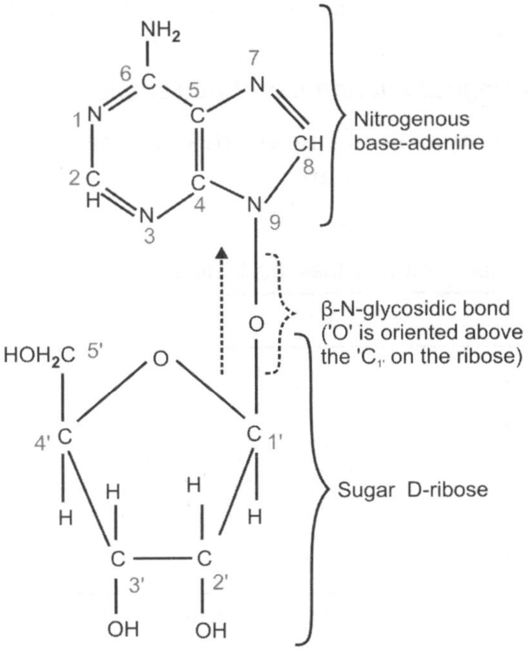


**Fig. 5.8:** *β*-N-glycosidic linkage in a nucleoside—adenosine *(Note the β orientation of the N-glycosidic bond)*

### *Naming of the Ribonucleoside*

Purine nucleosides have a suffix of *"sine"* and pyrimidine nucleosides have a suffix *"dine"*. The deoxyribonucleosides are denoted by adding a prefix of *"d"* to its name as shown in Table 5.1.

**Table 5.1:** Nitrogenous bases and their ribonucleoside and deoxyribonucleosides

| *Nitrogenous base* | *Ribonucleoside* | *Deoxy ribonucleoside* |
|---|---|---|
| Adenine | Adenosine | d-Adenosine |
| Guanine | Guanosine | d-Guanosine |
| Cytosine | Cytidine | d-Cytidine |
| Uracil | Uridine | d-Uridine |
| Thymine | Thymidine | d-Thymidine |

The nucleoside has two ring systems in it. Hence the carbon atoms of the pentose sugar are indicated with a prime *(′ )*. 2′ adenosine indicates 2nd carbon on the pentose sugar of adenosine (Fig. 5.8).

Phosphorylated nucleosides are called as *nucleotides.*

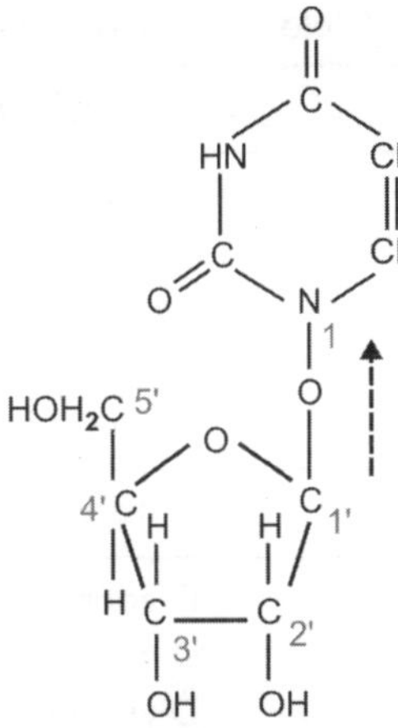

**Fig. 5.9:** β-N-glycosidic linkage in pyrimidine ribonucleoside-uridine

*Nucleotides:* Nucleosides + Phosphate

The hydroxyl group on the 5′ carbon of the sugar is phosphorylated by ester bond formation (Fig. 5.10).

The nucleotides are also called as *nucleoside monophosphates*. The nucleoside monophosphates of the ribonucleosides and deoxyribonucleosides are shown in Table 5.2.

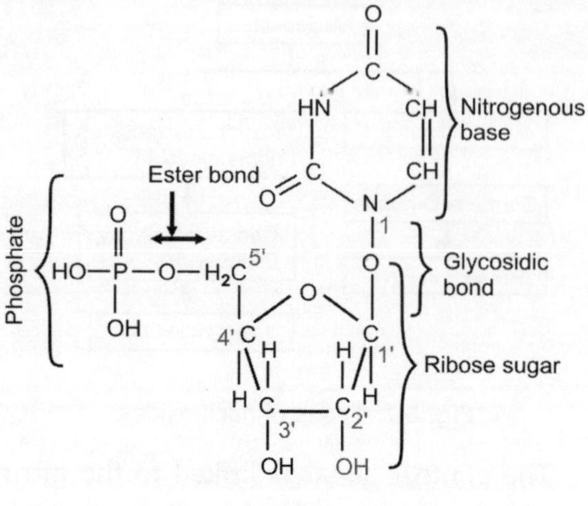


**Fig. 5.10:** Structure of a nucleotide—uridine monophosphate—UMP

### Higher Nucleotides

Addition of further phosphate groups to the existing phosphate of the nucleotide, forms nucleoside diphosphates (NDP) and nucleoside triphosphates (NTP) which are called as *higher nucleotides*. The phosphate groups are linked to one another by anhydride bond (Fig. 5.11).

*Biological importance*: The anhydride bonds found in the higher nucleotides are *high-energy bonds represented as "~"* (squiggle bond). On hydrolysis these bonds yield more than 7 kilocalories of energy. There is one high-energy bond in nucleoside diphosphates (N—P~P) and two high-energy bonds in nucleoside triphosphate (N—P~P~P). The energy released by the hydrolysis of these high-energy bonds is used for other metabolic reactions and functions of the living cells. Hence higher nucleotides, ATP/GTP, are called as the *universal energy currency*.

### Biological importance of nucleotides

- Polymers of nucleotides are the nucleic acids—DNA and RNA

**Table 5.2:** Ribonucleosides, deoxyribonucleosides and their nucleotides

| *Ribonucleoside* | *Ribonucleotide* | *-NMP* |
|---|---|---|
| Adenosine | Adenosine monophosphate | – AMP, Adenylic acid |
| Guanosine | Guanosine monophosphate | – GMP, Guanylic acid |
| Cytidine | Cytidine monophosphate | – CMP, Cytidylic acid |
| Uridine | Uridine monophosphate | – UMP, Uridylic acid |
| Thymidine | Thymidine monophosphate | – TMP, Thymidylic acid |
| *Deoxyribonucleoside* | *Deoxyribonucleotide* | *– d-NMP* |
| D-adenosine | D-adenosine monophosphate | – DAMP, D-adenylic acid |
| D-guanosine | D-guanosine monophosphate | – DGMP, D-guanylic acid |
| D-cytidine | D-cytidine monophosphate | – DCMP, D-cytidylic acid |
| D-uridine | D-uridine monophosphate | – DUMP, D-uridylic acid |
| D-thymidine | D-thymidine monophosphate | – DAMP, D-thymidylic acid |

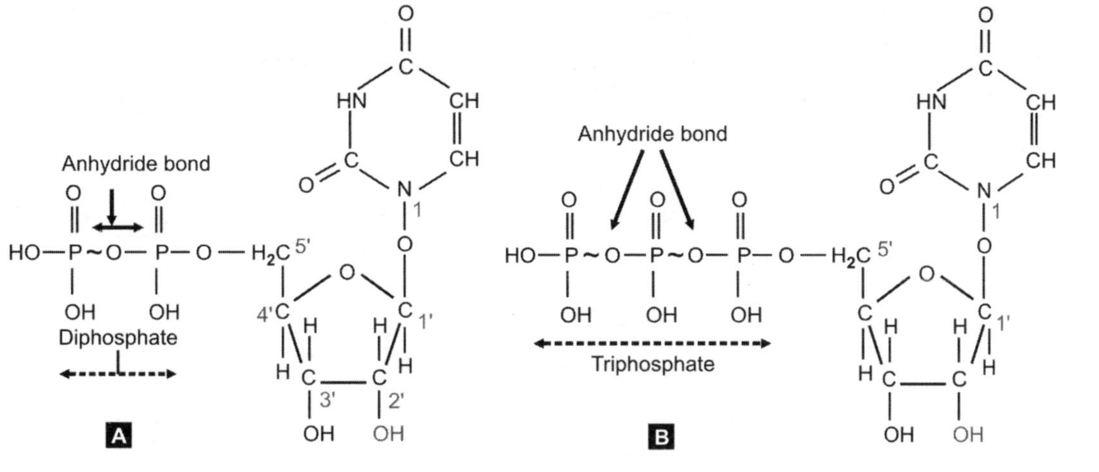


**Fig. 5.11:** Structure of (a) UDP (b) UTP

- Higher nucleotides like ATP/GTP are high-energy compounds.
- Nucleoside monophosphates (NMP) form part of coenzymes like FAD, $NAD^+$, $NADP^+$ and coenzyme A.
- Cyclic nucleotides act as second messengers in the signal transduction of hormones.
  *For example,* cyclic AMP (cAMP), cyclic GMP (cGMP).
- Nucleotide diphosphates act as carrier molecules.
  *For example,* UDP for mono-saccharide, CDP for nitrogenous bases.
- Nucleosides act as allosteric modulators of metabolic pathways.
  *For example,* Enzyme phosphofructokinase of glycolysis has AMP as positive modulator and ATP as negative modulator.
- Synthetic derivatives of nucleotides act as anticancer drugs. *For example,* 5-Fluorouracil.

### Formation of Phosphodiester Bond

Two nucleotides joined together by a phosphodiester bond forms a *dinucleotide.*

*Mechanism:* Dehydration between 3′-OH group of one nucleotide and the OH group of the phosphate on another nucleotide (Fig. 5.12).

When phosphodiester bonds link larger numbers of nucleotides, polymers are formed. 'n' number of nucleotides will be linked by 'n-1' number of phosphodiester bonds.

*Oligonucleotides* contain 2-10 numbers of nucleotides.

*Polynucleotides* are formed when the number of nucleotides is higher.

- RNA is a polyribonucleotide.
- DNA is a polydeoxyribonucleotide
- Two such polydeoxyribonucleotide chains form the DNA.

The *Polynucleotides* have a free 3′-OH group at one end—this is the *3′ end*. The other end has a free phosphate group at the 5′ end and is called as the *5′ end*. The sequence of bases is always written from the 5′ end to the 3′ end. This is *polarity* of the polynucleotide. The phosphodiester bond has definite direction i.e., *3′→5′ direction*. The backbone of the polynucleotide is made up of alternating phosphate and pentose sugar

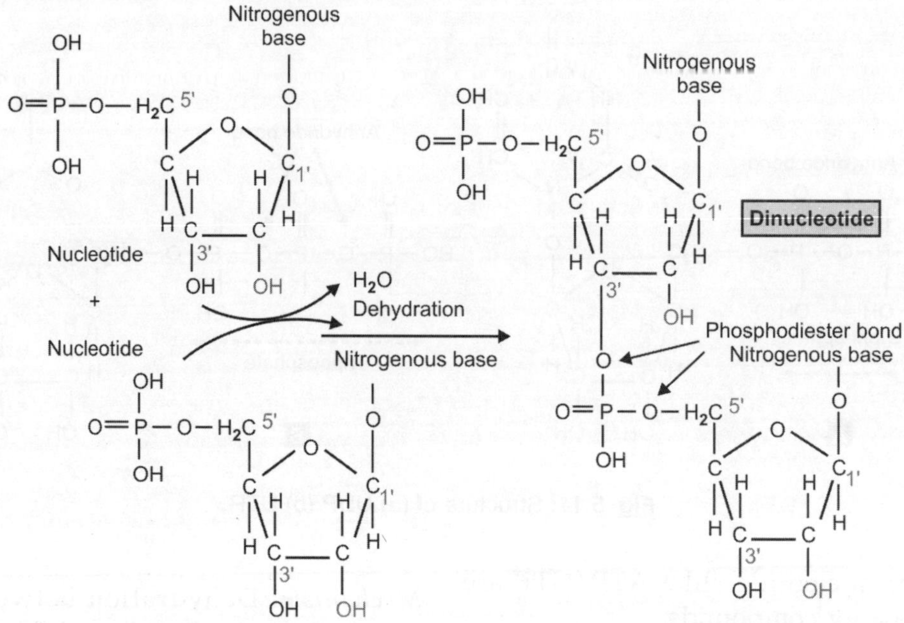


**Fig. 5.12:** Formation of a phosphodiester bond and dinucleotide

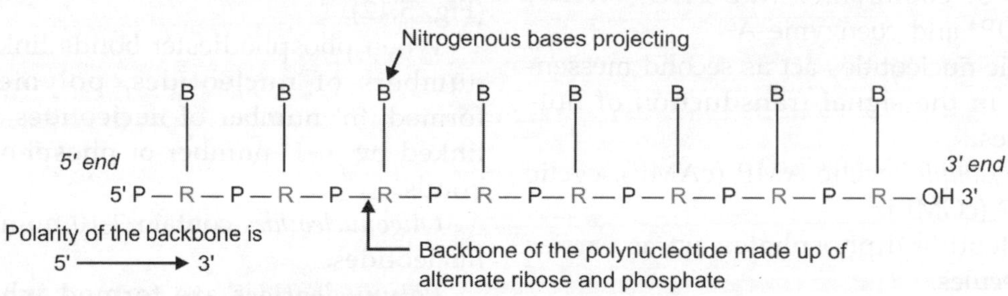


**Fig. 5.13:** Structure of an oligonucleotide

residues (P—R—P—R). The nitrogenous bases (B) project out from the backbone (Fig. 5.13).

## Nucleic Acids

This structure of DNA described below, is called as Watson and Crick model, named after the scientists who proposed it and were awarded the Nobel Prize for this work.

### *Structure of DNA*

- *Double stranded* – It has two polynucleotide chains (Fig. 5.14A).
- *Antiparallel*—One strand is in 3′ → 5′ direction and the other is in 5′ → 3′ direction.
- The two strands of the DNA are held together and stabilized by:
  - Hydrogen bonds between the nitrogenous bases of the two strands.
  - Hydrophobic interactions between the bases, which are stacked one above the other.
- Hydrogen bonds between the nitrogenous bases of the two strands give it a *ladder like appearance*. The sides are formed

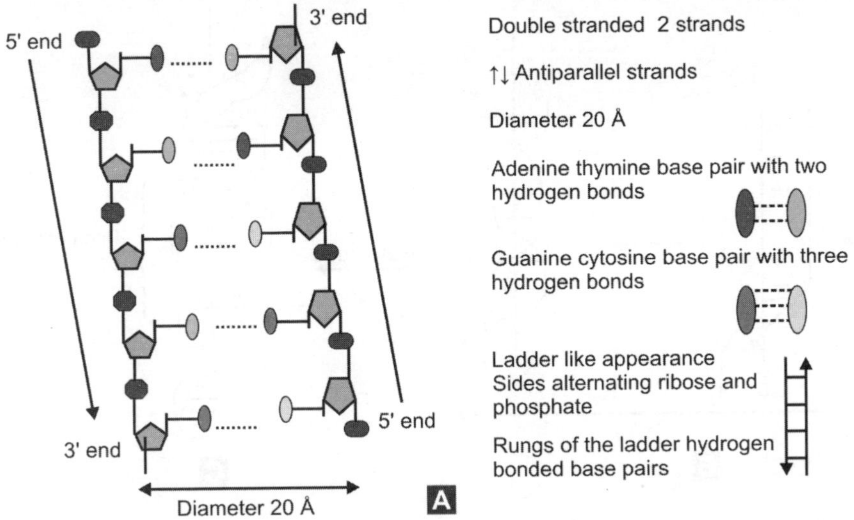


**Fig. 5.14A:** Structure of DNA *(For colour version see Plate 4)*

by alternate sugar and phosphate and the rungs are made up of the base pairs and the hydrogen bonds between them (Fig. 5.14A)

- *Chargaff's rule*: Adenine base pairs with thymine and guanine with cytosine on the two strands. This base pair rule is called as *Chargaff's rule.*
- *Base pairing*: Adenine base pairs only with thymine by two hydrogen bonds (A = T). And guanine with cytosine residue by three hydrogen bonds (G ≡ C). Hence GC pairs are stronger than AT pairs.
- *Helical*: This double stranded structure is twisted in a right-handed double helix like a spiral staircase. (Fig. 5.14B)
- *Diameter of the helix*: It is about 2.0 nanometers (20 Å). The distance between the two sides of the ladder is the diameter.
- *Pitch of the helix*: Each turn of the spiral helix is 3.4 nanometers and has 10 base pairs. This is called as the pitch of the helix (Fig. 5.14C).
- *Major groove and minor groove*: Formation of the double helix results in formation of two grooves on the surface—*major groove* (larger) and *minor groove* (smaller). *Importance of the grooves*: The DNA-protein interaction is along these grooves.
- *Each strand of the DNA serves as a template for* - (i) the synthesis of the antiparallel strand during the process of replication (duplication of DNA). (ii) specific regions of each strand is template for the process of transcription (synthesis of RNA) required for synthesis of proteins, formation of ribosomes and transport of amino acids to the site of protein synthesis.
- *Nucleosomes*: This double helical structure is coiled around proteins called histones. The proteins are organized into octomers. The histone-DNA complex is called as *nucleosome.*

## Structure of Ribonucleic Acids

### *General Feature*

- *Site of occurrence:* Cytoplasm, nucleus, mitochondria.

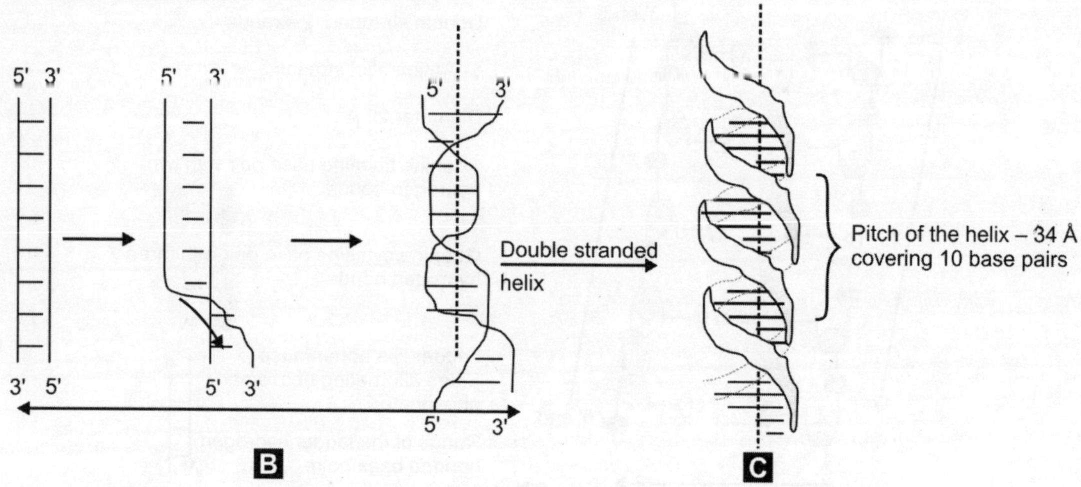


**Figs 5.14B and C:** (A) Ladder like double strand (B) Formation of double helix (C) Double helical structure

- *Half-life is short* unlike that of DNA.
- *Single strand* of ribonucleotides linked by 3′→5′ phosphodiester bonds.
- *Nitrogenous bases* present are adenine, guanine, cytosine and *uracil*.
- Does *not follow the Chargaff's rule*. However, certain regions of the RNA with base pair complementarity in the antiparallel direction do show base pairing to form a *stem*. Presence of modified bases that prevent base pairing in between this stretch results in a *loop*. (Figs 5.15A and B).
- RNA is a *product of the process of transcription* (synthesis of RNA) from selected regions of the DNA. Hence they are low molecular weight with limited number of ribonucleotides.
- There are three different types of RNA each with specific function in the process of translation (protein biosynthesis). They are:
  - *Messenger RNA/mRNA*: dictate the sequence of amino acids during the synthesis of proteins.

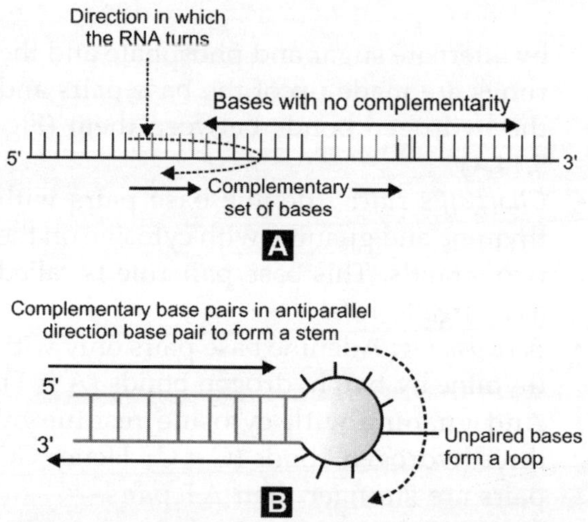


**Figs 5.15A and B:** (A) A stretch of RNA which can form an arm (B) Structure of an arm with a loop and a stem

  - *Transfer RNA / tRNA*: Transfers the activated amino acids from the cytoplasm to the ribosomes for the synthesis of proteins.
  - *Ribosomal RNA/rRNA*: rRNA along with proteins forms the organelle ribosome that is the site for protein biosynthesis.

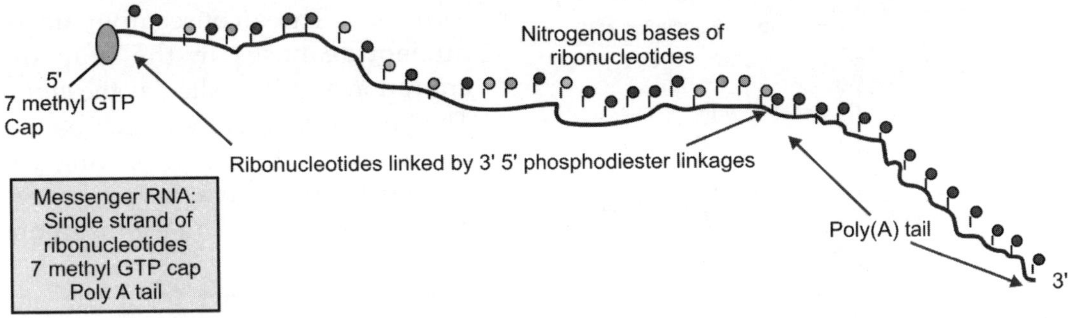


**Fig. 5.16:** Structure of messenger RNA *(For colour version see Plate 5)*

- Two other types of RNA that are found in the cells are—*heterogeneous nuclear RNA/hnRNA* and *small nuclear RNA/snRNA*.

### *Messenger RNA /mRNA (Fig. 5.16)*

- *Function:* Conveys the message from the DNA to the ribosome for the sequence of amino acids to be incorporated during protein biosynthesis.
- *Structure*: Single stranded, large molecules of varying sizes (according to the size of the gene which it codes for), having about 200 to $10^4$ nucleotides.
- *Primary structure:*
  - ➢ Single strand of ribonucleotides linked by phosphodiester bonds
  - ➢ The 5′ end of the mRNA has a *cap* made up of *7-methyl GTP*. This cap serves as a recognition site for the initiation of protein synthesis.
  - ➢ The 3′ end has a polyadenosine tail (about 20-30 adenosine residues)–*Poly (A) tail*. This protects the mRNA from degradation by ribonucleases.
- *Codon*: The sequence of three consecutive bases in the mRNA that code for an amino acid is called as the *codon.*
- *Cistron*: The stretch of codons in the mRNA that carries the genetic code for the synthesis of protein is called as a *cistron.*
- In prokaryotes, each mRNA carries more than one cistron—*polycistronic.* In eukaryotes, each mRNA carries only one cistron—*monocistronic.*

### *Transfer RNA/tRNA/soluble RNA/sRNA*

- *Function:* Transfers the activated amino acid from the cytoplasm to the particular codon of the mRNA, while it is being translated into a protein.
- *10-20%* of the total RNA in the cell is tRNA.
- *Number of bases:* Each tRNA has about *60 – 95 bases.*
- *Unusual bases:* Apart from the usual A, G, C and U bases, tRNA also contains modified bases such as dihydrouracil, methyl guanines, inosine, methyl inosine and pseudouridine to the extent of 25%.
- *Features of primary structure:*
  - ➢ 3′ end has a C-C-A sequence. This is the amino acid accepting site to form amino acyl tRNA.
  - ➢ 5′ end has purine nucleotide.
- *Secondary structure: Clover leaf model:* (Fig. 5.17) The sequence of bases in the primary structure of the tRNA allows for a large amount of hydrogen bonding between

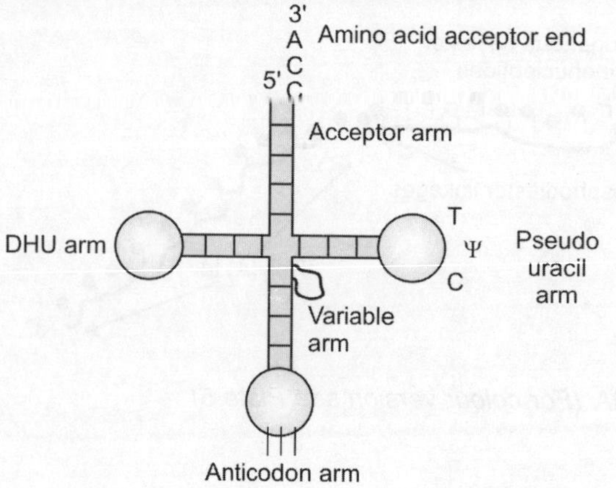


**Fig. 5.17:** Clover leaf model of tRNA

complementary bases in different anti-parallel stretches of the same polynucleotide strand (according to base pair rule). This results in the formation of four arms / stems.

- *Acceptor arm*—accepts the amino acid for transport to the ribosome.
- *Pseudouracil, (TψC) arm*—contains the unusual base pseudouracil, ψ, (C-C linkage between base and sugar) and thymine (pseudouridine arm). *Function*: This arm helps in binding the tRNA to the ribosome.
- *Anticodon arm*—has the anticodon, complementary to the codon on the mRNA. *Function:* The anticodon gives specificity to the amino acyl tRNA.
- *Dihydrouridine (DHU) arm*: Contains the unusual base dihydrouracil. (D arm). *Function*: It serves as the recognition site for the enzyme which adds the amino acid.
- *Loops:* (Fig. 5.16) Three of these arms have a stretch of modified, unpaired bases that cannot base pair. Hence these unpaired bases form a *loop.* Pseudouracil (ψ) arm, anticodon arm, and dihydrouridine arm have loops.
- *Anticodon:* Three consecutive, unpaired nucleotide bases in the loop of the anticodon arm are called as the *anticodon.* The anticodon gets hydrogen bonded to the codon on the mRNA during protein biosynthesis which helps bring the appropriate amino acid for incorporation into the protein.
- *Variable arm:* Apart from the above-mentioned four arms, the tRNA also has a *variable arm/extra arm.* The length of the variable arm varies from 3-21 bases.
- *Tertiary structure of tRNA*: The clover leaf secondary structure of the tRNA is bent into 'L'-shape: One arm of the L is made up of acceptor and pseudouridine (ψ) arms and the other arm of the L is made up of the anticodon and dihydrouracil arms. This tertiary structure is stabilised by hydrogen bonds and interactions between stacked adjacent bases (Fig. 5.17).

### *Ribosomal RNA /rRNA*

- *60-80%* of the total cellular RNA is made up of rRNA.
- *Bases*: In addition to A, G, C and U, methylated bases and pseudouridine are also present.
- *Types of rRNA*: Based on their sedimentation coefficients the different types are 5S, 5.8S, 18S and 28S.
- *Ribosomes*: rRNA is mostly in combination with proteins to form aggregates called as the *ribosomes.*
- *Subunits of ribosomes*: the ribosomes have two subunits (Fig. 5.18A)—a larger subunit and a smaller subunit. In prokaryotes –50S and 30S, and in eukaryotes 60S and 40S.
- *Sites on the ribosome*: Ribosome has the following sites:
  - *P site* / peptidyl site
  - *A site* / amino acyl site.

- *Polyribosomes*: One mRNA being simultaneously translated by a number of ribosomes is called as a *polyribosome or polysome* (Fig. 5.18B).

The differences between the two types of nucleic acids are given below (Table 5.3).

The DNA content of a cell (both nuclear and mitochondrial) is called as *Genome*.

Study of genome and its functions is called as *Genomics*.

The Watson and Crick model of DNA and the *semi-conservative mode of replication* suggest a simple mechanism by which genetic information can be precisely transferred from parent to daughter cells. This was soon extended and built upon to yield the central dogma of molecular genetics (Fig. 5.19) in which:

- DNA information is copied to form identical DNA by *replication*.

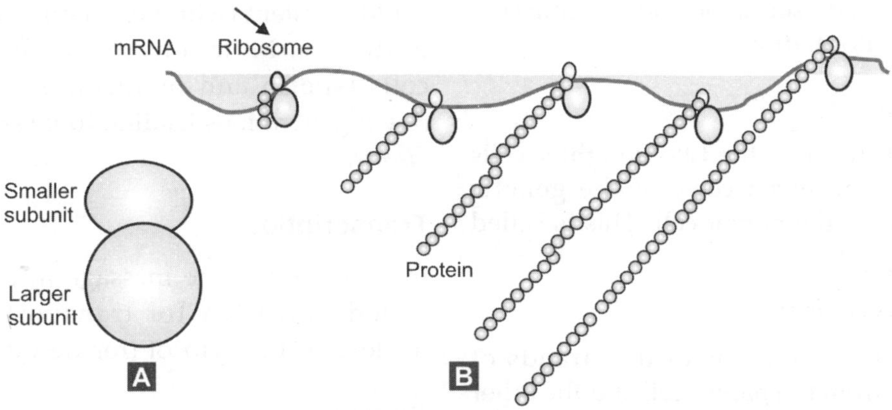


**Fig. 5.18:** (A) Ribosome (B) Polyribosome

**Table 5.3:** Differences between DNA and RNA

| Sl. No | Character | DNA | RNA |
|---|---|---|---|
| 1. | Location | Nucleus, mitochondria, chloroplasts | Cytoplasm, small amounts in nucleus, mitochondria |
| 2. | Size (number of bases) | Millions of base pairs | 200-$10^4$ bases |
| 3. | Number of strands | Double stranded, helical | Single stranded |
| 4. | Base complementarity | Interstrand base pair complementarity | Intrastrand base pair complementarity |
| 5. | Pentose sugar | 2-deoxy-D-ribose | D-ribose |
| 6. | Pyrimidine bases | Cytosine, Thymine | Cytosine, Uracil |
| 7. | Chargaff's rule | Followed | Not followed |
| 8. | Half life | Longer | Shorter than DNA |
| 9. | Function | Storage of genetic information and its transfer to the progeny. | Transfer of genetic information to the cytoplasm for protein biosynthesis |
| 10. | Hydrolysis by alkali | Not possible | Possible |

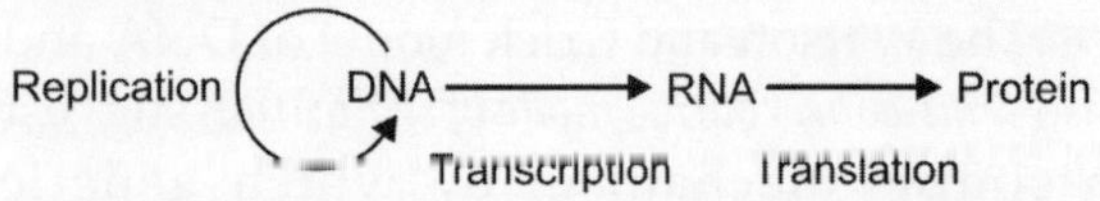


**Fig. 5.19:** Central dogma of molecular genetics

- DNA information is transcribed (coded) to RNA by *transcription.*
- RNA message is decoded to protein by *translation.*

These three major processes of the central dogma help in preservation and transmission of genetic information.

## Replication

During cell division, the two daughter cells formed, get an exact copy of the genetic information as the parent cell. This is called *DNA replication.*

### *Type of Replication*

In the daughter cell, one of the strands of the DNA is from the parent cell and the other is newly synthesised, complementary to the parent strand. So DNA of the daughter cell is partly conserved from the parent–*Semi conservative type of replication* (Fig. 5.20).

### *Enzyme*

*DNA polymerase* synthesizes DNA with fidelity and accuracy, proofreads for mistakes and edits it.

The DNA segment which codes for a protein is called as a *gene.* Each strand of the double stranded DNA in a chromosome codes for a different set of proteins. Hence, the twenty three pairs of chromosomes have large capacity for storage of genetic information.

### *Mutations*

Mistakes in replication lead to *change in the DNA code called as mutations.* Gene mutations could be silent (with no effect) or could lead to disease. Mutations in the reproductive cells (sperm and ovum) are transmitted down generations leading to *inherited genetic disease.*

## Transcription

In *transcription,* the message in the DNA is coded into RNA for transport from the nucleus to the cytosol (for decoding).

### *Enzyme*

DNA dependant RNA polymerase

### *Mechanism*

A limited region of the DNA is unwound and the two strands separated from one another. One strand is transcribed by polymerisation of ribonucleotides comple-

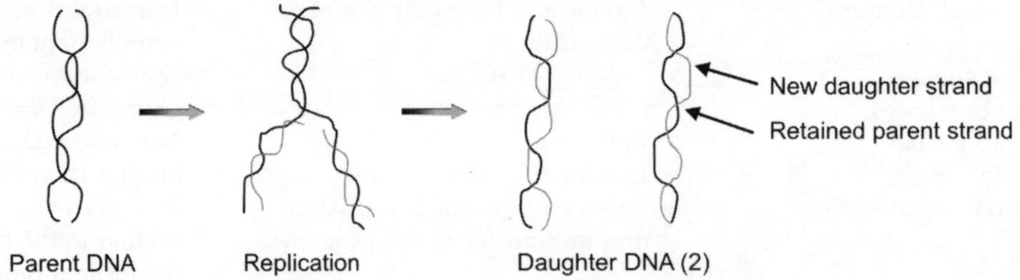


**Fig. 5.20:** Semi conservative type of replication

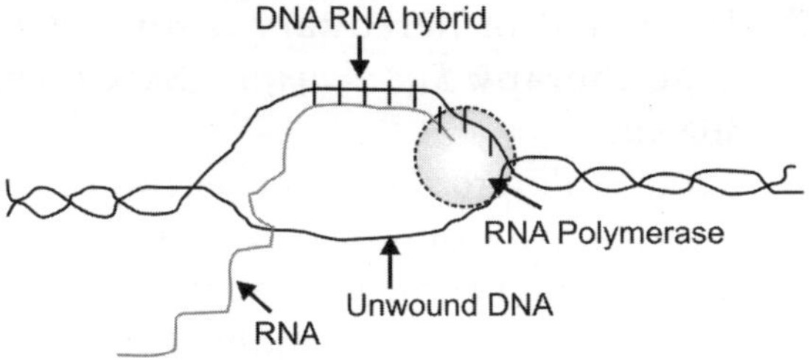


**Fig. 5.21:** Formation of RNA from DNA-transcription

mentary to the exposed unpaired bases. A transient DNA-RNA hybrid is formed (Fig. 5.21). The RNA is separated, undergoes modifications to form mature RNA which carries the code to the cytoplasm. The DNA goes back to its double helical structure.

*Codon:*

The genetic message in the RNA is read as a set of three consecutive bases which is called as the *triplet codon*. Each codon codes for and is specific for an amino acid.

## Translation

The messenger RNA transcribed from the DNA is decoded into a protein by translation.

*Site:* Ribosomes

### *Mechanism*

The ribosomes attach to the mRNA, the amino acids are assembled (as directed by the codon sequence in the mRNA) and linked by peptide bond formation (Fig. 5.22).

The nascent protein formed undergoes post translational modification to form the mature functional protein.

When there is a gene mutation, it is copied on to mRNA and later carried down to the translated protein resulting in a *mutant protein* with an altered/changed amino acid sequence. This could lead to disease. *For example,* sickle cell anaemia.

## Recombinant DNA Technology/ Genetic Engineering

The basic knowledge of DNA structure and associated molecular genetics along with techniques used in molecular biology (such as polymerase chain reaction, cloning, blotting, DNA finger printing and chimeric DNA formation), have been made use of in modern molecular medicine for analysis of—

- disease
- related genes
- their functions.

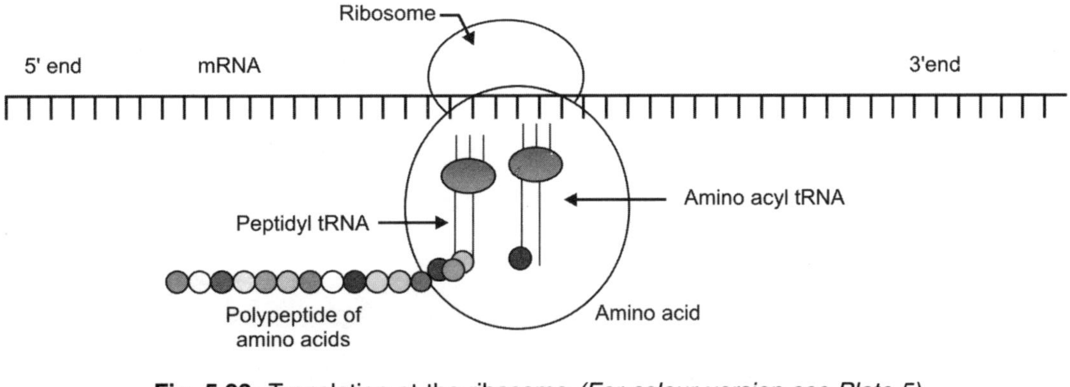


**Fig. 5.22:** Translation at the ribosome *(For colour version see Plate 5)*

Manipulation of genes from different segments of DNA and covalent recombination is termed as *recombinant DNA technology/ genetic engineering*.

### Important Tools of Genetic Engineering

- Enzymes—Restriction endonucleases
- DNA ligases
- Vectors

### Applications of Genetic Engineering

1. Gene cloning—amplification of genes.
2. Diagnosis of genetic disease. *For example* sickle cell anaemia.
3. Treatment of hereditary disorders by gene therapy *For example*, Sickle cell anaemia
4. Large scale production of useful human proteins like insulin (used in treatment).
5. Development of diagnostic kits of diseases like AIDS, hepatitis B.
6. In forensic medicine—identifying criminals and solving paternity disputes.
7. Evaluation of success of bone marrow transplants.
8. Production of animals, plants and vegetables of better quality.

# 6 Enzymes

In all living cells, life is possible due to the large number of metabolic reactions, which occur in a well coordinated manner. These reactions have to occur at body temperature with great speed and efficiency. Enzymes act as catalysts in biological systems. Hence they are called as *biocatalysts*. *Catalyst is a substance, which hastens the rate of the reaction without itself undergoing any change*. A specific enzyme catalyzes every reaction in a cell. In this chapter, the various aspects of enzymes including the classification, properties, factors affecting enzyme action, regulation, clinical importance and other uses have been discussed.

## DEFINITION

*Enzymes*: These are chemical catalysts that control various biochemical reactions in biological systems, without themselves being changed or utilised. Hence they are also called as *"biological catalysts"*.

*Substrate of an enzyme*: It is a specific molecule or a group of structurally related molecules on which the enzyme can act.

*Products:* These are molecules formed as a result of enzyme action on the substrate.

*Inhibitors:* These are molecules, which decrease/stop the rate of an enzyme-catalysed reaction.

## BIOMEDICAL IMPORTANCE

- Enzymes are biocatalysts which make life possible.
- End product and allosteric inhibition are of utmost importance to cellular economy, since a compound needs to be synthesised only when it is necessary.
- Enzymes are used in therapy.
- Estimation of serum enzyme levels are used for diagnosis and prognosis of disease.

## CHEMICAL NATURE OF ENZYMES

- Enzymes are proteins in nature (exception —*ribozymes* which are oligonucleotides with enzyme activity)
- They are either simple or conjugated proteins.
- When the enzyme is a conjugated protein— the protein part is called as *apoenzyme* and the non-protein part is called as the *coenzyme/activator*. Apoenzyme with the coenzyme forms the *holoenzyme*, which is biologically active.
  Apoenzyme (Protein) + Coenzyme/activator (Non-protein)→Holoenzyme (Active enzyme)
- Some hydrolytic enzymes have single polypeptide chain—*monomeric enzyme. For example, ribonuclease*. Many other enzymes have more than one polypeptide chains or subunits in each molecule—*oligomeric enzymes. For example, lactate dehydrogenase* has 4 polypeptide chains.
- Sometimes, several enzymes catalysing different, consecutive reactions of a metabolic reaction are located together. This is considered to be a *multienzyme*

*system*. Different enzyme activities occurring at different sites of the same macromolecule is considered to be a *multifunctional enzyme*.

*Examples:*
*Pyruvate dehydrogenase* and *α-ketoglutarate dehydrogenase* multienzyme complexes are similar, each having three enzyme activities.

*Fatty acid synthase multienzyme complex*—is dimeric, each polypeptide is made up of seven enzyme activities and an acyl carrier protein. Hence it is *multifunctional* as well.

*Significance of multienzyme/multifunctional systems*:

- Multienzyme systems enhance the efficiency of the overall reaction sequence by transferring the intermediates directly, from one active site to the next, thereby avoiding loss/dilution of intermediates in the free state.
- The intermediates are also protected from diversions into other reactions.

When separated from each other the catalytic activities of the aggregates are lost.

Enzymes maybe present either *inside or outside the cell.*

1. *Intracellular enzymes*: Most of the enzymes produced by a cell, function within the cell (endozymes).
   *For example,* enzymes of glycolysis.
2. *Extracellular enzymes*: These enzymes are liberated by the cell and catalyze reaction outside the cell or in the extracellular fluid (exozymes).
   *For example,* enzymes of digestion.

## Location of Enzymes in the Cell

Enzymes are distributed in the various subcellular fractions of the cell—nucleus, cytoplasm, mitochondria, microsomes and lysosomes.

Some processes take place in more than one compartment.

For example, urea cycle (five enzymes)—mitochondrial matrix (two enzymes) and cytosol (three enzymes)

Gluconeogenesis—mitochondria, cytosol and endoplasmic reticulum.

Shuttle mechanisms—mitochondria and cytosol.

## Classification of Enzymes

*Basis*: According to the IUB system of enzyme classification, enzymes are classified into *six classes* based on the type of reaction catalysed. Each class has been further classified into sub classes and sub sub classes. Accordingly, an enzyme has an enzyme code and a systematic code which is very complicated. In general, an enzyme is named: either (a) based on the substrate on which it acts, replacing the last three alphabets with *'ase'*.

*For example,* enzyme *lactase* acts on lactose, *maltase* acts on maltose.

Or (b) the name has two parts—

i. Name of the substrate
ii. Type of reaction that it catalyses

For example, *succinate dehydrogenase* is an enzyme, which dehydrogenates succinate.

The *six classes of enzymes* are as follows:

1. Oxidoreductases
2. Transferases
3. Hydrolases
4. Lyases
5. Isomerases
6. Ligases.

***1. Oxidoreductases***: Enzymes which catalyze oxidation-reduction reactions (redox reactions) belong to this class.

*Oxidation* is a process which involves any of the following:

i. Removal of hydrogen
ii. Removal of electrons
iii. Addition of oxygen

*Reduction* is

i. Addition of hydrogen
ii. Addition of electrons
iii. Removal of oxygen.

*Examples:*

i.

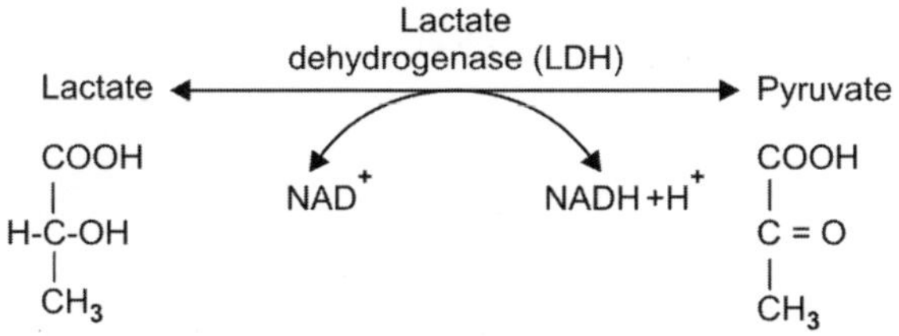


ii.

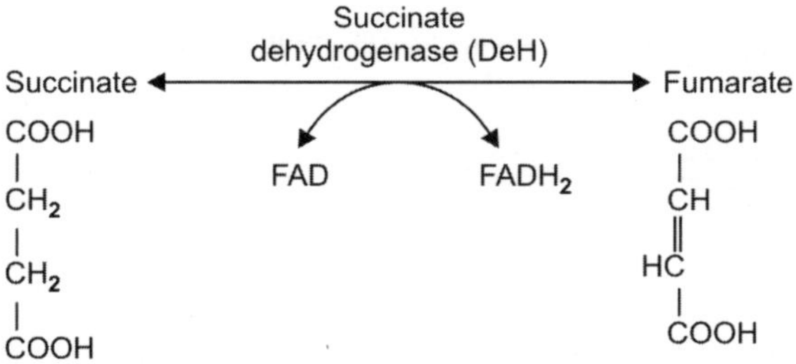


**2. *Transferases*:** Enzymes of this class catalyse the transfer of groups (other than hydrogen) from one substrate to another.

*Examples*:

i.

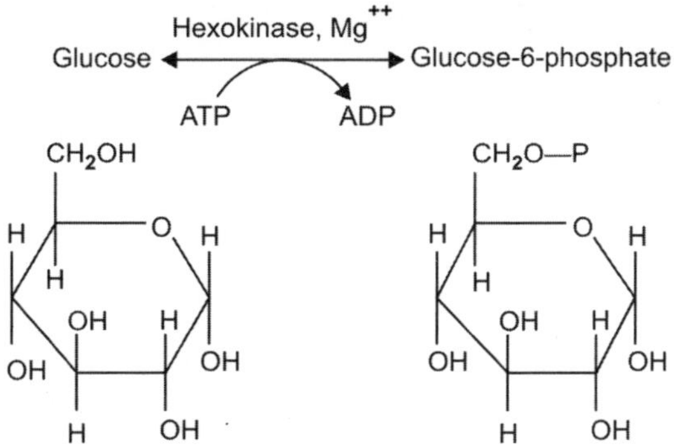

ii.

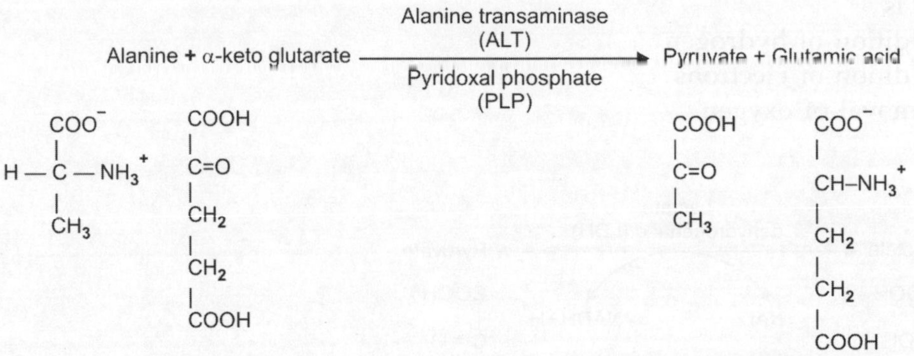


3. ***Hydrolases***: These enzymes help in breakdown of a covalent bond of a compound using a molecule of water.

*Examples:*

i. All enzymes of digestion belong to this class—*amylases, pepsin, nucleases*

ii. Lactose

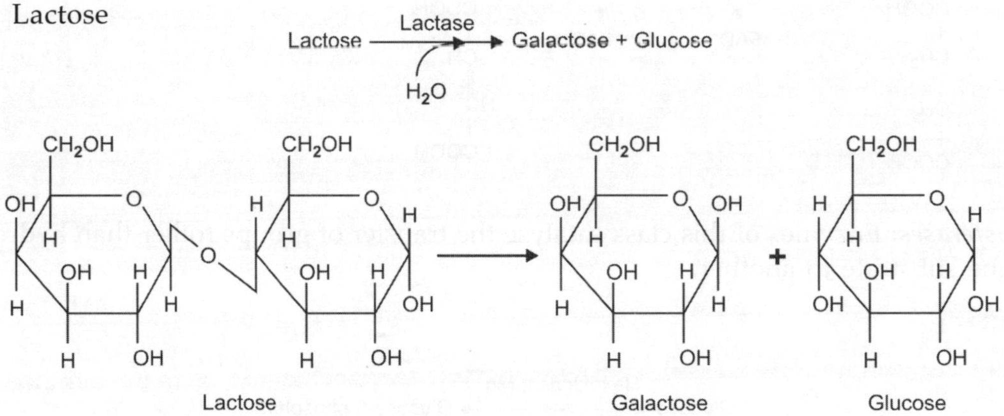


4. ***Lyases***: This class of enzymes helps in breaking a covalent bond in a substrate or catalyses removal of groups without using water.

*Examples:*

i.

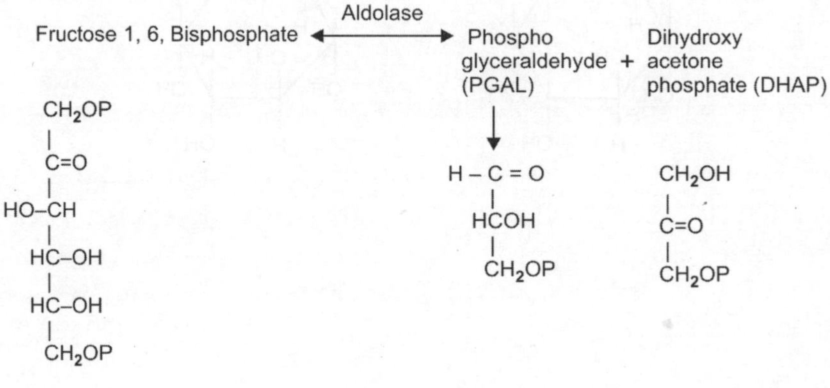

ii.

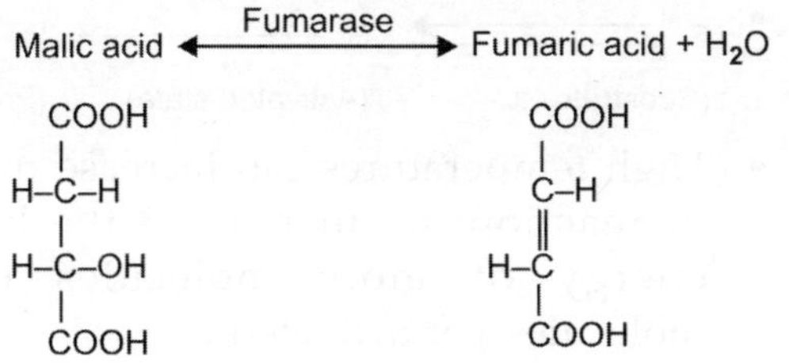


5. ***Isomerases***: Catalyze isomerization reactions and produce optical, geometric or positional isomers of the substrate.

*Examples:*

i.

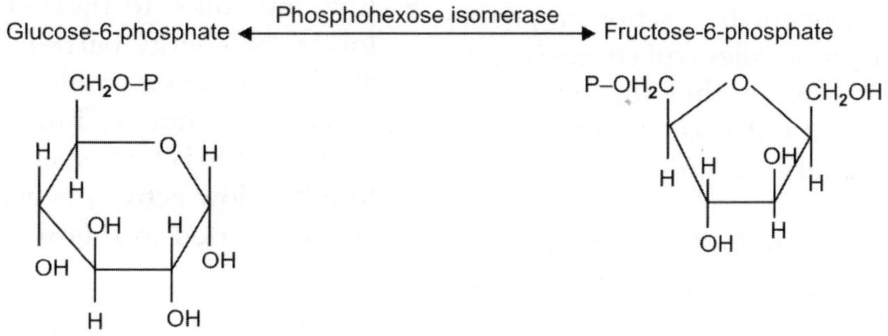


ii.

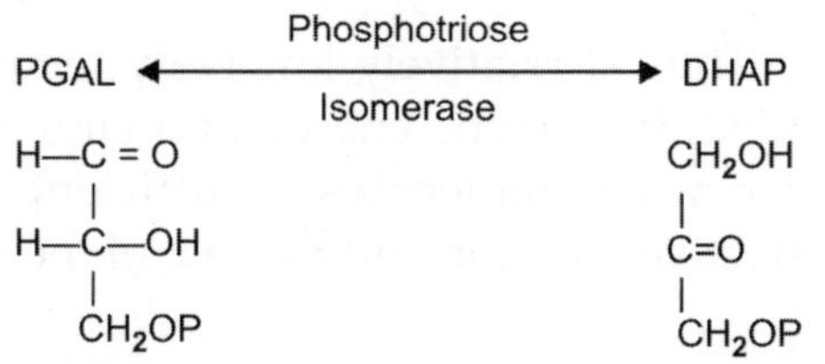


6. ***Ligases***: Help in linking of two substrates by formation of a covalent bond with input of energy. The energy needed is got by hydrolysis of ATP to ADP and Pi.

*Examples:*

i.

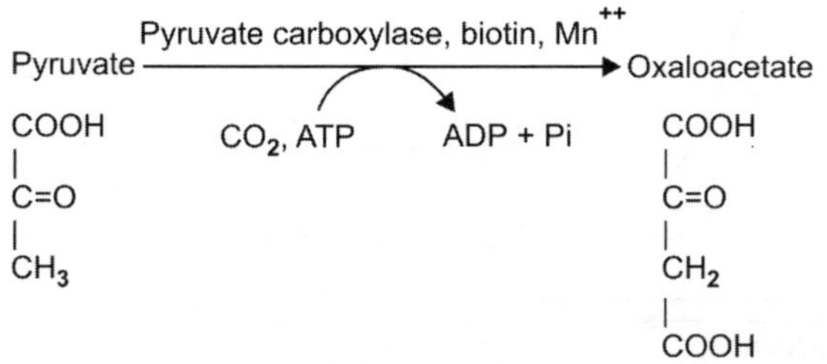

ii.

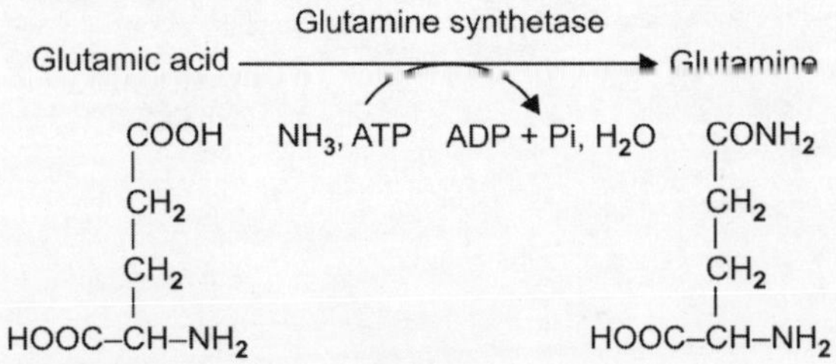


## Mechanism of Catalysis

- Every molecule has a small amount of energy called as *free energy.*
- The kinetic theory of the reaction proposes that, reacting molecules (substrates) must collide with one another (or approach each other), with at least the minimum amount of energy required to form a bond.
- When the substrate (S) molecules collide with one another, they get activated (S*). S* has higher energy than S.
- *Activation energy* is the difference in energy between the substrate (S) and the activated transient intermediate (S*). This higher energy intermediate is converted into either, the product (P) or the substrate (S).
- $$\underset{\text{(Substrate)}}{S} \rightleftharpoons \underset{\text{(Activated state)}}{S^*} \longrightarrow \underset{\text{(Product)}}{P}$$
- High temperatures can increase the rate of reaction; by increasing the kinetic energy of more molecules (more molecules get activated).
- But at relatively low temperature in biological systems, few reactant molecules possess sufficient kinetic energy for collision and overcome the energy barrier (Fig. 6.1). So rate of the reaction is slow.
- Enzymes bind to their substrates and lower the energy barrier. They provide alternate reaction pathway requiring less activation energy. Enzyme substrate complex (ES) is formed and gets activated to ES* at low activation energy. This ES* gets converted to product.

$$\underset{\text{(Substrate)}}{S} \xrightarrow{E} \underset{\text{(Complex)}}{ES} \longrightarrow \underset{\text{(Activated state)}}{ES^*} \rightleftharpoons \underset{\text{(Product)}}{P} + E$$

Thus at relatively low body temperature (37°C) the kinetic energy of a large number of reactant molecules is sufficient to have effective collision and the rate of the reaction

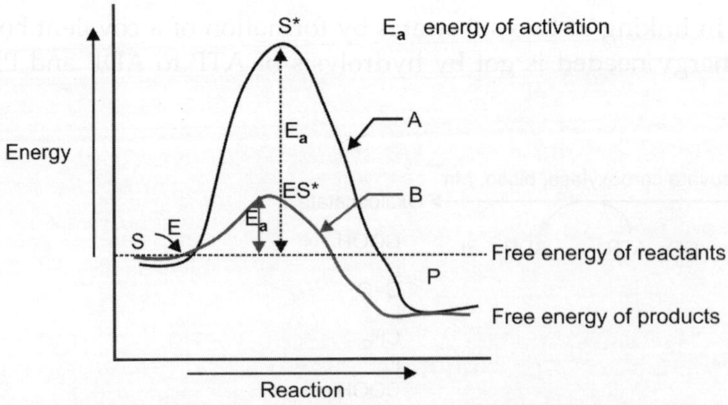


**Fig. 6.1:** Chemical reaction (A) Uncatalysed reaction (B) Enzyme catalyzed reaction needs less activation energy

is increased. So enzymes enhance the rate of reaction.

## Mechanism of Action of Enzymes

The substrate binds to enzyme on specific sites called as the *active sites* to form enzyme substrate complex which then gets converted to product and enzyme.

### *Active (catalytic) Sites*

- These are depressions or clefts on the surface of the enzyme.
- Made up of amino acids which are located far apart in the native conformation of the enzyme.
- These amino acids are brought close to one another by the three dimensional (3D) conformational changes that take place in presence of substrate.
- The amino acids at the active site either hold the substrate for catalysis (substrate binding residues) or bring about the catalysis (catalytic residues).

### *Models for Catalysis based on the Active Site*

1. *Lock and key theory:* This theory proposed by Emil Fischer states that active site is rigid. *Theory*: Enzyme is rigid like a lock and the substrate is like a key that fits into the lock (enzyme) for the catalysis to take place (Fig. 6.2). Hence it is also called as *Fischer's rigid template model*. This theory explains the following:

- Ordered binding of two or more substrates
- Kinetics of simple substrate saturation
- Absolute substrate specificity.

But has no explanation for

- Allosteric modulation.

2. *Induced fit theory*: Koshland proposed this theory in 1963 which states that active sites are formed on the surface of the enzyme only in the presence of substrate.

*Theory*: (Fig. 6.3) According to this theory,

- The enzyme has two conformations—
- In absence of substrate it exists in *native conformation*
- In presence of substrate it changes to a *active conformation*
- In presence of substrate, the native conformation undergoes a 3D conformational change resulting in an *active conformation*.

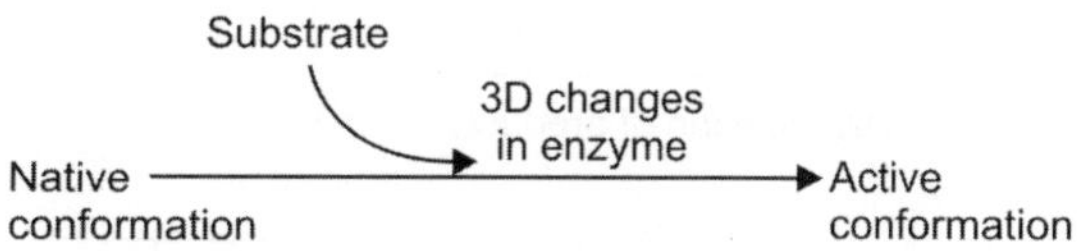


- In the active conformation of the enzyme, active sites are formed on the surface.
- The substrate molecule is held in the active site by the binding amino acid residues to form enzyme-substrate (ES) complex.
- The catalytic amino acid residues bring about the catalysis (either bond formation or bond breakdown) and enzyme-product (EP) complex is formed
- The EP complex then dissociates into enzyme and product

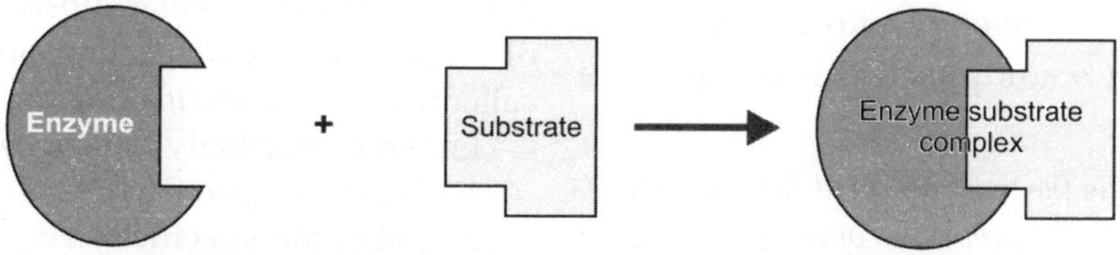


**Fig. 6.2:** Fischer's rigid template model of enzyme action

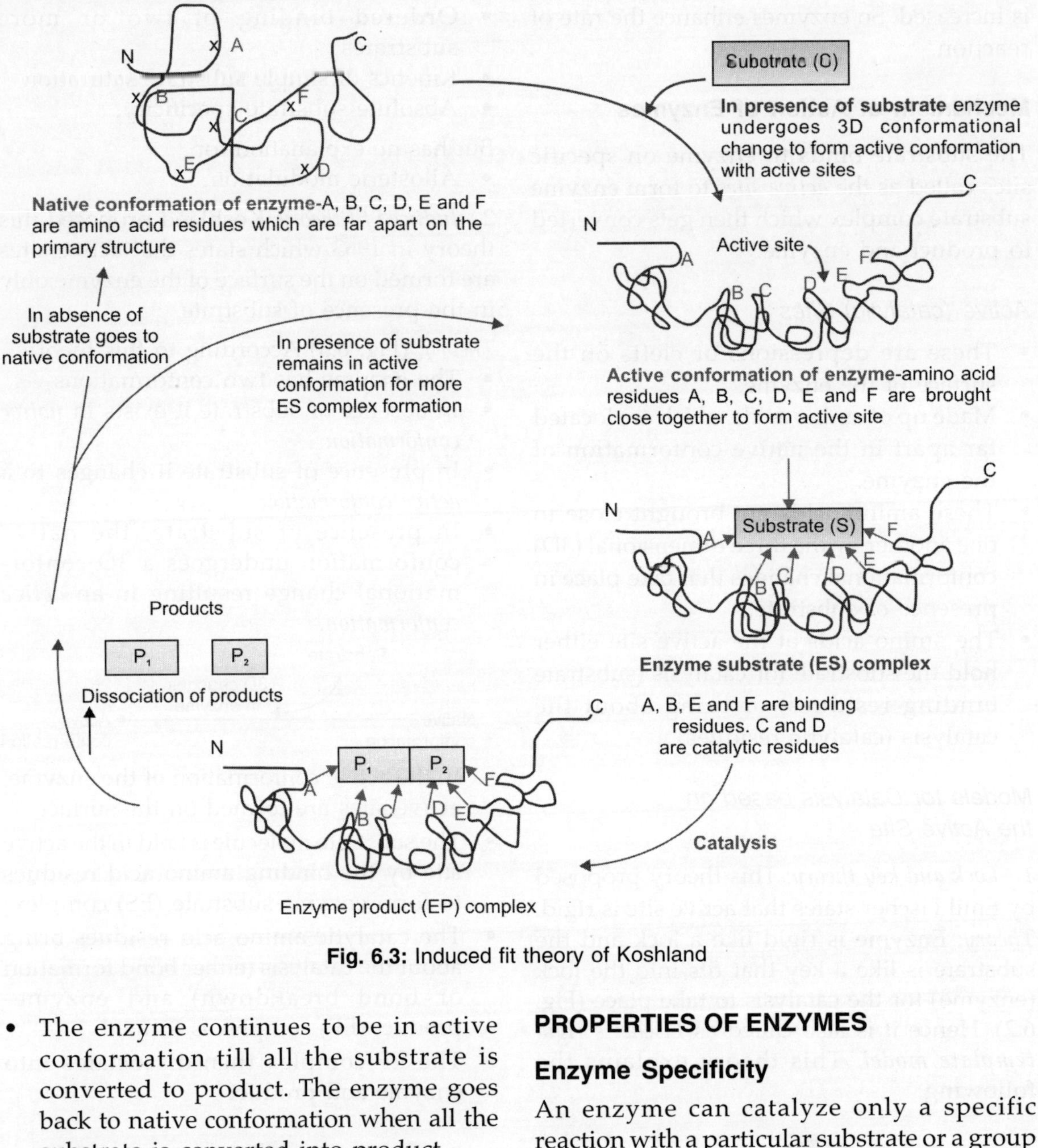


**Fig. 6.3:** Induced fit theory of Koshland

- The enzyme continues to be in active conformation till all the substrate is converted to product. The enzyme goes back to native conformation when all the substrate is converted into product.
- E + S →ES complex →EP complex → E + P

*Note: This is the most accepted theory since it explains most of the properties of enzymes.*

## PROPERTIES OF ENZYMES

### Enzyme Specificity

An enzyme can catalyze only a specific reaction with a particular substrate or a group of structurally related compounds. This is called as *enzyme specificity*.

Enzyme specificity is of two types:

1. Reaction specificity
2. Substrate specificity.

### Reaction Specificity

One enzyme catalyses only one of the various reactions which the substrate can undergo.

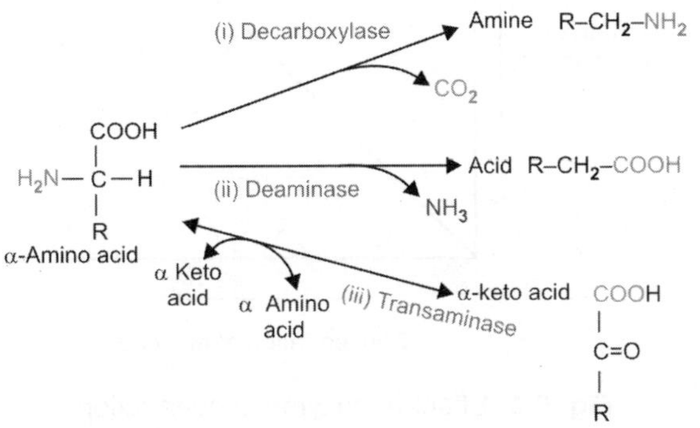


### Substrate Specificity

Different types of substrate specificity are known:

a. *Stereo specificity:*
Enzymes are specific in their recognition of substrates with respect to isomerism. A specific enzyme will catalyse the conversion of only one of the two optical isomers.

i. *DL-specificity:* For example, the enzymes of carbohydrate metabolism are specific for D-stereoisomers of sugars and those of amino acid metabolism are specific for L-amino acids.

ii. *Cis-trans specificity:* Enzymes which oxidise the substrate by removal of hydrogen, in either cis or trans positions are said to exhibit cis-trans specificity.

*Example:*

b. *Absolute substrate specificity:*
One enzyme will catalyse a specific reaction for a single substrate
*Examples:*

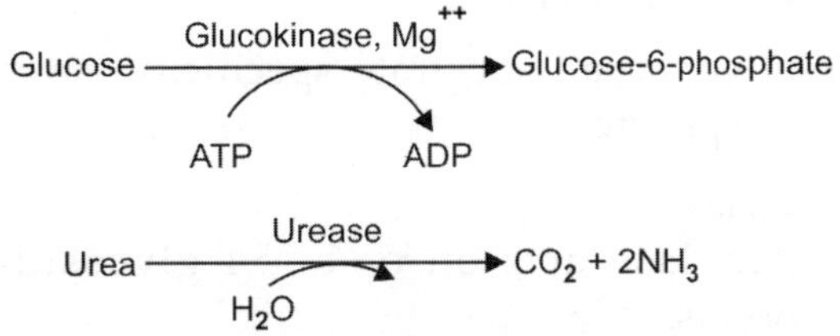


c. *Relative/broad substrate specificity:*
Enzyme catalyses a specific reaction with a group of structurally related compounds.
*Examples:*

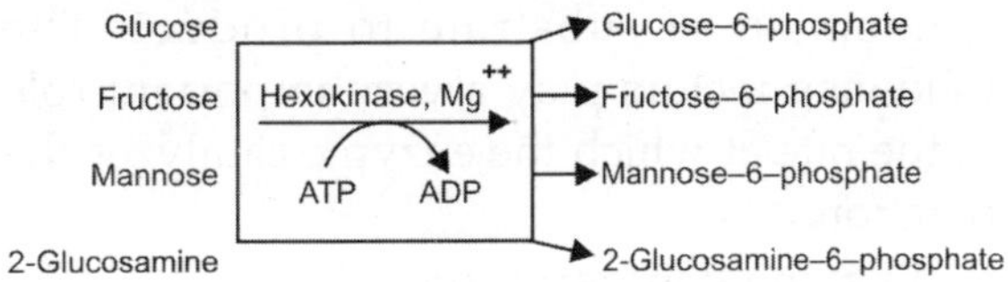


*Note: The configuration of all the 4 substrates around $C_3$, $C_4$, $C_5$ and $C_6$ is the same.*

### (i) Group specificity

Relative/broad substrate specificity may be group specific.

*Examples:*

a. Pepsin acts on those peptide bonds where C=O group is given by aromatic amino acids.

b. Trypsin acts on those peptide bonds where the C = O group is given by basic amino acids.

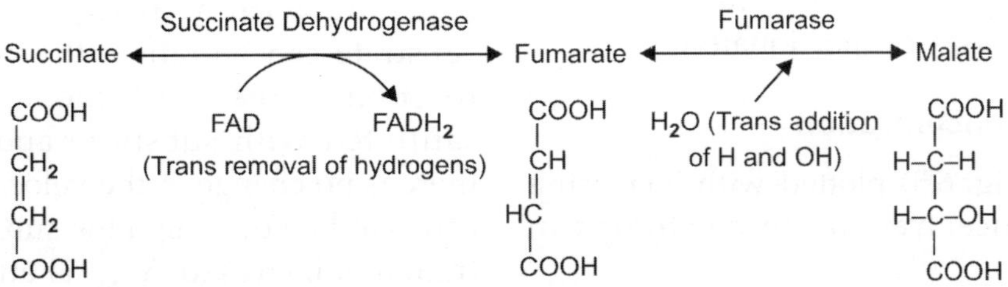

c. Chymotrypsin acts on those peptide bonds where the C = O group is given by aromatic amino acids/leucine.

*(ii) Bond specificity*

Relative/broad substrate specificity maybe bond specific:

*Examples*:

a. Amylase acts on α, 1 →4 glycosidic bonds.
b. Cellulase acts on β, 1 →4 glycosidic bonds.

## Factors Affecting Enzyme Action

Enzymes are biocatalysts catalyzing the conversion of substrate to product. The following factors play a very important role in the rate at which the enzyme catalyzes the reaction.

a. Enzyme concentration
b. Substrate concentration
c. Product concentration
d. Temperature
e. pH
f. Activators/coenzymes
g. Inhibitors/modulators.

## Enzyme Concentration

The rate of the reaction increases linearly with the increase in the enzyme concentration provided substrate concentration is not the limiting factor as shown in Figure 6.4.

*Reason*: As the enzyme concentration increases, the number of active sites for binding substrate also increases. Hence rate of reaction increases proportionately.

## Substrate Concentration

The graph (Fig. 6.5) plotted with increasing substrate concentration can be studied in three parts.

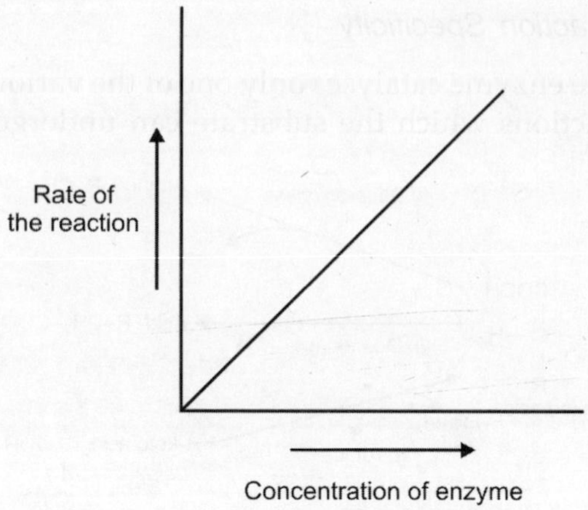


**Fig. 6.4:** Effect of enzyme concentration on enzyme action

i. At low concentrations of substrate the rate of reaction increases proportionately with the increase in substrate concentration (part (i) of Fig. 6.5A)
ii. Further increase in substrate concentration increases the rate of the reaction but not proportionally (part (ii) of Fig. 6.5A).
iii. At high substrate concentration maximal velocity is attained and the velocity of reaction remains unchanged however much the substrate concentration is increased. This is called as *maximal velocity or* $V_{max.}$ (part (iii) of Fig. 6.5A)

Figure 6.5(A) is a *hyperbolic curve.*

*Reason*: Substrate binds to the active site of the enzyme. When enzyme concentration is constant, with a steady increase in the substrate concentration, the velocity of the reaction increases till the active sites are saturated with substrate and there after, there is no change in the velocity. It remains constant however high the substrate concentration is increased. $V_{max}$ is attained.

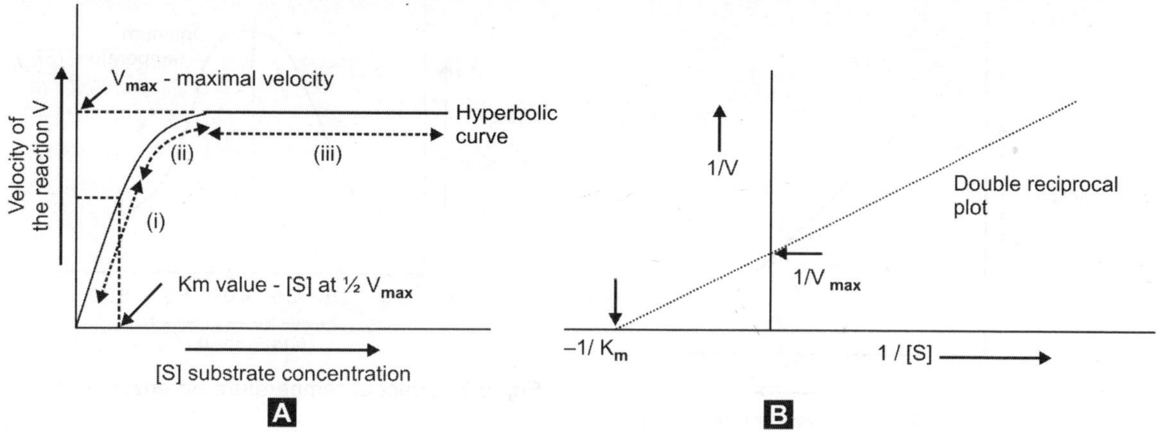


**Figs 6.5 A and B:** Effect of substrate concentration on enzyme action. (A) hyperbolic curve (B) double reciprocal plot

The substrate concentration [S] at half the maximal velocity ($\frac{1}{2}V_{max}$) is called as *Michaelis constant or $K_m$ value.*

$K_m$ value is characteristic of an enzyme at a definite pH and temperature. Normally, $K_m$ values lie in the range of $10^{-2}$ to $10^{-5}$ moles/litre for various enzymes.

### *Determination of $K_m$ Value*

The $K_m$ value obtained by the hyperbolic curve is not very accurate. Hence $K_m$ value is determined by the double reciprocal plot. (Fig. 6.5B)

### *Significance of $K_m$ Value*

- It indicates the affinity of the enzyme for the substrate—low $K_m$ *indicates* higher affinity of the enzyme for the substrate and vice versa. For example, hexokinase has a low $K_m$ for glucose, 0.05 mmol/L and glucokinase has a higher $K_m$ of 10 mmol/L. Hence hexokinase has higher affinity for glucose than glucokinase.
- For an enzyme with broad substrate specificity, the substrate with least $K_m$ is the natural substrate of the enzyme.
- Gives an indication of the type of inhibition—$K_m$ changes in competitive and remains unchanged in noncompetitive type of inhibitions.
- In a metabolic pathway, the enzyme which has highest $K_m$ for its substrate is the regulatory enzyme of the pathway.

## Product Concentration

Product formed as a result of enzyme action may accumulate and lower the rate of the reaction by occupying the active sites of the enzyme (Fig. 6.6). It is also possible that under certain conditions the high concentration of product favours the reverse reaction.

## Effect of Temperature

- Rate of a chemical reaction increases as the temperature is raised, since the number of enzyme molecules getting activated is more.

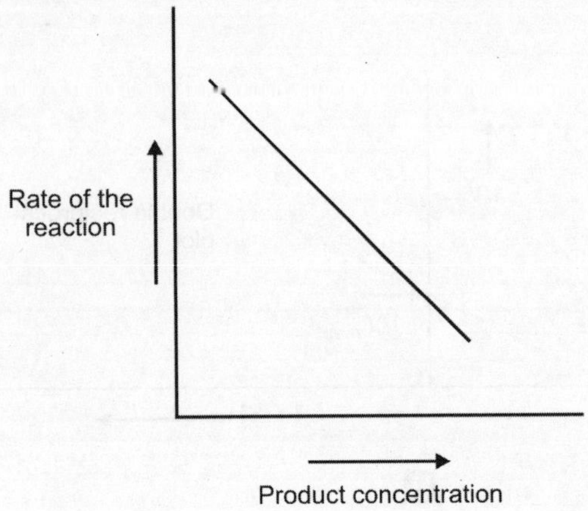


**Fig. 6.6:** Effect of product concentration on enzyme action

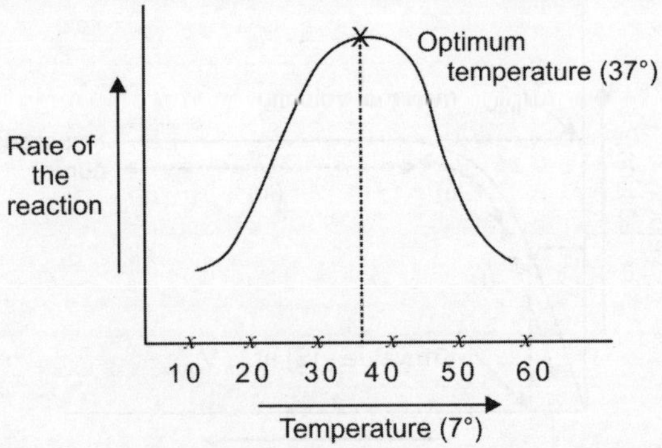


**Fig. 6.7:** Effect of temperature on enzyme action

- Each enzyme is most active at a specific temperature called as the *optimum temperature*. The optimum temperature of most enzymes ranges from 37 to 45°C.
- The activity of the enzyme progressively decreases when the temperature of the reaction is below or above the optimum temperature (Fig. 6.7). The curve plotted is called a *bell shaped curve*.
- Increase in temperature results in decrease/loss of activity due to denaturation (enzymes are proteins in nature). Low temperatures do not destroy the enzyme and the activity can be got back on heating to optimum temperature.
- *Exception*: Fat hydrolysing enzymes are active at –15°C. Ribonuclease and $T_{aq}$ polymerase are heat stable.

## Effect of pH

- Rate of an enzyme catalysed reaction is dependant upon pH. pH is the negative logarithm of hydrogen ion concentration (-ve log [$H^+$]).
- The enzyme activity is maximum at a particular pH which is called as *optimum pH*. The normal range of pH at which most of the enzymes are active is 4–9.
- Exceptions: Pepsin 1–2, arginase 8.9 – 9.9, acid phosphatase 4.–5.0, alkaline phosphatase–10–11.
- pH affects ionization of the active site of the enzyme which will exist only within a limited pH range and depends upon the pK value of the concerned groups forming ES complex.
- Ionization of the substrate, active site of the enzyme and coenzyme will be optimum for complex formation at the optimum pH. Hence, the velocity of the reaction will be maximum. At pH on either sides of the optimum pH the activity of the enzyme is lower (Fig. 6.8) and a bell shaped curve is seen.

## Effect of Activators/Coenzymes

- Many enzymes are conjugated proteins with a non protein part. The non protein part maybe inorganic or organic.
- The inorganic factors needed for enzyme action are usually called as *activators* and the organic factors are called as either *prosthetic group or coenzymes* as shown in the Figure 6.9.

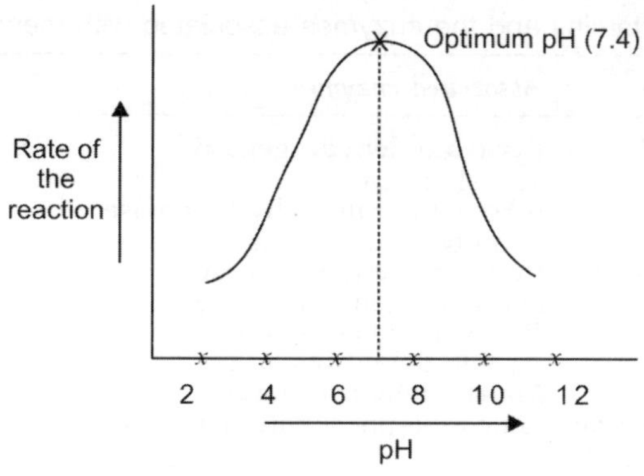


**Fig. 6.8:** Effect of pH on enzyme action

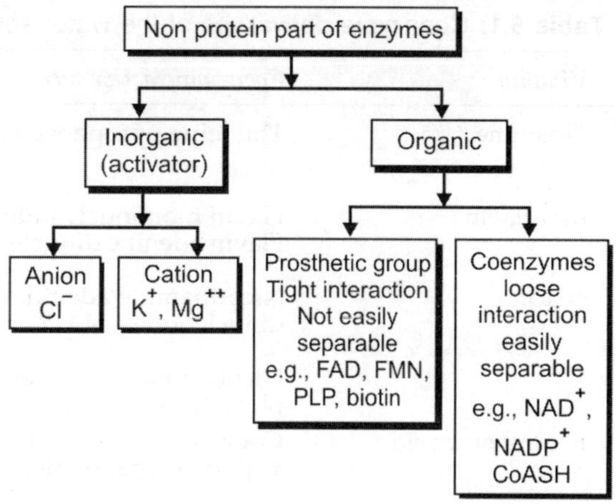


**Fig. 6.9:** Types of non protein part of enzymes—cofactors, prosthetic groups and coenzymes

*Note: Separation of the non protein part of the enzyme will inactivate it.*

*Activators*: These are usually metal ions.

- Depending on the type of interaction between the metal ion and the enzyme, the metal associated enzymes are divided into two categories—*a) metal activated enzymes*—metal ions are loosely bound to the enzymes. *For example,* amylase and $Cl^-$ ions *b) metalloenzymes*—metal ions are tightly bound to the enzymes. *For example,* carboxypeptidase and $Zn^{++}$ ions.
- Cofactors needed for enzyme action are of the following types (Fig. 6.10).

*Role of metal ions:*

Metal ions

- May help in forming and maintaining the active structural conformation of the enzyme.
- May help in formation of enzyme-substrate (ES) complex.
- May bring about structural changes in the substrate for easier interaction.
- May act as an electron donor/acceptor.
- May form a complex with the product and decrease the reverse rate of reaction.

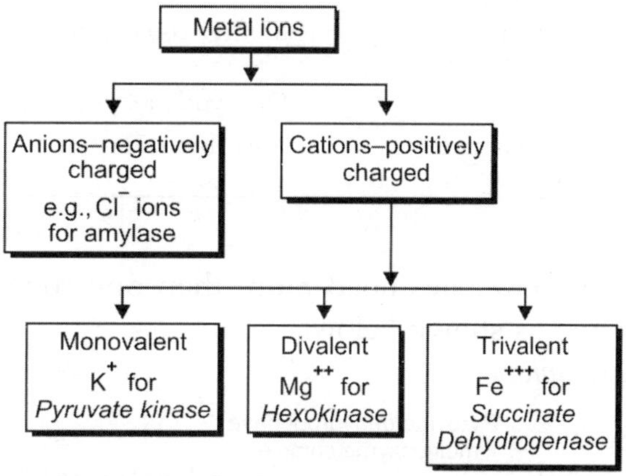


**Fig. 6.10:** Types of activators needed for enzyme action

*Coenzymes and prosthetic groups:*

- A *Coenzyme* is a low molecular weight, organic, heat stable substance which is essential for enzyme activity. Coenzymes and prosthetic groups are usually the derivatives of the water soluble group of vitamins (Table 6.1).

**Table 6.1:** Coenzyme derivatives of the water soluble vitamins and the enzymes associated with them

| *Vitamin* | *Coenzymic/active form* | *Associated enzymes* |
|---|---|---|
| Thiamine | Thiamine pyrophosphate –TPP | Pyruvate dehydrogenase<br>Transketolase<br>α-ketoglutarate dehydrogenase |
| Riboflavin | Flavin mononucleotide – FMN<br>Flavin adenine dinucleotide–FAD | L-amino acid oxidase<br>Pyruvate dehydrogenase<br>Succinate dehydrogenase |
| Niacin | Nicotinamide adenine dinucleotide – $NAD^+$ | Pyruvate dehydrogenase<br>Lactate dehydrogenase<br>Malate dehydrogenase |
| | Nicotinamide adenine dinucleotide phosphate – $NADP^+$ | Glucose-6-phosphate dehydrogenase |
| Pantothenic acid | Coenzyme A–CoASH<br>4′ phosphopantethiene | Pyruvate dehydrogenase<br>ACP of fatty acid synthase multienzyme complex |
| Pyridoxine | Pyridoxal phosphate–PLP | Enzymes of amino acid metabolism-transaminases, decarboxylases, deaminases |
| Biotin | Biotin | ATP dependant carboxylases |
| Folic acid | Tetrahydro folate–$FH_4$/THFA | One carbon metabolism |
| Vitamin $B_{12}$ (cobalamine) | Methyl cobalamine | Synthesis of methionine |
| | Deoxyadenosyl cobalamine | Methyl malonyl CoA isomerase |
| Ascorbic acid (vitamin C) | Ascorbic acid | Hydroxylation of proline residues in collagen |

Some enzymes need more than one coenzyme as shown below.
*Example:*

$$\text{Pyruvate} \xrightarrow[\substack{\text{Coenzymes: TPP, FAD,}\\ \text{NAD}^+\text{, lipoic acid,}\\ \text{Coenzyme A}\\ \text{Cofactor: Mg}^{++}}]{\substack{\text{Pyruvate dehydrogenase}\\ \text{multienzyme complex}}} \text{Acetyl CoA} + CO_2$$

Some enzymes need both activators and coenzymes for their action. *For* example, pyruvate carboxylase needs biotin and $Mn^{++}$ ions for its activity.
*Example:*

$$\text{Pyruvate} + \text{ATP} + CO_2 \xrightarrow{\substack{\text{Pyruvate carboxylase,}\\ \text{Biotin, Mn}^{++}}} \text{Oxaloacetate} + \text{ADP} + \text{Pi}$$

## Effect of Inhibitors/Modulators

Whenever the active site of the enzyme is not available for the binding of the substrate, the enzyme activity may be reduced. Substances which alter the enzyme activity (increase or decrease) are called as *modulators* and substances which decrease/stop the enzyme activity are called as *inhibitors*.

*Inhibitors*: These are chemical substances which decrease/stop the activity of a particular enzyme. For example,

(a) *Inorganic substance*—cyanide which inhibits cytochrome oxidase enzyme of electron transport chain.

(b) *Organic substance*—malonate which inhibits succinate dehydrogenase enzyme.

The phenomenon in which enzyme activity is decreased by the presence of inhibitor is known as *inhibition*. Inhibition can be classified as follows based on whether it is reversible or irreversible (Fig. 6.11).

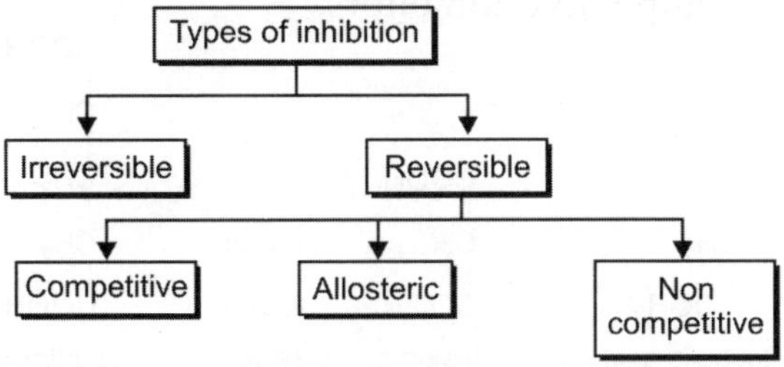


**Fig. 6.11:** Classification of inhibition based on whether it is reversible or irreversible

Inhibitors are classified based on whether they bind to active site or any other site (other than the active site) on the enzyme. (Fig. 6.12).

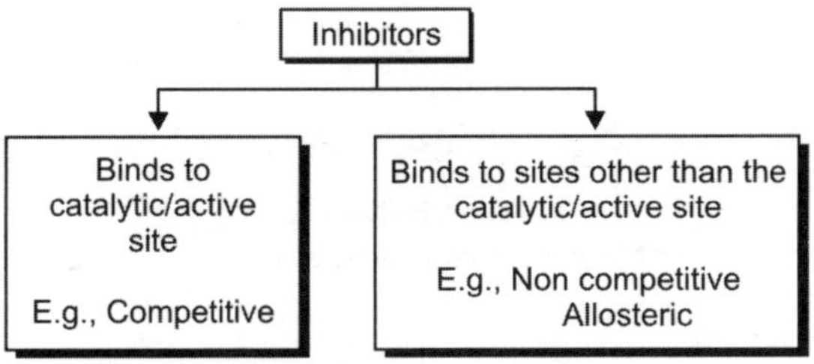


**Fig. 6.12:** Classification of inhibitors based on their site of binding

## 1. *Competitive Inhibition (Table 6.2)*

*Features:*

- The inhibitor is a structural analog (similar to) of the substrate.
- So inhibitor (I) competes with the substrate for the active site.
- Therefore, E + S →ES complex is not formed. Instead E + I →EI complex formation takes place. EI complex cannot be converted to product (Fig. 6.13). Hence the rate of the reaction will decrease.
- When [S] is increased, it pushes the inhibitor from the active site and combines with the enzyme to form ES complex which can be converted to product. (Fig. 6.13).

> In presence of substrate [S]:
> E + S → ES complex → EP complex → E + P
>
> In presence of competitive inhibitor (I) and [S] is low:
> E + S + I → EI complex (No formation of ES complex)
>
> When [S] is increased it is reversed:
> E I + ↑ [S] → ES complex + I
> (ES complex → EP complex → E + P)

**Fig. 6.13:** Mechanism of competitive inhibition

- *Kinetics:* In presence of competitive inhibitor $K_m$ value of the enzyme for the substrate is increased hence affinity for the substrate is decreased, $V_{max}$ remains unaltered (Fig. 6.14).

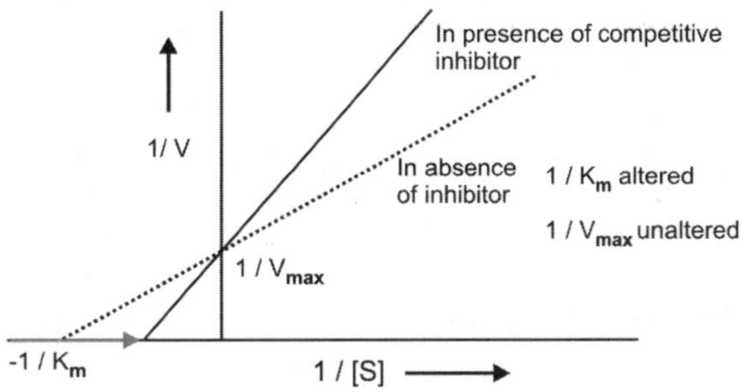


**Fig. 6.14:** Kinetics of competitive inhibition – altered $K_m$ value and unaltered $V_{max}$

*Examples:*

(a) The classical example for competitive inhibition is: *Malonate.* It inhibits succinate dehydrogenase enzyme of TCA cycle:

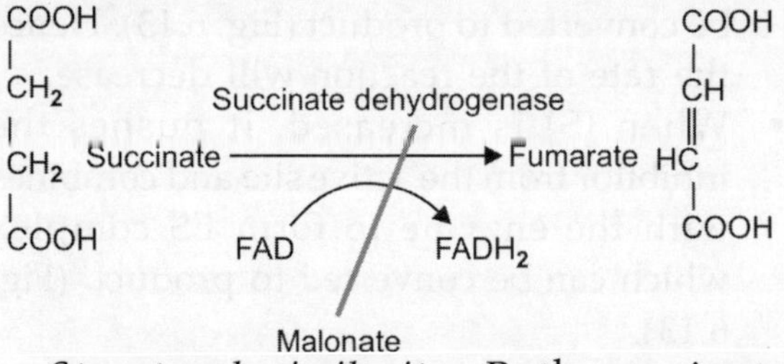


*Structural similarity*: Both succinate (substrate) and malonate (competitive inhibitor) have two carboxyl groups—dicarboxylic acids.

Other dicarboxylic acids which can inhibit the enzyme are oxalic acid and glutaric acid

Competitive inhibitors are of therapeutic importance since most of the drugs are competitive inhibitors.

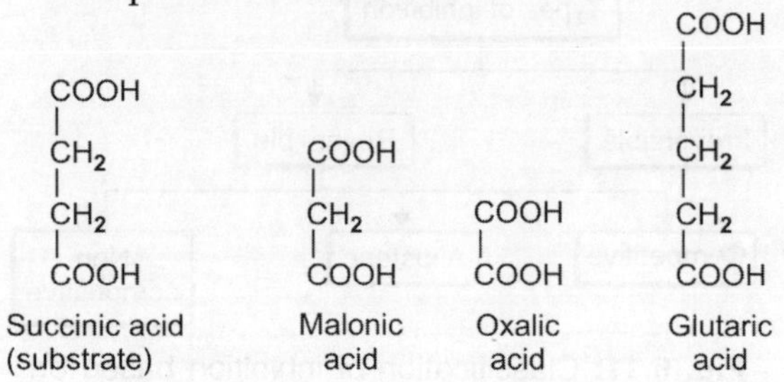


(Structural analogs-competitive inhibitors)

(b) *Dicumarol: Structural analog of vitamin K.* Glutamic acid residues in clotting factors II, VII, IX and X are activated by γ carboxylation. This is catalyzed by a vitamin K dependant carboxylase enzyme. Dicumarol inhibits the reactivation of vitamin K.

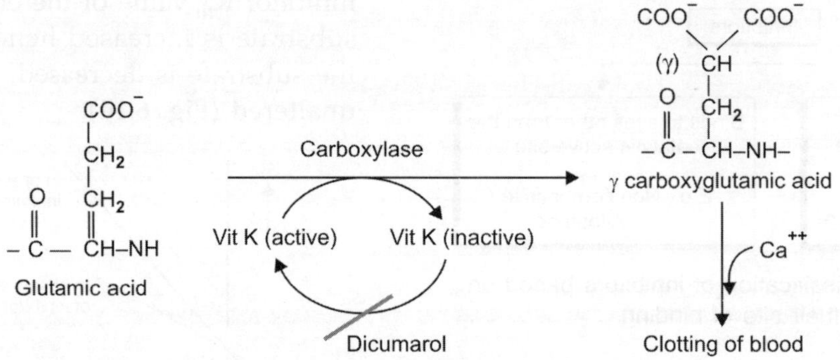


*Importance*: Dicumarol is used as an anticoagulant.

(c) *Aminopterin, amethopterin/methotrexate inhibit dihydrofolate reductase* enzyme which catalyses the following reaction in folic acid metabolism.

*Importance:* These drugs are used in cancer therapy. $FH_4$ is essential for the DNA metabolism and cell division. Therefore they can be used to control the division of cells in cancer.

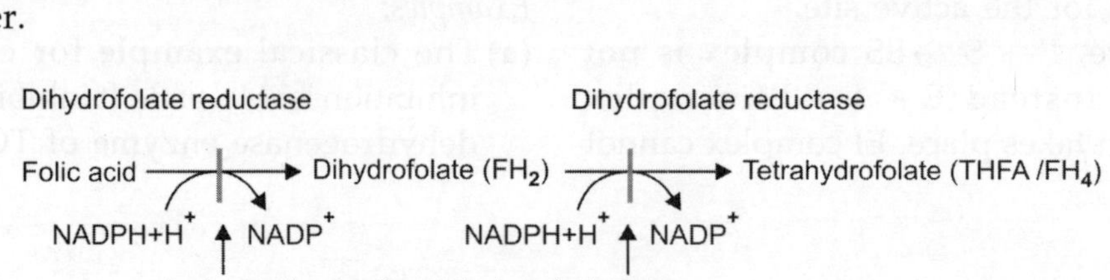

(d) *Sulphonamides:* Structural analog of para aminobenzoic acid (PABA). PABA is required by the bacteria for the synthesis of folic acid. Sulphonamides being structural analogs will inhibit the synthesis of folic acid in the bacteria.

*In bacteria :*

PABA + Pteridine + glutamic acid ——|——→ Folic acid

Sulphonamides

*Importance:* Sulphonamides are used as antibacterial/antibiotic drugs. Use of sulphonamides will destroy the intestinal bacteria. This drug is non toxic to the humans.

Certain other drugs that act as competitive inhibitors and are used in therapy are shown in Table 6.2.

## 2. Non Competitive Inhibition

*Features:*

- Inhibitor is not a structural analog of the substrate.
- It may bind to/modify amino acid residues at the active site or some other site on the enzyme.
- Inhibition cannot be overcome by increasing the concentration of the substrate.
- *Kinetics:* $K_m$ value of the enzyme remains unaltered. $V_{max}$ is decreased (Fig. 6.15.)

*Examples:* A variety of poisons are non competitive inhibitors.

(a) *Fluoride inhibits enolase* enzyme of glycolysis.

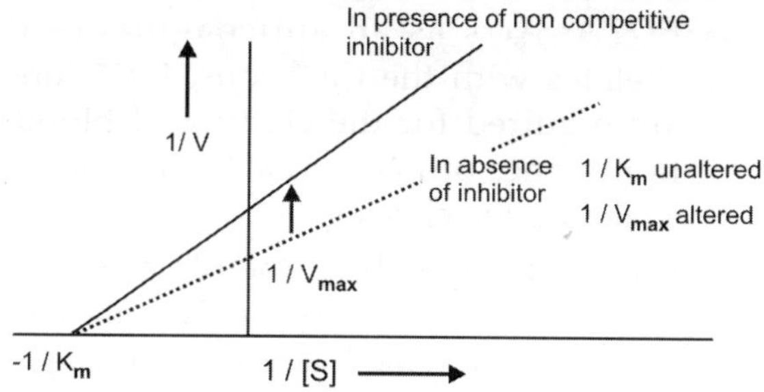


**Fig. 6.15:** Kinetics of non competitive inhibition

*Use:* Fluoride is used along with an anticoagulant for collection of blood sample for the estimation of blood sugar level. The anticoagulant prevents clotting of blood and fluoride prevents the loss of glucose by glycolysis in the RBCs.

Glucose → → → → → 2 phosphoglycerate —/—→ Phosphoenolpyruvate (Enolase, $Mg^{++}$; inhibited by Fluoride)

(b) *Cyanide: Respiratory poison*—complexes with metal ions, Cu and Fe of the enzyme cytochrome oxidase. This is an enzyme of the electron transport chain and generates ATP.

(c) *Oraganophosphorus compounds: Nerve poisons*—used as insecticides. *For example,* parathione, diisopropyl fluorophosphate. They modify the serine residues in the active site of enzymes (acetyl choline esterase) and bring about paralysis of the flight muscles of the insects.

**Table 6.2:** Drugs that are competitive inhibitors and used in therapy

| *S.No.* | *Drug* | *Enzyme inhibited* | *Therapy for/as* |
|---|---|---|---|
| (e) | Lovastatin | HMG CoA reductase | to lower cholesterol level |
| (f) | Trimethoprim | Dihydrofolate reductase | Antibacterial |
| (g) | Acyclovir | DNA polymerase | Antiviral |
| (h) | Penicillin | Transpeptidase | Antibacterial |
| (i) | Neostigmine | Acetyl choline esterase | Myasthenia gravis |

(d) *EDTA*: Acts as an anticoagulant—it chelates with the $Ca^{++}$ ions. $Ca^{++}$ ions are required for the clotting of blood. In presence of EDTA, $Ca^{++}$ is no longer available for clotting.

(e) *Iodoacetate*: Acetylates the SH group of cysteine residues in the active site and makes the enzyme inactive. *For example*, PGAL dehydrogenase of glycolysis.

(f) *Heavy metal ions*: $Pb^{++}$, $Hg^{++}$, As, $Ag^{++}$ bring about modifications of the SH groups on the enzymes and render them inactive resulting in *heavy metal ion poisoning*. BAL (British antilewisite; dimercaprol) has plenty of SH groups which will bind to the heavy metal ions and hence BAL is used as an antidote for *heavy metal ion poisoning*.

*Note: Non competitive inhibitors covalently modify some amino acid residues which are required for catalysis and render the enzyme inactive.*

*For example*, iodoacetate, diisopropyl fluorophosphate. These are irreversible. Reversible non competitive inhibition is also known e.g. trypsin inhibitors from soya bean and Ascaris.

### 3. Suicide Inhibition

*Features:*

- The inhibitor is a structural analog of the substrate.
- This inhibitor is converted to active form by the same enzyme which it later inhibits.
- This modified, active form binds to the enzyme and inhibits further reaction.
- It is irreversible type of inhibition and is also known as *mechanism based inactivation*.

*Examples:*

(i) *Allopurinol inhibits xanthine oxidase enzyme:*

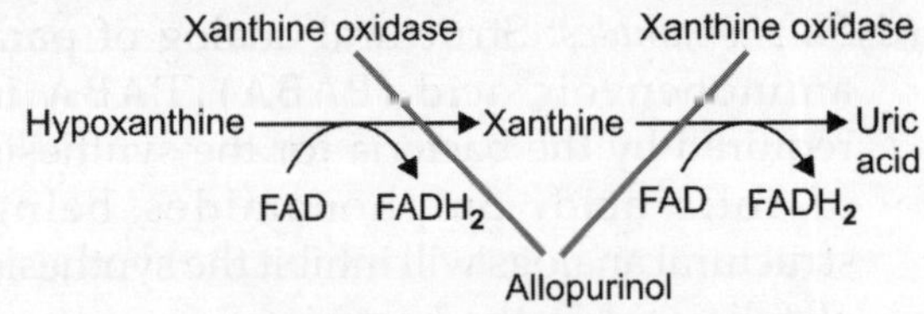


Uric acid is the end product of purine metabolism.

Increased production of uric acid produces gout.

*Importance:* Allopurinol is used as a drug for treatment of gout.

(ii) Di fluoro methyl ornithine (DFMO): Inhibits ornithine decarboxylase enzyme in trypanosomiasis (African sleeping sickness).

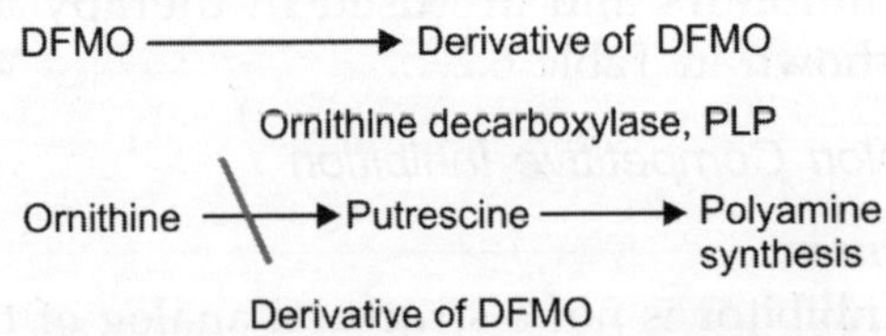


DFMO does not inhibit any enzyme. In the body, DFMO is converted into its derivative which binds to the coenzyme pyridoxal phosphate as well as the ornithine decarboxylase enzyme and inactivates them.

(iii) *5-fluorouracil*: It is a pyrimidine analog which is converted into its nucleotide derivative, fluorodeoxyuridylic acid (FdUMP) in the body which inhibits the enzyme thymidylate synthetase of nucleotide synthesis.

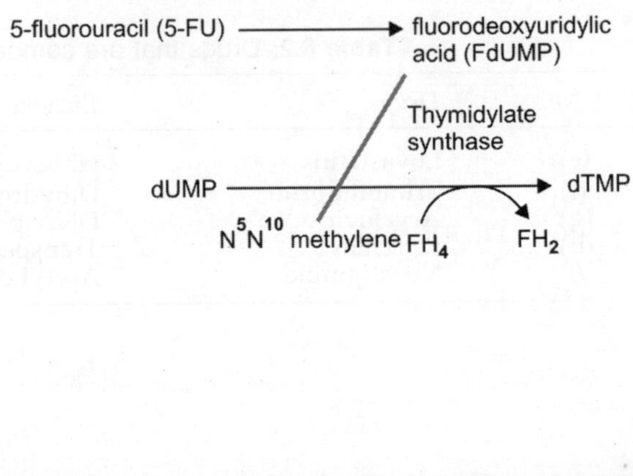

### 4. Allosteric inhibition:

*Features*:

- Allosteric enzymes have two sites which are different from each other.
  - *Catalytic site* which binds to the substrate and forms ES complex.
  - *Allosteric site* which is different from the catalytic site. Molecules which bind to the allosteric site are called as modulators/modifiers.
- When the modulator binds to the allosteric site, the enzyme undergoes three dimensional (3D) changes.
- These 3D changes results in alteration of the activity of the enzyme. It may either— increase the activity of the enzyme *positive modulator / allosteric activator*.

  - or it may decrease the activity of the enzyme, *negative modulator/allosteric inhibitor*.
    - The substrate cannot bind to the catalytic site as long as the negative modulator occupies the allosteric site on the enzyme.
    - In absence of the negative modulator (when the modulator concentration decreases in the cell) the 3D changes are reversed and enzyme goes back to its active conformation and binds to the substrate for catalysis. Hence it is called as *reversible type of inhibition*.

The examples of allosteric inhibition are given in Table 6.3.

When the negative modulator for the enzyme is the end product of the pathway then the inhibition is called as *feedback inhibition*.

*Feedback inhibition:*

*Features*:

- A metabolic pathway is made up of a number of steps each catalysed by a different enzyme (Fig. 6.16).

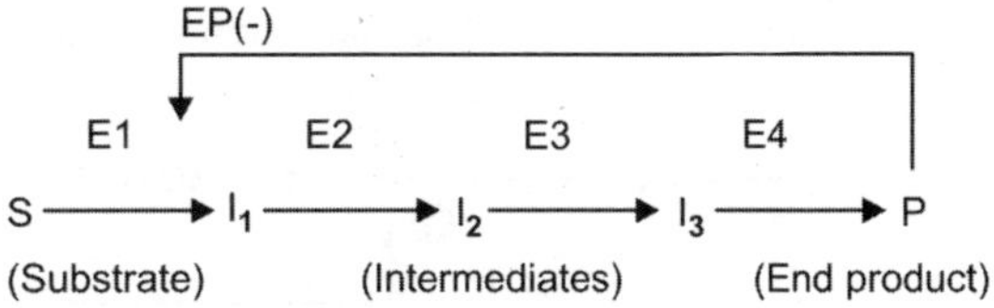


**Fig. 6.16:** Feedback inhibition in a metabolic pathway. E1, E2, E3, E4 are enzymes of the pathway

- When the concentration of the end product [P] increases, it inhibits one of the initial enzymes of the pathway. This is called as *feedback inhibition* or *end product inhibition*.
- When the concentration of [P] decreases, it dissociates from the enzyme and the enzyme goes back to its active conformation. Hence it is *reversible* (Fig. 6.17).

*Examples:* See in the box on page no. 188.

### Uses of Inhibitors

- Competitive inhibitors are used as anticancer, antibacterial, antiviral drugs.
- Used for elucidating mechanism of enzyme action.

**Table 6.3:** Allosteric enzymes and their modulators

| *Sl. No.* | *Allosteric enzyme/pathway* | *Modulators* Positive | *Modulators* Negative |
|---|---|---|---|
| (a) | Aspartate transcarbamoylase (ATC)/pyrimidine biosynthesis | ATP | CTP |
| (b) | Phosphofructokinase (PFK)/glycolysis | AMP | ATP |
| (c) | Acetyl CoA carboxylase/De novo synthesis of fatty acids | Citrate | Acyl CoA |
| (d) | Pyruvate carboxylase/gluconeogenesis | Acetyl CoA | ADP |

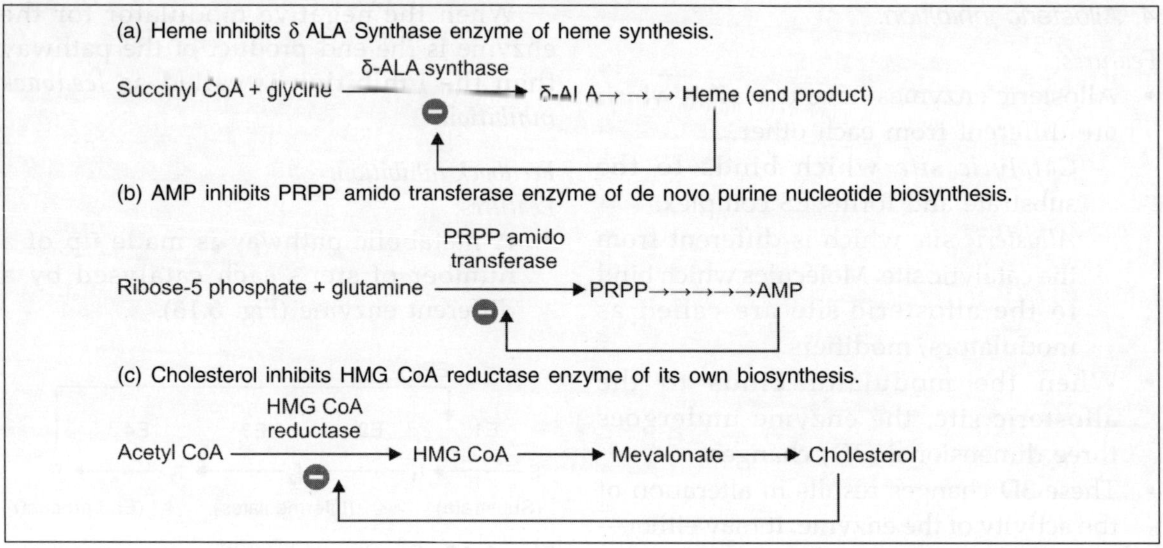


- Used in study of the active site of the enzyme.
- Used in establishing the relationship between function of a particular tissue and associated metabolic activity.

### Regulation of Enzyme Activity

In a living cell, there must be some regulation or control of the several reactions that occur. Regulation usually occurs at the first or one of the initial steps (committed step) in a pathway. It may be negative, positive or stimulatory.

- Regulation of enzyme activity is done as follows:
  (i) by regulating the catalytic efficiency of the enzyme
  (ii) or by regulating the concentration of the enzyme protein.

(i) *Regulation of catalytic efficiency of the enzyme*: can be done by the following:

➢ *Availability of the substrate, coenzyme and cofactor*. Decrease in any one of these factors leads to a decrease in catalytic efficiency of the enzyme.

➢ *Compartmentalisation of enzymes*: In metabolism of glucose, glycolysis is followed by TCA cycle. But glycolysis takes place in the cytosol and TCA cycle in the matrix of the mitochondria. Other examples for compartmentalisation are given in Table 6.4.

*Importance*: Compartmentalisation prevents overcrowding of metabolites in the subcellular compartments of the cell.

➢ *Allosteric / feedback regulation:*

In this type of regulation the metabolites or the end product themselves will regulate one or more enzymes of the metabolic pathway. These enzymes are said to be the rate limiting enzymes of the metabolic pathway. The details have been discussed under allosteric/feedback inhibition.

➢ *Regulation of enzyme activity by covalent modification:*

Enzymes can be regulated by covalent modification in two ways:

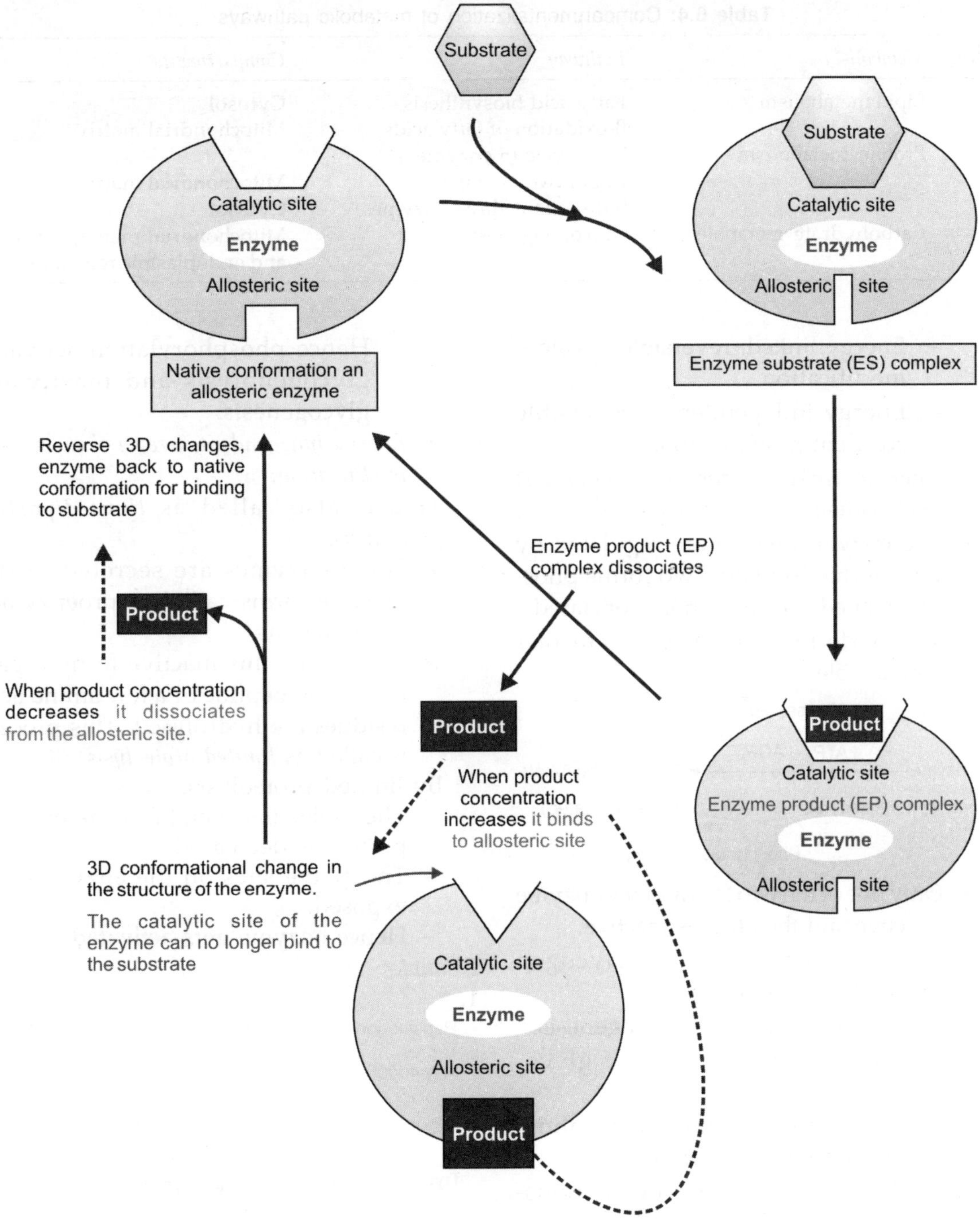


**Fig. 6.17:** Mechanism of feedback / allosteric inhibition

**Table 6.4:** Compartmentalization of metabolic pathways

| *Sl. No.* | *Metabolism* | *Pathway* | *Compartments* |
|---|---|---|---|
| (a) | Lipid metabolism | Fatty acid biosynthesis | Cytosol |
| | | β-oxidation of fatty acids | Mitochondrial matrix |
| (b) | Protein metabolism | Urea cycle (5 enzymes) | |
| | | Initial two enzymes | Mitochondrial matrix |
| | | Subsequent three enzymes | Cytosol |
| (c) | Carbohydrate metabolism | Gluconeogenesis | Mitochondrial matrix, cytosol and endoplasmic reticulum |

  - Energy linked, reversible covalent modification
  - Energy independent, irreversible covalent modification.
- *Energy linked, reversible covalent modification:*

➢ The enzymes which are regulated by this mechanism have two forms phosphorylated and nonphosphorylated.

➢ The two forms of the enzymes are inter convertible.

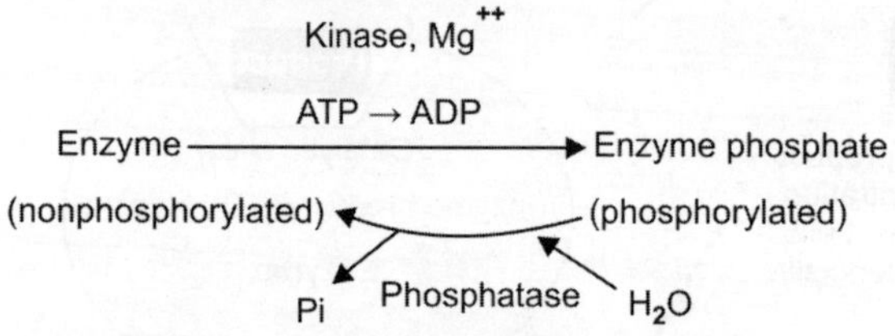


➢ Only one of the two forms of the enzyme is active and the other is inactive.

*Example:* Enzymes of glycogen metabolism (Fig. 6.18).

  - *glycogen synthase* of glycogenesis.
  - *glycogen phosphorylase* of glycogenolysis.
  - Both the enzymes are phosphorylated at the same time
  - Phosphorylated-glycogen phosphorylase is active and glycogen synthase is inactive
  - Non phosphorylated - glycogen synthase is active and glycogen phosphorylase is inactive
  - Hence phosphorylation activates glycogenolysis and inactivates glycogenesis.

- *Energy independent, irreversible covalent modification:*

This is also called as *limited/partial proteolysis.*

➢ Some enzymes are secreted in the inactive forms called as proenzymes or zymogens.

➢ To activate this inactive form, a part of the polypeptide or a few amino acid residues are hydrolysed. This process is called as *limited proteolysis*.

By limited proteolysis.

  - The molecular weight of the enzyme protein is decreased.
  - The active sites on the enzymes are exposed.
  - Hence enzyme gets activated.

*Examples:*

1.

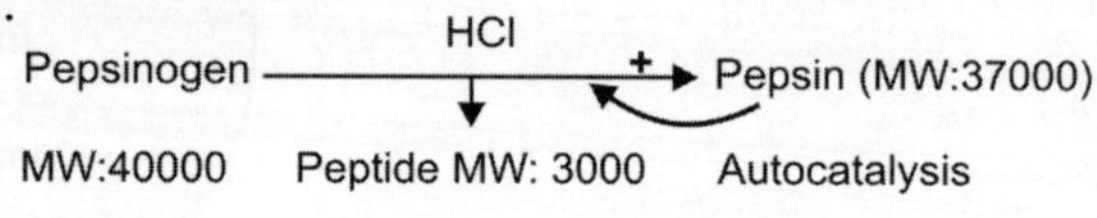


2.

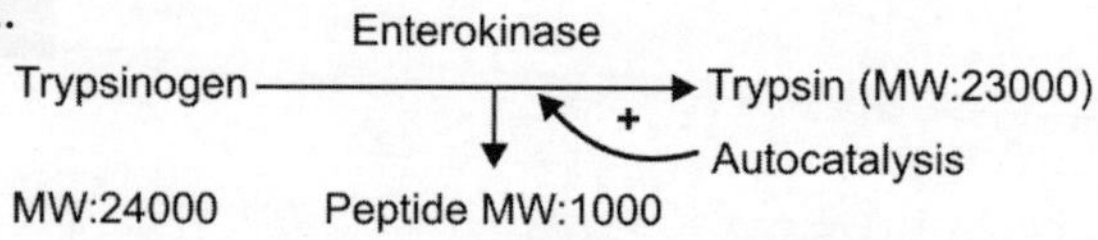


3. Blood clotting factors are activated by limited proteolysis in the *cascade phenomenon*.

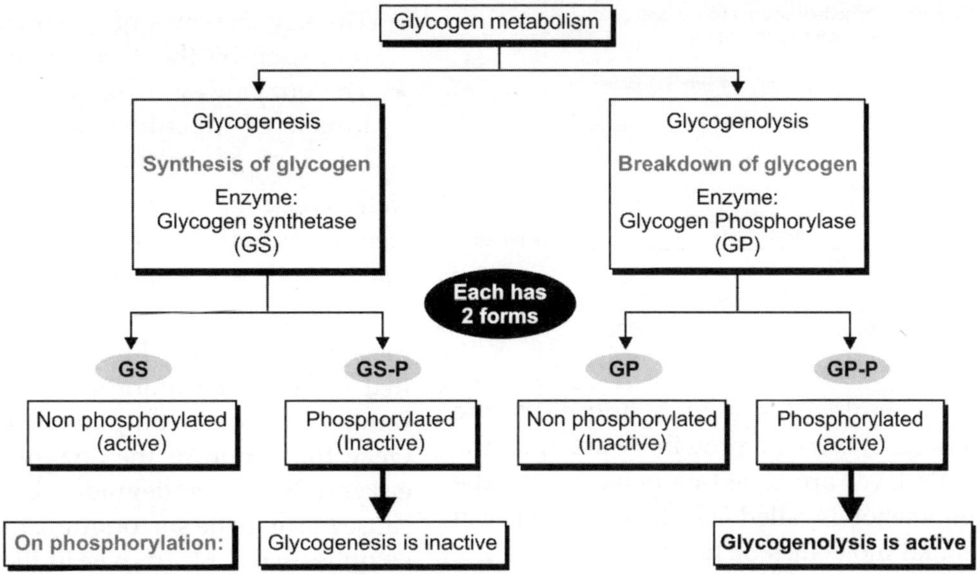


**Fig. 6.18:** Regulation of glycogen metabolism by covalent modification

Significance of secretion of the enzyme in proenzyme form:

- Prevents auto digestion of tissues.
- Prevents intravascular coagulation of blood.

*Autocatalysis:* Proenzymes are initially activated to active forms by either an acid or enzyme. Once small amounts of the active form of the enzyme is formed it catalyses its own formation. This phenomenon is called as *autocatalysis.*

(ii) *Regulation of enzyme activity by altering the concentration of the enzyme protein:*

When the concentration of the enzyme is more the number of active sites on the enzyme is more and hence the activity will be higher.

- *Synthesis of enzymes*: The concentration of the enzyme can be regulated by the following:
  - ➢ Increased enzyme production—*enzyme induction*
  - ➢ Decreased enzyme production—*enzyme repression*
  - ➢ Enzyme *degradation*
  - ➢ Enzymes could be of *two types based on their synthesis*:
- *Inducible enzymes*: Normal concentrations of these enzymes are very low in the cell but can be increased by certain molecules called as inducers.
- *Constitutive enzymes*: A fixed quantity of these enzymes is always produced in the cell. Concentration is not influenced by other molecules.

The concentration of the enzyme can be altered at the genetic level under the influence of inducers (increase) and repressors (decrease) (Fig. 6.19).

*Inducible enzymes and enzyme induction:* Depending on the need of the cell the concentration of some of the enzymes can be either increased or decreased by altering the rate of synthesis of the enzymes. Such enzymes are called as *inducible enzymes*.

*Mechanism:* Altered rate of transcription of the gene for the enzyme.

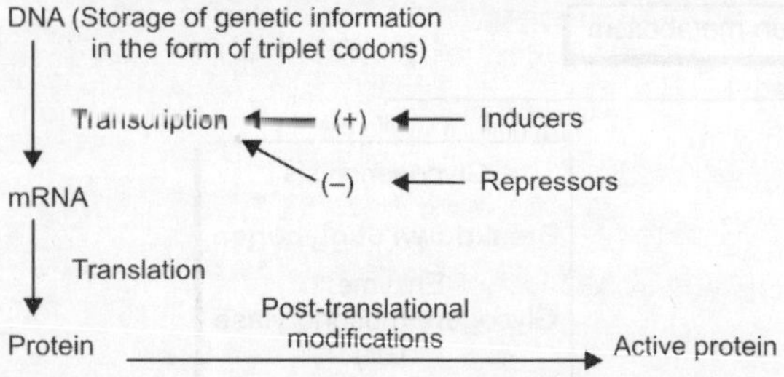


**Fig. 6.19:** Synthesis of proteins and site of action of inducers and repressors

The molecules which increase the synthesis of enzyme protein by acting at the genetic level are called as *inducers* and the phenomenon is called *induction*. Inducers can be small molecules or hormones.

*Example:*

➢ *Escherichia coli* bacteria need glucose and galactose for its growth. If these are present in the medium then the bacteria does not need lactase enzyme. However, if lactose is present in the medium lactase enzyme is needed to convert it into glucose and galactose. So lactose (inducer) induces the synthesis of the enzyme lactase (inducible enzyme).

➢ Insulin induces the glycogen synthase enzyme.

*Repressors and enzyme repression:* Small molecules which decrease the synthesis of enzyme protein by acting at the genetic level are called as *repressors* or *negative inducers* and the phenomenon is called *repression. Repressors* can be small molecules or hormones.

*Example*:

➢ *Salmonella typhimurium* can synthesise amino acid histidine. If histidine is present in the surrounding medium, the enzymes required for the synthesis of histidine are not produced in the pathogen. Histidine acts as the repressor molecule and the enzymes of histidine synthesis are said to be repressed.

➢ The key enzymes of gluconeogenesis are repressed by the hormone insulin.

➢ The enzymes glucokinase and pyruvate kinase of glycolysis are repressed by hormone glucagon.

*Note: When the concentration of the inducer/ repressor becomes less, it dissociates from the gene and the synthesis of the protein decreases / increases respectively.*

*Enzyme degradation:* Enzymes are produced in the body for metabolic reactions and removal of toxic compounds in the body. Once the action of the enzyme is over the enzyme has to be degraded before it could cause harm to the surrounding tissues. Hence enzymes have a turn over in the tissues and are degraded by specific mechanisms. The enzymes that regulate the metabolic pathways have a very high turnover. Defective enzymes that are synthesised are not useful to the body and are degraded before they cause any harm. Degradation of enzymes regulates the enzyme concentration in the cells.

*Isoenzymes or isozymes:* An enzyme catalysing a specific reaction will have small differences based on the tissue in which it is synthesised. Hence it is possible to have different molecular forms of the same enzyme. These multiple molecular forms of an enzyme are called as *isoenzymes or isozymes.* A study of the isoenzyme patterns can be useful in elucidating the normal metabolism and the altered metabolism in diseases of organs.

*Formation of isoenzymes:*

- Site of production of the same enzyme may be different with variable amounts of post-translational modifications. *For example,* alkaline phosphatase has five isoenzymes produced in different organs (liver, intestine, placenta, bone and kidney) with varying amounts of sialic acid content—*isoforms*.

- Isoenzymes may be products of different genes—*true isoenzymes. For example,* enzyme amylase hydrolyses starch. The two isoenzymes of amylase are salivary amylase and pancreatic amylase. These are products of different genes.
- Isoenzymes which are oligomeric proteins may be formed by different combinations of the constituent subunits—*Hybrid isoenzymes. For example,* Lactate dehydrogenase is a tetrameric protein with two types of subunits (H and M) and five isoenzymes ($H_4$, $H_3M$, $H_2M_2$, $HM_3$, $M_4$) are possible.
- Isoenzymes may be products of different alleles of the same gene—*Allozymes.* This is seen over a range of population. *For example,* allozymes of glucose-6-phosphate dehydrogenase enzyme are known.

### *Characteristic Features of Isoenzymes*

- They are physically distinct forms of an enzyme catalysing the same reaction.
- They differ from each other structurally, electrophoretically and immunologically.
- Isoenzymes of an enzyme may originate from different tissues, different cell types or from different sub cellular compartments.
- They differ from each other in their $K_m$, $V_{max}$ and optimum pH.
- Isoenzymes originating from different tissues have different affinities for their substrates.
- Usually they are oligomeric proteins with dissimilar monomers. Hence capable of existing in several forms. *Reason:* One tissue produces one monomer predominantly and the other monomer maybe produced by another tissue. These monomers combine in different proportions based on the rate of production in that tissue to construct an active enzyme.

*Example:*

*Lactate dehydrogenase (LDH):*

- *Reaction* catalyzed by the enzyme:

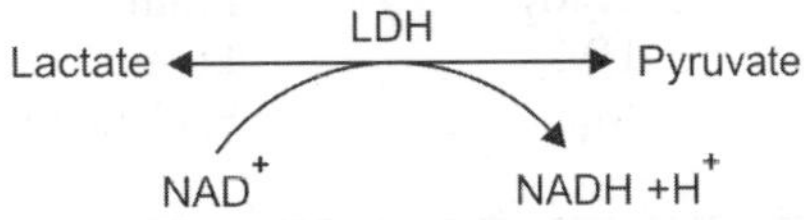


- *Structure of LDH*: Oligomeric protein with four monomers (tetrameric). The monomers are of two different types—H type from heart and M type from the muscle. They are products of different genes. Five isoenzymes can be separated by electrophoresis. Different combinations of the same or two different monomers give rise to these five isoenzymes (Table 6.5).
- The isoenzymes are $LDH_1$, $LDH_2$, $LDH_3$, $LDH_4$, and $LDH_5$.
- *Importance of LDH isoenzymes*: LDH isoenzymes are of *diagnostic importance* in *heart and liver disorders*. In normal healthy individuals $LDH_2$ is about 35% of the total serum LDH levels. In heart disorders like myocardial infarction, $LDH_1$ increases more than $LDH_2$. This is called as *flipped pattern*. Flipped pattern is seen 12-24 hours after the myocardial infarction. $LDH_5$ increases in liver diseases.
  LDH activity is very high in the RBCs.

*Creatine phosphokinase (CPK):*

- *Reaction* catalyzed:

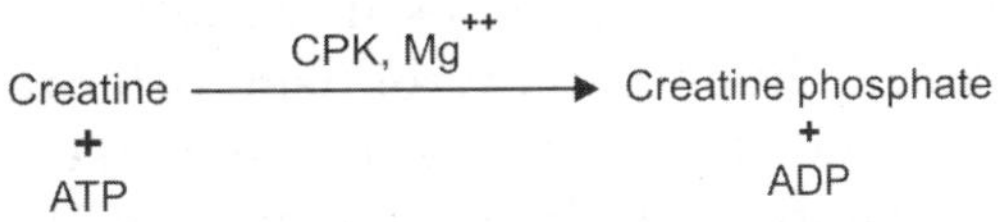


- *Structure of CPK*:

CPK is a dimeric protein—two monomers. The monomers are of two different types – B (brain) type and M (muscle) type:
Three isoenzymes are possible (Table 6.6).

**Table 6.5:** Comparison of the LDH isoenzymes

| *Isoenzyme* | *Structure* | *Tissue origin* | *Features* |
|---|---|---|---|
| $LDH_1$ | $H_4$ | Heart | Low Km for its substrate lactate, catalyses lactate → pyruvate conversion |
| $LDH_2$ | $H_3M$ | RBC | Normally high in the plasma (35%) |
| $LDH_3$ | $H_2M_2$ | Brain | |
| $LDH_4$ | $HM_3$ | Liver | |
| $LDH_5$ | $M_4$ | Skeletal muscle | Low Km for its substrate pyruvate, catalyses pyruvate to lactate conversion |

**Table 6.6:** Comparison of the isoenzymes of CPK

| *Isoenzyme* | *Structure* | *Tissue origin* | *Importance* |
|---|---|---|---|
| $CPK_1$ | BB | Brain | |
| $CPK_2$ | MB | Heart | Normal level is about 5% of the total plasma CPK. It increases to more than 20% in myocardial infarction |
| $CPK_3$ | MM | Muscle | |

### Alkaline Phosphatase (ALP)

*Action*: Alkaline phosphatase catalyzes the hydrolysis of phosphate from phosphate derivatives of aliphatic and aromatic compounds at an optimum pH of 9 to 10.
*Structure*: Monomeric protein.
*Isoenzymes*: Five isoenzymes are known based on the tissue of origin—liver, bone, intestine, placenta and kidney. They can be separated by electrophoresis since they have different electrophoretic mobilities due to a difference in the sialic acid content.
*Diagnostic importance*: Tissue specific isoenzymes increase in its disease. *Example:* Bone isozyme increases in rickets, osteomalacia, hyperparathyroidism, osteogenic sarcoma. Liver isozyme increases in cholestatic jaundice. Placental isozyme increases in pregnancy.

### Clinical Significance of Enzymes

Plasma contains certain enzymes in low levels. Enzymes in plasma could be of two different types based on their role (Fig. 6.20).

***Causes for abnormal increase in non functional enzymes in plasma:***

- Necrosis of various blood and tissue cells due to injury or disease. *For example,* ↑ CPK in myocardial infarction, muscular dystrophy.
- Blockage in the secretory pathway. *For example,* ↑ ALP in cholestatic jaundice.
- Cancers of tissues/organs leads to the ↑ in tissue/organ specific enzymes in plasma. (*Reason*: due to increased number of cells which synthesise the enzyme). *For example,* ↑ acid phosphatase in prostate cancers.

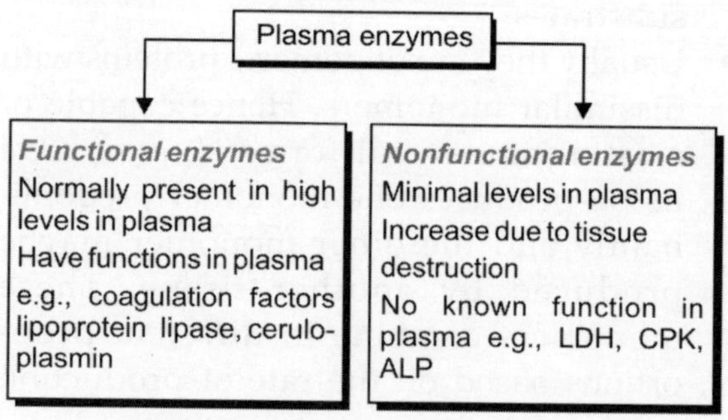


**Fig. 6.20:** Types of enzymes in the plasma

- Increased production of enzyme within the cell. *E.g.* ↑ Serum transaminase in patients treated with protein anabolic drugs.

Under all these conditions a quantitative estimation of the plasma enzyme levels gives an index of the disease and tissue destruction. The relation being the increase in plasma enzyme levels is directly proportional to the amount of tissue destroyed or damaged.

The enzyme activity is measured in *international units/litre* of plasma (IU/L).

An international unit is defined as activity of the enzyme which transforms one micromole of the substrate per minute/L under optimal conditions and defined temperature.

***Significance of serum enzyme assay in clinical practice:*** Single or repeated enzyme assays of selected enzyme or enzymes may

- Help in early detection of a disease/ damage to tissue
- Give information on the nature and extent of the disease process.
- Help in differential diagnosis of disease.
- Help in ascertaining prognosis and response to treatment.

Some of the enzymes of diagnostic and prognostic values are discussed in Table 6.7.

***Other uses of enzymes:***

- *Enzymes that function as therapeutic agents:*
  - Streptokinase/urokinase are used to dissolve intravascular clots in therapy of myocardial infarction.

**Table 6.7:** Enzymes of clinical importance with normal plasma levels

| *Diseases* | *Enzyme profiles* | *Normal levels* |
|---|---|---|
| Heart disorders | CPK<br>SGOT/AST<br>LDH | 20- 80 U / L<br>2- 20 U / L<br>60- 200 U / L |
| Liver diseases | SGPT/ALT<br>ALP<br>5´ Nucleotidase<br>γ-glutamyl transferase (GGT) | 3- 15 U/L<br>25- 90 U/L<br>2-10 U/L<br>10- 30 U/L |
| Bone disorders | ALP | 25- 90 U/L |
| Muscle diseases | CPK<br>Aldolase | <br>1.5- 7 U/L |
| GI tract | Amylase<br>Lipase | 50- 120 U/L<br>0.2- 1.5 U/L |
| *Malignancies*:<br>Prostate<br><br><br>Bone metastasis<br>Carcinoma of head of pancreas<br>Leukemia | <br>Acid phosphatase<br>ALP<br>ALP<br>$LDH_1$<br>Aldolase | <br>2.5- 12 U/L<br><br>25- 90 U/L<br>20- 90 U/L<br><br>1.5- 7 U/L |

- Trypsin and pepsin in therapy for poor digestion.
- Asparaginase is an anticarcinogenic drug for acute lymphocytic leukaemia.

- *Enzymes used in diagnostic kits in the clinical laboratory.*
  - Glucose oxidase is used for measurement of blood glucose levels by the GOD/POD method.
  - Immobilised glucose oxidase is also used in the strip test for detection of glucose in urine.
  - Urease is used for measurement of blood urea levels.
  - Peroxidase tagged antibody is used for colour development in ELISA tests.
  - Restriction endonucleases are used in recombinant DNA technology.

# 7 Vitamins

Vitamins are essential food factors which are organic compounds, occurring in small quantities in different natural foods. Vitamins are also required for proper utilisation of other food stuffs like carbohydrates, proteins and lipids. They are necessary for growth, differentiation, maintaining cellular functions, specific biological actions and thereby maintenance of good health. The prokaryotes do still have the ability to synthesise vitamins which have been lost during evolution. Hence, most of the vitamins cannot be synthesised by the human body in sufficient amounts.

## BIOMEDICAL IMPORTANCE

- Higher organisms have lost the ability to synthesise vitamins, hence, diet is the main source.
- Intestinal bacteria synthesise some vitamins like vitamin K, biotin and folic acid. Hence, vitamin supplements are advised along with antibacterials (antibiotics).
- Deficiency symptoms of vitamins like scurvy (vitamin C), beri beri (thiamine), rickets (vitamin D) and night blindness (vitamin A) led to the discovery of vitamins.
- Intake of the appropriate vitamin corrects the deficiency. Contrary to common belief, higher doses of vitamins do not boost health.

## CLASSIFICATION OF VITAMINS

Classification of vitamins is based on their solubility in fat or water (Fig. 7.1)

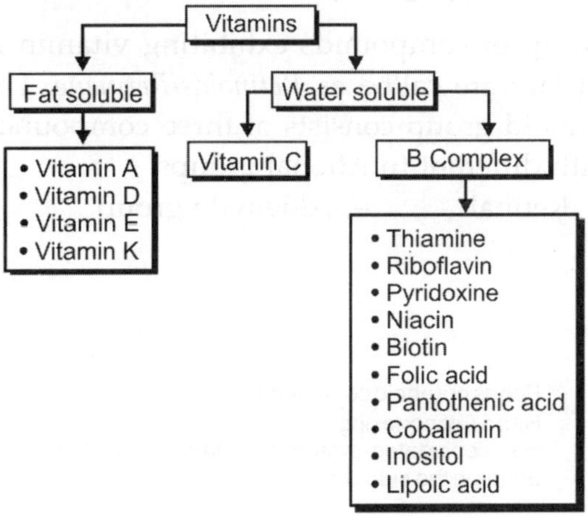


**Fig. 7.1:** Classification of vitamins

### Fat Soluble Vitamins

*General features:*

- Solubility – soluble in fat, insoluble in water
- Absorption from GI tract – is closely linked with dietary lipid/fat absorption, requirement of bile salts for formation of micelles.
- Transport in blood – is by specific proteins found in the chylomicrons
- Storage – liver/adipose tissue

- Excess intake – shows toxicity symptoms
- Deficiency – results when stored vitamins are depleted completely or fat digestion and absorption is impaired.
- Treatment – single large dose of the vitamin usually cures the deficiency.

**Vitamin A (Fig. 7.2)**

Group of compounds exhibiting vitamin A activity are called as *Retinoids/Retinene*. The retinoid group consists of three compounds with different functional groups:

a) Retinal – Aldehyde group
b) Retinol – Alcohol group
c) Retinoic acid – Carboxylic group

*Dietary Sources*

- *Animal sources:* Dairy products, egg yolk, liver, fish liver oils—cod and shark liver
- *Vegetable sources*: β-carotene containing yellow vegetables like carrot, papaya, pumpkin and green leafy vegetables. 12 microgram β-carotene ≡ 1 microgram vitamin A.

*Chemistry of Vitamin A*

Vitamin A is a *fat soluble vitamin*. It exists in active form in animal tissues and in inactive pro-vitamin form in the plant tissues. The pro-vitamin form is called as β-carotene (Fig. 7.2 D). Each β-carotene can give rise to two molecules of vitamin A.

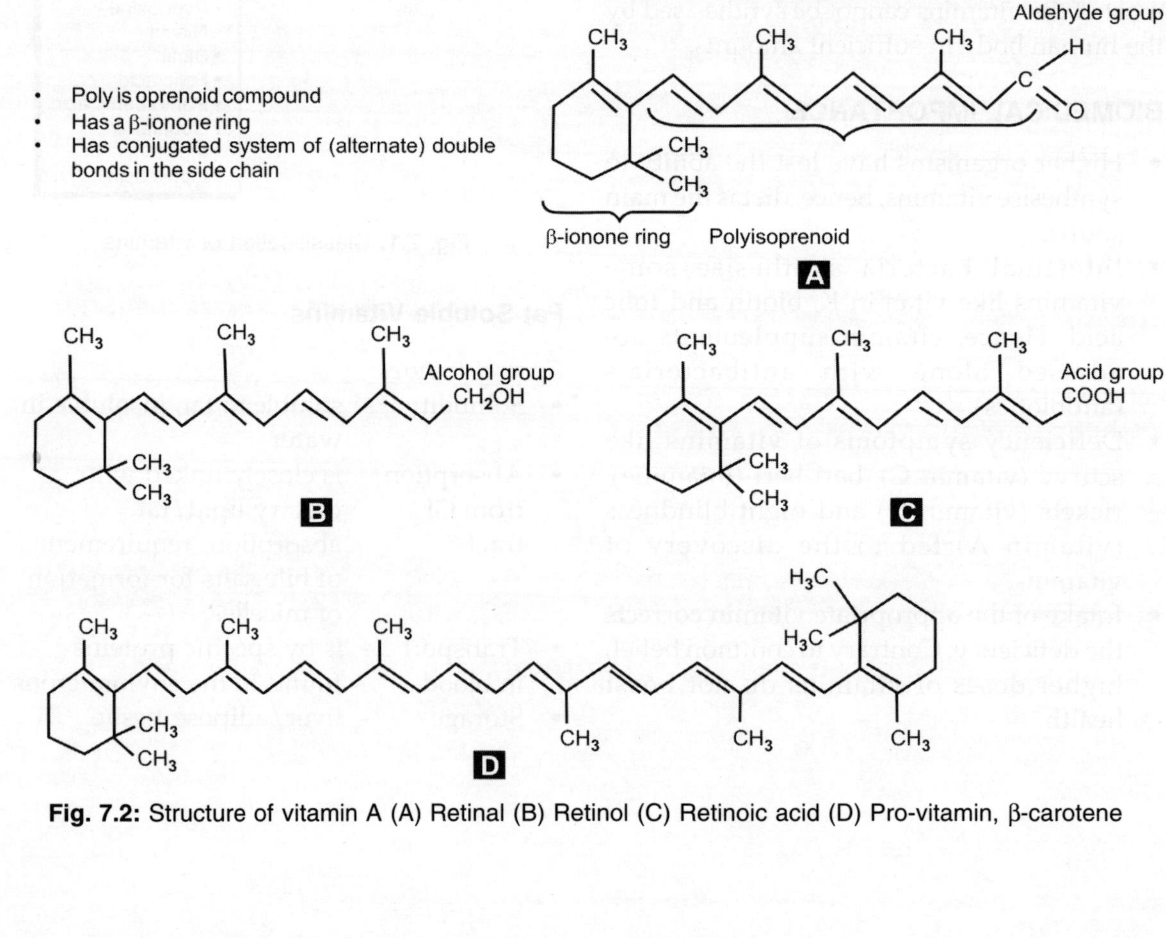


**Fig. 7.2:** Structure of vitamin A (A) Retinal (B) Retinol (C) Retinoic acid (D) Pro-vitamin, β-carotene

*Interrelationship between β-carotene and the other retinoids:*

The retinol and retinal forms are easily interconvertible. But once retinoic acid is formed it cannot be converted back to either form as shown in Figure 7.3.

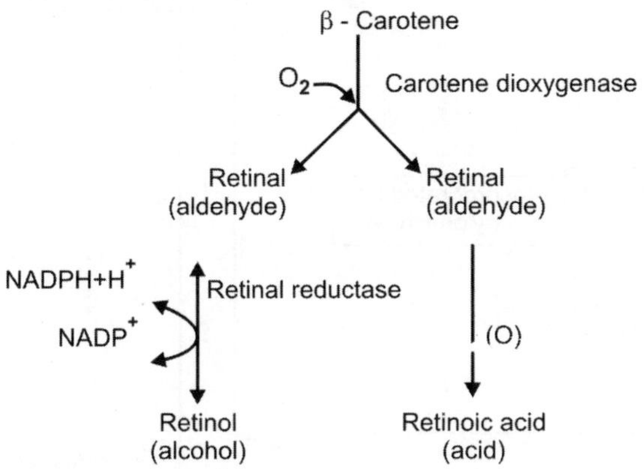


**Fig. 7.3:** Formation of vitamin A from β-carotene and conversion to other retinoids

*Isomerism in Vitamin A*

Retinal has two forms.

*All trans form*: All the double bonds in the side chain are in trans configuration (Fig. 7.4A). This is an intermediate in the visual cycle.

*11-cis form*: The double bond in the 11th position of the side chain is in cis configuration. So there is a bend in the side chain (Fig. 7.4B). This form of retinoid is the prosthetic group of *rhodopsin*, the visual pigment present in the photosensitive rod cells of the retina.

*Significance*: Both these isomers play an important role in vision and the visual cycle.

*Absorption of vitamin A in the GI tract:* (Fig. 7.5).

*Major site of absorption*: Small intestine

*Dietary vitamin A*: β-carotene and retinol esters.

*Requirement*: Dietary fat (since it is fat soluble) and bile salt. Decreased absorption is seen in all conditions where digestion and absorption of dietary fat is impaired. *Conditions*: Obstructive jaundice, steatorrhoea, sprue.

*In the intestinal lumen*:

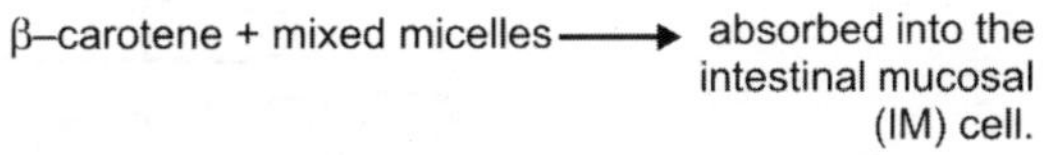


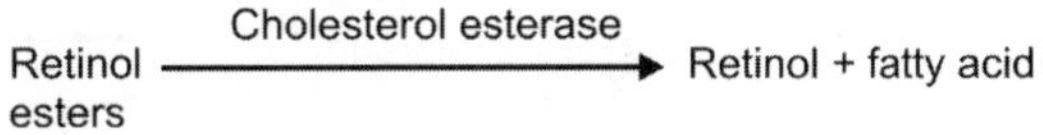


*Absorption*: Into the intestinal mucosal cells along with the dietary fat.

*Changes in the IM cell*:

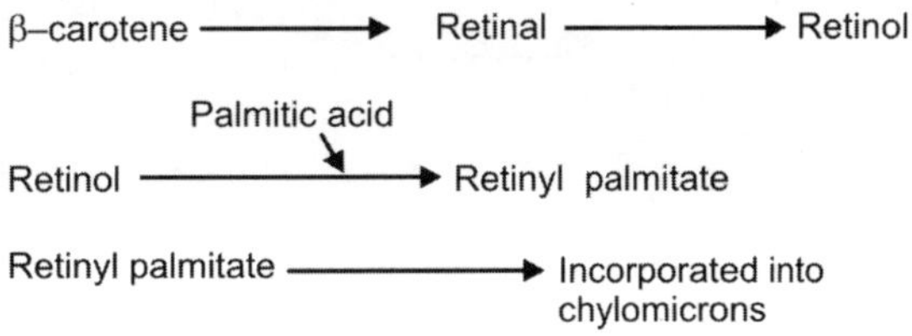


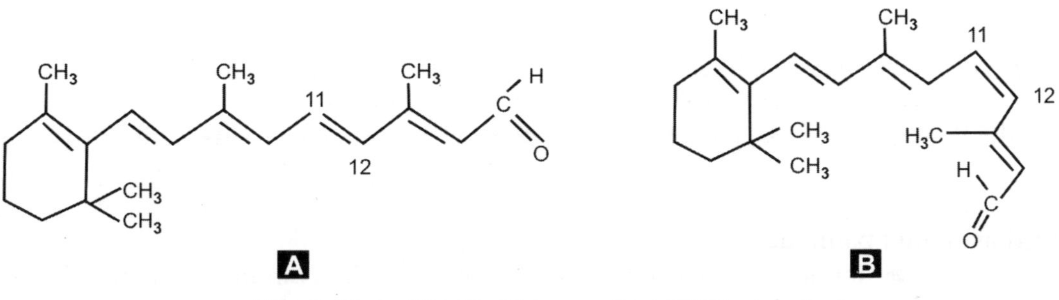


**Figs 7.4A and B:** (A) All *trans* retinal (B) 11-*cis* retinal

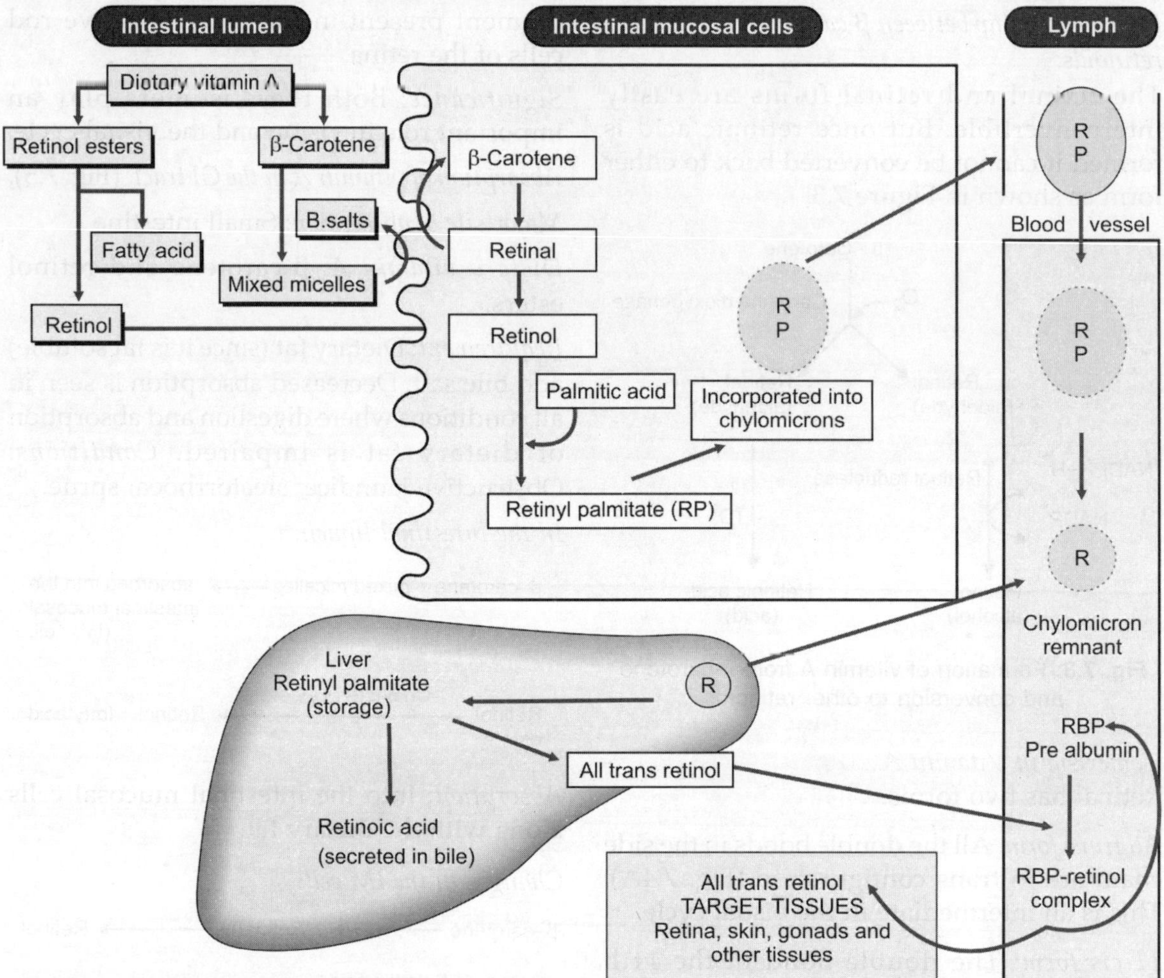


**Fig. 7.5:** Absorption and transport of vitamin A

*Transport of vitamin A:* Chylomicrons transport retinyl palmitate via lymph into the blood to the liver and other target organs.

*In liver*:

(a) Stored as retinyl palmitate

(b) Retinol ⟶ Retinoic acid (secreted in the bile)

*Transport of stored vitamin A to the target organs*:

*In liver the stored* :

Retinyl palmitate ⟶ Retinol + Palmitic acid

↓ Enters blood

*In blood*:

[Retinol + Retinol binding protein (RBP) ⟶ Retinol—RBP complex]

Retinol—RBP complex ⟶ Transports the retinol to the target organs which have receptors for retinol—retina, skin, gonads and other tissues.

*Excretion of vitamin A*:

(a) In bile: as retinoic acid

(b) In urine: as retinol epoxides and uronides.

### Biochemical Role of Vitamin A

- Retinal → vision
- Retinoic acid → growth and differentiation of tissues
- Retinol → reproductive system
- β-carotene → antioxidant property
- *Role in vision*: Retina the innermost layer of the eye has the photosensitive cells—rods (Fig. 7.6) and cones. The cells have visual pigments rhodopsin and photopsin. These visual pigments play a vital role in vision. Rhodopsin is needed for visibility in dim light and photopsin in bright light. Both these visual pigments have 11-*cis* retinal as their prosthetic group.

*Structure of the visual pigments*: They are *conjugated proteins*—containing carbohydrates (glycoproteins) and 11-*cis* retinal as prosthetic group. The apoprotein is called as opsin. They are embedded in the folds of the membranes (discs) of the rods and cones (Fig. 7.6).

Opsin + 11-*cis* retinal ⟶ Rhodopsin
*(Glycoprotein)* *(photosensitive prosthetic group)*

When *light falls on the retina,* the 11-*cis* retinal is isomerised to all trans retinal as shown in the following equation.

$$\text{Opsin} - 11\text{-}cis \text{ retinal} \xrightarrow{\text{Light}} \text{Opsin} - \text{all } trans \text{ retinal}$$

The opsin undergoes a series of conformational changes and gets detached from the all *trans* retinal—*photolysis* (Fig. 7.7). The

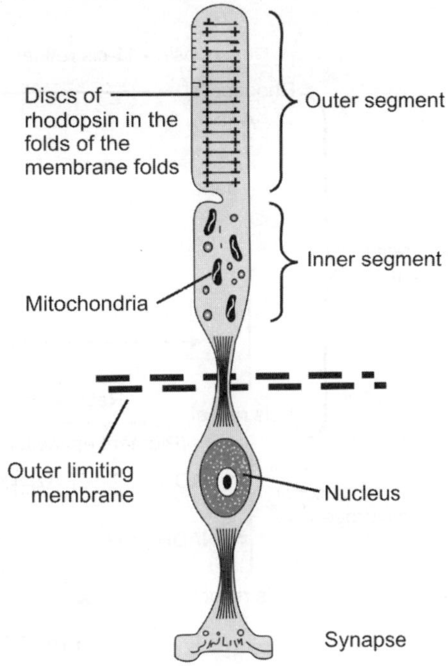


**Fig. 7.6:** Structure of rod cell

rhodopsin is later regenerated. These changes involving photolysis and regenera-tion of rhodopsin are called as *Wald's visual cycle* (Fig. 7.7) after the scientist who identified the role of rhodopsin in vision.

*Other biochemical functions of vitamin A:*

- *Role in the health of epithelial tissue*: Vitamin A in the form of retinoic acid is needed for the maintenance of healthy epithelial tissue. It stimulates the growth, differentiation and maintenance of normal living, secretory epithelial tissue.
  **Mechanism:** Retinyl phosphate acts as a carrier of phospho-hexoses and oligosaccharides for the synthesis of glycoprotein component of the membranes and mucus of the secretory epithelial cells.
- *Role in skeletal system:* Vitamin A as retinoic acid is also needed for the sulphation of the bone matrix mucopolysaccharides.

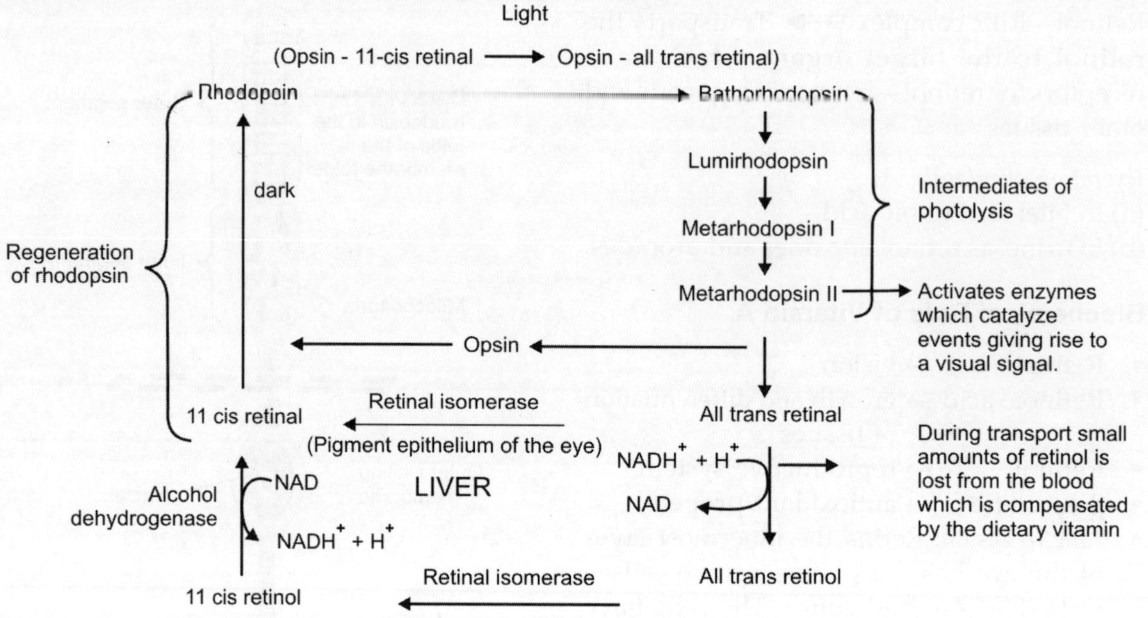


**Fig. 7.7:** Wald's visual cycle

- *Role in reproductive system:* Retinol acts like a hormone and influences the expression of certain genes in the gonads.
- *Role in growth promotion:* Retinol and retinoic acid forms act like steroid hormones and control the synthesis of specific growth promoting protein factors at the genetic level.
- *Anti-oxidant and anti-carcinogenic activity of vitamin A:* Retinol and retinoic acid are said to be anti-carcinogenic and β-carotene is anti-oxidant. The anti-carcinogenic property is because of the anti-oxidant nature of the molecules. The conjugated system of double bonds in the poly-isoprenoid side chains are known to trap and inactivate the free radicals generated in the tissues during metabolism before it causes any harm to the surrounding tissues. Hence, deficiency of vitamin A is known to be associated with epithelial cancers. The anti-oxidant property of β-carotene is useful in protection against myocardial infarction.
- *Role in iron metabolism:* Retinol/retinoic acid promote transcription of the gene for *transferrin* (iron transport protein in plasma). Hence, there is an increased mobilisation of iron from liver to blood and its transport to the bone marrow (the site of RBC synthesis).

### Mechanism of Action of Retinol/Retinoic Acid

Retinol/retinoic acid have hormone like action and affect the transcription of genes for specific proteins in the reproductive system and for promoting growth. The mechanism is like that of steroid hormones which have intracellular, cytosolic, mobile receptors. The receptor—vitamin A complex will bind to specific regions of the DNA and promote the transcription and translation of specific genes to proteins which bring about the metabolic effect (Fig. 7.8).

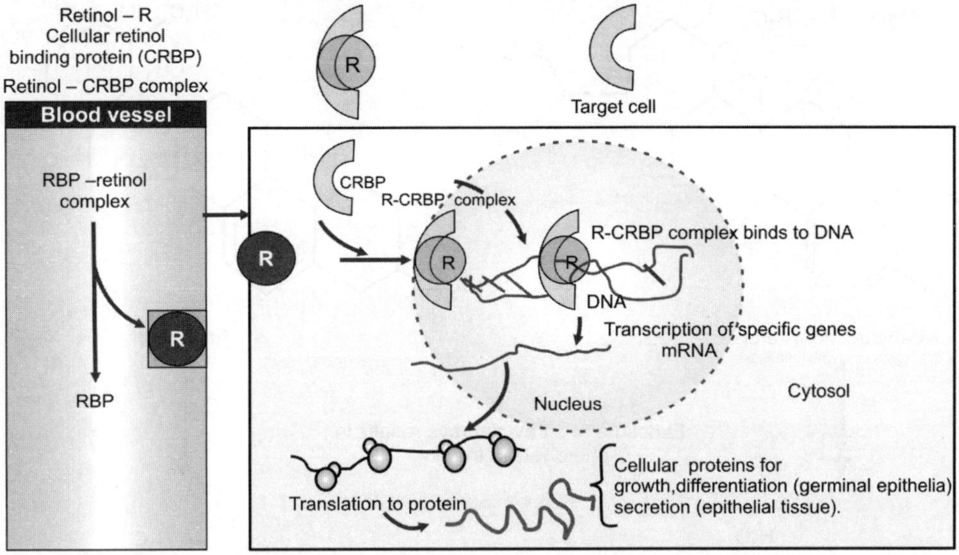


**Fig. 7.8:** Mechanism of hormone like action of retinol *(For colour version see Plate 5)*

### Daily Requirement

| | | |
|---|---|---|
| Adults | – | 750 μg/day |
| Pregnancy | – | 1000 μg/day |
| Lactation | – | 1200 μg/day |
| Infants and children | – | 400-600 μg/day |

### Deficiency Manifestations of Vitamin A

- *Ocular manifestations:*
  - ➢ Nyctalopia – night blindness.
  - ➢ Xerophthalmia – conjunctiva becomes dry, thick, wrinkled, keratinised and looses transparency.
  - ➢ Bitot's spots – grayish white triangular spots on the conjunctiva.
  - ➢ Keratinisation of lacrimal gland epithelia (dryness of the eye).
  - ➢ Keratomalacia – softening of the cornea.
- *Dermatological manifestations:*
  - ➢ Atrophy of the secretory epithelia.
  - ➢ Phrynoderma-follicular hyperkeratosis in the skin (thick, wrinkled, toad like skin as in essential fatty acid deficiency).
  - ➢ Keratinising hypermetaplasia – keratinization of the secretory epithelia of respiratory system, gastrointestinal system and genitourinary system.
- *Manifestations in other systems:*
  - ➢ Urinary calculi and prone to infection-reduced immunity.
  - ➢ Growth retardation due to retarded skeletal growth.
  - ➢ Reproductive failure.
  - ➢ Retarded growth of nervous system due to impaired growth of vertebrae and cranium.
  - ➢ Microcytic anaemia.

### Hypervitaminosis

*Symptoms*: Nausea, irritability, headache, abnormal calcifications, painful swelling over bones, dizziness, blurring of vision, loss of hair, dermatitis.

### Vitamin D

*Synonyms:* Cholecalciferol, antirachitic vitamin (cures rickets), sunshine vitamin.

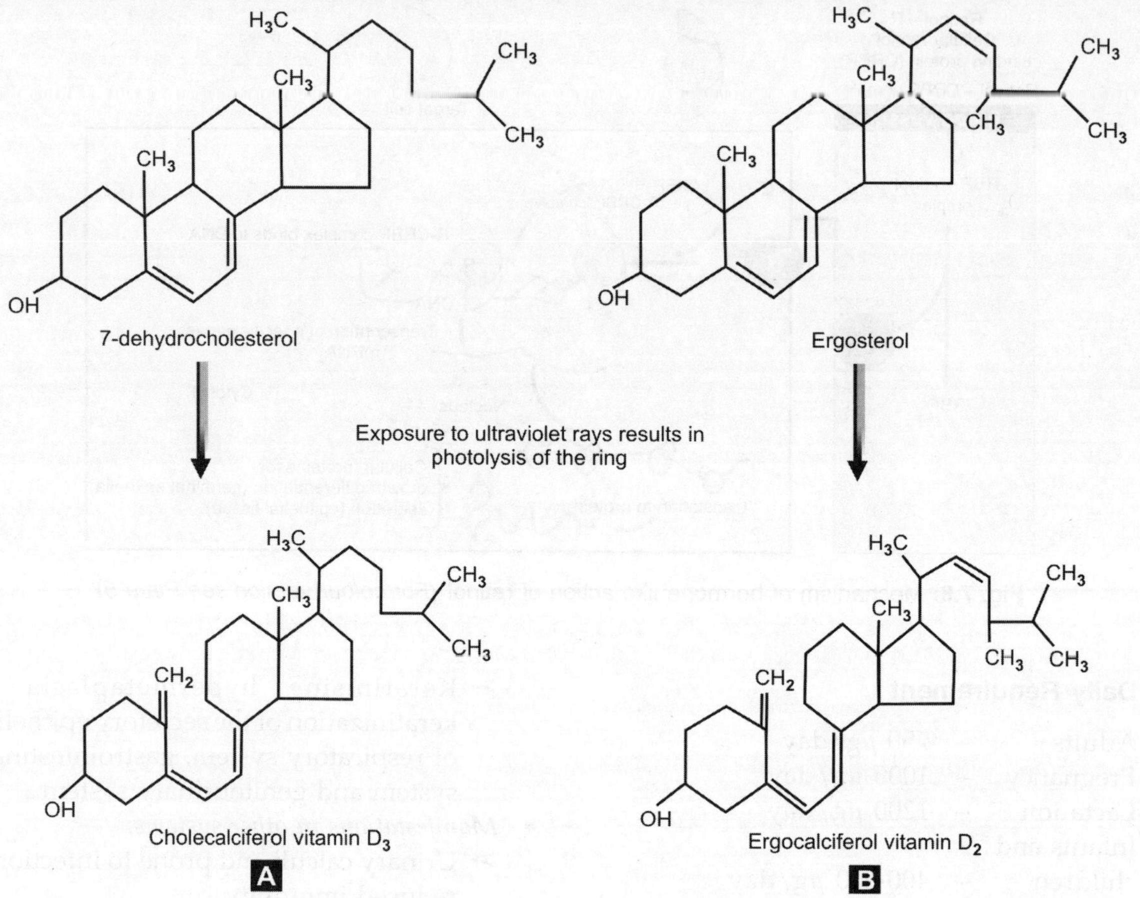


**Figs 7.9A and B:** Formation of (A) Vitamin $D_3$ from 7-dehydrocholesterol (B) Vitamin $D_2$ from ergosterol

## Sources

1. Synthesis by exposure to sunlight.
2. Animal sources—fish liver oils flesh of oily fishes—herring, sardine, salmon, egg, milk
3. Plant sources—ergots/yeast—are absorbed only after irradiation.

### *Chemistry of Vitamin D*

Vitamin D is a steroid derivative as shown in Figure 7.9.

### *Formation of Vitamin D*

It is derived from pro-vitamins on exposure to sunlight by photolysis of the sterol ring structure (Fig. 7.9). Hence, it is also called as *sunshine vitamin*. The pro-vitamins of vitamin D are *7-dehydrocholesterol* (an intermediate in biosynthesis of cholesterol) in animals and *ergosterol* in plants (yeast/ergots).

## Absorption, Metabolism and Transport

*Major site of absorption*: Small intestine—ileum and jejunum.

*Requirement*: Dietary fat (since it is fat soluble) and bile salt (for emulsification). Decreased absorption is seen in all conditions where digestion and absorption of dietary fat is impaired.

*Conditions*: Obstructive jaundice, steatorrhoea, sprue.

*Sources of cholecalciferol:*

- Diet
- Synthesis in the body (skin and intestinal mucosa) from 7-dehydrocholesterol on exposure to sunlight.

*Transport of vitamin $D_3$*: By chylomicrons to liver.

Activation of vitamin $D_3$ (Cholecalciferol):

Activation involves 2 hydroxylation reactions – one in liver and the other in the kidney (Fig. 7.10)

*End product*: 1, 25 dihydroxycholecalciferol – *calcitriol* (since it has three hydroxyl groups) is the active form of vitamin $D_3$.

*Regulation of activation:* A decrease in plasma calcium levels (hypocalcaemia), stimulates the parathyroid gland to release parathormone

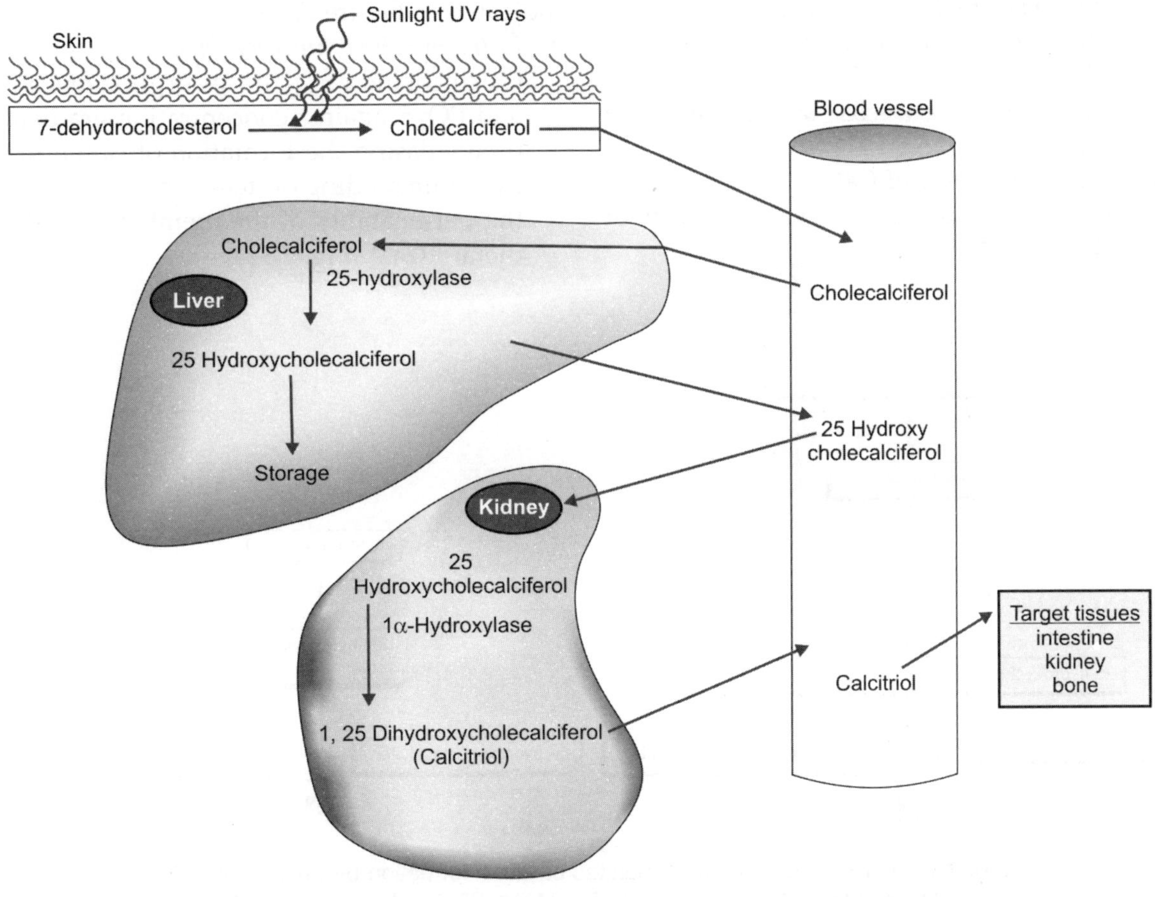


**Fig. 7.10:** Activation of vitamin $D_3$ (cholecalciferol) to calcitriol

which activates the *25-hydroxycholecalciferol-1α hydroxylase* enzyme of the kidney tubular cells (Fig. 7.11A).

*Alternative pathway:* An increase in plasma calcium levels (hypercalcaemia), activates enzyme *24-hydroxylase* in the renal cells (Fig. 7.11B). 24, 25-dihydroxycholecalciferol is formed in the renal tubular cells. It plays a very important role in the formation of bone by osteoblasts.

### Biochemical Functions of Vitamin D

The stimulus for synthesis of calcitriol is hypocalcaemia. So its effect is to raise the serum calcium level back to normal–*hypercalcemic effect.*

Calcitriol, the active form, acts on three target organs:

- Intestinal mucosal cells – site of absorption of dietary $Ca^{++}$
- Bone – reservoir of $Ca^{++}$
- Renal tubular cells – reabsorption of filtered $Ca^{++}$

*In the intestinal mucosal cells:* it increases the absorption of dietary calcium (Fig. 7.12).

### Mechanism of Action of Calcitriol

Calcitriol acts like a steroid hormone. It binds to intracellular, cytosolic, mobile receptors. The receptor—calcitriol complex will bind to specific regions of the DNA and promote the transcription and translation of specific genes to proteins (Fig. 7.12) which either bind to $Ca^{++}$ for retention (like osteocalcin of bone) or are involved in active transport of $Ca^{++}$ as in the intestinal mucosal cells, renal tubular cells and bone cells.

The action of calcitriol on the target organs is indicated in Figure 7.13.

- *In the intestinal mucosal cells:*
  Calcitriol increases the absorption of $Ca^{++}$ and $PO_4^{\ 2-}$ against concentration gradient by increasing the formation of calbindin, a calcium binding protein, which increases the permeability of the membrane to the dietary $Ca^{++}$ (Fig. 7.12).

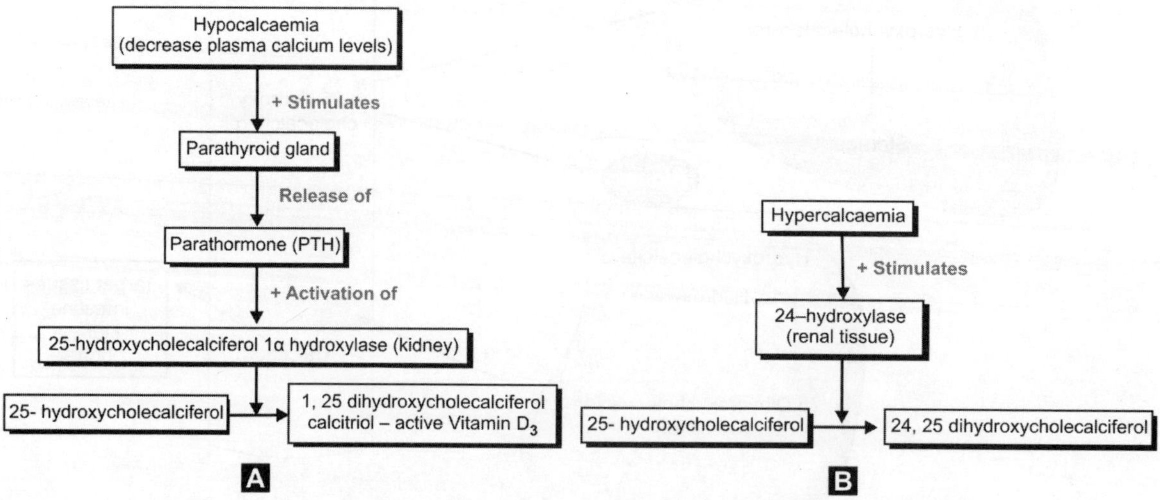


**Fig. 7.11:** (A) Hypocalcaemia stimulates calcitriol formation by release of PTH (B) Hypercalcaemia favours formation of 24, 25-dihydroxycholecalciferol

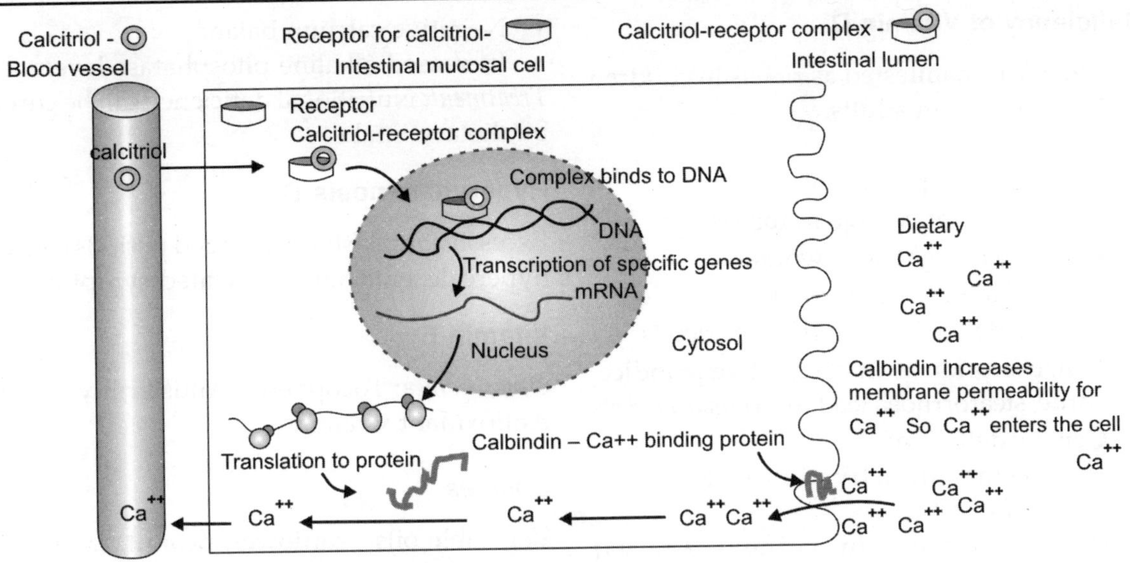


**Fig. 7.12:** Influence of calcitriol on the absorption of dietary calcium from the intestinal mucosal cell *(For colour version see Plate 6)*

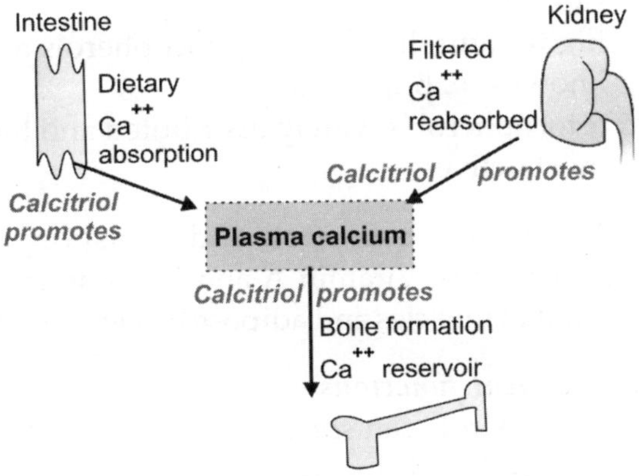


**Fig. 7.13:** Action of calcitriol on the target organs

- *In bone cells:*
  Calcitriol induces the formation of osteocalcin and enzyme alkaline phosphatase which increase the retention of $Ca^{++}$ and mineralization of the bone tissue.

When serum calcium levels are low it also mobilizes calcium from the bone to the blood.

- *In renal tubular cells:*
  There is increased reabsorption of the filtered $Ca^{++}$ and $PO_4^{2-}$ from the glomerular filtrate back into the plasma. Hence $Ca^{++}$ and $PO_4^{2-}$ are retained in the plasma.

Since the mechanism of action of calcitriol is like that of a steroid hormone it is considered as a *hormone* and cholecalciferol as *prohormone*.

***Daily requirement:***

| | |
|---|---|
| Pre school children, Pregnancy, Lactation, Old age | 400 International units (IU) or 10 μg/day |
| Older children, Adults | 200 IU or 5 μg/day |

### Deficiency of Vitamin D

Deficiency is manifested as *rickets* in children and *osteomalacia* in adults.

*Causes:*

- Inadequate exposure to sunlight
  - ➢ Workers of cottage industries
  - ➢ Regular night shift workers
  - ➢ Purdah system of clothes
- Dietary deficiency—*nutritional rickets*
- Malabsorption as in obstructive jaundice, sprue, steatorrhoea lead to *secondary rickets*
- Genetic defects of:
  - ➢ Enzymes of activation in kidney—*Renal rickets*
  - ➢ Reabsorption in kidney—*Fanconi syndrome*
  - ➢ End organ resistance—*Vitamin D resistant rickets.*

*Clinical features of rickets:* Defects are associated mainly with the skeletal system.

- Decreased mineralisation of bones
- Soft pliable bones
- Decreased growth of bone

  Bone deformities seen are:
  - ➢ Bent bow shaped bones of the legs
  - ➢ Knock knees
  - ➢ Rickety rosary
  - ➢ Pigeon chest
  - ➢ Delay in closure of fontanellae
  - ➢ Thinning of the dental enamel
  - ➢ Pain and swelling at the wrist, ankle and sternal ends of ribs.

*Osteomalacia:* Seen in adults

Inadequate mineralisation leads to soft bones with aches and pains—*osteoporosis*, bending of the vertebral column and bow legs. The bones are prone to easy fractures.

*Biochemical parameters:*

- Decreased serum calcium and phosphate levels
- Negative calcium balance
- Increased alkaline phosphatase levels.

*Treatment:* Nutritional deficiency can be cured by dietary intake.

### Hypervitaminosis D

Excessive intake for prolonged periods causes hypercalcaemia and associated symptoms.

### Vitamin E

*Other names*: Tocopherol, Antisterility factor, Antioxidant vitamin.

#### Sources

Vegetable oils—safflower, peanut, rapeseed, maize, wheat germ, soyabean, sunflower, cotton seed, palm.

***Chemistry:*** (Fig. 7.14)

- It has a chromane ring and a isoprenoid side chain
- Eight naturally occurring tocopherols are known—α, β, γ, δ, etc.
- α-tocopherol is widely distributed and the most active

***Metabolism:*** Absorption and transport is same as that of Vitamin A and D. Vitamin E is stored in muscles and adipose tissue as well.

***Biochemical functions***

- Tocopherols can *scavenge free radicals* that are produced during metabolism. This is the *antioxidant action* of vitamin E (Fig. 7.15).

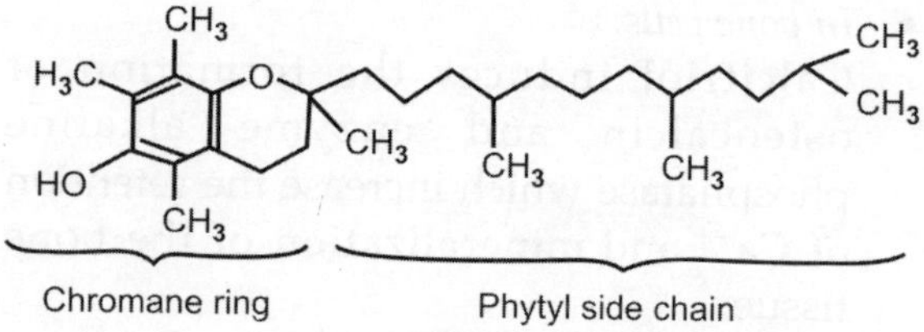


**Fig. 7.14:** Structure of α-tocopherol

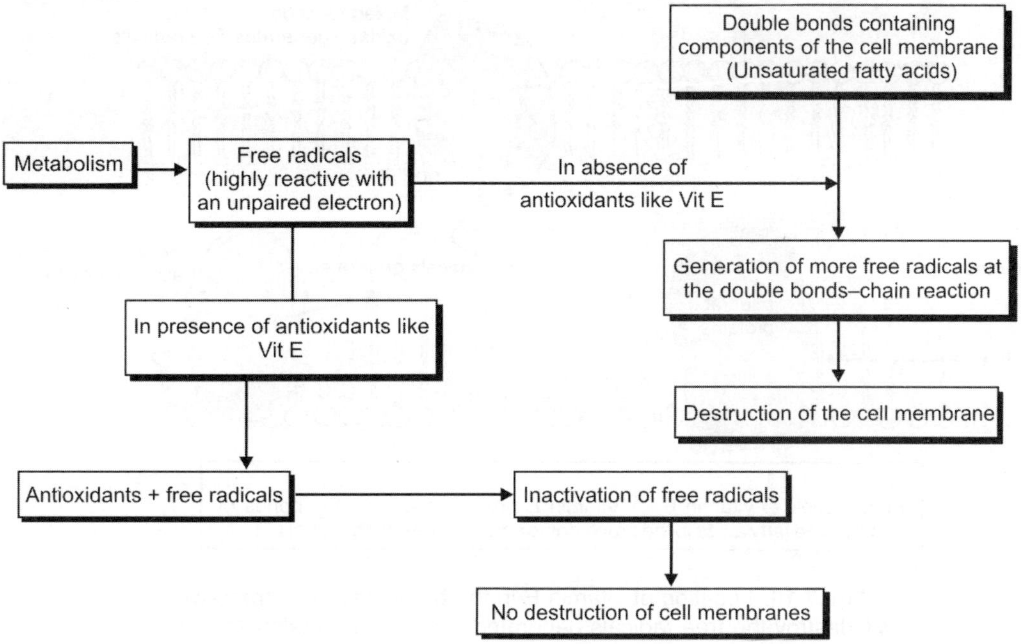


**Fig. 7.15:** Role of vitamin E in scavenging free radicals—antioxidant activity

This is possible because α-tocopherol is located:

- In the cell membrane in between the unsaturated membrane lipids and the free radical generating enzyme systems of the cell membranes (Fig. 7.16).

Vitamin E scavenges the nascent free radicals generated by the enzyme systems before they attack the unsaturated membrane lipids and preserve the integrity of the cell membranes and organelles. This is the basis for the biochemical role of vitamin E.

- Vitamin E preserves the structure of the cell membranes and integrity of the organelle and tissues and prevents the following: In,
  - ➢ RBC – prevents haemolysis
  - ➢ Muscles – prevents muscular dystrophy
  - ➢ Liver – prevents necrosis due to toxins generated
  - ➢ Cardiac muscles – prevents myocardial infarction
  - ➢ Mitochondria – preserves oxidative phosphorylation and ATP generation
- *Aging* is additive effect of the peroxidative changes of free radicals generated in the body during metabolism. Vitamin E by its antioxidant action *delays the aging process.*
- Vitamin E *prevents the action of mutagens* on the DNA—*antimutagenic.*
- The *action of peroxidative industrial pollutants* in the atmosphere like $O_3$, $H_2O_2$ and $NO_2$ on the respiratory system is controlled by vitamin E.
- Tocopherol *protects metal ions* (selenide and non heme iron), *metalloproteins* and *enzymes* containing sulphydryl groups from getting oxidized by free radicals.
- *Antisterility factor*: It prevents sterility in the lower animals by preserving and

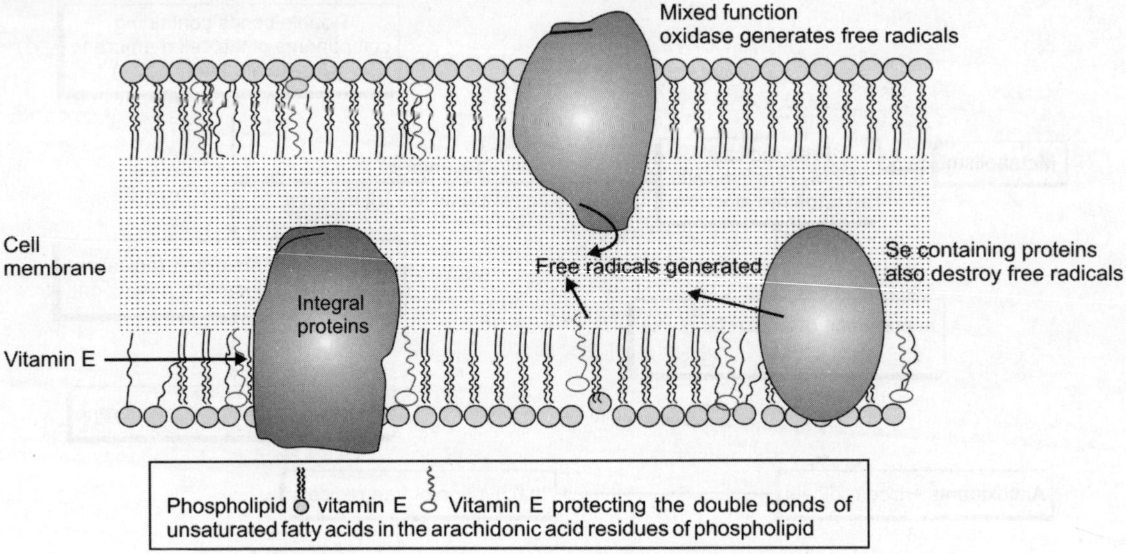


**Fig. 7.16:** Location of vitamin E in the bilipid cell membrane which will destroy the free radicals generated *(For colour version see Plate 6)*

maintaining the germinal epithelium of the gonads.

- It *delays the onset of cataract* in the eyes.

Vitamin E works in association with other vitamins like A, C, provitamin A (β-carotene) and also with the mineral selenium which is a constituent of the enzyme glutathione peroxidase. Glutathione peroxidase is an enzyme which destroys free radicals.

***Daily requirement:***

| | | |
|---|---|---|
| Adult male | - | 10 mg/day |
| Adult female | - | 8 mg/day |
| Pregnancy | - | 10 mg/day |
| Lactation and old age | - | 12 mg/day |

### *Deficiency Manifestations of Vitamin E*

In rats:

- Failure to reproduce due to degeneration of the reproductive organs both in male and female.
- Blood – haemolysis
- Liver – hepatic necrosis
- Muscle – dystrophy
- Kidney – degeneration of the renal tubular epithelium and creatinuria

In humans no deficiency has been reported. However, sometimes haemolytic anaemia in infants fed on commercial food and anorexia and muscle pain in women has been attributed to vitamin E deficiency.

*Causes of deficiency could be*: Obstructive jaundice, steatorrhoea, sprue, pancreatitis.

## Vitamin K

*Synonyms*: Coagulation vitamin

### *Sources*

- Synthesis by intestinal microflora
- Green leafy vegetables.

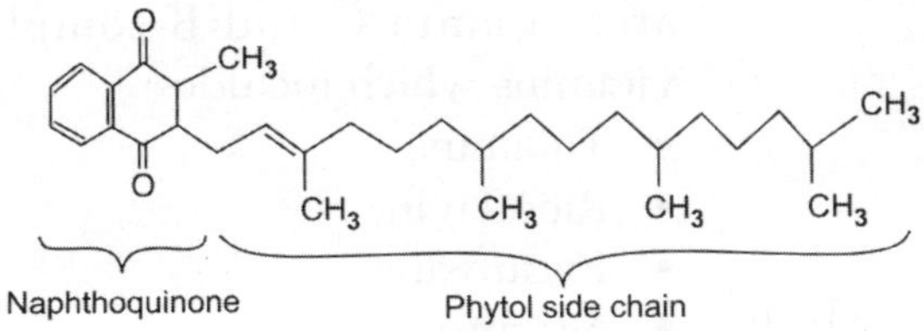


**A** Phylloquinone – $K_1$

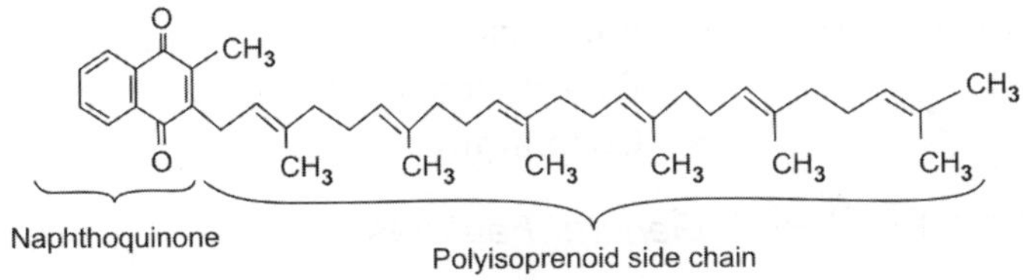


**B**
Menaquinone – $K_2$ – from fish meal, intestinal microflora

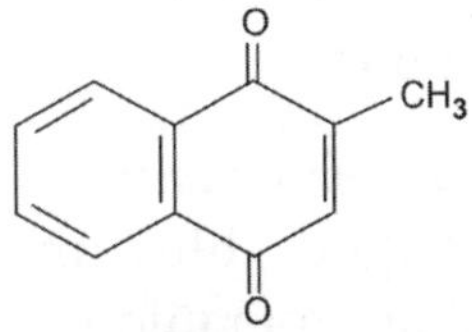


**C**
Menadione – $K_3$ – water soluble, synthetic and used in clinical practice

**Figs 7.17A to C:** Structure of vitamin E from various sources (A) Plant source (B) Animal source (C) Synthetic

*Chemistry*: Figure 7.17

- Naphthoquinone derivative
- Polyisoprenoid side chain

### *Metabolism of Vitamin K*

Same as that of vitamin A and D.

### *Biochemical Role*

*Role in blood clotting*: Glutamic acid residues in clotting factors II, VII, IX and X are activated by γ-carboxylation. This is catalyzed by a vitamin K dependant carboxylase enzyme. Dicumarol and warfarin inhibits the reactivation of vitamin K.

Hence, these are used as anticoagulants in therapy in cases of thrombosis.

*Role in other tissues:* Vitamin K acts as a coenzyme for the γ- carboxylation of proteins in other tissues as well—bone, spleen, placenta, kidneys. In bone, this reaction helps in retention of calcium by osteocalcin of the osteoblasts leading to mineralisation of bone.

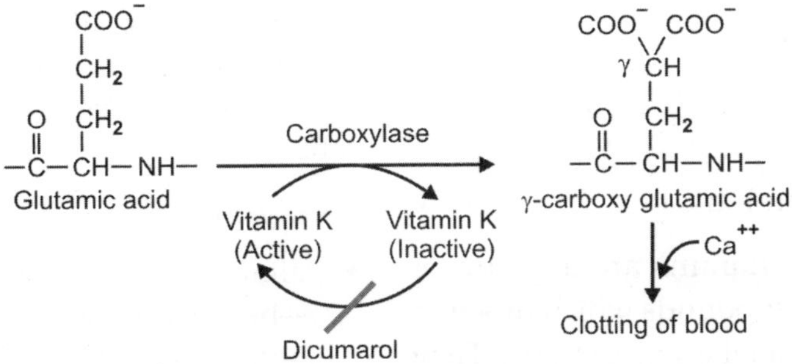

**Daily Requirement: 50-100 μg/day**

*Deficiency of Vitamin K*

*Conditions:*

- malabsorption of dietary lipids
- prolonged therapy with antibiotics which will kill the intestinal microflora
- diarrhoea associated with GI tract infections.

*Clinical manifestations:*

- Prolonged bleeding time—profuse bleeding from minor injuries
- Delayed clotting time
- Easy bruises
- Echymotic patches
- Internal bleeding

In the new born babies—very little vitamin K crosses the placental barrier. As a result, there is a very low vitamin K store in the foetal liver. The intestinal microflora is not developed to synthesise sufficient vitamin K. The breast milk has small quantities of vitamin K which is not sufficient for the infant. Hence, *haemorrhagic disease of the newborn* is often seen. This can be treated with menadione, the water soluble vitamin K.

*Hypervitaminosis:* Toxicity is seen when there is an increased dose of menadione in treatment of the newborn which may result in:

- Haemolysis
- Hyperbilirubinemia
- Kernicterus
- Brain damage.

## Water Soluble Vitamins

The water soluble vitamins are a group of unrelated organic compounds which function as coenzymes for metabolic reactions. They are: vitamin C and B complex group of vitamins which includes:

- Thiamine
- Riboflavin
- Pyridoxine
- Niacin
- Biotin
- Folic acid
- Pantothenic acid
- Cobalamin.

*General Features*

- *Solubility:*
  - soluble in water, insoluble in fat
- *Absorption from GI tract:*
  - not dependant on dietary fat and bile salts
- *Transport in blood:*
  - does not require carrier proteins/ lipoproteins (*exception:* vitamin $B_{12}$)
- *Storage:*
  - no storage (exception: vitamin $B_{12}$ in liver)
- *Excess intake:*
  - no toxicity (exception-vitamin C)
- *Deficiency:*
  - decreased dietary intake
- *Treatment:*
  - inclusion of the vitamin in the diet.

## Vitamin C

*Other names:* Ascorbic acid, antiscorbutic factor.

Dietary sources: Rich source – amla (Indian goosebery), citrus fruits, guava.

Green leafy vegetables – coriander, amaranth, cabbage.

*Chemistry*

- Vitamin C is a carbohydrate, sugar lactone —hexuronic acid. The L isomer of the sugar is active (Fig. 7.18).

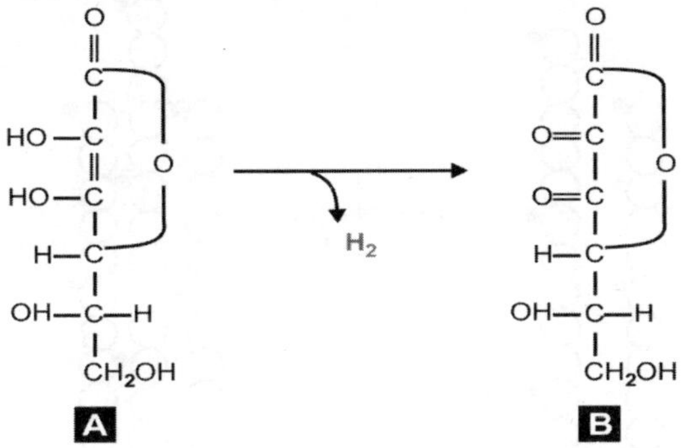

Fig. 7.18: Oxidation of vitamin C (A) L-ascorbic acid (B) Dehydroascorbic acid

- It is very easily oxidised by oxygen to dehydroascorbic acid.
- It is a strong reducing agent, due to the presence of enediol (double bond with two hydroxyl groups) Fig. 7.18 (A).
- Both L-ascorbic acid and dehydroascorbic acid have antiscorbutic property.

### *Biosynthesis of Ascorbic Acid*

Ascorbic acid can be synthesised in the lower animals by the uronic acid pathway. The higher animals and human beings cannot synthesise vitamin C due to the absence of the enzyme *L-gulonolactone oxidase*. Hence, it should be included in the diet. Fruits and vegetables are a rich source of vitamin C.

### *Metabolism of Vitamin C*

*Absorption* : At small intestine by a carrier mediated active process.
*Transport* : By blood to the various tissues
*Excretion* : In urine.

### *Biochemical Functions of Vitamin C*

Vitamin C plays two important roles in metabolism:

- Coenzyme for hydroxylation reactions
- As an antioxidant.

Other roles of vitamin C are:

- Role in oxidation-reduction reactions.
- In immune response.

*Role of vitamin C as a coenzyme for hydroxylation reactions:*

- *Hydroxylation of proline and lysine residues in collagen molecules:* Collagen is a protein, of the supporting tissues of mesenchymal origin—like bone, dentine, capillaries and connective tissue. Immature collagen has large amounts of proline and lysine residues.
- During post-translational modification of the immature collagen, these amino acids are hydroxylated by hydroxylase enzymes which need vitamin C as coenzyme.
- The hydroxylated amino acid residues help in crosslinking of the immature collagen fibrils into mature collagen fibrils Figure 7.19 (discussed in Chapter 14).
  *Significance*: Failure of hydroxylation results in weak crosslinking and delay in maturation of collagen—leading to fragility of blood vessels, spongy bleeding gums, weakness of bones and delay in wound healing process.

*Role in catecholamine synthesis:*

- In adrenal medulla, vitamin C is required for the synthesis of catecholamines-adrenaline (epinephrine), nor adrenaline (norepinephrine).
  *Significance:* catecholamines are neuro-transmitters

*Role in carnitine biosynthesis:*

- Carnitine (β-hydroxy, γ-trimethyl ammonium butyrate) is a carrier molecule present in the inner membrane of the mitochondria for transport of the activated fatty acids from cytosol into the mitochondrial matrix for β-oxidation and release of energy.

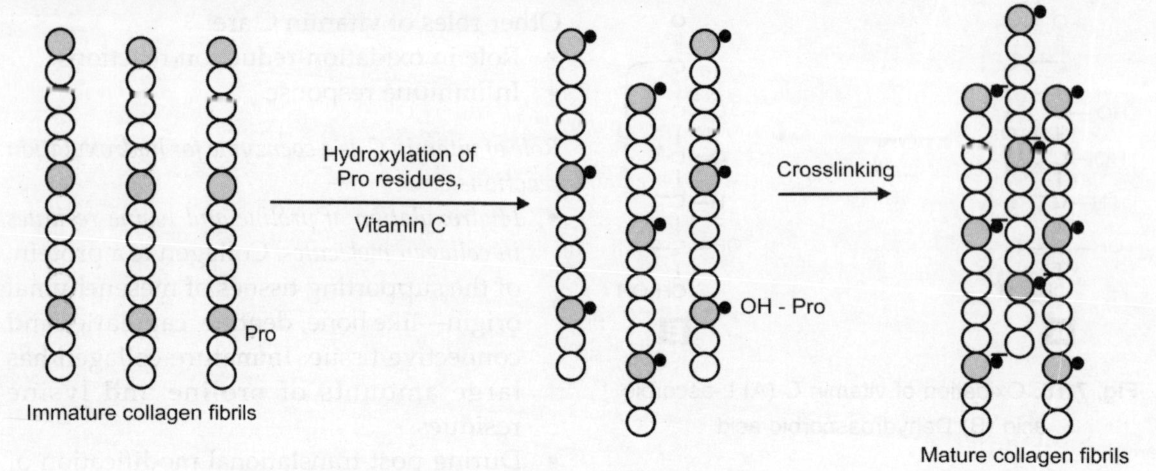


**Fig. 7.19:** Maturation of collagen fibrils

*Significance:* In vitamin C deficiency fatigue is due to decreased energy production from fatty acids.

*Role in serotonin production*

- Serotonin (5-hydroxy tryptamine) is produced from tryptophan in the brain, mast cells, platelets and intestinal mucosa. The hydroxylation reaction needs vitamin C as coenzyme.

*Role in steroid synthesis*

- Synthesis of bile acids, steroid hormones from cholesterol needs vitamin C for the hydroxylation reactions.

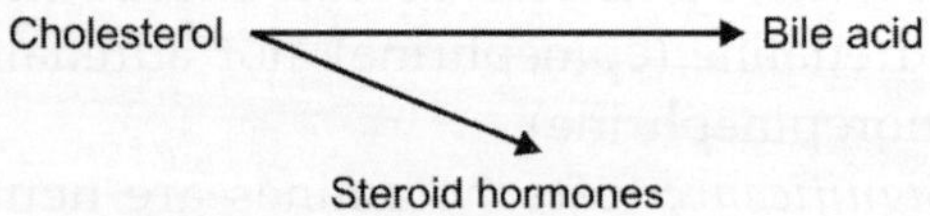


*Role of vitamin C as an antioxidant*

- Vitamin C acts as a natural antioxidant. It destroys free radicals due to its reducing property and helps to prevent cancer. (Vitamin C is known to prevent bladder cancers caused by aniline dyes)
- *Role of vitamin C as an antioxidant in the eyes:* Vitamin C prevents cataract formation in the eye.

*Role of vitamin C in oxidation-reduction reactions:*

- *Role in iron absorption:* Vitamin C reduces the dietary ferric iron to ferrous iron which can be absorbed from the intestine.

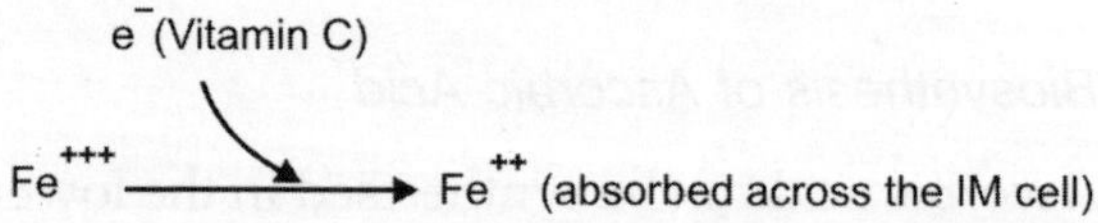


*Significance*: Iron is needed for synthesis of haemoglobin which transports respiratory gases.

*Role of vitamin C in haemoglobin metabolism:*

- Haemoglobin has an atom of $Fe^{++}$. When this $Fe^{++}$ is oxidized to $Fe^{+++}$ methaemoglobin is formed. Vitamin C helps in reduction of methaemoglobin to haemoglobin.

*Reduction by vitamin C*

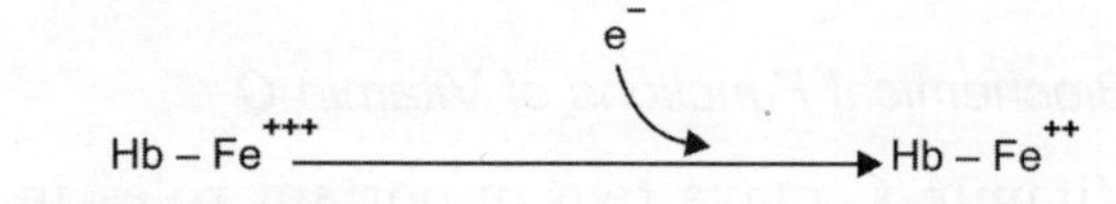


*Significance*: Methaemoglobin has no oxygen transport capacity.

*Role of vitamin C in folic acid metabolism:*

- Vitamin C helps in reducing folic acid to its active form tetrahydrofolate ($FH_4$) which is a one carbon carrier molecule required for the synthesis of the nucleic acids.
  *Significance*: $FH_4$ is one of the maturation factors for RBC formation.
- *Role in immune system:*
  - Ascorbic acid stimulates formation of antibodies and phagocytosis in leucocytes.

*Daily requirement:*

| | |
|---|---|
| Adults | 60–80 mg/day |
| Pregnancy, Lactation, Old age | 100–125 mg/day |

*Deficiency manifestations:* Deficiency of vitamin C results in *scurvy.*

*Symptoms due to defective maturation of collagen:*

- Haemorrhagic tissues due to capillary fragility.
- Spongy, bleeding, painful, swollen gums.
- Loosening of teeth.
- Weak bones susceptible to fractures—leading to *osteoporosis.*
- Painful swollen joints.

*Symptoms due to defects in other metabolism:*

- Anaemia with lassitude and fatigue (iron deficiency, loss of blood by haemorrhage, folic acid deficiency, formation of methaemoglobin).

*Uses in therapy:*

- Large doses in common cold
- In treatment of tissue infections
- Decreases incidence of cancer
- Delays aging.

## Thiamine

*Synonyms*: Vitamin $B_1$, aneurine, antiberiberi factor.

### Sources

*Cereals:* Outer colored layer of cereals-*Aleurone* (close to the husk—as in whole wheat, hand pounded rice), parboiled rice.

*Legumes:* Peanuts, dry peas, beans, nuts, lentils, soya beans.

*Non-vegetarian sources:* yeast, liver, eggs, lean meat.

### Structure

Thiamine has two ring systems attached by a methylene bridge. The two ring systems are substituted pyrimidine and substituted thiazole (Fig. 7.20A). It is a sulphur containing vitamin.

### Metabolism

Absorption is either by simple diffusion or by active transport mechanism in the small intestine.

*Coenzyme/active form of thiamine*

The transfer of pyrophosphate group from ATP as shown in the following equation forms *thiamine pyrophosphate (TPP) (Figure 7.20B).*

Thiamine + ATP —[Thiamine pyrophospho transferase (Liver enzyme)]→ TPP + AMP

### Physiological Functions of Thiamine

1. The reactions in which TPP acts as coenzyme are mostly from carbohydrate metabolism.

Pyruvate —[pyruvate dehydrogenase multienzyme complex]→ Acetyl CoA + $CO_2$

Coenzymes: TPP, FAD, $NAD^+$, lipoic acid, Coenzyme A, $Mg^{++}$

α-ketoglutarate —[α-ketoglutarate dehydrogenase complex]→ Succinyl CoA + $CO_2$

Coenzymes: TPP, FAD, $NAD^+$, lipoic acid, Coenzyme A, $Mg^{++}$

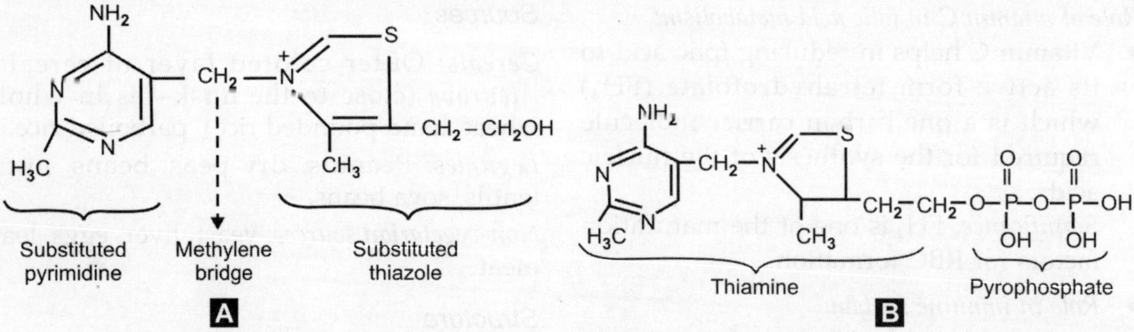


**Fig. 7.20:** (A) Structure of thiamine (B) Structure of active form thiamine pyrophosphate –TPP

(a) Oxidative decarboxylation reactions: of pyruvate, α-ketoglutarate, α-keto-butyrate, α-keto acid intermediates of branched chain amino acid metabolism.

*Note:* Both these reactions are catalysed by multienzyme systems in the mitochondria made up of 3 enzymes, 5 coenzymes and 1 cofactor.

(b) Transketolase enzyme catalysed reactions of HMP shunt pathway:

i. [Ribose 5 phosphate + Xylulose 5 phosphate]

*Transketolase*, $Mg^{++}$, TPP ↕

[Sedoheptulose 7 phosphate + Glyceraldehyde 3 phosphate]

ii. [Xylulose 5 phosphate + Erythrose 4 phosphate]

*Transketolase,* $Mg^{++}$, TPP ↕

[Fructose 6 phosphate + Glyceraldehyde 3 phosphate]

2. TPP also helps in nerve impulse transmission in the peripheral neurons either by affecting the ion translocation across stimulated nerve membrane or by affecting the synthesis of neurotransmitter acetyl choline.

*Importance*: Since thiamine is required for carbohydrate metabolism, the requirement increases with the increase in the carbohydrate content of the diet.

### Dietary Requirement

| | | |
|---|---|---|
| Adults | : | 1.4 mg/day |
| Pregnancy | : | 1.3 mg/day |
| Lactating woman | : | 1.5 mg/day |
| Children | : | 0.6 – 1.3 mg/day |

*Note*: *Dietary requirement increases with increased carbohydrate intake or increased calorie intake.*

### Deficiency of Thiamine

*Causes:*

- $B_1$ deficient diets—consumption of polished rice as in diets of South Asian countries.
- Intake of thiamine antagonists—e.g., pyrithiamine and oxythiamine.
- Chronic alcoholism.
- Old age
- Sea faring fishermen due to consumption of raw fish containing *thiaminase* enzyme, which destroys thiamine.

Deficiency of thiamine results in *"Beriberi"*. In general the *deficiency manifestations* include:

- Anorexia, loss of appetite, fatigue
- Loss of body weight
- Poor growth

- Nausea, headache
- Vomiting, dizziness
- Muscle weakness, numbness, pricking sensation in lower limbs.

In adults there are two types of beriberi—

(i) *Wet beriberi/ "edematous beriberi"*—characterised by:
  - Extensive edema in face, trunk, legs and serous cavities.
  - Cardiovascular manifestations—heart enlargement and congestive cardiac failure.
  - Breathlessness, palpitation.

(ii) *Dry beriberi/"neuritic beriberi"*—characterised by:
  - Central nervous system manifestations
  - Wasting of muscles—paralysis *(Chastek paralysis)*
  - Peripheral neuritis
  - Sensory disturbances

*Infantile beriberi*: Seen in infants age (2-10 months)

*Cause:* Mothers with inadequate thiamine intake.

*Wernicke-Korsakoff syndrome:*

*Cause:* Genetically defective transketolase activity (decreased affinity for TPP) or chronic alcoholism.

*Symptoms:*

- Haemorrhagic lesions of the third ventricle of the brain.
- Incoordinated eye movements due to paralysis of the eye muscles.
- Muscular incoordination—abnormal stance and gait.
- Mental symptoms—functional derangements, confusion, depression, anxiousness, loss of memory, insomnia.

*Biochemical alterations:*

- Decreased thiamine levels in the blood
- Decreased transketolase activity in the erythrocytes
- Increased pyruvate and α-ketoglutarate levels in blood
- Increased lactate in blood (conversion of accumulated pyruvate by LDH).

## Riboflavin

*Synonyms:* Vitamin $B_2$

### Dietary Sources

*Good sources:* Whole grains, legumes, dhal and green leafy vegetables yeast, eggs, milk and meat.

*Fair sources:* Root vegetables and fruits.

Other compounds identical to riboflavin are: Lactoflavin in milk, hepatoflavin in liver, ovoflavin in egg.

### Structure of Riboflavin

Riboflavin has a dimethylisoalloxazine ring system attached to ribitol (alcohol form of the pentose sugar ribose). (Fig. 7.21).

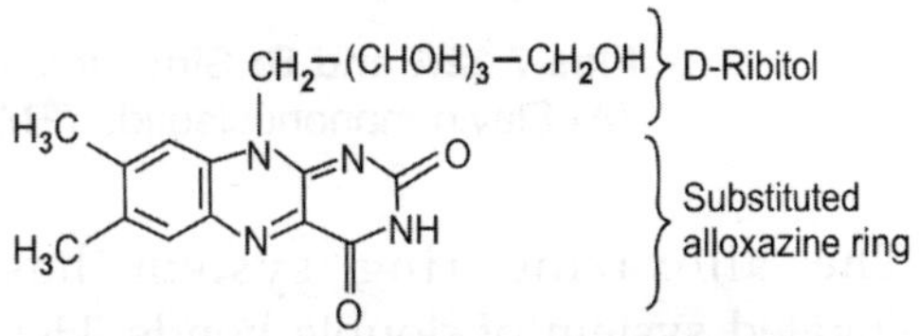


**Fig. 7.21:** Structure of riboflavin (D-ribitol + substituted alloxazine ring)

### Coenzyme/active Form of Riboflavin

There are two coenzymic forms of riboflavin – (Fig. 7.22)

- Flavin mononucleotide (FMN)
- Flavin adenine dinucleotide (FAD)

These coenzymic forms of riboflavin are tightly bound to proteins to form conjugated proteins called *flavoproteins.* Some flavoproteins require metal ions (Mo or Fe) as well —*metalloflavoproteins*.

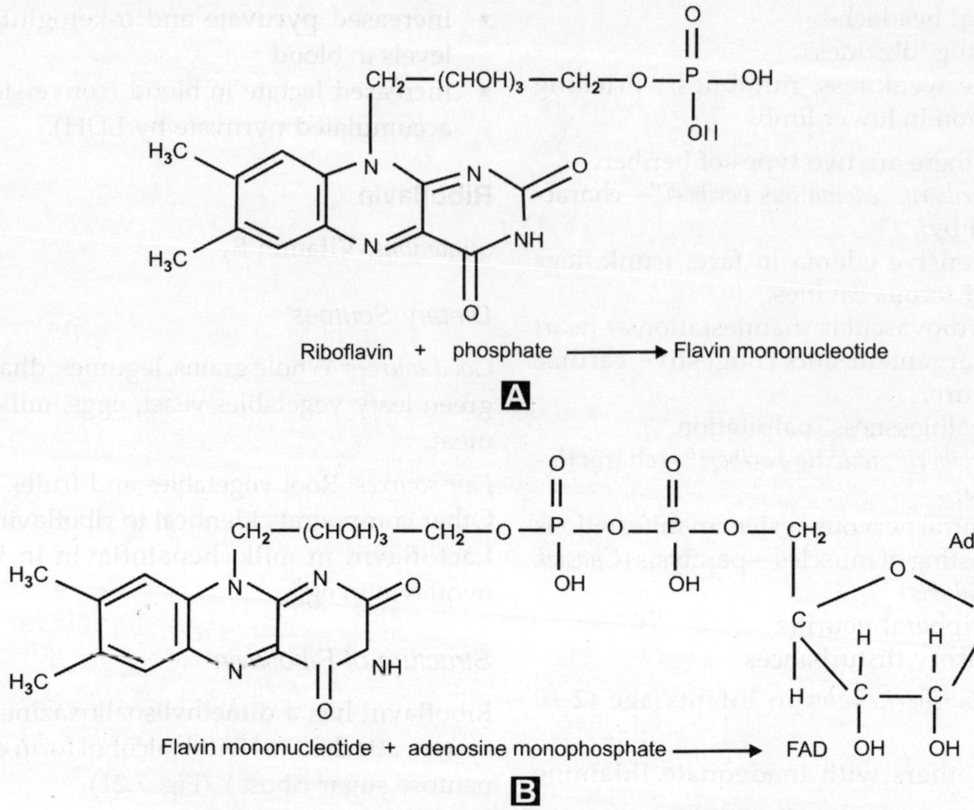


**Figs 7.22A and B:** Structures of the coenzymic/active forms of riboflavin (A) Flavin mononucleotide (FMN) (B) Flavin adenine dinucleotide (FAD)

The alloxazine ring system has a conjugated system of double bonds. Hence, it emits a strong yellowish-green fluorescence when exposed to UV rays.

*Formation of the coenzymic forms of riboflavin:* Riboflavin is converted to its coenzymic forms in the small intestinal cells, which have the enzymes *flavokinase* and *FAD synthase* as shown in the following equation:

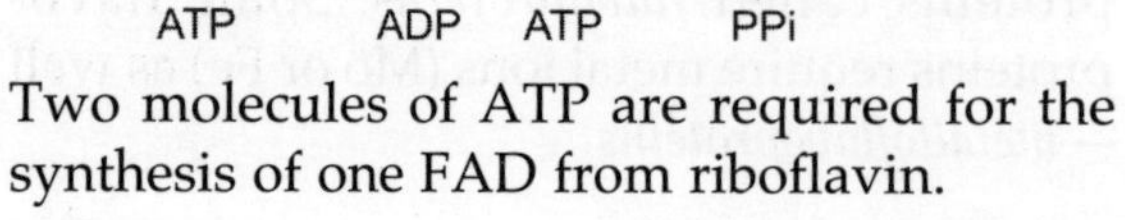


Two molecules of ATP are required for the synthesis of one FAD from riboflavin.

### *Biochemical Functions of Riboflavin*

1. *Coenzymes of riboflavin take part in oxidation-reduction (redox) reactions* (Fig. 7.23). They act as acceptor of hydrogen atoms from substrates. The coenzyme forms get reduced and the substrate gets oxidised as shown in the following equation:

   Two nitrogen atoms in the alloxazine ring accept the hydrogen atoms.

*Examples:* FMN mononucleotide dependant enzymes:

- NADH dehydrogenase of the respiratory chain
- L-amino acid oxidase.

$$AH_2 + \text{Enzyme-FAD} \longrightarrow A + \text{Enzyme-FADH}_2$$

(Reduced substrate) (Oxidised coenzyme) (Oxidised substrate) (Reduced coenzyme)

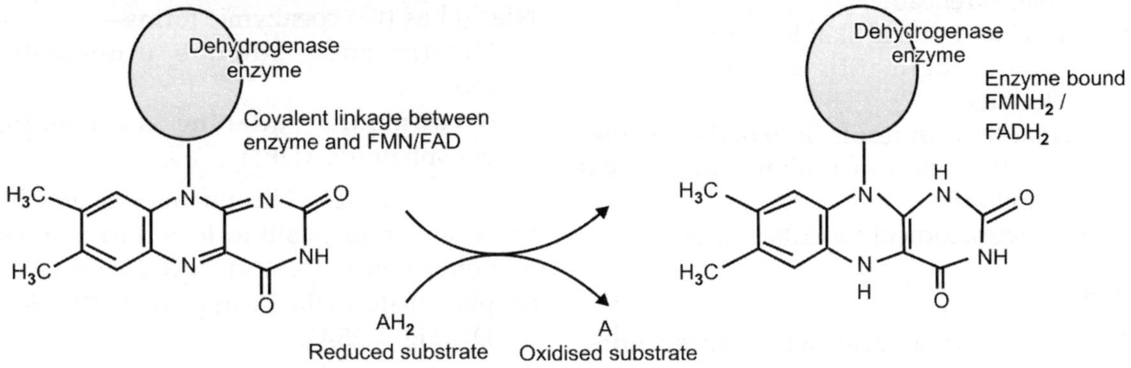


**Fig. 7.23:** Role of riboflavin in redox reaction

*FAD dependant enzymes:*

- Final common metabolic pathway:
- *Succinate dehydrogenase and α-ketoglutarate dehydrogenase complex* from TCA cycle
- Carbohydrate metabolism: *Pyruvate dehydrogenase* multienzyme complex (oxidative decarboxylation reaction)
- Lipid metabolism—*Acyl CoA dehydrogenase* in β-oxidation of fatty acids
- Amino acid metabolism – *D-amino acid oxidase*
- Nucleotide metabolism – *Xanthine oxidase* of degradation of purine bases

2. Riboflavin helps in proper maintenance of ectodermal tissues of skin and oral mucosa.

### Dietary Requirements

| | |
|---|---|
| Adults | - 1.3 –1.7 mg/day |
| Pregnancy | - 1.5 mg/day |
| Lactation | - 1.6 mg/day |
| Infants and growing children | - 0.7 – 1.6 mg/day |

### Deficiency of Riboflavin

Plants, micro-organisms and intestinal microflora synthesise riboflavin. Hence dietary deficiency is rare.

*Causes*

- Deficiency of other vitamins is accompanied by riboflavin deficiency — beriberi, pellagra.
- Transient deficiency is seen in phototherapy for neonatal jaundice due to photolysis of circulating riboflavin.
- Intake of riboflavin antagonists — galactoflavin.
- Prolonged intake of drugs like chlorpromazine.

***Symptoms:*** Associated with the riboflavin deficiency:

- Oral lesions of the mucous membranes
  - ➢ Glossitis—inflamed magenta coloured tongue
  - ➢ Cheilosis—fissures of the tongue
  - ➢ Angular stomatitis—inflammation of the corners of the mouth.

- Facial skin lesions:
  - ➢ Seborrhoeic dermatitis—thickening and atrophy of skin at nasolabial folds, ear, forehead.
- Scrotal and vulval skin lesions:
  - ➢ Scrotal dermatitis and urinogenital lesions.
- Ocular lesions: results in blood shot eyes
  - ➢ Proliferation of bulbar conjunctival capillaries
  - ➢ Circumcorneal vascularisation.

## Niacin

*Synonyms:* Nicotinic acid, pellagra preventing factor.

### Dietary Sources

*Rich sources:* Dried yeast, liver, meat, fish, peanut, rice polishing, whole cereals and legumes.

Milk and egg are poor sources of niacin but rich sources of tryptophan (from which niacin can be synthesised).

*Poor sources:* Fruits and vegetables.

Intestinal microflora synthesises small amounts of niacin.

### Structure of Niacin

Niacin is pyridine-3-carboxamide. There are two forms of niacin—nicotinic acid and nicotinamide (Fig. 7.24).

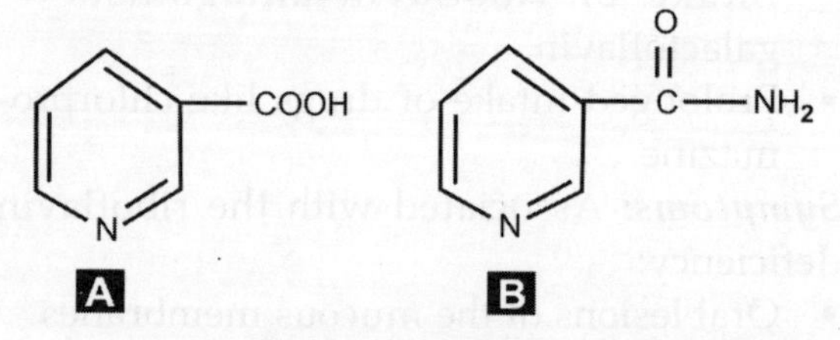


**Fig. 7.24:** Structure of niacin (A) Nicotinic acid (B) Nicotinamide

### Metabolism of Niacin

Absorption is from small intestine, circulated to various tissues by blood where it is converted to its coenzymes. Niacin is excreted from the body as N-methylnicotinic acid by transmethylation reaction.

*Coenzyme/active form of niacin:*

Niacin has two coenzymic forms—

- Nicotinamide adenine dinucleotide ($NAD^+$).
- Nicotinamide adenine dinucleotide phosphate ($NADP^+$).

*Structure of $NAD^+$:* Nicotinamide is attached to ribose-5-phosphate to form nicotinamide mononucleotide (NMN). NMN is linked to the phosphate on the 5′carbon of AMP to form $NAD^+$ (Fig. 7.25A).

*Structure of $NADP^+$:*

3′ hydroxyl group on the AMP part of $NAD^+$ is phosphorylated to form $NADP^+$. (Fig. 7.25B).

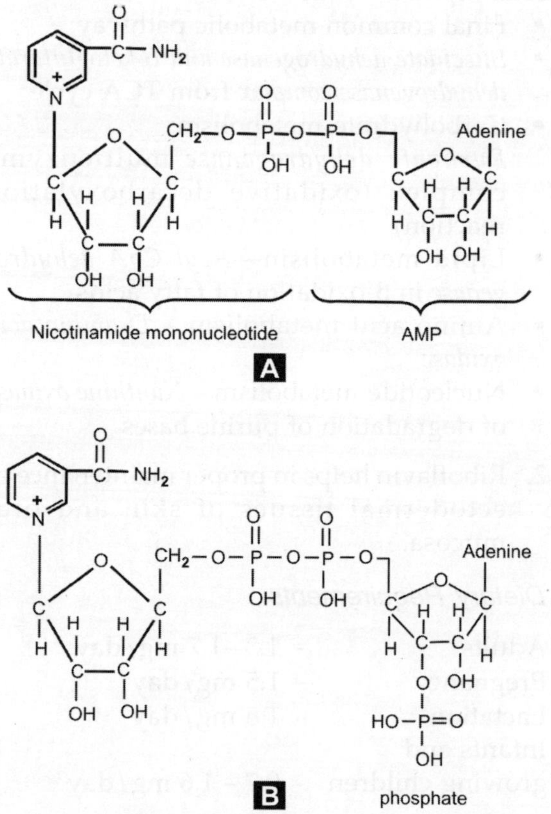


**Figs 7.25A and B:** Structure of (A) $NAD^+$ (B) $NADP^+$

The nitrogen atom of the pyridine ring has a positive charge hence both NAD/NADP are indicated as $NAD^+/NADP^+$.

*Formation of $NAD^+$ and $NADP^+$:* Niacin is formed from the amino acid tryptophan. Sixty milligrams tryptophan gives rise to one milligram niacin in the liver.

*Biochemical Functions of Nicotinamide Coenzymes:*

1. Coenzymes of nicotinamide take part in oxidation-reduction (redox) reactions.

*Mechanism of action of nicotinamide coenzymes: (Fig. 7.26)*

- Enzymes that are dependant on nicotinamide nucleotides are called as *pyridine dependent dehydrogenase* (Table 7.1).
- The nicotinamide coenzymes are bound to the apoenzymes of the dehydrogenases with loose and transient interactions.
- These enzymes catalyze the removal of two hydrogen atoms from the substrate (2 protons - $H^+$ and 2 electrons - $e^-$).
- While the coenzyme is bound to the apoenzyme a hydride ion (a proton with 2 electrons – $H^-$) is transferred from the substrate to the pyridine ring of the coenzyme. $NAD^+$ becomes NADH.
- The other proton ($H^+$) is left free in the medium.
- Hence $NAD^+$ becomes NADH + $H^+$—*reduced coenzyme.*
- The oxidised substrate and the reduced coenzyme (NADH/NADPH) are immediately released from the apoenzyme.
- NADH/NADPH now donate its reducing equivalents to mitochondrial electron transport chain/other substrates and get oxidised back to $NAD^+/NADP^+$ forms.

**Table 7.1**: Pyridine dependent dehydrogenases

| *$NAD^+$ dependant enzymes* | *$NADP^+$dependant enzymes* |
|---|---|
| *Glycolysis:* PGAL dehydrogenase<br>Pyruvate dehydrogenase multienzyme complex<br>Lactate dehydrogenase | *HMP shunt*:<br>Glucose-6-phosphate dehydrogenase<br>6-phosphogluconate dehydrogenase |
| *TCA Cycle*:<br>Isocitrate dehydrogenase (mitochondrial)<br>α-ketoglutarate dehydrogenase multienzyme complex<br>Malate dehydrogenase | Isocitrate dehydrogenase (cytosolic) |
| *B-Oxidation:*<br>β-hydroxy acyl CoA dehydrogenase | *Fatty acid Synthesis:*<br>β-ketoacyl reductase<br>Enoyl reductase<br>Malic enzyme<br>Fatty acid desaturase<br>*Cholesterol synthesis:*<br>HMG CoA reductase<br>Squalene synthase |
| *Amino acid metabolism:*<br>L-glutamate dehydrogenase | L-glutamate dehydrogenase<br>Dihydrofolate reductase |

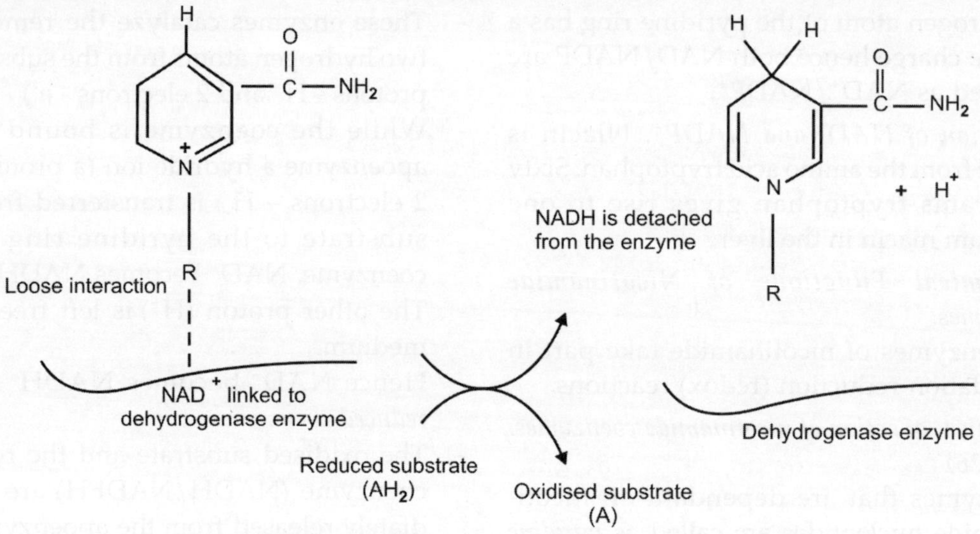


**Fig. 7.26:** Mechanism of action of nicotinamide coenzymes

2. Nicotinic acid produces transient vasodilatation of blood vessels and histamine release in the skin. This results in tingling sensation, burning and itching of the skin.
3. Niacinamide has cholesterol lowering effect in plasma (it decreases the release of fatty acids from the adipose tissue, which decreases the acetyl CoA content in the cells—the starting material for cholesterol biosynthesis).
4. Niacin lowers plasma levels of Lp (a). Increased Lp (a) is indicative of susceptibility to coronary heart diseases.

### *Dietary Requirement*

| | |
|---|---|
| Adults | : 15-19 mg/day |
| Pregnancy | : 17 mg/day |
| Lactation | : 19 mg/day |
| Infants and children | : 8-14 mg/day |

### *Deficiency of Niacinamide*

In man deficiency of niacinamide produces *"Pellagra"* meaning rough skin.

*Causes:*

- Dietary deficiency—diets containing mainly maize/jowar (deficient in tryptophan, high leucine content which depresses the synthesis of tryptophan)
- *Deficiency of pyridoxine (vitamin $B_6$):* PLP (coenzyme of vitamin $B_6$) is needed for formation of niacin from tryptophan.
- *Intake of drugs:* Isoniazid (anti tuberculosis drug) inhibits formation of PLP.
- Inborn errors: *Hartnups disease* – defective absorption of tryptophan from intestine.
- Intake of niacin antagonists: acetyl pyridine, pyridine sulphonamide, aminonicotinamide

*Symptoms of pellagra:* Pellagra is called as the "3D" disease since it is characterised by dermatitis, diarrhoea and dementia.

- *Dermatitis: Erythema* is seen in the skin exposed to sunlight—feet, ankle, neck, axilla, forearms and fingers. Increased pigmentation round the neck is called as *Casal's necklace.*
- *Diarrhoea:* Mainly due to inflammation of the mucous membrane of the gastro-

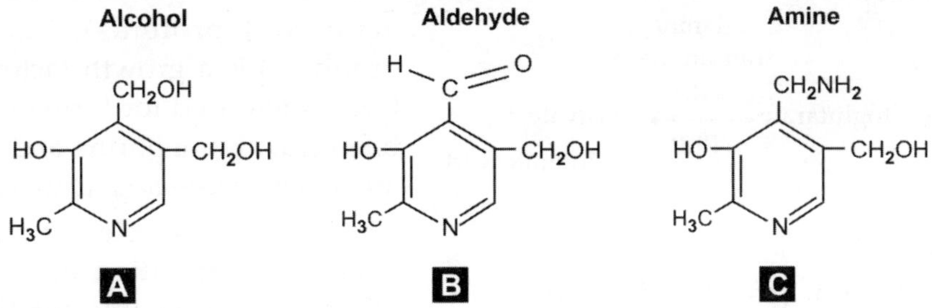


**Figs 7.27A to C:** Structure of (A) pyridoxine (B) pyridoxal (C) pyridoxamine

intestinal tract. Diarrhoea results in weight loss associated with nausea and vomiting.

- *Dementia:* Neurological symptoms like depression, headache, delirium and memory loss are associated with pellagra. Posterolateral tract degeneration may lead to ataxia and spasticity.
- *Other dermatological symptoms:* glossitis (black tongue in dogs), stomatitis.
- Unless treated with the vitamin the disease may be fatal.

## Pyridoxine

*Synonyms* : Vitamin $B_6$.

### Sources

*Rich vegetable sources:* Dried yeast, rice polishing, wheat germs, cereals, legumes, oil seeds and green leafy vegetables.

Milk, egg, fish and meat are also fair sources.

Pyridoxine can be synthesised by the intestinal bacteria.

### Structure of Pyridoxine

Pyridoxine is 2-methyl, 3-hydroxy, 4, 5-hydroxymethyl pyridine. Vitamin $B_6$ is a group of structurally related pyridine compounds, which have similar biological effects (Fig. 7.27).

### Coenzyme/Active form of Pyridoxine

Pyridoxal phosphate (PLP) is the active form of pyridoxine.

*Formation of PLP:* PLP is formed by the following reaction:

$$\text{Pyridoxal} + \text{ATP} \xrightarrow[\textit{kinase}]{\textit{Pyridoxal}} \text{Pyridoxal phosphate} + \text{ADP}$$

Pyridoxal kinase can also phosphorylate pyridoxamine and pyridoxine to pyridoxamine phosphate (PMP) and pyridoxal phosphate. PLP and PMP are interconvertible during transamination reaction.

PLP is the prosthetic group of many enzymes of amino acid metabolism. It is bound to the ε-amino group of lysine residue of the apoenzyme by schiff base formation. (Fig. 7.28)

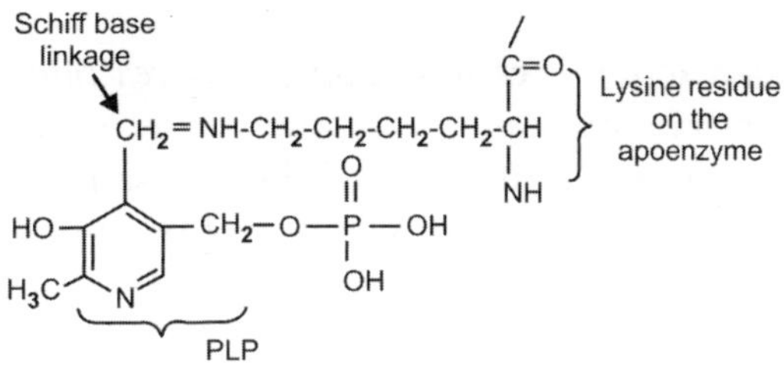


**Fig. 7.28:** Enzyme bound PLP

### Biochemical Role of Pyridoxine

1. *Role of PLP in transamination reaction:* PLP is the prosthetic group of the *amino transferase/transaminase* enzymes (see Fig. 4.16) which catalyze transamination as shown in the following reaction:

$$\text{Alanine} + \alpha\text{-ketoglutarate} \xrightleftharpoons[\text{PLP}]{\text{Alanine transaminase (ALT)}} \text{Pyruvate} + \text{Glutamic acid}$$

*Mechanism*

- The α-amino group of alanine is transferred to the aldehyde group of PLP to form PMP and the oxo (=O) group of PLP is transferred to the α-position of alanine to form pyruvate (Figure 7.29A).
- Next, the amino group from PMP is transferred to α-position of α-ketoglutarate forming glutamic acid and the oxo (=O) group of α-ketoglutarate is transferred to the prosthetic group to form PLP. Figure 7.29B.

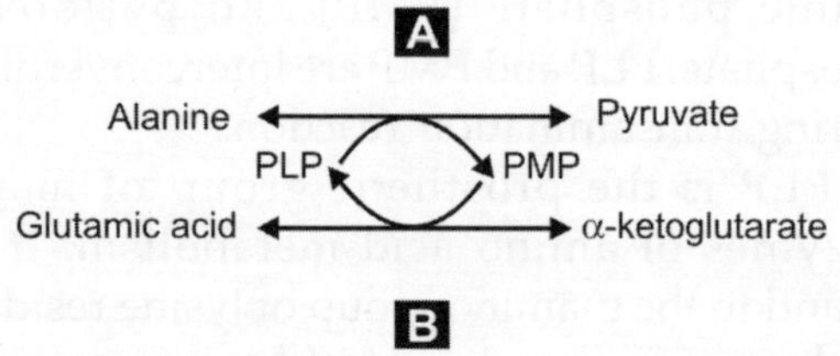


**Figs 7.29A and B:** (A) Transfer of amine group from alanine to PLP (B) Transfer of amine group from PMP to α-ketoglutarate

This reaction is called as *ping pong reaction* or *bisubstrate reaction*.

Clinical significance of the transaminase enzymes are dealt with under general reactions of amino acid metabolism and enzymes chapters.

2. *Role of PLP in decarboxylation reactions:* PLP is the prosthetic group of *decarboxylase* enzymes which catalyze decarboxylation reactions of amino acids. The products are amines which are called *biogenic amines* or *bioamines*.

   Biogenic amines have a wide range of functions. GABA, dopamine, and serotonin are neurotransmitters (refer chapter on amino acids and proteins). Putresceine (from ornithine) is a growth factor. Taurine (from Cys) is required for formation of conjugated bile acids. Epinephrine and nor epinephrine are *adrenal medullary hormones*.

   *Importance:* Deficiency of $B_6$ results in decreased formation of neurotransmitters which is responsible for the neurological symptoms.

3. *Role of PLP in non-oxidative deamination of hydroxy/sulphur containing amino acids:* Hydroxy amino acids serine and threonine undergo non-oxidative deamination (dehydration/desulfuration followed by deamination). The enzymes are called as *dehydratases* (serine dehydratase and threonine dehydratase) and *desulphydrases* (cysteine desulphydrase).
4. *Role of PLP in glycine metabolism:* The following reaction in glycine metabolism is a PLP dependent reaction.

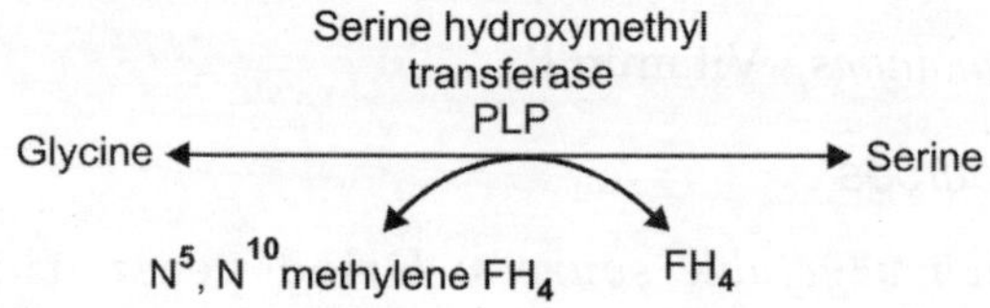


*Importance:* synthesis of serine (nonessential amino acid) and $FH_4$ derivative for further $C_1$ reactions.

5. *Role of PLP in metabolism of sulphur containing amino acids:* PLP is coenzyme for the enzyme cystathionine synthase and cystathionine lyase (cystathionase) enzymes of the metabolism of sulphur containing amino acids as shown in the following reactions:

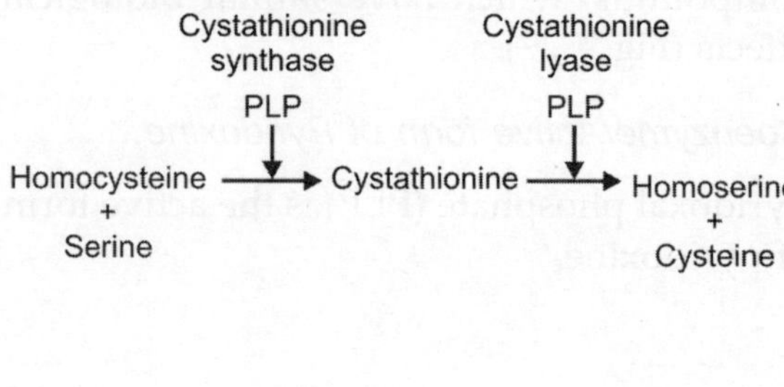

*Importance:* Deficiency of pyridoxine increases the homocysteine levels in blood. Increased homocysteine levels in blood are closely associated with myocardial infarction. Pyridoxine is used to maintain low homocysteine levels in therapy.

6. *Role of PLP in synthesis of niacin from tryptophan: Fig. 7.27.*

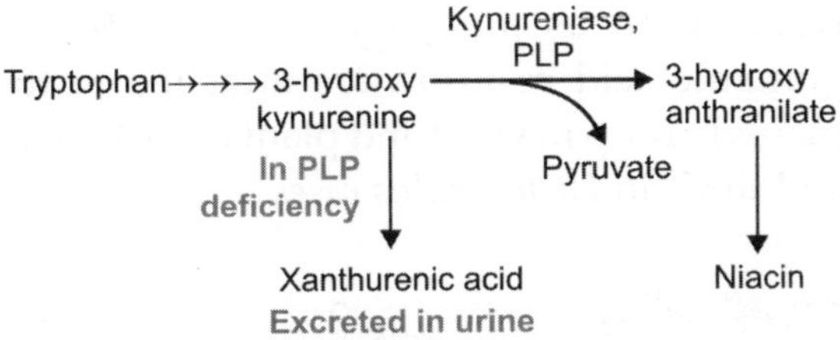


*Importance*: Deficiency of vitamin $B_6$ results in deficiency of niacin and pellagra like symptoms. Alternatively, kynurenine is converted to xanthurenic acid which is excreted in the urine.

7. *Role of PLP in heme synthesis:* PLP is coenzyme for the enzyme δ-ALA synthase the first reaction in heme synthesis as shown in the following reaction:

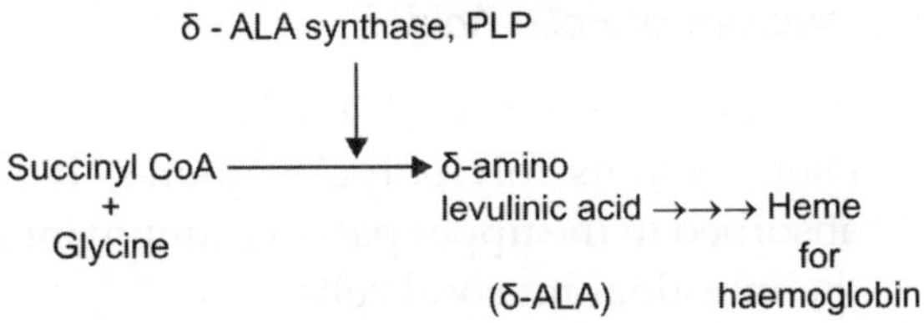


*Importance:* Deficiency of pyridoxine is associated with haematological symptoms–microcytic anaemia.

8. *Role of PLP in glycogenolysis:* Glycogen phosphorylase the enzyme of glycogenolysis requires PLP. Glycogen phosphorylase converts 90% of the glycogen to glucose-1-phosphate. Hence more than 70% of the body's PLP is concentrated with this enzyme in muscle/liver.

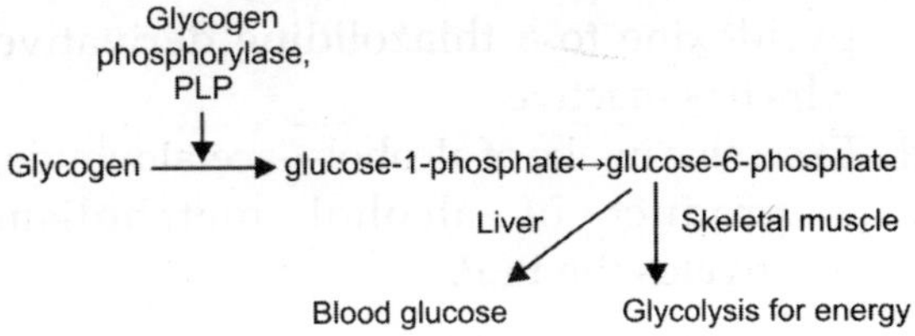


9. *Role of PLP in sphingolipid metabolism:* PLP is coenzyme for the enzymes involved in formation of sphingosine as shown in the following reaction. Sphingosine is the alcohol present in the sphingophospholipids and glycolipids.

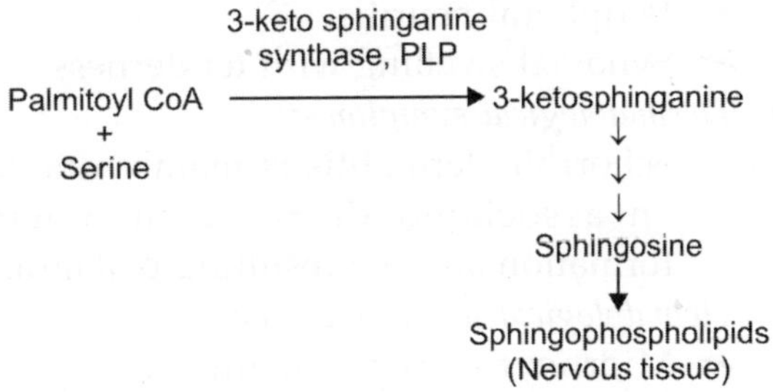


*Importance:* Deficiency of Vitamin $B_6$ is associated with demyelination of nerves resulting in peripheral neuritis.

### *Daily Requirement*

| | | |
|---|---|---|
| Adults | : | 1-2 mg/day |
| Pregnancy and lactation | : | 2.5 mg/day |

### *Deficiency of Pyridoxine*

*Causes:*

a. Administration of pyridoxine antagonists—methyl pyridoxine and isoniazide (antituberculosis drug which binds to pyridoxine to form pyridoxal hydrazone which is rapidly excreted in urine).
b. Intake of oral contraceptives and steroid drugs.
c. Intake of drugs—pencillamine which is used in the treatment of *rheumatoid arthritis and Wilson's disease,* changes

pyridoxine to a thiazolidine derivative which is inactive.

d. Excessive intake of alcohol—acetaldehyde, a product of alcohol metabolism inactivates the PLP.

### Deficiency Manifestations

Can be studied under three groups:

- *Neurological symptoms are due to decreased neurotransmitter production and demyelination of nerves:*
  - ➢ Convulsions due to decreased GABA and catecholamines.
  - ➢ Peripheral neuritis.
  - ➢ Synovial swelling with tenderness.
- *Dermatological symptoms:*
  - ➢ Seborrhic dermatitis is mainly due to an associated decrease in niacin formation and the resulting pellagra.
- *Hematological manifestations:*
  - ➢ Microcytic hypochromic anaemia (decrease in heme synthesis)
  - ➢ Impaired antibody formation is also seen.
- *Biochemical findings:*
  - ➢ Increased xanthurenic acid excretion in urine-xanthurenic aciduria
  - ➢ Homocystinuria.

## Folic Acid

*Synonyms*: Pteroyl glutamic acid (PGA). The groups of compounds which are related to folic acid are called as folacin.

### Sources

*Rich sources*—green leafy vegetables (spinach, cabbage, ladies finger, curry leaves, mint), yeast. Intestinal bacteria can synthesise folic acid.

*Moderate sources*—cereals, pulses (bengal gram, green gram, black gram) and oil seeds.

*Poor source*—milk, coconut, whole cereals.

### Chemistry (Fig. 7.30)

Folic acid is made up of three components:

- Pteridine ring—the colored part } Pteroyl group
- Para amino benzoic acid (PABA) } Pteroyl group
- Glutamic acid (amino acid)

The number of glutamic acid residues in naturally occurring folic acid can vary from three to seven in number. Hence they are referred to as *folylpolyglutamate*. Seven glutamic acid residues are attached to the pteroyl group in yeast and plants, five in liver and three in *Lactobacillus casei*.

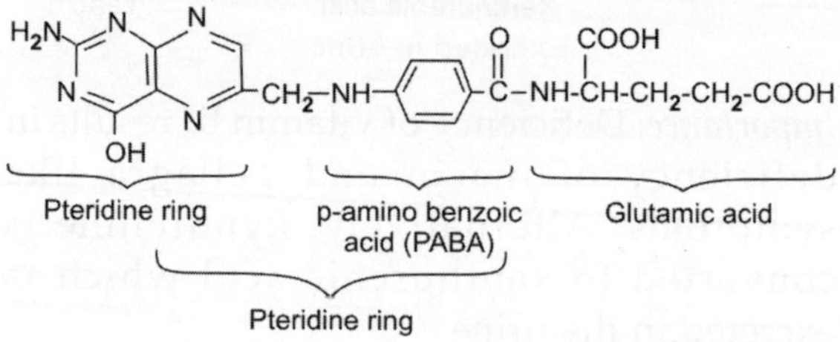


**Fig. 7.30:** Structure of folic acid—pteroyl monoglutamate

### Metabolism of Folic Acid

*Digestion and absorption: (Fig. 7.31)*

- Diet contains folylpolyglutamates. It is absorbed in the upper part of jejunum into the intestinal mucosal cells.
- In the lysosomes of the IM cells is present folylpolyglutamate hydrolase enzyme which hydrolyses the excess glutamate residues to form folic acid.
- Folic acid converted to tetrahydrofolate ($FH_4$) by the dihydrofolate reductase enzyme and then methylated to form $N^5$ methyl $FH_4$.
- In blood, $N^5$ methyl $FH_4$ is the major circulating form. It is transported by two β-globulins.

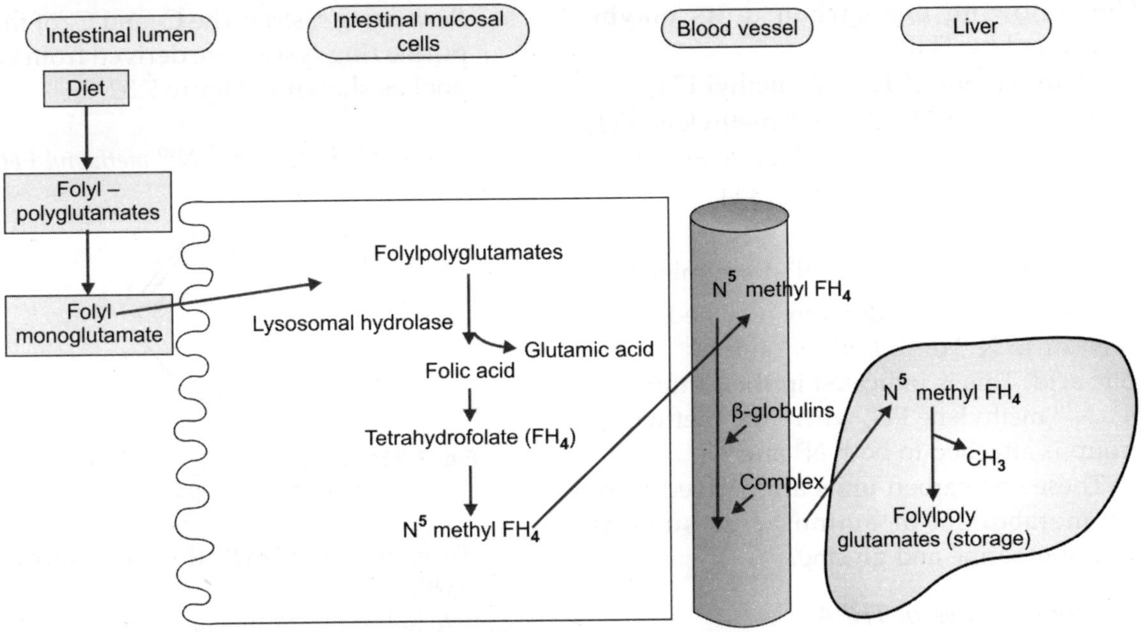


**Fig. 7.31:** Metabolism of folic acid

- In the liver, $N^5$ methyl $FH_4$ is converted back to folylpolyglutamate after transfer of the methyl group.
- In the tissues, there are different tissue specific folic acid transport molecules. The folic acid-transporter complex enters the cells by endocytosis.

### *Conversion of Folic Acid to its Active/Coenzyme Form*

Folic acid is converted to its active coenzyme form tetrahydrofolate as shown in the following reaction by a two step process catalysed by dihydrofolate reductase enzyme. $NADPH + H^+$ is the reducing equivalent used.

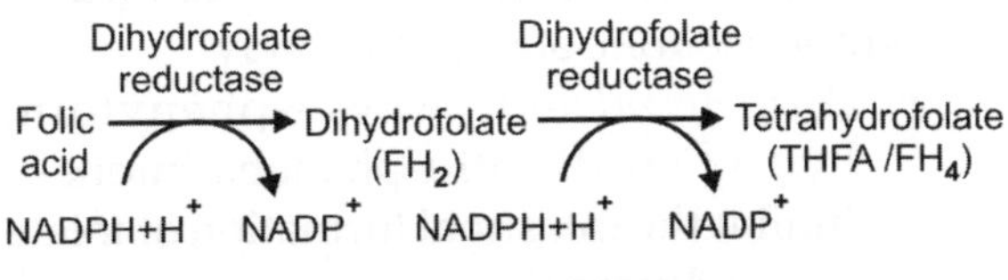


*Importance:* This enzyme is a target for cancer therapy.

### *Role of Tetrahydrofolate (THFA/$FH_4$)*

THFA acts as a carrier of one carbon unit during metabolism—it carries the one carbon unit from one reaction to another, *e.g.*

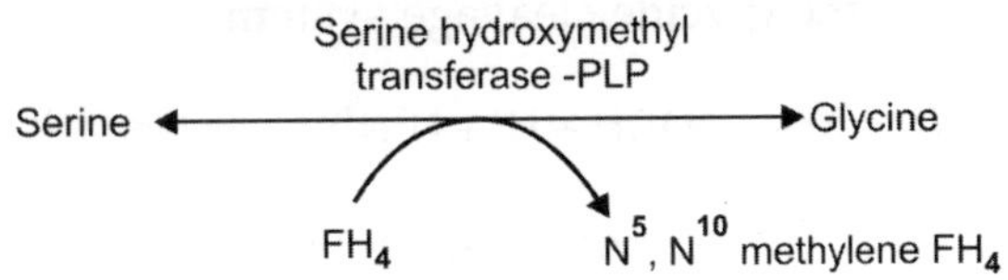


Methylene group is transferred to deoxyuridylic acid (dUMP) to form thymidylic acid (dTMP) by enzyme thymidylate synthase as shown in the following reaction:

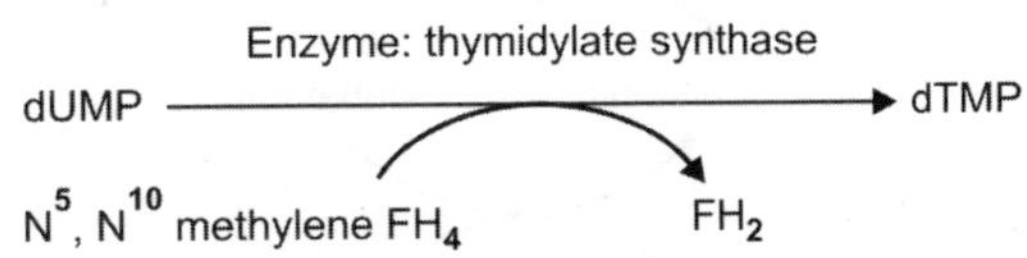

The following one carbon units maybe transferred by THFA.

- Methyl group ($CH_3$); $N^5$-methyl $FH_4$
- Methylene ($CH_2$); $N^5$, $N^{10}$-methylene $FH_4$
- Methenyl (–CH); $N^5$, $N^{10}$-methenyl $FH_4$
- Formyl (–CHO); $N^5$-formyl $FH_4$, $N^{10}$-formyl $FH_4$
- Formimino (–CH=NH); $N^5$- formimino $FH_4$

These one carbon units are attached either to $N^5$ or to $N^{10}$ or to both $N^5$ and $N^{10}$ of the folic acid. This is indicated in the formula as $N^5$, $N^{10}$ methylene $FH_4$ where the methylene group is attached to both $N^5$ and $N^{10}$.

These one carbon units are derived from the metabolism of amino acids such as histidine, serine and glycine.

### *Biochemical Role of THFA*

1. Transfer of one carbon unit as in the following reactions:
   a. Serine hydroxymethyl transferase: *Reaction shown earlier.*
   b. Glycine cleavage system:

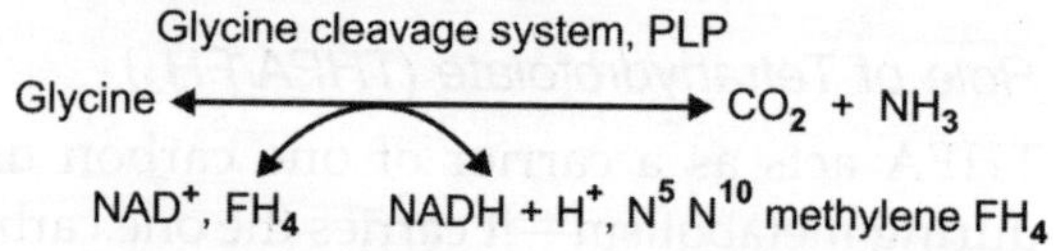


   c. *Synthesis of glycine:* It is the reverse of the glycine cleavage system.

$$CO_2 + NH_3 + NADH + H^+ + N^5N^{10}\text{ methylene } FH_4$$
$$\downarrow$$
$$\text{Glycine} + NAD^+ + FH_4$$

   d. *Homocysteine methyl transferase:*

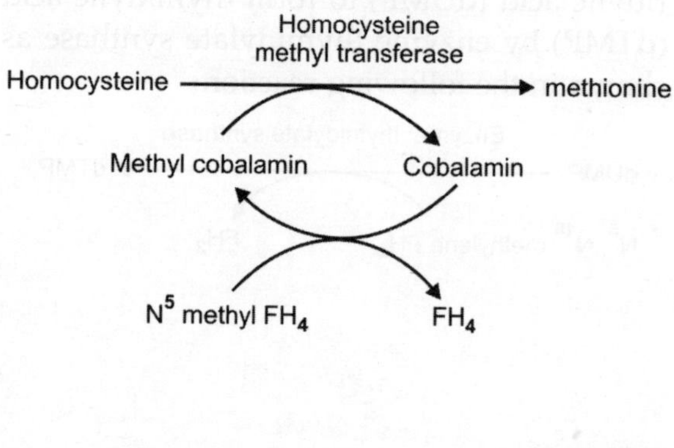


   e. *Purine ring system:* The $C_2$ and $C_8$ of the purine ring system are derived from $C_1$ pool as shown in Figure 7.32.

$C_2$ - $N^{10}$ *formyl* $FH_4$, $C_8$ – $N^5$ $N^{10}$ *methenyl* $FH_4$

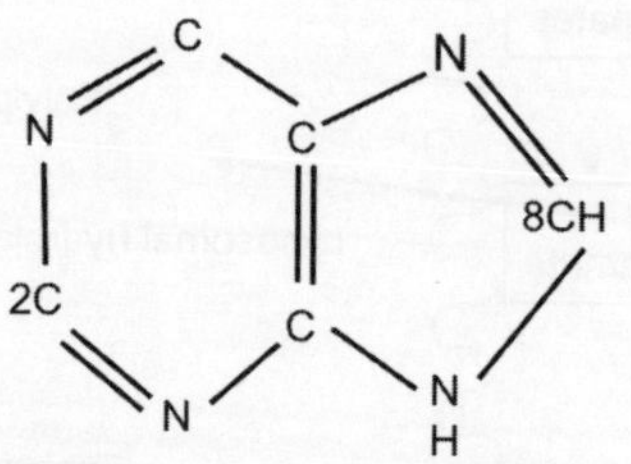


**Fig. 7.32:** Purine ring – $C_2$ and $C_8$ are from one carbon pool

   f. *Formation of dTMP:* Reaction shown earlier.

THFA helps in synthesis of purines and pyrimidines by transfer of $C_1$ groups in all tissues. Purines and pyrimidines are required during mitosis for DNA synthesis in all cells including maturing blood cells—erythropoiesis (synthesis of RBCs) and leucopoiesis (synthesis of WBCs).

*Inter-conversion and fate of the $FH_4$ derivatives:* See in the box on page no. 229.

### *Deficiency of Folic Acid*

*Causes:*

- Dietary deficiency—diets with less green leafy vegetables, non vegetarian diets
- Defective absorption as seen in disorders related with the GI tract—*celiac disease, sprue*
- Increased requirement is not compensated —pregnancy, haemolytic anaemia.
- *Drugs:* Destroy intestinal microflora which synthesise folic acid in the body.
  - ➢ *Anticonvulsant drugs*—phenytoin, hydantoin, dilantin, phenobarbitone—inhibit the intestinal formation of folate monoglutamate.
  - ➢ *Folate antagonists:* methotrexate

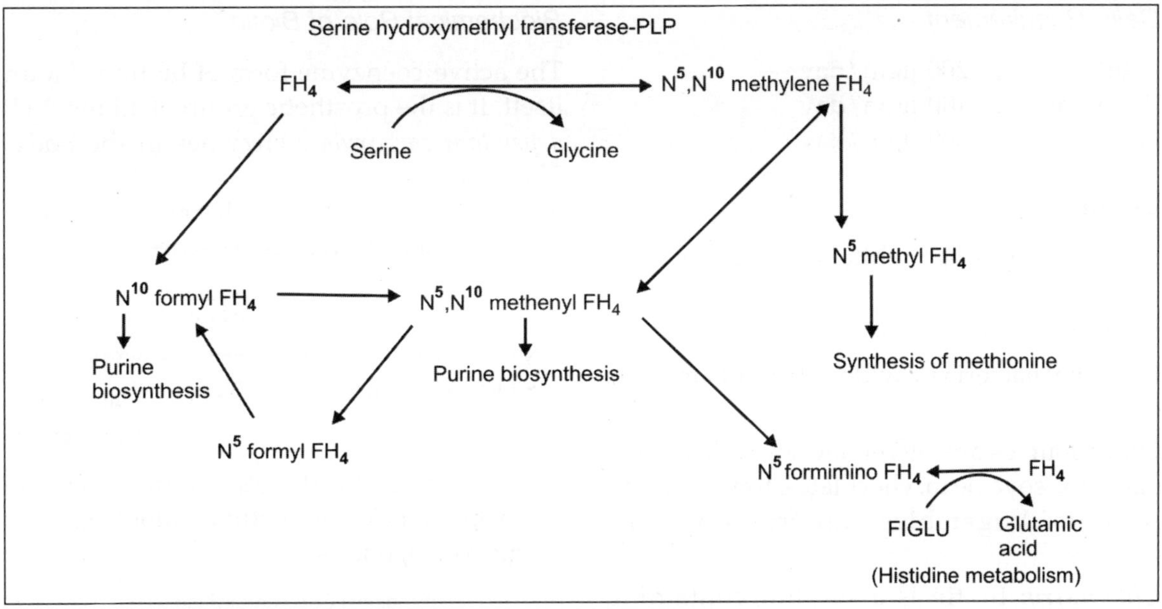


- *Antibiotics/antibacterials*—sulphonamide group of drugs —decrease the intestinal bacterial synthesis of folic acid by competitive inhibition.
- *Trimethoprim*–decrease the formation of THFA.
- *Antimalarial* drugs.
- *Anti cancer drugs*—aminopterin, amethopterin are competitive inhibitors of dihydrofolate reductase enzyme.

### Deficiency Manifestations

THFA derivatives are mainly required for the synthesis of DNA as explained above. In folate deficiency, the process of mitosis is arrested due to decrease in the synthesis of DNA. This is very evident in the blood cells. Even though folic acid is required in all the tissues, the tissues first affected are the RBC, leucocytes and intestinal epithelium.

- *Anaemia*—Macrocytic anaemia is the characteristic blood picture in folate deficiency. This is because the DNA is not synthesised but haemoglobin is synthesised in the RBC precursors. The nucleus is not well developed but the cytoplasm is well developed. Hence the RBCs are said to be macrocytes.
- Reticulocytosis
- Howell Jolly bodies and Cabot rings (remnants of nucleus destruction) are seen.
- Abnormal nucleated RBCs are seen in the peripheral blood smear which is destroyed in the spleen resulting in anaemia.
- Leucopenia – ↓ WBCs
- Thrombocytopenia – ↓ platelets
- Homocysteinaemia – ↑ Homocyteine levels in blood. This increases the risk of coronary heart diseases.
- *Congenital birth defects:* Deficiency of folic acid leads to fetal neural tube defects resulting in malformations like *spina bifida.*
- Vomiting, diarrhoea and abdominal pain are associated with folate deficiency.

### Daily Requirement

| | | |
|---|---|---|
| Adults | : | 200 μgm/day |
| Pregnancy | : | 400 μgm/day |
| Lactation | : | 300 μgm/day |

## Biotin

*Synonyms*: Anti egg white injury factor.

### Sources

Intestinal bacteria provide sufficient amounts of biotin.

Other sources are—liver, kidney, yeast, milk, egg yolk, soya bean, chocolates, tomatoes and peanuts. Vegetables and fruits are fair sources.

***Chemistry:*** Biotin is a sulphur containing vitamin with an imidazole ring fused to a1 thiophene ring with a valeric acid side chain (Fig. 7.33).

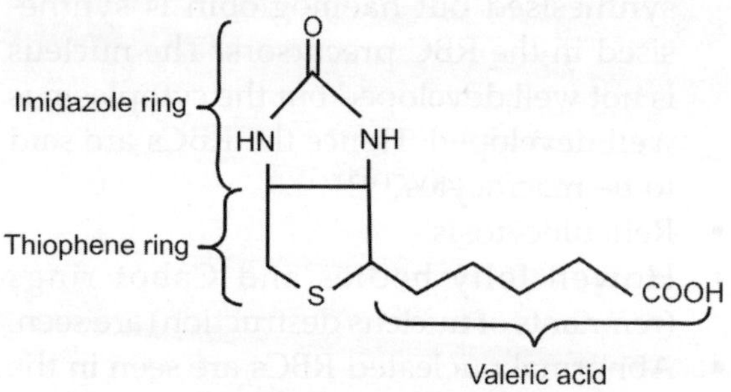


**Fig. 7.33:** Structure of biotin—Imidazole + Thiophene + Valeric acid

The carboxyl group of the valeric acid is linked to the ε-amino group of lysine residue on the apoenzyme.

### Metabolism

| | | |
|---|---|---|
| Synthesis | : | intestinal microflora |
| Absorption | : | small intestine |
| Transport | : | by blood to various parts of the body |

In liver it is present in the form of biocytin (adenylated biotin).

### Biochemical Role of Biotin

The active/coenzyme form of biotin is biotin itself. It is the prosthetic group of all the ATP *dependant carboxylase* enzymes in the body. They are:

- *Pyruvate carboxylase*: Which converts pyruvate to oxaloacetate in gluconeogenesis.

$$\underset{\text{(3C compound)}}{\text{Pyruvate + ATP + }CO_2} \xrightarrow[\text{Mn}^{++}\text{, Biotin}]{\text{Pyruvate carboxylase,}} \underset{\text{(4C compound)}}{\text{Oxaloacetate + ADP + Pi}}$$

- *Propionyl CoA carboxylase:* Which converts propionyl CoA to methyl malonyl CoA in gluconeogenesis.

$$\text{Propionyl CoA + ATP + }CO_2 \xrightarrow[\text{Mn}^{++}\text{, Biotin}]{\text{Propionyl CoA carboxylase}} \text{Methylmalonyl CoA + ADP + Pi}$$

- *Acetyl CoA carboxylase* which converts acetyl CoA to malonyl CoA for de novo biosynthesis of fatty acid.

$$\text{Acetyl CoA + ATP + }CO_2 \xrightarrow[\text{Mn}^{++}\text{, Biotin}]{\text{Acetyl CoA carboxylase,}} \text{Malonyl CoA + ADP + Pi}$$

- *β-methylcrotonyl CoA carboxylase* of leucine metabolism:

$$\beta\text{-Methyl crotonyl CoA} \xrightarrow{\beta\text{-methyl crotonyl CoA carboxylase, biotin}} \beta\text{-methyl glutaconyl CoA}$$

The carboxylase enzymes bring about fixation of $CO_2$. The mechanism of carbon dioxide fixation involves the following steps: (Fig. 7.34).

- Biotin is bound by an amide linkage to the ε-amino group of a lysine residue on the apoenzyme of the carboxylase enzyme to form the holoenzyme which is active.

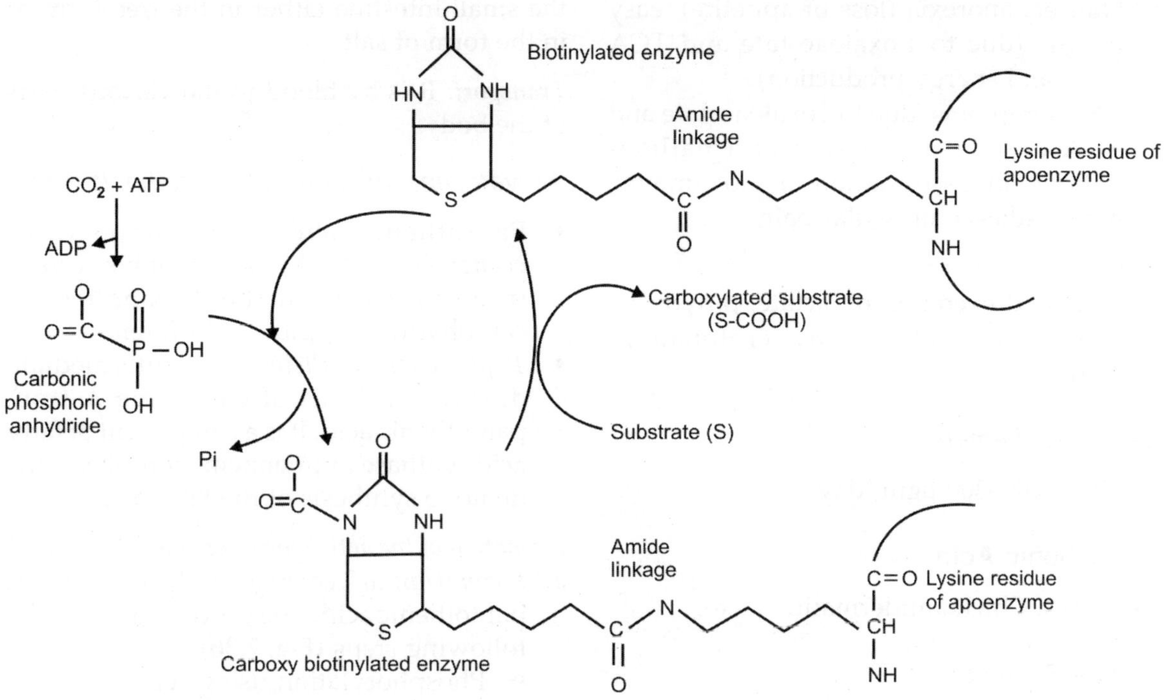


**Fig. 7.34:** Mechanism of carbon dioxide fixation—biotin dependant carboxylation

- Hence, biotin is attached to the apoenzyme of the carboxylase by a long, flexible arm of one nitrogen and nine carbon atoms.
- $CO_2$ in the form of $HCO_3^-$ combines with the biotinylated enzyme in presence of ATP to form carboxy biotinylated enzyme.
- The carboxyl group is then transferred from the biotin to the substrate (Pyruvate/ Propionyl CoA/Acetyl CoA/β-methyl crotonyl CoA) to form the carboxylated product (oxaloacetate/methyl malonyl CoA/malonyl CoA/β-methyl glutaconyl CoA) respectively.

***Deficiency of Biotin:***

*Causes*: Deficiency is rare in humans.

- Biotin deficient diets
- Intake of antibiotic sulfa group of drugs destroys the biotin synthesizing intestinal microflora.
- Intake of biotin antagonist—egg white contains a glycoprotein *avidin* (biotin antagonist) which binds to biotin and makes it unavailable for absorption. One avidin molecule combines with four molecules of biotin. Consumption of 4-10 raw egg whites/day makes the individual prone to biotin deficiency.
- Genetic deficiency of enzyme *holocarboxylase synthase* which catalyses binding of biotin to the apocarboxylase enzyme.
- *Leiner's disease*—seen in breastfed infants with diarrhoea.

*Cause:* Low biotin content of human milk with loss of biotin due to diarrhoea.

*Deficiency Manifestations*

- Epithelial tissue manifestations—dermatitis, atrophic glossitis, alopecia (loss of hair).

- Nausea, anorexia (loss of appetite), easy fatigue (due to ↓oxaloacetate and TCA cycle and energy production).
- CNS symptoms (due to ↓oxaloacetate and TCA cycle)—depression and hallucinations, inco-ordination of movements, hyperesthesia, muscular pain.

*Biochemical findings:*
- Increased excretion of lactate, β-hydroxy propionate, and β-methyl crotonate in urine.

### Daily Requirement

Adults : 200–300 μgm/day

## Pantothenic Acid

*Synonyms*: Chick antidermatitis factor

### Dietary Sources

- Synthesised by the normal intestinal microflora.
- Good sources—yeast, liver, organ meat, eggs, whole cereals and legumes.
- Poor sources—fruits and vegetables.

***Chemistry***: Contains β-alanine and dihydroxy, dimethyl butyric acid (pantoic acid) (Fig. 7.35).

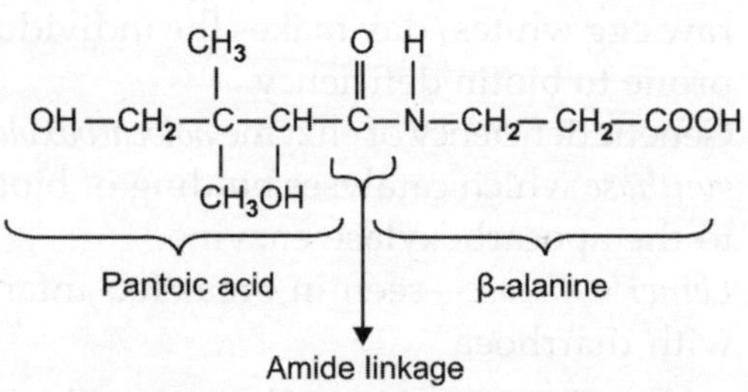


**Fig. 7.35:** Structure of pantothenic acid

### Metabolism

*Site:* Pantothenic acid is released from the dietary sources by the action of intestinal phosphatases. Pantothenate is absorbed from the small intestine either in the free form or in the form of salt.

*Transport:* It is by blood to the various parts of the body.

### Active/coenzymic form of Pantothenic Acid

- Pantothenic acid is a component of *coenzyme A* (CoA) which plays a very important role in the metabolism of carbohydrates, lipids and proteins.
- *4′ phosphopantotheine* is an intermediate during synthesis of coenzyme A from pantothenic acid. It is a component of fatty acid synthase multienzyme complex of the de novo synthesis of the fatty acids.

*Formation of the active forms of pantothenic acid:*

a. *Formation of coenzyme A*: It is from pantothenic acid. The process involves the following steps (Fig. 7.36).
   - ➢ Phosphorylation using ATP
   - ➢ Cysteinylation—addition of cysteine. Energy: ATP →ADP + Pi
   - ➢ Decarboxylation - loss of $CO_2$ from the attached cysteine residue and formation of 4′ phosphopantotheine
   - ➢ Adenosyl transfer—transfer of AMP group from ATP
   - ➢ Phosphorylation of the adenosyl group using ATP.

- Synthesis of one molecule of coenzyme A needs four high energy bonds. CoA has a free thiol/sulphydryl group at one end of the molecule. Hence, it is abbreviated as HS-CoA/CoA-SH.

*Biochemical Role of Pantothenic Acid:*

1. This thiol group carries the acyl group by forming a thioester linkage. The thioester linkage in the acyl CoA is a high energy bond.

The CoA-SH acts as a

- Carrier of acyl group to an acceptor: *For example,* formation of acetyl choline from choline and acetyl CoA.

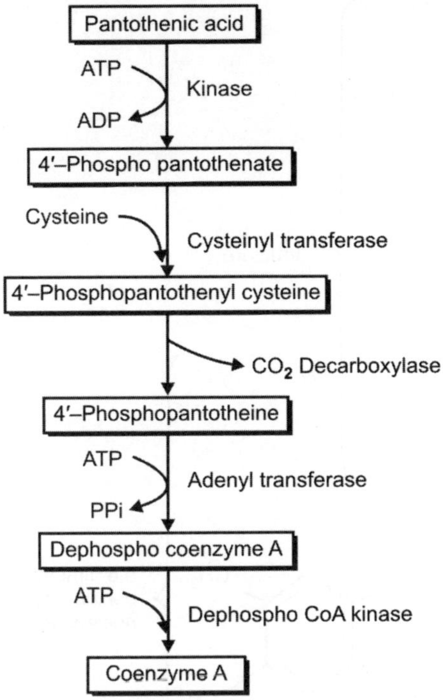


**Fig. 7.36:** Formation of coenzyme A

- Acceptor of acyl group during metabolism:
  *For example,* acetyl group from pyruvate to form acetyl CoA.

*Examples*:
- Acetyl CoA formation from pyruvate
- Succinyl CoA from α-ketoglutarate

2. *Acyl carrier protein (ACP):* ACP is a component of the fatty acid synthase complex and carries 4′-phosphopantotheine as a prosthetic group. The free SH group of 4′-phosphopantotheine acts as a carrier of the malonyl/acyl group to form malonyl/acyl ACP. The long flexible 4′-phosphopantotheine enables the acyl group to be transferred from one enzyme site to another on the multienzyme complex. This prevents the loss of acyl group by diffusion till the process is complete.

### *Deficiency of Pantothenic Acid*

*Causes:*
- Deficient diets
- Intake of pantothenate antagonists—ω-methyl pantothenate, pantoyl taurine.

*Deficiency manifestations in humans:*
- Gastrointestinal symptoms—abdominal cramps, fatigue, fatty liver.
- CNS symptoms—restlessness, growth failure, sleep disturbances.
- Paraesthesia of feet (burning, lightning pain)—burning feet syndrome with incoordination of movements (staggering gait).
- Skin manifestations—dermatitis, alopecia, graying of hair
- Anaemia— ↓porphyrin and haemoglobin synthesis.
- Adrenal necrosis— ↓ decreased steroid synthesis

### *Daily Requirement*

4–6 mg/day.

## Vitamin $B_{12}$

*Synonyms:* Cobalamin, extrinsic factor of Castle, antipernicious anaemia factor, Lactobacillus lactis Dorne factor, Cobamides.

### *Sources*

Vitamin $B_{12}$ is totally absent in plant sources.
*Rich sources:* Liver (site of storage)
*Good sources:* Meat, fish, egg and curds (source for vegetarians).

### *Chemistry (Fig. 7.37)*

- It has a tetrapyrrole (I, II, III, IV) ring structure called as the corrin ring.

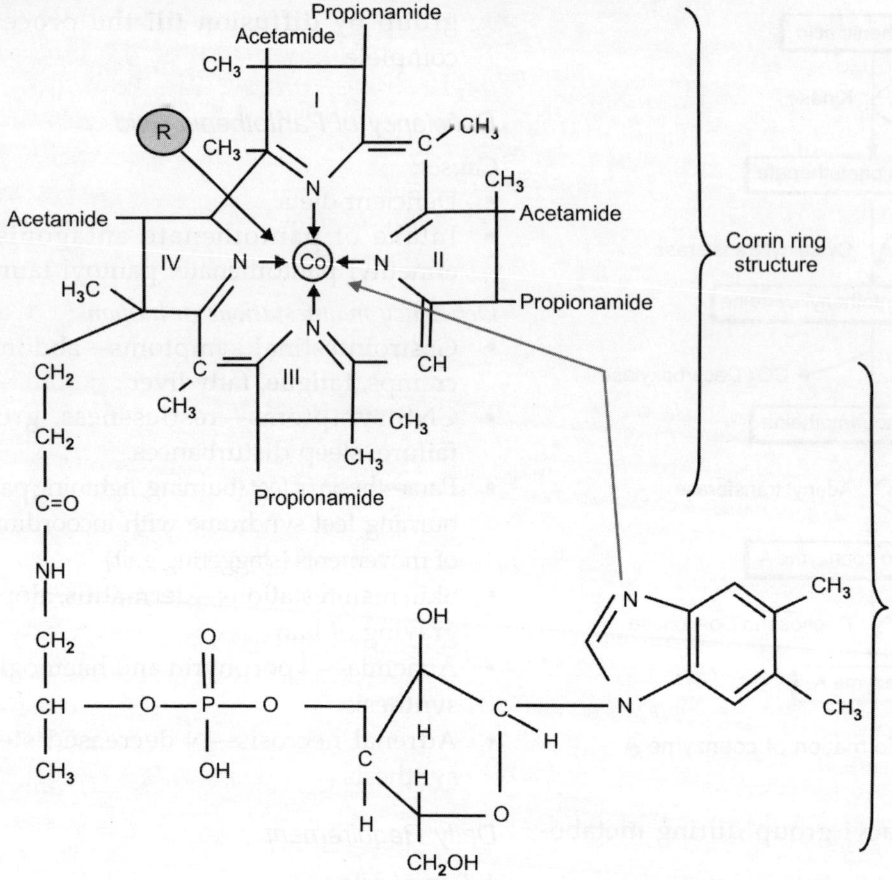


**Fig. 7.37:** Structure of cobalamin

- The four pyrroles are linked by 4 bridges. Three methane bridges and the one direct bridge.

The corrin ring has an atom of cobalt (Co) in the centre. The atom of Co is coordinated with the nitrogens of the four pyrrole rings. The fifth valency is covalently linked to a 5, 6 dimethyl benzimidazole ring (a nitrogenous base). Based on the group satisfying the sixth valency, $B_{12}$ is called as in Table 7.2.

**Table 7.2:** Types of cobalamin

| *R Group* | *Type of cobalamin* | *Importance of the cobalamin* |
|---|---|---|
| Cyanide, $CN^-$ | Cyanocobalamin | Therapeutic preparations |
| Hydroxo, $OH^-$ | Hydroxocobalamin | Injectible form |
| Deoxyadenosine | Deoxyadenosylcobalamin | Storage form, active form |
| Methyl, $CH_3$ | Methylcobalamin | Circulatory form, active form |
| Nitro | Nitrocobalamin | Cobalt compounds with $B_{12}$ like activity |
| Chloro | Chlorocobalamin | |

### *Metabolism of Vitamin $B_{12}$: (Fig. 7.38)*

*Form in the diet*: In the diet vitamin $B_{12}$ is found as complexes of methyl cobalamin and deoxyadenosylcobalamin with proteins.

*Release of vitamin $B_{12}$*: It is in the gastric lumen by the action of HCl and pepsin or trypsin.

*Site of absorption*: Microvillous surface of the ileum part of the small intestine.

*Requirement for absorption:* HCl and Intrinsic factor of Castle.

Intrinsic factor is a dimeric glycoprotein secreted by gastric parietal cells.

Before it is absorbed, vitamin $B_{12}$ is dissociated from protein complexes by HCl and pepsin. The free $B_{12}$ now complexes with Castle's intrinsic factor to form a vitamin $B_{12}$–intrinsic factor complex.

One intrinsic factor binds two molecules of vitamin $B_{12}$. Vitamin $B_{12}$ –Castle's intrinsic factor complex is transported to the ileum.

*In the ileum:* The microvillous surface of the intestinal mucosal cells of the ileum has specific receptors for this complex. The complex will bind to these receptors in presence of divalent cations and the whole complex is internalized. In the ileal cells the complex dissociates and the vitamin $B_{12}$ is transferred to transcobalamin II.

*Transport:* Transcobalamin I and transcobalamin II, both globulins, transport the vitamin to the tissues. Methylcobalamin is the predominant form in blood.

*In the tissues:* The transcobalamin II-vitamin $B_{12}$ complex is internalised by a specific cell surface receptor mediated mechanism in all the tissues including the liver. The complex dissociates in the cytoplasm and the cobalamin is converted into either of its active forms–methylcobalamin or deoxyadenosyl cobalamin.

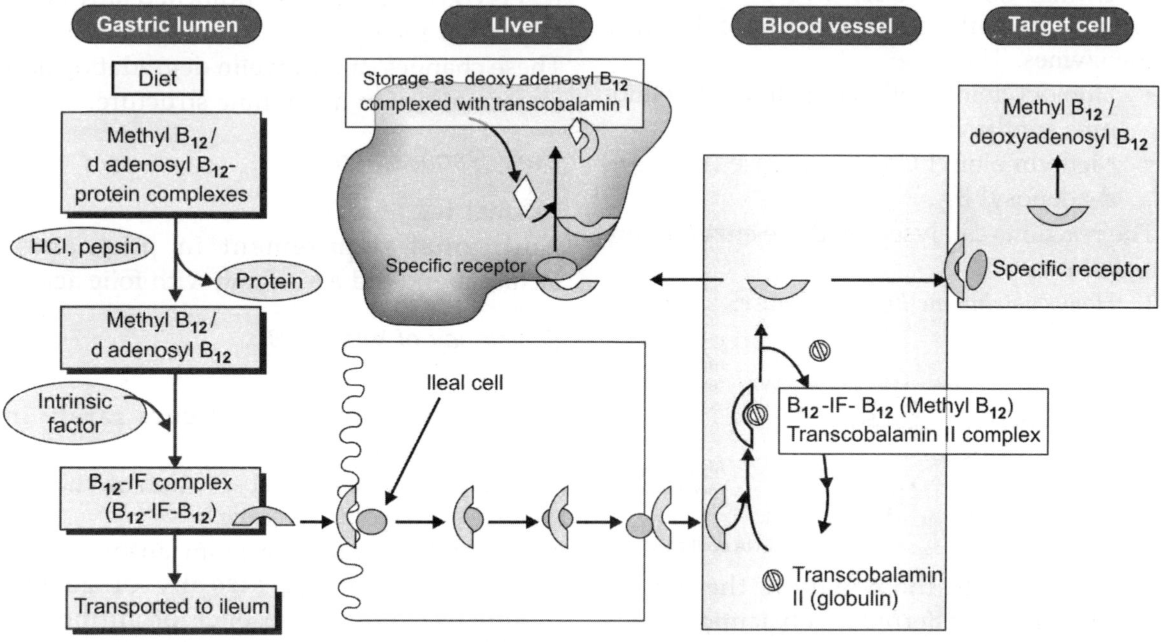


**Fig. 7.38:** Metabolism of cobalamin

*Storage:* Vitamin $B_{12}$ is stored in the liver cells in form of deoxyadenosyl cobalamin. Deoxyadenosyl cobalamin complexes with a protein–transcobalamin 1/transcorrin, for storage. Vitamin $B_{12}$ is the only water soluble vitamin which is stored in the body.

About two milligrams of vitamin $B_{12}$ is stored in the whole liver which is sufficient for about two years. Cobalamin is capable of being secreted in the bile and undergoing enterohepatic circulation because of its water soluble property.

### Coenzymic/Active Forms of Cobalamin

There are two active forms of cobalamin:

- Methyl cobalamin—methyl $B_{12}$
- Deoxyadenosyl cobalamin—d-adenosyl $B_{12}$

These coenzymes are also called as *cobamide coenzymes*.

### Biochemical Functions of Vitamin $B_{12}$

In animal tissues, only two enzymes are known to contain/require the cobamide coenzymes. They are:

- Homocysteine methyl transferase (methionine synthase) - methyl $B_{12}$.
- Methylmalonyl CoA isomerase (mutase)-d-adenosyl $B_{12}$.

The reactions catalysed by these enzymes are as follows:

1. *Homocysteine methyl transferase*:

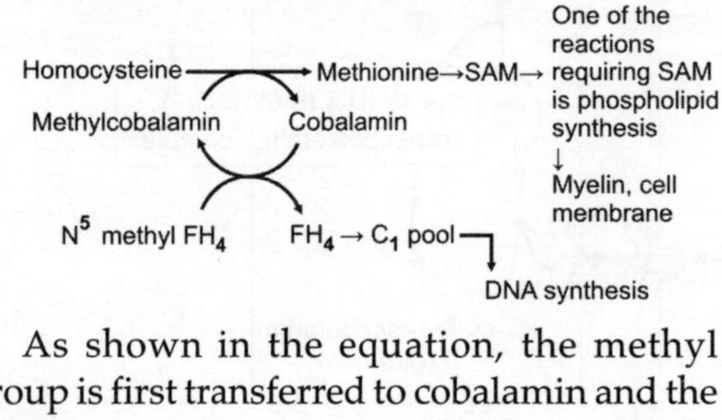


As shown in the equation, the methyl group is first transferred to cobalamin and the THFA is released for further action. Methylcobalamin then transfers the methyl group to homocysteine to form methionine. In deficiency of cobalamin, the methyl group is trapped as methyl THFA which cannot be used for any other reaction. This leads to folate deficiency as well. Hence it is called as *folate trap*.

2. *Methylmalonyl CoA isomerase:* This is an enzyme of gluconeogenesis.

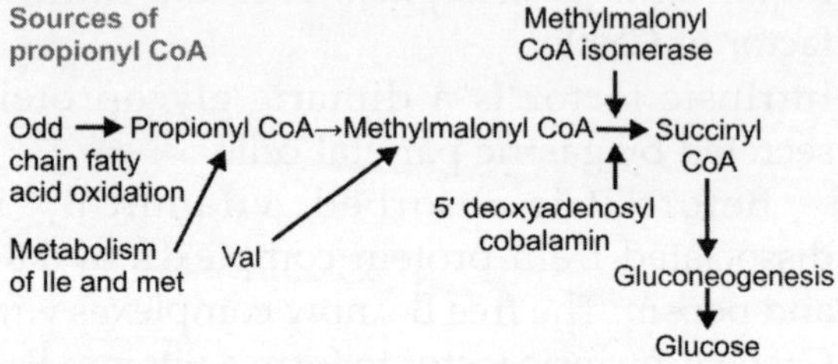


In deficiency of vitamin $B_{12}$, the propionate/propionyl CoA gets converted to methyl malonyl CoA which accumulates. The accumulated methylmalonyl CoA interferes in fatty acid synthesis and myelin formation. It gets converted to other branched fatty acids which are incorporated in membrane lipids. These changes cause myelin degradation and abnormalities in membrane structure.

### Daily Requirement

Normal 1–2 μgm/day.
Additional requirement in pregnancy, lactation and old age, along with folic acid.

### Deficiency of Vitamin $B_{12}$

*Causes:*

- Dietary deficiency—Occurs rarely in vegetarians.
- Failure of intestinal absorption of vitamin $B_{12}$ – *"Pernicious anaemia"*
- Pernicious anaemia is an autoimmune disorder causing atrophy of gastric mucosa due to presence of antibodies either to parietal cells or to intrinsic factor.

*Causes of pernicious anaemia:*

- Gastric atrophy causes intrinsic factor as well as $B_{12}$ deficiency.
- Decreased absorption—Surgical procedures of stomach and small intestine.
- Congenital defects of Intrinsic factor maybe at the level of:
  - Synthesis
  - Secretion
  - Defect in intestinal receptor of Intrinsic factor
  - Presence of auto antibodies against parietal cells.
- Presence of fish tapeworm—*diphillobothrium latum.* This tapeworm is present in fish. The population which eats this live fish will get deficiency of the vitamin $B_{12}$ due to nonavailability of the vitamin for absorption.
- Old Age—Unable to absorb the vitamin $B_{12}$.
- Pregnancy—The requirement is higher and so diet should contain more vitamin $B_{12}$.
- Genetic defects of enzymes which synthesise deoxyadenosyl cobalamin also produce vitamin $B_{12}$ deficiency.

### *Manifestations of Vitamin $B_{12}$ Deficiency*

The deficiency manifestations can be grouped into two:

- Macrocytic anaemia
- Neurological symptoms
- *Macrocytic anaemia:* Due to deficiency of both folic acid and $B_{12}$.
  *Symptoms:*
  - ➢ Megaloblastic (large basophilic cells-megaloblasts) macrocytic (abnormally large RBCs) anaemia
  - ➢ Abnormal and immature RBCs in the bone marrow and peripheral circulation. (Megaloblasts with reticulated nuclei, less haemoglobin, high RNA: DNA ratio).
  - ➢ ↓RBC count and haemoglobin content.
  - ➢ Hyperplastic bone marrow.

*Cause:* Deficiency of vitamin $B_{12}$ causes folate trap, leads to non availability of folate for synthesis of DNA and decreases the maturation of the red cells leading to anaemia.

*Absence of cobalamin:*

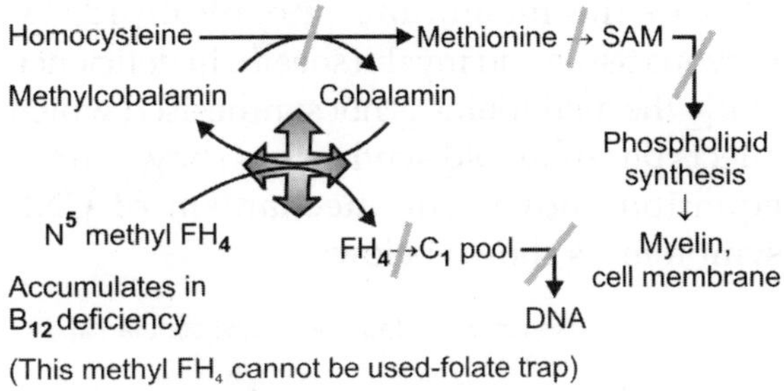


*Accumulates in $B_{12}$ deficiency (This methyl $FH_4$ cannot be used-folate trap).*

### *Neurological Symptoms:*

"Subacute Combined degeneration" due to

- Progressive demyelination of nerves
- Degeneration of the spinal cord
- Neuronal dysfunction.

This affects the central nervous system like cerebral cortex, fibres of dorsal column and pyramidal tract of the spinal cord—both sensory and motor functions.

*Symptoms:*

*Early stages of deficiency*:

- Symmetrical paresthesia of extremities (numbness and tingling sensations of hands and feet).
- Altered deep senses and tendon reflexes.
- Loss of equilibrium
- Unsteady gait
- Romberg's sign—loss of balance when eyes are closed.
- Babinski's sign is positive—extensor plantar reflex.

*At later stages:*

- Loss of memory
- Confused state
- Psychosis

*Reason:* Methylcobalamin is needed for synthesis of methionine which is a methylating agent for synthesis of phospholipid, phosphatidyl choline, an important component of the membrane phospholipids of nervous tissue and myelin sheath. In deficiency of $B_{12}$, the methionine is not synthesised which affects phospholipid synthesis as shown in the equation above. The mechanism of CNS symptoms is shown below:

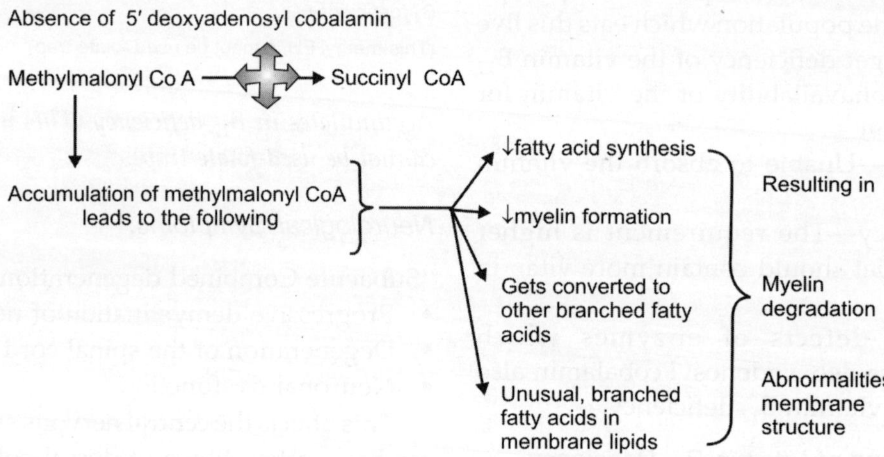


## Choline

### Sources

*Good sources:* Rice polishings, vegetables, milk, egg and liver.

It can be synthesised in the body from serine.

### Chemistry (Figure 7.39)

It is N-trimethyl ethanolamine (hydroxyethyl trimethylamine).

### Biochemical Functions of Choline

It is a nitrogenous base found in phospholipids-phosphatidyl choline (Lecithin).

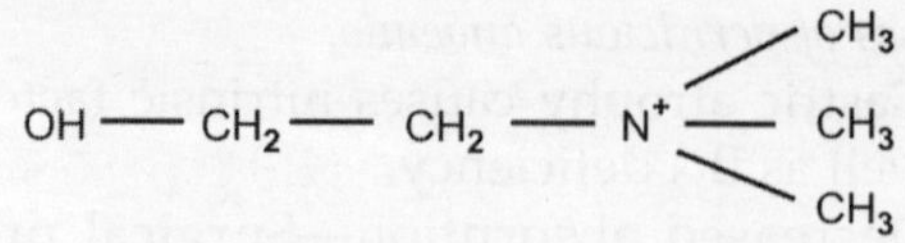

**Fig. 7.39:** Structure of choline

a. Lecithin is a very important constituent of the cell membrane of all tissues.
b. It is a constituent of the myelin sheath of the myelinated nerves.
c. It is a constituent of the lipoprotein complexes which are involved in transport of fat and cholesterol from the liver.

Choline deficiency involves the deposition of fat in the liver leading to fatty liver. Choline prevents fatty liver (lipotropic factor) and liver cirrhosis.

1. Choline is used in the synthesis of acetyl choline (Ach)—a neurotransmitter (Fig. 7.40).
2. Choline can donate three methyl groups to the $C_1$ pool for transmethylation reactions.

### Deficiency of Choline

*Symptoms:*

- Demyelination of nerves
- Fatty liver

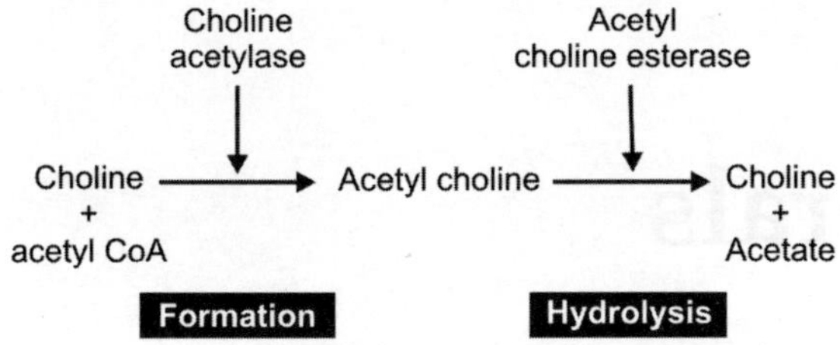


**Fig. 7.40:** Metabolism of acetyl choline

- ↓synthesis of acetylcholine—relaxation of muscles.

### Inositol

***Chemistry:*** *(Fig. 7.41)* It is hexahydroxy cyclohexane.

Myo-inositol is the main isomer which is found in tissues.

*Sources:* yeast, cereals, milk.

#### *Biochemical Functions of Inositol*

- It is non nitrogenous and found in the phospholipid—phosphatidyl inositol, important constituent of the cell membrane, heart muscle.
- In response to extracellular signals the following reaction takes place:

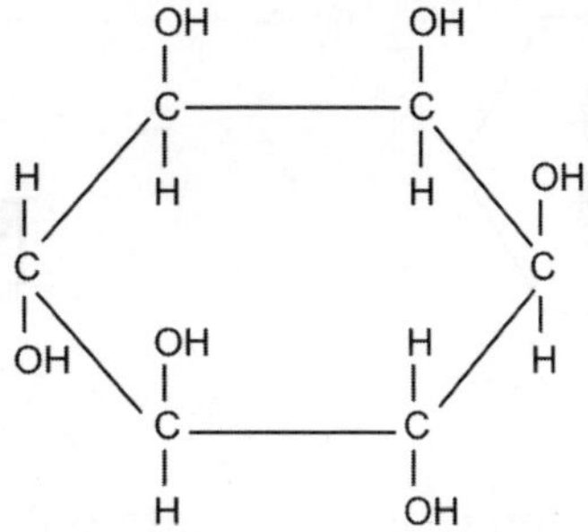

**Fig. 7.41:** Structure of inositol

Phosphatidyl inositol →1, 2-DAG + Inositol triphosphate ($IP_3$)

Both, $IP_3$ and 1, 2-DAG act as second messengers in the target tissues.

- Phytol—inositol hexaphosphate is found in the vegetables which complexes with $Ca^{++}$ and $Fe^{+++}$ and decreases their absorption from the GI tract.
- It prevents fatty liver—*lipotropic factor.*
- Inositol is known to be a requirement for growth of fibroblasts in tissue culture.
- Diabetic peripheral neuropathy is known to have a link with inositol metabolism.

# 8 Minerals

Although minerals constitute only a small proportion (4%) of the body tissue, they are essential as structural components and in many vital processes. Some form hard tissues, such as bones and teeth; some are present in the fluids and soft tissues as integral parts of organic compounds in the body, such as iron in Hb, iodine in thyroxine, cobalt in vitamin $B_{12}$, sulphur in amino acids and in the vitamins, thiamin and biotin. Various minerals act as cofactors, a requirement for activity of enzymes.

## BIOMEDICAL IMPORTANCE

A deficiency of any one of the minerals may result in disorders like anaemia (deficiency of iron), cretinism and goiter (deficiency of iodine), tetany (decreased levels of plasma calcium).

Increased intake may also be harmful. *For example,* increased intake of sodium (as NaCl) may cause hypertension in susceptible individuals, fluoride may cause fluorosis.

Minerals may be classified as:

1. *'Macrominerals'*, which are required in amounts greater than 100 mg/day and include sodium, potassium, chloride, calcium, phosphorus, magnesium and sulphur.
2. *'Microminerals or trace elements'*, which are required in amounts less than 100 mg/day. This includes iron, copper, iodine, manganese, zinc, molybdenum, selenium, cobalt and fluorine.

## CALCIUM

Calcium contributes 1.5-2.2% of the human body by weight.

In a normal adult approximately 1200 g of calcium is present.

### *Sources*

Milk, cheese, dark green leafy vegetables, soft fish bones, dried legumes and figs.

### *Recommended Dietary Allowance (RDA)*

| | | |
|---|---|---|
| Adult man/woman | : | 800 mg |
| Pregnant/lactating woman | : | 1.2 g |
| Infant | : | 300-500 mg |

### *Functions of Calcium*

1. It is a constituent of bones and teeth.
2. It is required for blood coagulation.
3. Calcium is involved in the regulation of the activity of many enzymes, e.g., phospholipase $A_2$, glycogen phosphorylase.
4. Calcium helps in the transmission of nerve impulses.
5. Calcium is essential for the regulation of both skeletal and smooth muscle contraction.
6. Calcium plays a role in permeability of membranes and gap junctions. Calcium reduces neuromuscular excitability by reducing membrane permeability.
7. Calcium acts as a second messenger in mediating the cellular effects of many hormones such as vasopressin and thyroid releasing hormone.

### Absorption

About 40% of ingested calcium is absorbed actively. Absorption occurs mainly in the duodenum and first half of jejunum, with the help of a calcium binding protein and a $Ca^{++}$ dependent ATPase.

*Factors affecting absorption*

1. *Ca: P ratio*: If the amount of either calcium or phosphorus in the food exceeds double the amount of the other {1:2-2:1} intestinal absorption of both are reduced due to the formation of insoluble complexes. Absorption is maximum when food contains equal amounts of readily available calcium and phosphorus.
2. *Sugars*: Sugars and organic acids produced by microbial fermentation of sugars in the intestine increase the solubility of calcium salts and enhance their absorption.
3. *Proteins*: A high protein diet increases calcium absorption.
4. *Fatty acids*: Presence of fatty acids decreases calcium absorption because of the formation of insoluble calcium soaps. So failure of fat absorption due to any reason lowers calcium absorption.
5. *Oxalates*: Present in spinach and colocasia, form insoluble calcium oxalates and decrease absorption of calcium.
6. *Phytates* form insoluble calcium phytates.
7. *Vitamin D and parathormone* (via vitamin D) both increase the absorption of calcium. Calcitriol acting as a hormone induces the formation of calcium binding protein which increases calcium absorption (refer Vit D).

### Calcium Balance:

*Calcium equilibrium:* Amount absorbed is equal to amount excreted.

*Positive calcium balance*: Intake exceeds excretion, occurs during pregnancy and growth.

*Negative calcium balance*: Excretion of calcium exceeds intake, occurs in old age osteoporosis and prolonged bed confinement.

### Distribution in the Body

99% of calcium is present in bone,
1% in ECF.

*Plasma calcium level*: 9-11 mg %.

Present in two forms:

- *Ionized calcium*: (45% of total), about 5 mg% in blood, biologically active fraction.
- Nonionized calcium:
  - *Nondiffusible:* (50%), 4.5-5.5 mg%, protein bound, primarily to albumin.
  - *Complexed:* Nonionic (5%), 0.4-0.6 mg%, complexed to citrate, phosphate and other anions.

### Excretion

- About 200-300 mg/day lost in urine.
- About 600 mg (unabsorbed & lost from GIT)/ day in feces
- About 20 mg in sweat.

### Regulation (Fig. 8.1)

Mainly by three hormones

1. Calcitriol or vitamin D } ↑ calcium level
2. Parathormone (PTH) }
3. Calcitonin ↓calcium level

*Calcitriol or Vitamin $D_3$*

a. Increases intestinal absorption of calcium (refer vitamin D for mechanism).
b. Increases calcium mobilisation from bones.
c. Increases reabsorption of calcium from renal tubules.

*Parathormone*

– Secreted by parathyroid glands.
– Increases blood calcium in many ways.

a. Main action is by bone resorption or mobilisation of calcium from bone.

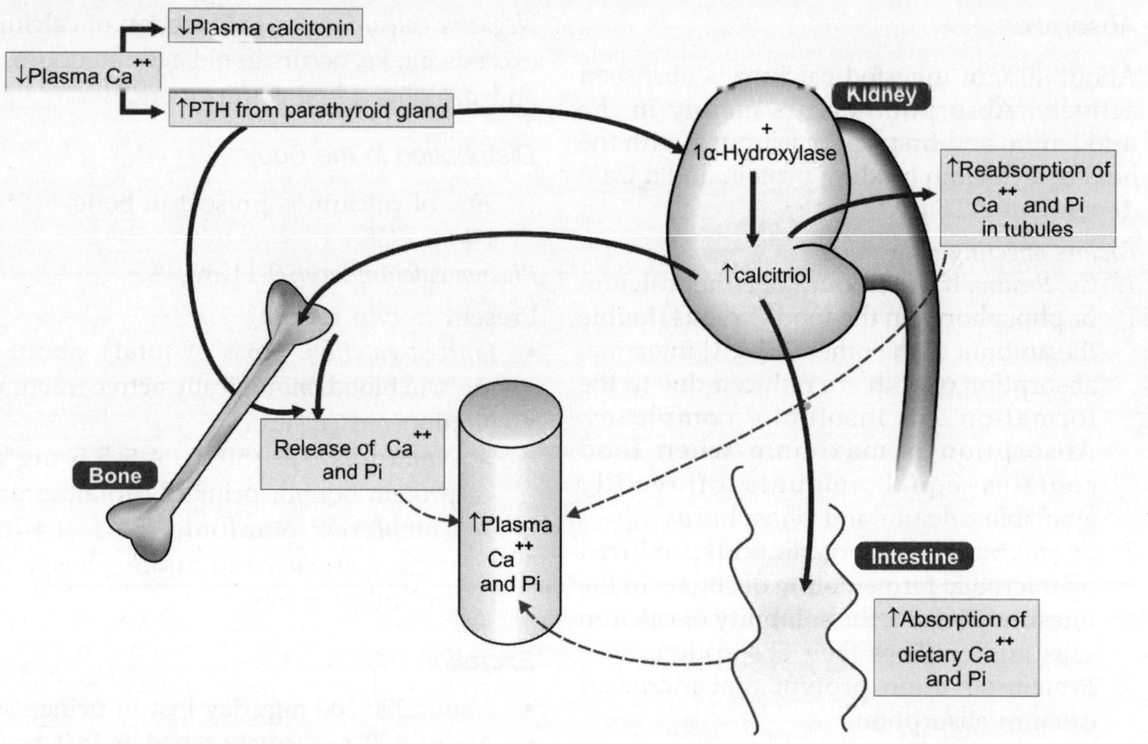


**Fig. 8.1:** Interrelationship between vitamin D, PTH and calcitonin in regulation of plasma calcium

b. Increases intestinal absorption of calcium indirectly by increasing 1α-hydroxylase activity, which increases synthesis of calcitriol.
c. Promotes reabsorption of calcium by the kidney tubules.

*Calcitonin*

- Secreted by thyroid parafollicular cells.
- Lowers serum calcium level.
- Decreases mobilisation of calcium from bones, particularly in growing children.

*Sex hormones: Oestrogens and testosterone*

Promote retention and deposition of calcium in bones.

### Disorders

*Hypocalcaemia*: Causes rickets in children and osteomalacia in adults.

There is:

- Decreased Ca and $PO_4$ absorption.
- Low serum $PO_4$.
- Demineralisation of bones—negative $Ca^{2+}$ balance.

*Causes:*

- Dietary deficiency
- Hypoparathyroidism
- Deficiency of vitamin D
- Malabsorption.

*Rickets* is characterised by:

- Stunted growth
- Poor quality of bones and teeth
- Malformation of bones: bowlegs, knock knees, rachitic rosary, pigeon chest.
- Fall in serum $Ca^{2+}$ level below 4.5 mg % due to hypoparathyroidism causes tetany,

characterised by neuromuscular tremors, carpopedal spasm and laryngeal spasm.

*Hypercalcaemia*: Increase in serum calcium level.

Occurs due to

- Hypervitaminosis D
- Malignancy
- Hyperparathyroidism due to parathyroid adenoma or ectopic PTH secreting tumours.

Characterized by–

- Demineralisation of bones
- ↑ serum $Ca^{2+}$
- ↓ serum Pi
- ↑ urinary Ca and $PO_4$
- Metastatic calcification of soft tissues like kidney, pancreas, muscle, arterial walls.

## PHOSPHORUS

Phosphorus contributes to 1% by weight of the human body.

### *Sources*

Milk, cheese, cereals, legumes, nuts and organ meats.

### RDA

| | |
|---|---|
| Adult man/woman – 800 mg | |
| Infants. | 1200 mg |
| Pregnant and lactating women. | |

### *Functions*

1. *Constituent of bones and teeth:* 80% of the phosphorus in the body is present in bones and teeth.
2. *Energy transfer*: Energy produced by the metabolic processes is stored as high energy phosphates like ATP and creatine phosphate and this energy is then utilised for the energy requiring processes.
3. *Buffers* ($H_2PO_4^-/HPO_4^{2-}$). Help to maintain the physiological pH.
4. *Regulation of enzymes:* Many enzymes are regulated by reversible phosphorylation and dephosphorylation, for example glycogen phosphorylase, glycogen synthase, pyruvate dehydro-genase.
5. Constituent of many biological molecule
   - Phosphoproteins
   - Nucleic acids
   - Phospholipids
   - Lipoproteins
6. Phosphorylation of glucose is required for its utilisation. Most intermediates in carbohydrate metabolism are in phosphorylated form.

### Absorption

- Absorbed as inorganic phosphates from the middle third of jejunum.
- 70% of dietary intake is absorbed.
- All factors affecting calcium absorption also affect absorption of phosphorus.

### Excretion

- 70% of the phosphate in glomerular filtrate is actively reabsorbed from proximal tubules.
- 0.8–1.4 g/day is excreted in urine
- Small amount is excreted in feces.

### Regulation

1. *Vitamin D:* Increases serum phosphate level by
   - Stimulating intestinal absorption of phosphorus
   - Increasing mobilisation from bones
   - Increasing reabsorption in kidney tubules.

2. *Calcitonin:* Lowers phosphate level by
   - Increased excretion of phosphate in urine
3. *Parathormone:* Lowers serum phosphate level by decreasing reabsorption from kidney tubules.

## Deficiency of Phosphorus

- Results in *rickets in children* and *osteomalacia in adults.*
- Characterised by poor bone mineralisation.
- Occurrence of bone fractures.

### *Conditions*

- Hyperparathyroidism
- Persistent vomiting
- Deficiency of vitamin D
- Kidney or liver disorder (Vitamin D is not converted to its active form).
- Alcoholism
- Malabsorption
- Fanconi syndrome characterised by increased excretion of phosphorus.

### *Symptoms*

- Loss of appetite
- Weakness
- Muscle tremors
- Demineralisation of bones, bone pain
- Loss of calcium
- Extremely low levels causes seizures and coma.

### *Treatment*

All symptoms can be relieved by administration of phosphorus.

## IRON

### *Sources*

Liver, meat, egg yolk, green leafy vegetables, cereals, legumes.

### *RDA*

Adult man: 10 mg
Adult woman: 28 mg
Pregnant and lactating women: 40 mg
Children: 10–15 mg.

## Functions

Compounds which contain iron can be classified as heme compounds and nonheme iron compounds.

1. Heme compounds (where iron is a component of the porphyrin ring).
   - Haemoglobin and myoglobin: Participate in transport and storage of oxygen.
   - Enzymes like cytochromes, catalases and peroxidases participate in electron transport and tissue oxidations.
2. Non-heme iron compounds (these compounds contain iron in their prosthetic group but no porphyrin).
   - These include iron—sulfur proteins, e.g., succinate dehydrogenase, NADH dehydrogenase, cytochrome C reductase.

## Absorption

- Only 10% of dietary iron is absorbed.
- Absorbed from upper part of duodenum.

### *Factors Affecting Iron Absorption*

1. *HCl in the stomach:* Liberates free $Fe^{3+}$ from nonheme iron compounds present in food.
2. Vitamin C, glutathione and cysteine present in food help in iron absorption by reduction of $Fe^{3+}$ to less polymerisable and more soluble $Fe^{2+}$.
3. Vitamin C and amino acids help in iron absorption by forming soluble iron ascorbate and iron amino acid chelates.

4. Oxalates, phytates, phosphates and carbonates of food reduce iron absorption by forming insoluble complexes of iron.

Heme in the food is absorbed as such and degraded in the intestinal mucosal cells to release iron.

### *Regulation of Iron Absorption*

Iron is the only nutrient, the levels of which is regulated at the point of absorption from the intestine.

$Fe^{2+}$ binds to a receptor on the brush border of the mucosal cells. Absorbed iron is converted to $Fe^{3+}$ by *ferroxidase (ceruloplasmin)* in the mucosal cell and is bound to *apoferritin* to form *ferritin* depending on the demand.

If iron absorbed greatly exceeds the demand, a large amount of apoferritin is synthesised, to trap the iron within the mucosal cells as ferritin to prevent its transfer to the capillary bed. Once iron is so incorporated, this iron is not available to the body. The life span of mucosal cells is very short, i.e., 3-5 days in human adult and iron bound as ferritin is lost in feces, when the mucosal cells are sloughed off.

On the other hand, if there is iron deficiency, the receptor mediated uptake is more and virtually no ferritin is synthesised and most of the iron in the mucosal cell is transferred to the plasma.

### Transport

In the plasma it is reoxidised to $Fe^{3+}$ form and is incorporated into apotransferrin, a β-globulin to form *transferrin*. It is a glycoprotein consisting of a single polypeptide chain in MW 79,570. Each molecule has two binding sites for iron and two for carbonate ions which are needed for stable binding configuration. Normally, 100 ml of blood contains 300 mg of transferrin which may bind upto 360 mg of iron. This is known as total iron binding capacity (TIBC).

Normally, in a healthy adult about 1/3 of iron binding sites are occupied *i.e.* 70-170 mg iron is bound/100 ml of blood.

TIBC is decreased in liver diseases because synthesis of transferrin is decreased. Plasma iron level is decreased in iron deficiency anaemia.

### Utilisation of Iron

Required by all tissues in small amount, mainly developing erythrocytes draw iron from transferrin for haemoglobin synthesis.

*Uptake of iron:* Mainly by erythropoietic tissues of bone marrow by binding of transferrin to specific receptors in plasma membrane.

The receptor-transferrin complex is internalised, where it releases the bound iron and changes into receptor—apotransferrin complex.

This complex returns to the membrane, where receptor is retained and apotransferrin is released back into the plasma.

### Storage

- The iron storage protein is *ferritin* found in liver and spleen.
- Ferritin is a large protein having 24 subunits arranged spherically.
- Iron is taken up by apoferritin in $Fe^{3+}$ form.
- Each molecule of ferritin can bind large amounts of iron. In a normal individual, about 2000 atoms of iron are bound to ferritin, i.e., upto 20% weight.
- Ferritin acts as a *'sink'* of iron in times of excess and a *'source'* of iron in times of iron deprivation. When necessary $Fe^{3+}$ form is converted to $Fe^{2+}$ form by a ferritin reductase of liver cells and released into plasma.

- Normal plasma ferritin level
  Males: 3 – 10 mg/dl
  Females: 2 – 12 mg/dl
- Abnormal level of ferritin in plasma is an indicator of iron status of the body.
- The water insoluble amorphous, microscopic aggregates of iron around ferritin spheres, containing upto 37% of iron, formed during iron overload is called *haemosiderin.*
- The iron in haemosiderin is less accessible than ferritin and its formation is probably a secondary protective mechanism against iron overload.

## Distribution of Iron

About 4 g of iron is present in the body, which is distributed as follows:

| | |
|---|---|
| Haemoglobin | – 3 g (70%) |
| Myoglobin | – 200 mg (5%) |
| Heme containing enzymes | – about 300 mg (7%) |
| Transferrin | – 130 mg (3%) |
| Ferritin | – 0.6 g (15%) |

## Excretion

Only 1-1.5 mg of iron is excreted daily, mostly in feces.

## Disorders

Deficiency of iron

- *Results in hypochromic, microcytic anaemia.*

### Causes for Deficiency

- Dietary deficiency.
- Hookworm infestation.
- Deficiency in absorption, due to lack of HCl in stomach.
- Resection of GI tract; absorption is impaired.
- Sprue, steatorrhoea; absorption is impaired.
- Excessive loss of blood.
- Repeated pregnancy.

*Characterised by*

- Low serum iron.
- Low haemoglobin level.
- Poor tissue cytochromes.

### Iron Overload

1. *Haemochromatosis*

   *Cause:* Occurs due to excessive intestinal absorption of iron due to genetic disorder. There is excessive deposition of haemosiderin in soft tissues characterised by:
   - High serum iron
   - Pancreatic fibrosis, leading to diabetes mellitus
   - Browning of skin due to iron deposition.
   - Hepatic fibrosis.
   - Degeneration of cardiac muscles leading to congestive heart failure.

   Since there is browning of skin and diabetes mellitus, it is also called as *bronze diabetes.*

### Treatment

Phlebotomy, repeated bleeding and removal of blood.

2. *Siderosis or haemosiderosis:* Occurs secondary to other problems.
   *Causes:*
   - Repeated blood transfusion.
   - Hereditary haemolytic anaemia.
   - Thalassemia.
   - In African Bantu people due to habitual cooking in iron pots.
   - Alcoholics.

Other trace elements are shown in Table 8.1:

**Table 8.1:** Sources, RDA, functions and metabolism of some trace elements

| *Nutrient* | *Sources* | *RDA in adults* | *Functions* | *Metabolism / Disorder* |
|---|---|---|---|---|
| Magnesium | Green leafy vegetables, whole grains, nuts, legumes | 300-400 mg | Constituent of bones and teeth, affects neuromuscular excitability. Essential cofactor for many enzymes e.g., all ATP requiring enzymes, phosphoglucomutase, enolase | Deficiency: Observed in intestinal and renal disorders and alcoholism. Symptoms: Neuromuscular hyperirritability. |
| Manganese | Green leafy vegetables, cereals, legumes, meat, fish | 5 mg | Essential cofactor for many enzymes, e.g., phosphatases, arginase, carboxylases, mitochondrial superoxide dismutase (SOD) | Liver is the most active organ of metabolism. Storage: In liver and kidney. |
| Copper | Organ meats, shell fish, nuts, legumes | 2-5 mg | Helps in utilisation of iron for Hb synthesis. Required as cofactor for many enzymes like ferroxidase, cytosolic SOD, cytochrome oxidase, lysyl oxidase. Plays a role in bone formation and maintenance of myelin sheath. | Mainly present bound to ceruloplasmin (ferroxidase) in serum. Deficiency: Impaired Hb synthesis may cause anaemia in infants. *Wilsons' disease:* Hepatolenticular degeneration, defect in incorporation of $Fe^{2+}$ into ceruloplasmin. |
| Zinc | Cereals, legumes | 15 mg | Required as cofactor for many enzymes, e.g., DNA and RNA polymerase, carbonic anhydrase, cytosolic SOD, alcohol dehydrogenase, carboxypeptidase; required for storage and release of insulin and vitamin A metabolism. Necessary for proper growth and reproduction, maintenance of skin, normal brain function, taste, wound healing | Metabolised in liver, pancreas, muscles, bones and other organs. Deficiency causes *Achrodermatitis enteropathica* |
| Iodine | Iodised salt, seafood | 100-200 mg | Synthesis of thyroid hormones | Storage: Thyroid gland Deficiency: Nutritional deficiency causes *'goitre'* with enlarged thyroid |
| Molybdenum | Organ meats, milk. | 2.5–5 mg | Required for activity of xanthine oxidase, aldehyde oxidase and sulphite oxidase. | Excretion: In urine, deficiency unknown. |
| Selenium | Marine fish and shell fish, grains | <3 ppm | Promotes growth, essential component of the enzyme glutathione peroxidase, acts synergistically with vitamin E in defense: antioxidant action. | Deficiency: Growth retardation, aspermatogenesis, cataracts. |
| Fluoride | Water | 1 ppm | Necessary for development of teeth and bone, prevents dental caries | Disorders : high intake of fluoride causes *'fluorosis'* – mottling and discoloration of teeth. |

# 9 Nutrition

The science of nutrition deals with the processes by which components of food are made available to an organism for meeting energy requirements in order to maintain the organism in an optimum functional state. The science of nutrition is concerned primarily with the following aspects.

a. Foods that provide essential nutrients.
b. Quantity of each needed.
c. Nutritional purpose provided by the food.
d. Effects of their deficiency.

## Biomedical Importance

The common nutritional disorders observed due to the deficiency of nutrients are, *kwashiorkor* (protein), *xerophthalmia* (vitamin), *anaemia* (iron) and *starvation* (energy). Excess intake of nutrients (carbohydrate & lipids) are associated with *obesity, diabetes mellitus* and *coronary heart disease* (CHD).

## NUTRIENTS

Are those chemical substances essential for the growth and maintenance of normal cells. About 45 to 50 chemical entities are known to be essential. These are divided into six main categories.

1. Carbohydrates – Glucose; fibre (2)
2. Lipids – Linoleic acid, Linolenic acid (2)
3. Proteins – Essential amino acids (10)
4. Minerals – Macroelements (8)
   – Trace elements (11)
5. Vitamins – Fat soluble (4)
   – Water soluble (10)
6. Water

## Recommended Dietary Allowances (RDA)

RDAs are the levels of intake of essential nutrients considered to be adequate to meet the known nutritional needs of practically all healthy persons. They are established for 'healthy populations' and population groups rather than for specific individuals.

*Note: RDA is published by National Academy of Sciences in the United States as the major standard for adequacy of nutrient intake.*

## ENERGY

Is recognized as a prime requirement for humans and other living organisms. A major proportion of the energy (55-60%) is derived from carbohydrates. Lesser amounts are obtained from fats (30-40%) and proteins (10-15%).

Energy value of food is measured in calories. The measurement of energy requirements of the body under various physiological conditions and the determination of energy values of foods is called *Calorimetry*.

### Calorie

One *calorie* represents the amount of heat required to raise the temperature of 1 gram of water by 1°C (15° to 16°C). Since the calorie is too small a unit, it is conveniently expressed as *kilocalories* and is represented by *kcal* or simply *Cal* with capital *C*.

1 kcal = 4.118 kJ

1000 kcal = 4.18 MJ

1 MJ = 239 kcal

In the SI unit energy value is expressed as *megajoules* (MJ).

## CALORIC VALUE OF FOODS

The calorific value (energy content) of a food is calculated from the heat released by the total combustion of food in a calorimetre. The energy value of these principal food stuffs carbohydrates, fats and proteins, measured in a bomb calorimeter and in the body are given in Table 9.1.

*Note: Bomb calorimeter is an instrument in which food is combusted completely in the presence of oxygen and the total energy released is determined.*

**Table 9.1:** Calorific value of principal food stuffs

| *Food Stuff* | *Energy Value (Cal/g)* | |
|---|---|---|
| | *Bomb Calorimetre* | *Body* |
| Carbohydrate | 4.1 | 4.0 |
| Fat | 9.4 | 9.0 |
| Protein | 5.4 | 4.0 |

Carbohydrates and fats are completely oxidised in the body. Hence their energy values measured in the bomb calorimetre or in the body are almost the same. Proteins however are not completely burnt in the body as they are converted to products such as urea, creatinine and ammonia. Due to this reason calorific value of protein obtained in the body is less than that of the bomb calorimetre.

The nutrients namely vitamins and minerals have no calorific value although they are involved in several important body functions, including generation of energy from carbohydrates, fats and proteins.

## RESPIRATORY QUOTIENT (RQ)

The RQ of food stuffs is the ratio of the volume of $CO_2$ produced to the volume of $O_2$ consumed in the oxidation of food stuff.

1. ***Carbohydrates:*** These are completely oxidized and their RQ is close to 1, as represented below for glucose.
   $C_6H_{12}O_6 + 6O_2 \rightarrow 6\ CO_2 + 6\ H_2O$
   $RQ = CO_2/O_2 = 6/6 = 1$
2. ***Fats:*** These have relatively low RQ since they have a low oxygen content. Hence they require more oxygen for oxidation. The RQ for the oxidation of tristearin is given below.
   $2\ C_{57}H_{110}O_6 + 163O_2 \rightarrow 114\ CO_2 + 110\ H_2O$
   $RQ = CO_2/\ O_2 = 143/\ 163 = 0.7$
3. ***Proteins:*** The RQ of protein cannot be represented by any specific formula since the chemical nature of the protein is highly variable. By indirect measurements the RQ of the protein is found to be around 0.8.
4. ***Mixed Diet:*** The RQ of the diet is dependent on the proportion of carbohydrates fats and proteins. For a normally ingested mixed diet RQ is *0.82*.
5. When the rate of utilisation of fat increases in relation to carbohydrates, RQ falls, e.g., in diabetes mellitus. RQ is lowest when ketolysis is very active.

## ENERGY REQUIREMENT OF A NORMAL PERSON

Man consumes energy to meet the fuel demands of the three major processes in the body.

1. Basal metabolic rate (BMR).
2. Physical activity.
3. Specific dynamic action (SDA).

Besides, additional energy supply is needed during growth, pregnancy and lactation.

### 1. Basal Metabolic Rate

May be defined as *the energy required by an individual to maintain life at complete physical, mental, emotional and digestive rest (post absorptive state)*. Thus basal metabolism measures the energy required for sustaining the basic life processes. About 50% of the energy of basal metabolism is used for ion transport. Remainder is expanded for maintaining functions such as muscle tone, heartbeat, respiration and body temperature.

#### Units of BMR

BMR is expressed as *kilocalories (C) per sq.m. of the body surface area per hour i.e., C/sqm/hr*. It may also be expressed as percent above or below the normal for persons of the same age and sex.

*Nomograms* of the body surface area (directly in $m^2$) from height and weight are readily available in literature.

Body surface area may be calculated by the simple formula devised by Dubois and Dubois.

$$A = H^{0.725} \times W^{0.425} \times 71.84$$

A = Surface area in $cm^2$.
H = Height in cm.
W = Weight in kg.

*Note: To convert body surface area into square metres $(m)^2$ divide the above value $(cm)^2$ by 10,000.*

#### Normal Values

Adult man — 35-38 C/sqm/hr
Adult woman — 32-35 C/sqm/hr

Some authors represent BMR as C/day. For an adult man, BMR is around 1600 C/day while for an adult woman it is 1400 C/day. This is important for calculating the energy requirement per day.

#### Measurement of BMR

BMR is measured directly by the heat evolved, or indirectly by the volume of oxygen consumed per unit time. It was the most important test of thyroid function at one time. At present the determination of BMR has only theoretical importance. Benedict Roth Apparatus (closed circuit) is commonly used to measure BMR.

*Preparation of the subject:* The subject should be awaken, at complete physical and mental rest, in a post absorptive state and in a comfortable surrounding (25°C).

*Procedure:* The subject breathes in oxygen from a metal cylinder through a mouth piece for 6 min. The $CO_2$ produced is absorbed in soda lime. The volume of oxygen consumed is recorded.

BMR may be calculated as follows. Let the oxygen consumed in 6 minutes be *X litres*.

Calorific value of oxygen–4.8

[When 1L of $O_2$ is utilised, 4.8 kcal of heat is generated].

Heat produced in 6 min = $4.8 \times X$

Heat produced in 24 hrs = $4.8 \times X \times 10 \times 24$ kcal

= $4.8 \times X \times 10 \times 24$ kcal

BMR = $(4.8 \times X \times 10 \times 24 \text{ kcal})/A$

[where A = Surface area in sqm]

*Note:*

- *Basal metabolism is expressed as C/sq m of the body surface area /hr only for clinical purposes.*
- *For all practical purposes BMR is empirically taken as 24 C/kg body weight.*

***Factors Affecting BMR:*** Several factors affect the BMR and are listed below.

1. *Surface area:* BMR is proportional to the surface area.
2. *Age:* The BMR is higher in infants and children. This is because of more energy

required for rapid growth that is reflected in a higher basal metabolic rate.

3. *Sex*: Men have a marginally higher (5%) BMR than women. This is due to the higher proportion of lean muscle mass in men.
4. *Physical activity:* BMR is increased in persons with regular exercise (athletes) due to the higher proportion of lean muscle tissue, which has a higher BMR, than the adipose tissue.
5. *Hormones*: Thyroid hormones, sex hormones, growth hormones and to a lesser extent epinephrine and cortisol increase the BMR.
6. *Fever*: Causes an increase in BMR. Elevation of more than 10% is observed for every 1°C rise in body temperature.
7. *Starvation:* BMR is decreased in starvation. This may be an adaptation by the body.
8. *Environment:* In cold climates the BMR is higher compared to warm climates.
9. *Diseased status*: BMR is increased in various infections, leukaemias, polycythaemia, cardiac failure, hypertension etc.

BMR is marginally lowered in Addison's disease. It is also decreased in starvation.

*Significance:* Determination of BMR was of considerable value in the diagnosis of endocrine disorder, particularly of the thyroid gland. At present it is of importance only in calculating the caloric requirement of an individual and planning of diets.

### 2. Physical Activity

The physical activity of an individual is highly variable and depends on the duration and intensity of muscular activity. The expenditure of energy for the various physical activities is given in Table 9.2.

**Table 9.2:** Energy required for various activities

| *Activity category* | *C/kg/hr* |
|---|---|
| Sleeping, reclining | 0.9 |
| *Very Light* | |
| Sitting or standing, studying, talking | 1.5 |
| *Light* | |
| Laboratory work, typing, driving a car, table tennis etc. | 2.9 |
| *Moderate* | 4.3 |
| Walking 3.5 – 4 mph, gardening, cycling, carpentry etc. | |
| *Heavy* | 8.4 |
| Climbing, swimming, running, playing football etc. | |

For the calculation of energy requirement the activity level may be divided into 3 groups with regard to their physical activity and the requirement of energy.

Sedentary work 30% of BMR
*(teachers, office workers, doctors)*

Moderate work 40% of BMR
*(housewives, students)*

Heavy work 50% of BMR
*(agricultural labourers, miners)*

### 3. Specific Dynamic Action (SDA)

Consumption of food is associated with the expenditure of energy. This fact is well established. This phenomenon of *extra heat production by the body, over and above the calculated caloric value when a food is metabolised by the body is known as specific dynamic action (SDA).* It is also known as *calorigenic action or thermogenic action* of food. Considerable amount of research has demonstrated that SDA for the three major food stuffs are as follows:

Proteins - 30%
Carbohydrates - 6%
Fats - 4%

Thus from 25 g of proteins only 70 kcal would be available as usable energy, the remaining 30 kcal being expanded as SDA. However, when proteins are administered as a mixture with carbohydrates and fats the SDA is only about 6-10%. Fats are mostly involved in reducing the SDA.

- It is generally believed that SDA of foods is due to the energy required for digestion, absorption, transport, metabolism and storage of foods in the body. The SDA of proteins is primarily due to the energy required for the deamination of amino acids and the formation of urea.
- SDA of carbohydrates is attributed to the energy expanded for the conversion of glucose to glycogen.
- SDA of fat may be due to its storage, mobilisation and oxidation.

*Significance:* For every 100 g of protein consumed, energy available for doing useful work is 30% less than the calculated value. Hence for a mixed diet extra 10% calories should be provided to account for the loss of energy as SDA.

*Note: It is a common experience that consumption of a protein rich diet makes us feel warm and comfortable in cold weather. This is due to the high SDA of proteins.*

***Recommended Energy Allowances:*** The value of total and basal metabolism is of greatest importance in estimating the calorific requirements of human subjects. The recommended daily energy requirements as proposed by the food and nutrition board of the National Academy of Sciences—National Research Council is given in Table 9.3.

*Note: Due allowance may have to be made to the above mentioned overall energy requirement based on the nature and duration of physical activity. (Box 9.1)*

The recommendation by authorities in the field of nutrition is that energy intake should balance energy expenditure to obtain and maintain an appropriate body weight for height throughout the entire lifecycle.

**Table 9.3:** Recommended daily energy requirement for various age groups

| *Category (Age in years)* | *Energy Needs (C/day)* |
|---|---|
| Infants (0.5 – 1.0) | 80 – 135 |
| Children (below 10) | 900 – 3300 |
| Adults | |
| Males | |
| 11 – 50 | 2000 – 3100 |
| 51 – 75 | 2000 – 2450 |
| Females | |
| 11 – 54 | 1500 – 2400 |
| 51 – 75 | 1400 – 2200 |
| Pregnancy | + 300 |
| Lactation | + 500 |

- A negative balance caused by an inadequate energy intake to balance energy expenditure is undesirable since, if prolonged, it can lead to loss of body weight, increased susceptibility to infections, ketosis, acidosis and other pathological manifestations that can be life threatening.
- On the other hand, a prolonged positive energy balance, can lead to an increase in body weight and obesity. Obesity in turn is associated with decreased longevity and the development of chronic health disorders such as cardiovascular diseases, hypertension, gallbladder disease, etc.
- Major food sources of energy are carbohydrates, lipids and proteins although protein serves as a minor source of energy. Current opinion is that carbohydrates should supply about 65% of the total energy needs (with not more than 10% coming from total sugars such as sugar cubes, sweets, etc), fats should supply about 20% and the energy from proteins should not be less than 10%.

**Box 9.1:** Calculation of energy requirement

*Calculation of energy requirement for a 60 kg person doing moderate work*

| | | | | |
|---|---|---|---|---|
| Energy for BMR | — | 60 kg × 24 | = | 1440 C |
| Energy for physical activity | = | 40% of BMR | = | 576 C |
| | = | $\frac{40}{100} \times 1440$ | = | 2016 C |
| Allowance for SDA | = | 10% of energy | | |
| | = | $\frac{10}{100} \times 2016$ | = | 201.6 C |
| Total energy required | | | = | 2217.6 C |
| Rounded to the nearest multiple of 50 | | | = | **2250 C** |

## NUTRITIONAL ASPECTS OF CARBOHYDRATES, FATS AND PROTEINS

***Carbohydrates:*** Serve as a major source of energy in the human diet (65-75%) yielding 4 C/g.

*Role of carbohydrate in the body*:

- Principal source of energy.
- Central nervous system is dependant on a continuous supply of glucose and cannot use any other energy substrates than glucose until some capacity to use ketone bodies develop.
- Necessary for the oxidation of fats. Excess catabolism of fats coupled with the deficiency of carbohydrates leads to ketosis.
- Energy for muscle contraction is chiefly obtained from glycogen.
- Pentoses (ribose) are the constituents of some essential compounds such as nucleic acids and coenzymes. These are produced during the metabolism of glucose.
- Glucuronic acid derived from glucose is an important conjugating agent used in detoxification.

### Sources

- Sugars, grain flour, legumes, potatoes and other root vegetables.
- Sugars are present mainly as sucrose in jams, jellies, syrups, honey and fruits.
- Lactose is present in milk and dairy products.

Carbohydrate content of some common foods are given in Table 9.4.

**Table 9.4:** Carbohydrate content of some common foods

| | |
|---|---|
| Sugar | 99.5% |
| Chocolates | 85 % |
| Cereals | 75 % |
| Cookies | 70 % |
| Cakes | 60 % |
| White bread | 20 % |
| Cooked rice, sphagetti | 23-30% |
| Potatoes | 15-77% |
| Fresh fruits, carrots | 7-11% |
| Milk | 2-4 % |
| Many vegetables, liver | 2-5% |

### *Distribution of Dietary Carbohydrate*

About 55% of dietary carbohydrates is obtained from starch, 35% from sucrose, 5% from lactose and the remaining 5% from maltose, glucose and other sugars.

Pentoses in the food is of limited nutritional value. Cellulose, gums, pectins and other non-digestible residues are not used by humans and form the *bulk or fibre* of foods.

RDA: *350–500 g/day*

The amount however may vary widely depending on physical activity, food preferences, ethnic customs and other factors.

*[Note: Minimum carbohydrate intake should be 100 g/day, to prevent ketosis. Excessive breakdown of body protein, loss of cations (especially $Na^+$) and dehydration will also offset the undesirable metabolic responses associated with high fat diets and fasting].*

Nutritional authorities currently recommend decreasing the use of sugars, especially sucrose, and increase the consumption of complex carbohydrates. The principal reason of this is the mounting evidence incriminating sucrose in the development of obesity and a number of related pathological conditions, such as hypertension, cardiovascular disease, maturity onset diabetes, decreased longevity and dental caries. It also lacks the essential vitamins and minerals. Jaggery may be preferred over sucrose since it is a good source of iron.

### Dietary Fibre

The indigestible or non-available carbohydrate that cannot be digested constitute the dietary fibre, that gives bulk to the feces. Pathological conditions such as cancer of the colon and coronary heart disease may be related to the lack of dietary fibre. The beneficial effects of dietary fibre are as follows:

1. *Prevention of constipation:* Fibre helps to maintain the normal motility of the GI tract and prevents constipation.
2. *Elimination of bacterial toxins*: Fibre absorbs large quantities of water and also the toxic compounds produced by the intestinal bacteria that lead to increased faecal mass and its easier expulsion.
3. *Prevention of colon cancer*: Pathological conditions such as diverticular disease and cancer of the colon are related to the lack of fibre.
4. *Improvement of glucose tolerance*: Fibre improves the glucose tolerance by decreasing the rate of glucose absorption from the intestine.
5. *Reduction of plasma cholesterol*: Fibre binds the bile acids and cholesterol in a nonabsorbable complex and thus increases the fecal excretion. This in turn results in decreasing the plasma and tissue cholesterol levels.
6. *Satiety value*: Intake of dietary fibre gives a sensation of satiety (stomach fullness) without consumption of excess calories. Table 9.5. gives a list of the various dietary fibres, the chemical nature and possible physiologic effects of dietary fibre.

Thus it is wise to include fibre in the daily diet of normal adults in the form of whole grains, cereals, legumes, fruits and vegetables. **RDA** – 30 g/day.

*Disadvantages of using excess dietary fibre:* Excessive intake of fibre (ingestion of huge quantities of rice bran) may lead to several complications. Digestion and absorption of protein is adversely affected.

1. The absorption of Ca, P and Mg is decreased.
2. Intestinal bacteria ferment some fibers causing flatulence and often discomfort.

### Lipids

Fats in the form of triglycerides yield approximately 9 C/g. About 20-30% of the caloric intake may be obtained from fat. Other non caloric functions of triglycerides are:

1. Supply of polyunsaturated fatty acids.

**Table 9.5:** Various dietary fibres, their chemical nature and possible physiologic effects

| *Fibre* | *Chemical Nature* | *Physiologic Effect* |
|---|---|---|
| Cellulose | Polymer of glucose | Promotes peristalsis and increases bowel action |
| Hemicellulose | Pentoses, hexoses and uronic acids | Increases bile acid excretion. |
| Pectins | Partially esterified rhamnogalacturans | Absorbs water, slows gastric emptying, increases bile acid excretion. |
| Lignin | Aromatic alcohol | Antioxidant, increases bile acid excretion |
| Mucilage | Branched arabino glycans | Binds bile acids and increases their excretion; thus lowers the cholesterol level. Improves glucose tolerance. |

2. As carriers of fat soluble vitamins A, D, E and K to aid their absorption in the intestine.
3. High satiety value.
4. Fats increase the taste and palatability of food and is the favoured cooking medium.

Dietary fat may be divided into two types:

a. *Visible fat:* Fats consumed as such, *example:* butter, ghee, oils.
b. *Invisible fat:* Fat present as part of the other food items, e.g. egg yolk, fish, meat, cereals.

A significant fraction of the fatty acids ingested should be contributed by long chain PUFA. The PUFA present mainly in vegetable oils and fish oils belong to the group of essential fatty acids. (For the functions of essential fatty acids refer Chapter 3).

***RDA*** = 50 – 60 g.

*Note: Not more than 30% of the total daily energy requirement should be supplied by fat. 10% of the dietary energy should be supplied as PUFA. Although PUFA has beneficial effects, excessive amounts (more than 30% of total fat) may be undesirable since it may lead to the formation of free radicals. Possible adverse effects are as follows:*

- Increased requirement for vitamin E.
- Increased tendency of gall stone formation.
- Increased incidence of tumour formation (in some experimental animals).

*Dietary sources:* Oil, butter, ghee.

*Cholesterol and heart diseases:* The risk of coronary artery disease is greater in people with hypercholesterolaemia. Hence the dietary intake should be restricted by restricting the amount of foods rich in cholesterol, such as egg yolk, liver, brain and kidney.

*Note:*

- *Vegetables, cereals and pulses do not contain any cholesterol.*
- *Plant sterols inhibit cholesterol absorption.*
- *Saturated fats raise serum cholesterol while unsaturated fats lower it.*
- *High fibre content also reduces serum cholesterol.*

### Proteins

Are of prime importance since they form the building blocks for body tissues. Only 10-15% of the total energy is derived from proteins. When enough carbohydrates are present in the diet, the amino acids are not used for yielding energy.

### Role of Protein in the Body

- Proteins are essential structural components of all cells.

- Building new cells and tissues in growing children, during pregnancy, in athletic training and after injury.
- Maintaining tissues already built.
- Manufacturing functional and regulatory substances such as blood proteins, enzymes and some hormones.
- Forming milk—supplying energy.
- Protein is also needed to synthesise the plasma proteins which are essential for maintaining osmotic balance, transporting substances through the blood and maintaining immunity.
- In the fasting state the breakdown of body protein is enhanced and the resulting amino acids are utilised for glucose production, synthesis of NPN substances and in the synthesis of the essential secretory and plasma proteins.
- Turn over of body protein is a normal process and an essential factor of what is called *Nitrogen balance*.
- Proteins thus are the only source of essential amino acids. If any of these essential amino acids is missing from the diet the body cannot synthesise new protein to replace the protein lost due to normal turnover.

### *Nitrogen Balance: Nitrogen Output = Nitrogen Input*

Proteins undergo constant turnover. Nitrogen balance is a comparison between the intake of nitrogen (in the form of protein) and the excretion of nitrogen (in the form of undigested protein in the faeces and urea, ammonia and other nitrogenous compounds in the urine).

The normal adult will be in *Nitrogen Balance* OR *Equilibrium,* when the amount of nitrogen consumed equals the amount of nitrogen excreted.

### *Negative Nitrogen Balance: Nitrogen Output > Nitrogen Input*

Results from an inadequate dietary intake of protein, since the amino acids utilised for energy and biosynthetic reactions are not replaced. It is observed in the following conditions:

i. Injury - There is net destruction of the tissue

ii. Illness, iii. Starvation - Increased catabolism of body protein stores

iv. Old age - Inadequate protein intake, both quality and quantity wise.

### *Positive Nitrogen Balance: Nitrogen Output < Nitrogen Input*

Is observed when there is a net increase in the body protein stores, e.g.,

- In a growing child
- In a pregnant woman
- In convalescing adults.

### *Sources*

Widely distributed in foods of both animal and plant origin (Table 9.6).

**Table 9.6:** Sources of proteins

| *Animal origin* | *Plant origin* |
|---|---|
| Red meats | Cereals |
| Milk | Grains |
| Milk products | Legumes |
| Poultry | Vegetables |
| Egg | Nuts |
| Cheese | |
| Fish | |

### *Nutritive Value*

Not all food proteins have the same ability to meet the needs of the body for tissue formation, maintenance and other functions performed by dietary proteins. Thus the effectiveness of a food protein in supporting growth in the young organism, and in tissue maintenance and repair in the adult, depends primarily on both quantity (amount) and the quality (amino acid composition) of the protein.

A protein should have all the essential amino acids in proper proportion and should satisfy the amino acid requirement for growth/maintenance of total body function. Nutritionally the proteins may be classified into two types (Table 9.7).

**Table 9.7:** Nutritional classification of proteins

| | |
|---|---|
| *Complete proteins* *For example,* Egg albumin, casein | Have all the essential amino acids. Promotes good growth |
| *Incomplete Proteins* *For example,* Gelatin and corn | Completely lacks one or more essential amino acids. Do not promote growth. |

*Assessment of Nutritive Value:* It is usually expressed in comparison with some arbitrarily designated food protein of high quality such as whole egg protein. The digestibility of the protein and the efficient absorption of its constituent amino acids are other factors that can affect its utilisation. Different methods for assessing the nutritive value of protein are as follows:

a. *Biological value (BV):* One of the commonly used method to determine the BV involves the determination of the amount of food protein nitrogen (N) ingested and the nitrogen excreted in the urine and the faeces. The biological value of the test protein is then expressed as the percentage of the absorbed nitrogen retained by the subject. It is calculated as follows:

$$\text{BV} = \frac{\text{N retained} \times 100}{\text{N absorbed}}$$

$$= \text{Food N} - \frac{\text{Faecal N} + \text{Urinary N}}{[\text{Food N} - \text{Faecal N}]} \times 100$$

*Limitation:* The BV does not take into account the nitrogen losses that occur during digestion. Hence other indices are employed.

b. *Coefficient of digestibility:* The digestibility of the protein may be calculated by the analysis of faecal and urinary nitrogen when the subject is given an appropriate nitrogen free diet.

$$= \text{Food N} - \frac{[\text{Faecal N} - \text{Faecal N on protein free diet}] \times 100}{[\text{Food N}]}$$

c. *Net protein utilisation (NPU):* It is defined as the percentage of the dietary protein that is retained.

$$\text{NPU} = \frac{\text{N retained}}{\text{N intake}} \times 100$$

NPU score of a particular protein is a measure of two things.

i. How well the protein is digested.
ii. How well it is utilised once the amino acids have been absorbed into the system.

d. *Protein Efficiency Ratio (PER):* It is defined as the weight gain observed per gram of the protein consumed.

$$\text{PER} = \frac{\text{Gain in weight of growing animal}}{\text{Protein consumed}}$$

NPU, BV and PER of some common foods are listed below (Table 9.8).

e. *Chemical score:* Is another method of expressing the nutritive quality of a protein and is based on its amino acid composition. Essentially, this value is the amount of the most limiting amino acid

**Table 9.8:** Nutritive value of some proteins as assessed by BV, NPU and PER

| *Protein* | *NPU* | *BV* | *PER* |
|---|---|---|---|
| Egg | 87 | 94 | 4.5 |
| Milk | 81 | 84 | 3.0 |
| Legumes | 45-55 | 60-70 | 1.5-2.0 |
| Cereals | 47-60 | 58-68 | 1.5-2.2 |

in the test protein expressed as the percentage of the amount present in egg protein.

$$= \frac{\text{mg of limiting amino acid / g test protein}}{\text{mg of same amino acid / g egg protein}} \times 100$$

- Limiting amino acid is defined as that amino acid which limits the weight gain when a protein is supplied to an animal. Limiting amino acids of some of the proteins are given in Table 9.9.

This value does not take into account differences in the digestibility of the test proteins nor the intestinal absorption of amino acids derived from them. However, the chemical scores agree grossly with values as determined biologically and have advantage over the latter of requiring less time.

**Table 9.9:** Limiting amino acid in some of the proteins

| *Protein* | *Limiting amino acid* |
|---|---|
| Rice | Lys; Thr |
| Wheat | Lys; Thr |
| Corn | Trp; Lys |
| Bengal gram | Cys; Met |
| Tapioca | Phe |

### Complementary Effect of Proteins

The proteins of animal origin have the highest nutritive value unlike the proteins of vegetable origin. Cereal proteins have low content of lysine and corn proteins are lower in tryptophan. However, when cereals are employed as mixtures with other proteins, particularly milk, meat or eggs, an excellent source of proteins result. This is referred to as the *complementary effect*. The essential amino acids present in one protein balance the deficits in other proteins of lower biological value. For this reason vegetarians should combine food such as cereals with legumes (soya beans, peas or beans) or with milk products to obtain a more complete protein source. Protein content of some of the foods are given in Table 9.10.

**Table 9.10:** Protein contents of some foods

| *Food* | *Content* | *Percentage* |
|---|---|---|
| Fish, cheese, peanuts | High | 18-28 |
| Nuts, egg, shell fish | Moderate | 10-18 |
| Bread, cereals, chocolates | Low | 6-10 |
| Fruits, vegetables | Very low | 1-5 |

*Note: Enrichment of wheat flour and other cereal grain products with lysine has been suggested as a means of substantially increasing the nutritive value, making them nearly equivalent to the more expensive animal protein.*

Heat processing (cooking) generally improves the nutritive value of proteins. *For example:*

- Egg white is said to be more easily digested.
- Soya beans have a higher nutritional value after cooking.
- Phaseolin, a protein of navy bean has a greater nutritive value if cooked.

*Note: Dry heat has a deleterious effect particularly on the cereal proteins. Lysine becomes less available; example, toasted and puffed cereals.*

**RDA:**

Adults : 0.8-1.0 g/kg body weight
Infants : 2.0-2.2 g/kg body weight
Pregnancy : + 30 g
Lactation : + 20 g

*Note: Nutritional aspects of vitamins, minerals and water balance are discussed separately (Chapters 7, 8 and 12).*

***Balanced Diet:*** It is defined as the diet which contains different types of foods and possesses the principal nutrients *viz,* carbohydrates, fats, proteins, vitamins and minerals in a proportion to meet the requirements of the body. The basic composition of a balanced diet is highly variable as it differs from country to country.

***Prescription of diet:*** While prescribing the balanced diet the following general rules are to be remembered.

1. The ideal body weight is to be preserved.
2. Protein requirement is roughly calculated as 1g/kg body weight for adults and 2 g/kg body weight for children.
3. The caloric requirement mainly depends on the health status, physical activity and occupation. The caloric requirement of a person is taken as 30-35 kcal/kg body weight. For sedentary habits, lower value (30 kcal/kg body weight) and for moderate muscular activities, higher values (35 kcal/kg body weight) are prescribed. The average caloric requirement for an adult between 55-65 kg is around 1700-2300 kcal/day. Extra allowance of 300 kcal is necessary during pregnancy and lactation.
4. Expenditure of energy by SDA has to be taken into account.
5. Balanced diet should include calories from carbohydrates, proteins and fats in the ratio 60 : 20 : 20.

### Planning of a Balanced Diet

As a guide for the selection of nutritious food and for the planning of daily meals, most Nutritionists suggest the following general food plan.

It is based on the daily use of four basic food groups in quantities and variety that will supply the nutrients known to be essential for human nutrition and in balanced amounts that will satisfy the recommended daily allowances for all normal healthy individuals.

**I.** *Milk group:*

| | |
|---|---|
| Children | 3-4 cups |
| Adults | 2 or more cups |
| Pregnant women | 4 or more cups |
| Nursing mothers | 6 or more cups |

Cheese and ice cream or other dairy products can replace part of the milk.

* Provides high quality proteins, calcium and riboflavin.

**II.** *Meat group:* 2 or more servings daily.

Beef, veal, pork, lamb, poultry fish, eggs. Dry beans, peas and nuts may be used as alternatives.

* Supplies excellent protein, thiamin, vitamin $B_{12}$ and minerals.

**III.** *Vegetable/fruit group:* 4 or more servings daily.

Dark green or yellow vegetable every alternate day—for vitamin A.

A citrus fruit or vegetable daily—for vitamin C.

* Contributes vitamin A, ascorbic acid, minerals and fibre.

**IV.** *Bread/cereal group:* 4 or more servings daily.

Whole grains, enriched, or restored,

* Furnishes carbohydrate and protein. If enriched, or restored provides thiamin, niacin, riboflavin and iron.

Each of the four food groups has been designed to contribute specific nutrients essential for a complete and balanced diet.

The above foods may be taken as 3 meals a day which includes—

i. A moderate breakfast.

ii. Lunch
iii. Hearty evening meal.

This food plan along with the variety of selections within each of the food groups and additional carbohydrate or fat as required to meet the individual energy needs, forms a foundation for a well balanced diet. The above food plan with suitable modifications, may be used for therapeutic purposes.

Recent Guidelines Set up by the Nutrition Board are as Follows:

1. Eat a well balanced diet selected from a wide variety of foods to enhance an adequate intake of all essential nutrients.
2. Do not overeat. Maintain proper body weight according to height. If overweight, eat less food, particularly sugars, fats and oils. Exercise regularly.
3. Use salt in moderation.

---

Calculation of the amount of carbohydrate, protein and fat for a 60 kg man doing moderate work. The total energy expenditure for this subject is 2250 C. Carbohydrate, fat and protein should be provided in the ratio of 65: 20: 15

Accordingly, the energy to be obtained from the principal foods may be calculated as follows:

- First, the protein requirement is calculated
  60 x 1g = 60 g (RDA = 1g/kg wt)
  This provides 240 C of energy (60 × 4 = 240).
- Calculate the fat requirement $\frac{20}{100} \times 2250 = 450$
  20% of the energy requirement = 450C
- The remaining energy should be supplied by carbohydrates = 1560 C 2250 – (240 + 450)

Compute the amount of each food component that should be consumed.

| | | |
|---|---|---|
| Carbohydrate — | 390 g | (1560 ÷ 4) |
| Fat — | 50 g | (450 ÷ 9) |
| Protein — | 60 g | (240 ÷ 4) |

---

Accordingly he should consume 390 g of carbohydrate, 50 g of fat and 60 g of protein, in order to meet his energy requirement.

Apart fom this the diet should include recommended allowances of vitamins, minerals and water.

*Nutritional disorders:* Some of the important nutritional disorders are as follows:

### Proteins Energy Malnutrition (PEM)

Sometimes called protein calorie malnutrition is the most common nutritional disorder of the developing countries. PEM is widely prevalent in infants and preschool children. The two extreme forms of PEM are *Kwashiorkor* and *Marasmus.*

*Kwashiorkor*: It is a nutritional disease chiefly caused by protein deficiency. Literally it means a disease the child gets when the next baby is born.

- It is predominantly found in children between 1-5 years of age.
- The disease occurs in children after weaning and is related to the deficiency of protein and apparently other essential nutrients in the food that is taken. *For example,* yams, bananas, potatoes, maize, cassava etc. The proteins are poor in quality and also low in amount.

*Symptoms:* A child with this disease often has an enlarged abdomen. Rash occurs at various sites of the skin and cunjunctivitis may appear. Oedema is a characteristic finding. The child loses his appetite and becomes anaemic. Serum proteins are lowered considerably (Albumin < 2 g/dl). Loss of melanin pigment giving the hair a reddish appearance is another characteristic finding.

*Treatment:* A protein intake of 3-4 g/kg body weight is suggested. Treatment may be monitored by measuring plasma albumin concentration.

*Marasmus:* Is mainly caused due to calorie deficiency. Literally this term mean *"to waste"*.

**Table 9.11:** Characteristic features of marasmus and kwashiorkor

| *Features* | *Kwashiorkor* | *Marasmus* |
|---|---|---|
| 1. Oedema | Lower legs, face or generalised | Absent (none) |
| 2. Wasting | Less obvious | Loss of subcutaneous fat |
| 3. Growth retardation in terms of body weight | Less | Severe |
| 4. Mental changes | Usually present | Usually absent |
| *Biochemical changes* | | |
| 1. Serum albumin | Low | Normal or slightly decreased. |
| 2. Urinary urea / creatinine | Low | Normal or decreased. |
| 3. Anaemia | Common; iron or folate deficiency may be associated | May be observed |
| 4. Liver biopsy | Fatty infiltration | Normal |

- It mainly occurs in children under one year of age. Early weaning tends to lead to marasmus.
- Symptoms of marasmus include growth retardation, muscle wasting, anaemia and weakness.
- Plasma albumin levels may be normal or slightly reduced and hence the child does not have oedema.
- Characteristic symptoms of the two conditions are given in Table 9.11.

*Obesity:* An abnormal increase in the body weight due to excessive fat deposition. In the modern world where millions of people are suffering from serious malnutrition, many in the effluent societies are suffering from obesity. Excessive food intake associated with reduced energy expenditure is the major cause of obesity. The two types of obesity are known as follows:

1. Juvenile onset—Increased number of adipocytes.
2. Adult onset—Enlarged adipocytes.

Hormonal related disorders like Cushing's syndrome, hypothyroidism, insulinoma also cause obesity. Recently it is believed that obesity is related to a low proportion of brown adipose tissue.

*Biochemical changes in obesity:* Significant metabolic changes take place in obese individuals that are reflected in plasma concentration. Most characteristic is the altered pattern of lipid metabolism. The concentration of β lipoproteins and free fatty acids are increased. The concentration of insulin in circulation is also increased markedly.

*Treatment*: Obesity is treated with a weight reducing diet and increasing the energy expenditure by exercise such as walking, swimming, jogging or bicycle riding. The normal diet should be modified based on the basic four groups. Diets containing 1200 kcal daily for women and 1500 to 1800 kcal daily for men are acceptable. Obesity leads to increased incidence of cardiovascular diseases, diabetes etc.

Several other nutritional disorders such as xerophthalmia, rickets, beriberi, scurvy are discussed under vitamins (Chapter 7), goitre fluorosis etc under minerals (Chapter 8).

**Parenteral Feeding:** When a patient is unconscious or unable to take food by mouth it is necessary to provide nutrients by special feeding. The solution containing the nutrients is infused by way of subclavian vein directly into the superior vena cava. This procedure is known as *total parenteral nutrition or intravenous hyperalimentation*.

# 10 Hormones and Neurotransmitters

The word *Hormone* is derived from a Greek word that means to "arouse activity". Classically it is defined as a chemical substance present in very low concentration in the blood, and has a regulatory effect on the metabolism of at least one specific organ or tissue at a site distant from the site of production.

- Hormones can also act on adjacent cells (paracrine action) and on the cell in which they are synthesised (autocrine action) without entering the systemic circulation.
- Recently the term hormone is used for any compound in an organism that generates a signal to cause some sort of alteration at the cellular level.
- Hormones initiate their biologic effects by binding to specific *receptors* located on or within the *target cell*. (*Target cell is defined by its ability to bind selectively a given hormone via a receptor).*

## MECHANISM OF HORMONE ACTION

Cells sense and respond to multiplicity of signals from their environment. In multicellular organisms many signal molecules such as hormones originate in organs distant from their site of action. Hence they have to be carried in the blood to their effector *target* cells. Signals generated in these ways are sensed and processed by *signal transduction.*

Receptors sense and transduce the specific hormone signal by *transmembrane coupling* to effector enzyme systems, generating low molecular weight molecules termed *second messengers.* The second messengers influence the activity of key protein kinases that ultimately determines the type of biological response obtained.

A *"receptor"* is defined as a protein that specifically binds a hormone and produces a biologic action after binding. Thus receptors are the *first* site of action of the hormones at the cell. They are highly specific for the individual hormones. All receptors have two functional domains:

1. *Recognition domain*—Binds the hormones.
2. *Coupling domain*—Brings about the signal transduction.

It is the coupling of hormone binding to signal transduction that provides the first step in the amplification.

Binding of the hormone to the receptor results in the formation of a *Hormone Receptor Complex*. Based on the mechanism of action the hormones are classified into 2 types.

***Type I***: Includes the protein hormones and catecholamines that bind to receptors located on the external surface of the plasma membrane of the target cells. They do not enter the target cells.

***Type II***: Includes the lipid soluble steroid hormones calcitriol and thyroid hormones which bind to the cytosolic receptors (*Internal receptors*). Table 10.1 lists some of the important hormones belonging to the two groups.

**Table 10.1:** Hormones mediating their action through Type I and Type II receptors

| *Type I* | *Type II* |
|---|---|
| Glucagon | Androgens |
| Insulin | Oestrogens |
| Calcitonin | Glucocorticoids |
| Follicle stimulating hormone (FSH) | Mineralocorticoids |
| Parathyroid hormone (PTH) | Calcitriol |
| Antidiuretic hormone (ADH) | Thyroid hormones ($T_3$ and $T_4$) |
| Leuteinizing hormone (LH) | Retinoic acid |

In both cases the effects of the hormones are mediated by the *Hormone Receptor Complex*. However, the cell surface and the internal receptors mediate the hormone action by different mechanisms.

Complexing of a hormone with a cell surface receptor stimulates the production of a *Second Messenger* within the cell. The second messenger produced modifies the activity of specific cellular enzymes (Fig. 10.1).

In contrast, complexing of a hormone with an intracellular receptor, acts directly on the *DNA* and influences the transcription of specific genes.

*Note: More than one kind of receptor may be present for certain hormones, thus, bringing about different metabolic effects. Example, catecholamine receptors (adrenergic receptors) are of two types, $\alpha$ and $\beta$.*

### Mechanism of Action of Type I Hormones

Since these hormones do not enter the cells, *Second Messengers* are used to transmit the hormonal message into the cell. The hormone binding signal of most plasma membrane

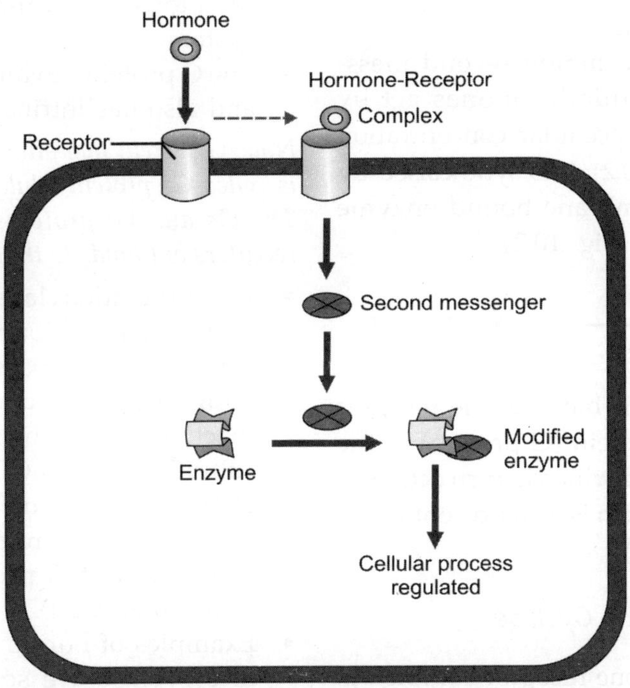


**Fig. 10.1:** Mechanism of peptide hormone action

receptors is transduced to the interior of the cells by the binding of hormone receptor complex [HR] to a series of membrane localised GDP / GTP binding proteins known as G proteins. It involves the following steps.

- The hormone binds to receptors on the surface of a target cell (hormone itself is the first messenger).
- Binding of the hormone to the receptor triggers the formation of second messenger within the cell.
- The *Second Messenger* modifies the activities of specific cellular enzymes.
- The modified enzymes mediate changes in the cellular process.

The various second messengers are:

- cAMP
- $Ca^{++}$
- Inositol Triphosphate ($IP_3$)
- Diacylglycerol (DAG)
- Nitric oxide (NO)
- Tyrosine kinase

- *cAMP*: Is a very common second messenger. Most peptide hormones act by increasing the intracellular concentration of cAMP (Table 10.2). It is synthesised by the action of membrane bound enzyme adenylate cyclase (Fig. 10.2).

$$\text{ATP} \xrightarrow[\text{Cyclase}]{\text{Adenylate}} \text{cAMP} + \text{PPi}$$

When the hormone binds to the receptor a GTP dependent protein (G protein) in the membrane is activated which in turn activates adenylate cyclase which is also present in the membrane (Fig. 10.2).

### *Activation of Adenylate Cyclase*

Binding of the hormone receptor complex to the G protein activates adenylate cyclase.

**Table 10.2:** Hormones using cAMP

| |
|---|
| TSH |
| Vasopressin |
| Glucagon |
| PTH |
| ADH |
| Calcitonin |

Adenylate cyclase is a membrane bound enzyme found in the plasma membrane of most mammalian cells.

- The active form of adenylate cyclase consists of a complex of G protein *(Guanine nucleotide binding protein)* that carries bound GTP.
- The G protein constitutes the communication link between the hormone receptor complex and the adenylate cyclase. It consists of 3 subunits, namely, α, β, γ. β and γ are common to all cAMP dependent receptor systems and α is specific for each system.
- The G protein can bind either GDP or GTP and also has intrinsic GTPase activity.

*Note: G protein that stimulates adenylate cyclase is called Gs protein while Gi protein inhibits it. The Gs and Gi proteins work with different receptors but binds to the same adenylate cyclase.*

- After the adenylate cyclase is activated by Gs or is inhibited by Gi protein the complex hydrolyses GTP to GDP liberating phosphate and dissociates into an inactive cyclase and the specific G protein subunit which is GDP bound.
- Overall activity of adenylate cyclase at any given moment depends on the proportion of G protein that carry GTP rather than GDP.
- Examples of hormones inhibiting adenylate cyclase are somatostatin, α-adrenergics and angiotensin II.

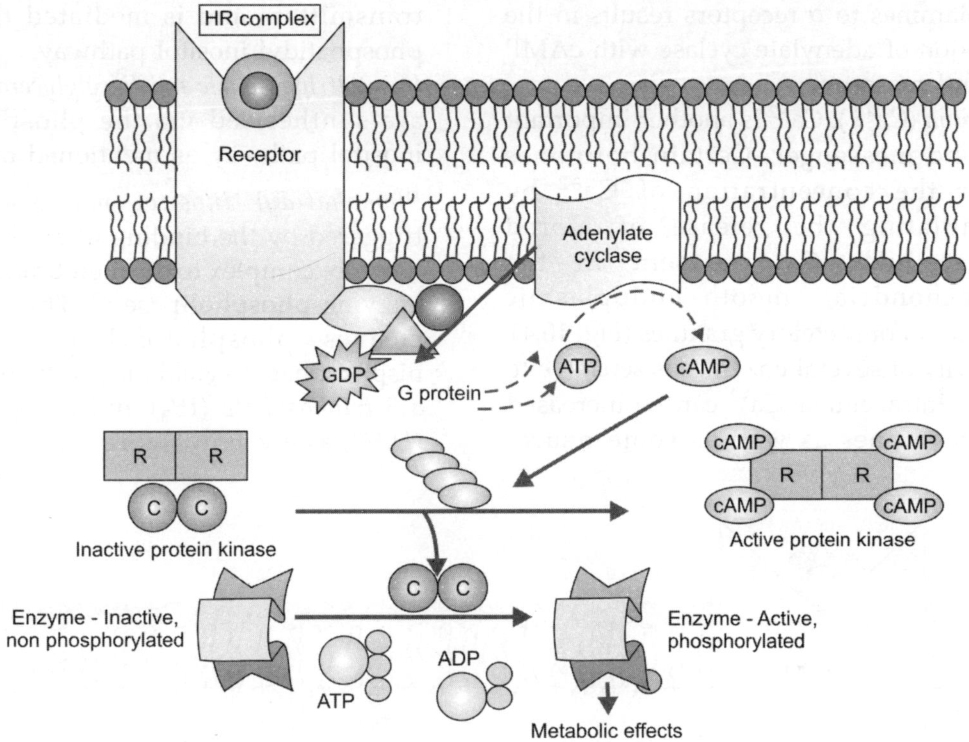


**Fig.10.2:** cAMP mediated peptide hormone action
*(For colour version see Plate 7)*

### Action of cAMP in Cells

cAMP acts by stimulating protein kinases. Protein kinases consist of catalytic and regulatory subunits (Fig. 10.2). cAMP binds to the regulatory subunit and causes it to dissociate thus activating the catalytic subunit.

Activated protein kinase catalyzes the phosphorylation of specific enzymes by reactions of the following type.

$$\text{Enzyme + ATP} \xrightarrow{\text{Protein kinase}} \text{Enzyme - P + ADP}$$

The enzymes may be either *activated* or *inhibited* by phosphorylation. Examples of enzymes which are activated and inhibited by phosphorylation: Glycogen phosphorylase, and acetyl CoA carboxylase are *activated* by phosphorylation. While glycogen synthase and pyruvate dehydrogenase are *inhibited* by phosphorylation.

The action of protein kinases is opposed by the action of phosphoprotein phosphatases which dephosphorylate the enzymes.

$$\text{Enzyme - P} \xrightarrow{\text{Phosphatase}} \text{Enzyme + Pi}$$

cAMP is inactivated by phosphodiesterase which catalyzes the conversion of cAMP to AMP. The activity of phosphodiesterases may also be regulated by hormones.

$$\text{cAMP} \xrightarrow{\text{Phosphodiesterase}} \text{5'AMP}$$

Catecholamines bind to membrane receptors on target cells. There are two types of catecholamine receptors (adrenergic receptors). They are $\alpha$ and $\beta$. Binding of catecholamines to α receptors results in the stimulation of adenylate cyclase with cAMP as second messenger.

- *Calcium* ($Ca^{+2}$): $Ca^{+2}$ is another important second messenger. Certain hormones alter the concentration of $Ca^{+2}$ by stimulating the release of stored intracellular $Ca^{+2}$ present in the mitochondria, smooth endoplasmic reticulum or secretory granules (Fig. 10.4). Activity of several enzymes is sensitive to $Ca^{+2}$. Intracellular $Ca^{+2}$ can be increased by hormones as well as some neuro-transmitters and is mediated through phosphatidyl inositol pathway.
- *Inositoltriphosphate and diacylglycerol*: These are synthesised via the phosphatidyl inositol pathway as mentioned below.

*Phosphatidyl inositol pathway*: It is triggered by the binding of the hormone receptor complex to the membrane bound enzyme phospholipase C. This enzyme cleaves phosphatidyl inositol 4,5 bisphosphate to yield the products *inositol 4,5 triphosphate* ($IP_3$) and *diacylglycerol (DAG)* as given in Figure 10.3.

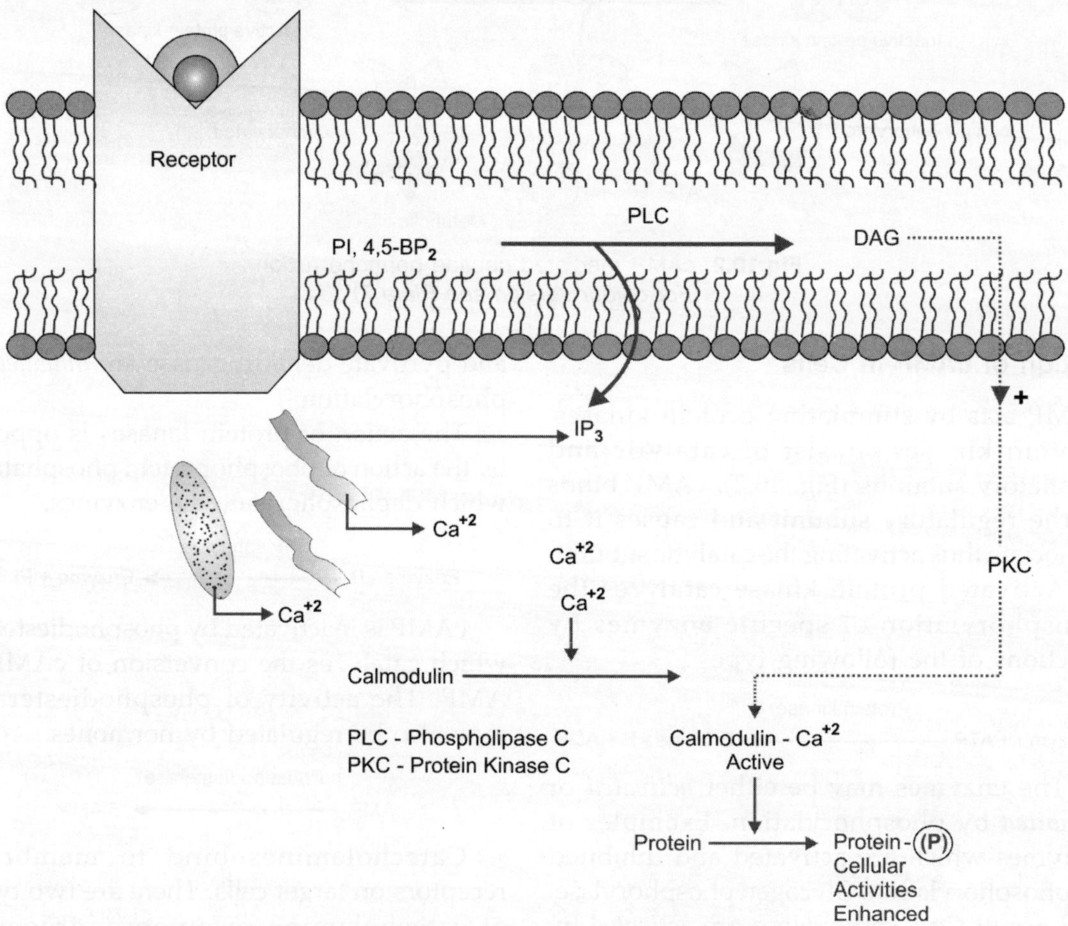


**Fig.10.3:** Formation of $IP_3$ and DAG

$IP_3$ passes through the cytoplasm and stimulates the endoplasmic reticulum to release $Ca^{+2}$.

*Note: $IP_3$ is a short lived second messenger. It is rapidly degraded to inositol 4, 5 bisphosphate which is inactive.*

*Diacyl glycerol*: The diacylglycerol generated by the action of phospholipase C as mentiond above is retained in the membrane. It also functions as a second messenger. It activates the key signaling enzyme PKC (protein kinase C) which phosphorylates a wide range of target signal transduction proteins on serine or threonine residues.

Changes in the intracellular $Ca^{+2}$, affect cellular metabolism largely by means of the calcium binding protein *calmodulin*. In this case $Ca^{+2}$ acts as a third messenger.

Some of the cellular processes mediated by $Ca^{+2}$ are the following:

- Muscle contraction
- Blood clotting cascade
- Membrane excitability

*Calmodulin*: Is the major $Ca^{+2}$ binding protein in the cell. It is also called as $Ca^{+2}$ dependent regulatory proteins. Calmodulin has 4 $Ca^{+2}$ binding sites.

- Binding of $Ca^{+2}$ to all 4 sites triggers a conformational change that activates it.
- Active calmodulin can now complex with a number of cellular enzymes stimulating their activity (Fig. 10.4).

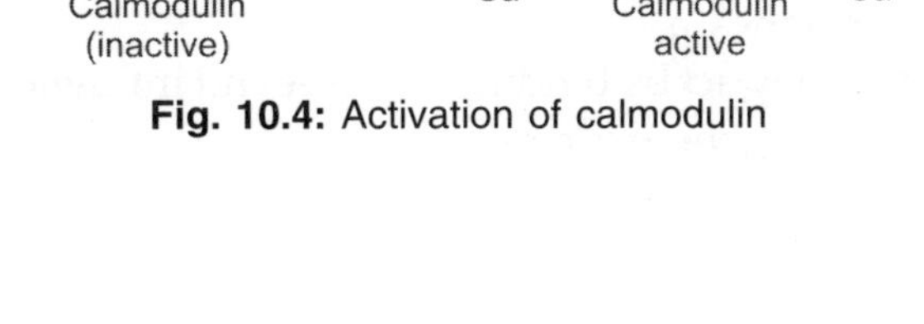


**Fig. 10.4:** Activation of calmodulin

*Synergism between messengers*: The two second messengers ($IP_3$ and DAG) synergistically act to cause increased phosphorylation of proteins. DAG activates protein kinase C by a process requiring $Ca^{+2}$ whereas $IP_3$ elevates $Ca^{+2}$ to activate calmodulin dependent protein kinase. The elevated $Ca^{+2}$ may also act by additional mechanisms that do not involve calmodulin.

Typical example is the binding of epinephrine to α adrenergic receptors; leads to the dissociation of phosphatidyl inositol 4, 5 bisphosphate to $IP_3$ and DAG by the enzyme phospholipase C. $IP_3$ or DAG then stimulates the release of $Ca^{+2}$ from the endoplasmic reticulum (Fig. 10.4).

The hormones utilising phosphatidyl inositol pathway are given in Table 10.3.

*cGMP*: cGMP also acts as a second messenger like cAMP and is synthesised by the membrane bound enzyme guanylate cyclase.

$$\text{GTP} \xrightarrow{\text{Guanylate cyclase}} \text{cGMP} + \text{PPi}$$

- Two kinds of guanylate cyclase are present: Mitochondrial and cytosolic. It is present mainly in the cardiac atrial tissues, platelets and smooth muscles. It is linked to a specific biological signal through a membrane receptor.

**Table 10.3:** Hormones mediating their action through $Ca^{+2}$

| |
|---|
| Epinephrine |
| Norepinephrine |
| Thyrotropin releasing hormone |
| Cholecystokinin |
| Vasopressin |
| Oxytocin |
| Gonadotropin releasing hormone |
| Acetylcholine |
| Gastrin |

- The increased cGMP activates cGMP dependent protein kinase (PKG) which in turn phosphorylates a number of smooth muscle proteins resulting in vasodilation and relaxation of smooth muscle.
- Antinatriuretic factor, ANF (*Atriopeptins*) mediates its action through cGMP. It is produced in the cardiac atrial tissues and causes natriuresis, vasodilation and inhibition of aldosterone. cGMP is also involved in visual system, platelet aggregation and smooth muscle relaxation.

*Note: The drug viagra (sildenafil) enhances and prolongs the response of cyclic GMP.*

*Nitric oxide (NO)*: Nitric oxide is a unique second messenger. It is a potent vasodilator and exerts its effect by activating a soluble cytosolic type of guanylate cyclase and thus increases the cGMP.

- Nitric oxide is formed when the amino acid arginine is broken down into NO and citrulline by nitric oxide synthase.

$$\text{Arginine} \xrightarrow[\text{Synthase}]{\text{Nitric oxide}} \text{Citrulline + NO}$$

- Action of cGMP is terminated by the phosphodiesterases.
- *Tyrosine kinase*: Some receptors produce their biologic effect by phosphorylating the tyrosine groups of proteins. These receptors are generally for anabolic peptides such as insulin and many growth factors. Some have intrinsic tyrosinase activity, e.g., insulin receptor.

Such receptors are characterised by 3 domains:

- ➢ Extracellular domain—binds the hormone.
- ➢ Hydrophobic transmembrane domain
- ➢ Intracellular domain—has tyrosine kinase activity.

These receptors may be composed of single polypeptide chain (*EGF receptor*) or multiple subunits (*Insulin receptor*). After binding of the hormone, the signal is passed to tyrosine kinase which phosphorylates proteins. The hormones which stimulate tyrosine kinase is given in Table 10.4.

**Table 10.4:** Hormones mediating their action through tyrosine kinase

| |
|---|
| Insulin |
| Epidermal growth factor (EGF) |
| Platelet derived growth factor (PDGF) |
| Somatomedin |

Hormones such as EGF which are involved in growth control, differentiation and inflammatory response mediate their action through *protein kinase cascade* involving several protein kinases.

***Mechanism of action of lipid soluble hormones:*** Unlike the protein hormones the steroid hormones and thyroid hormones enter the cells by passive diffusion. They complex with the intracellular receptors forming hormone receptor (HR) complex. Receptors for steroid hormones are present in the *cytosol* whereas the thyroid hormone receptors are in the *nucleus*.

- HR complex binds to the chromatin and stimulates the expression of certain genes. Thus protein composition of the cell is altered which influences cellular metabolism (Fig.10.5).

### *Steroid Hormone Receptors*

Steroid hormones, calcitriol and thyroid hormones exert their metabolic effects by interacting with specific intracellular receptors. The receptors for all these hormones are closely related.

- They act by binding to DNA and influence the gene expression.

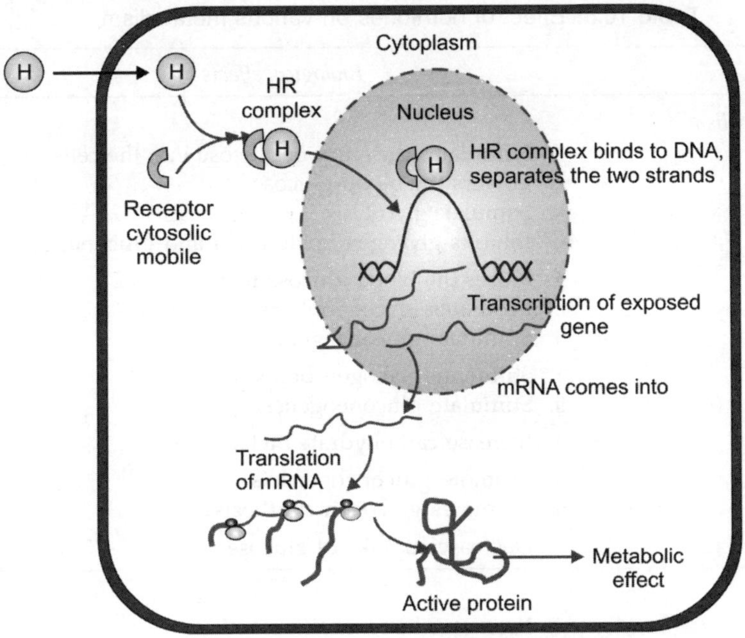


**Fig. 10.5:** Action of steroid hormone and thyroid hormones; *H – hormone*
*(For colour version see Plate 7)*

- The receptors may be made up of one or multiple subunits. Each subunit has 2 specific areas essential for biologic activity.
- One for binding of DNA and the other for binding of the specific hormone.
- Binding to the specific gene increases synthesis of mRNA.
- Binding to hormone response element (HRE) alters structure of chromatin and allows much greater transcription of gene.

***Regulation of hormone synthesis:*** Hormone levels are regulated mainly by *feedback inhibition.* Once the hormone has produced its effect it results in further inhibition of hormone synthesis *(negative feedback inhibition).* A lack of feedback inhibition results in profound increases in hormone synthesis and is usually detrimental.

***Degradation of the hormones:*** For any message to be useful it must be active only for finite time. Similarly hormones which are messages must be destroyed in a timely fashion. It may be specific or non-specific.

- Many hormones are non-specifically destroyed by the liver and kidneys via proteases and others may be excreted into the urine.

***Regulation of receptors:*** Elevated levels of hormones typically cause reductions in the number of their own receptors, a process called *down regulation.* Most hormones that bind to plasma membrane receptors will cause down regulation to occur. Steroid hormone receptors are also recycled and can undergo down regulation but the mechanism is poorly understood.

Effect of the various hormones on the metabolism are listed in Table 10.5. *The various hormonal disorders are listed in Table 10.6.*

**Table 10.5:** Effect of hormones on various metabolism

| *Hormone* | *Biological effects* |
|---|---|
| *On carbohydrate metabolism* | |
| Insulin | i. Stimulates the entry of glucose into the cells<br>ii. Lowers the blood glucose<br>iii. Stimulate glycolysis & glycogenesis<br>iv. Inhibits glycogen breakdown and gluconeogenesis |
| Glucagon | i. Raises the blood glucose levels<br>ii. Promotes glycogenolysis<br>iii. Stimulate gluconeogenesis |
| Catecholamines | i. Stimulate glycogen breakdown<br>ii. Stimulate gluconeogenesis |
| $T_3$ & $T_4$ | i. Increase carbohydrate metabolism |
| Glucocorticoids | i. Promote gluconeogenesis<br>ii. Promotes glycogen synthesis |
| Growth hormone | Decreases uptake of glucose |
| *On lipid metabolism* | |
| Insulin | i. Promotes lipogenesis<br>ii. Inhibits mobilisation of fatty acids in the adipose tissue<br>iii. Enhances activity of lipoprotein lipase |
| Glucagon | i. Promotes lipolysis in adipose tissue<br>ii. Inhibits lipogenesis<br>iii. Stimulates ketogenesis<br>iv. Stimulates mobilisation of fatty acids from adipose tissue |
| Catecholamines | i. Stimulate lipolysis in adipose tissue<br>ii. Inhibit the release of insulin |
| $T_3$ & $T_4$ | i. Increase the rate of lipid metabolism |
| Growth hormone | i. Stimulates mobilisation of fatty acids from adipose tissue<br>ii. Inhibits the responses of the target tissues to insulin |
| *On protein metabolism* | |
| Insulin | i. Stimulates synthesis of proteins<br>ii. Enhances cellular uptake of amino acids |
| Glucagon | i. Inhibits protein synthesis in the liver, thus provides amino acids for gluconeogenesis. |
| *Catecholamines* | |
| $T_3$ & $T_4$ | Increase the rate of protein metabolism |
| Glucocorticoids | Promote protein degradation |
| Growth hormone | Increases transport of amino acids into muscle cells and also increases protein synthesis |

*Contd...*

*Contd...*

| *On calcium homeostasis* | *Site* | |
|---|---|---|
| PTH | Kidney | i. Increases reabsorption of $Ca^{+2}$<br>ii. Stimulates production of 1, 25 dihydroxy cholecalciferol, |
| | Bone | Resorption of bone $Ca^{+}$ |
| | Intestine | Exerts indirect effect. 1, 25 dihydroxy cholecalciferol synthesised in the kidney stimulates intestinal reabsorption of $Ca^{+2}$ |
| Calcitonin | Kidney | Resorption of bone $Ca^{+2}$ and tubular reabsorption of Pi |
| Calcitriol | Intestine | Stimulates absorption of $Ca^{+2}$ and Pi by the intestinal cells, stimulates uptake of $Ca^{+2}$ and promotes bone formation. Along with PTH, stimulates resorption and excretion of phosphate. |

| *Action of anterior pituitary hormones* | |
|---|---|
| ACTH (Adrenocorticotropic hormone) | Stimulates synthesis and secretion of adrenocortical steroids |
| TSH (Thyroid stimulating hormone) | Stimulates synthesis and secretion of thyroid hormones. |
| GH (Growth hormone) | Stimulates skeletal and soft tissue growth |
| FSH (Follicle stimulating hormone) | In females, stimulate growth of the graafian follicles and oestrogen production |
| LH (Luteinizing hormone) | In females stimulates synthesis of progesterone |
| Prolactin | Stimulates and maintains lactation. |
| *Action of posterior pituitary hormones* | |
| ADH (Antidiuretic hormone) | Concerned with regulation of water balance. Stimulates the kidneys to retain water |
| Oxytocin | i. Causes contraction of the uterus and induces labour<br>ii. Stimulates ejection of milk from the breast |

## NEUROTRANSMITTERS

Neurotransmitters are the molecules that act as chemical signals between the nerve cells. More than 30 neurotransmitters are known. Most of them are low molecular weight compounds containing a positively charged nitrogen atom. Each neurotransmitter has a characteristic excitatory or inhibitory effect on a specific type of nerve fibre, although some may be excitatory in certain parts of the nervous system and inhibitory in other.

### Classification

Neurotransmitters may be classified based on the chemical composition (Table 10.7). Many of them are derived from common amino acids. Some of the important examples are acetylcholine, norepinephrine, γ-amino-

**Table 10.6:** Hormonal disorders

| *Disorder* | *Defect* | *Symptoms* |
|---|---|---|
| 1. Cushing's syndrome | Over production of ACTH – leading to synthesis of corticosteroids | Characterised by<br>• Hypertension<br>• Oedema<br>• Osteoporosis |
| Addison's disease | Deficiency of ADH (mineralocorticoids) | Characterised by<br>• Loss of appetite<br>• Muscle weakness<br>• Impaired cardiac function<br>• Hypoglycaemia<br>• Hyponatraemia<br>• Hypokalemia |
| 2. Diabetes insipidus | Insufficient levels of ADH or defect in the receptors of the target cells. | Characterised by<br>• Polyuria |
| 3. Goitre | Iodine deficiency in the diet. | • Abnormal increase in the size of the thyroid gland |
| a. Simple endemic goitre | Excess of thyroid hormones. | • Characterised by |
| b. Hyperthyroidism | Failure in the autoregulation of $T_3$ & $T_4$ synthesis. | • Increased BMR<br>• Loss of weight<br>• Sweating<br>• Exophthalmos |

butyric acid (GABA), dopamine and serotonin. Presently the peptides are gaining a lot of importance as neurotransmitters.

**Table 10.7:** Some of the physiologically important neurotransmitters

| *Group* | *Examples* |
|---|---|
| Ester of amino alcohol | Acetylcholine (ACh ) |
| Amines | Norepinephrine, Epinephrine, dopamine, serotonin |
| Amino acids | Glutamate, GABA, Glycine |
| Peptides | Endorphins |
| Purines | ATP; Adenosine |
| Gases | Nitric oxide |

Functionally they may be grouped into excitatory (stimulators) and inhibitory neurotransmitters (Table 10.8).

**Table 10.8:** Excitatory and inhibitory neurotransmitters

| *Excitatory* | *Inhibitory* |
|---|---|
| Acetylcholine | GABA |
| Glutamate | Glycine |
| Norepinephrine | Adenosine |
| Epinephrine | Dopamine |
| ATP | |

## NEUROTRANSMISSION

Involves the transmission of nerve impulses. During the transmission of nerve impulse, neurotransmitter is ejected from the axon into the synaptic cleft and then passes into the dendrites of the next nerve cell. This process is extremely rapid requiring less than 1 msec. Synthesis of the neurotransmitter occurs in the cell body of the nerve cell (Fig. 10.6).

## Nerve Impulse (Fig. 10.7)

Consists of a wave of depolarisation that travels along the nerve cell membrane from the dendrites to the axon. When a nerve cell is stimulated a local area of the membrane becomes slightly more permeable to $Na^+$. As a result $Na^+$ flows into the cell reducing the membrane polarisation (*depolarisation*). If this depolarisation exceeds a specific value called threshold potential (*20 mv*) a nerve impulse or action potential is triggered.

When the action potential reaches the end of the axon it causes the neurotransmitter molecules to be released into the synaptic clefts and binds to specific receptor sites on the post synaptic cells triggering the appropriate physiologic response. Nerve cells contain voltage dependent sodium channels that open very rapidly when a depolarising change in voltage is applied. When they open they allow the inward passage of huge number of $Na^+$ ions from the ECF which swamps the resting voltage and drives the membrane potential to positive values. This reversal of voltage from the –ve value to the + ve value is the *"action potential"*.

Neurotransmitters alter the activity of various ion channels to cause changes in the membrane potential.

- Excitatory neurotransmitters cause a depolarising change in voltage and an action potential is more likely to occur.
- Inhibitory neurotransmitters hyperpolarize the membrane and an action potential is less likely to occur.

## MECHANISM OF ACTION

Like the hormones neurotransmitters also bind to specific receptors and open or close the ion channels. Directly or indirectly they cause changes in ion flow across the membrane until the potential reaches the threshold value.

- Receptors that directly control the opening of an ion channel are called *'ionotropic'* while those that control the opening of the ion channel through second messenger systems are termed *'metabotropic'* receptors.

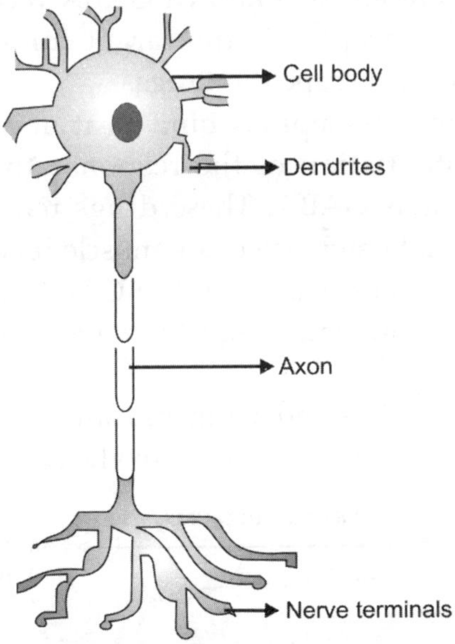


**Fig. 10.6:** Structure of a nerve cell

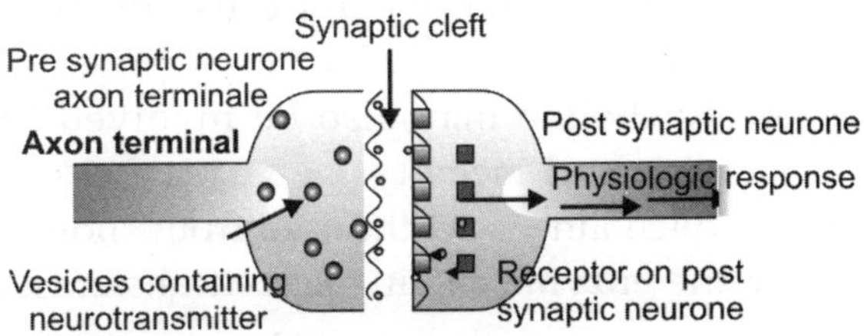


**Fig. 10.7:** Transmission of nerve impulse

### Ionotropic Receptors

Contain an ion channel within their structure. These are transmembrane proteins with several subunits, surrounding a pore through the membrane. When the neurotransmitter binds, there is a change in the three dimensional structure of the complex which allows the flow of ions through it. *For example,* Nicotinic acetylcholine receptor and the glutamate and γ-aminobutyrate (*GABA*) receptors.

### Metabotropic Receptors

Are coupled to second messenger pathways and act more slowly than ionotropic receptors. All known metabotropic receptors are coupled to G proteins. Typically they couple either to adenylate cyclase, altering the production of cAMP or to the phosphatidyl inositol pathway which alters the calcium fluxes, e.g. adrenergic receptors. A brief note on some of the important neurotransmitters are given below.

*Acetylcholine (ACh)*: The first chemical neurotransmitter identified was acetylcholine. It is a neurotransmitter of the parasympathetic, autonomic nervous system and of the sympathetic ganglia producing the effects such as slowing of heart rate, bronchoconstriction and stimulation of intestinal smooth muscle.

- Acetylcholine acts at the neuromuscular junctions where motor neuron contact skeletal muscle cells and cause them to contract.
- Acetylcholine may also be involved in learning and memory.
- Acetylcholine is synthesised from choline by the enzyme choline acetyltransferase.

$$\text{Choline} + \text{Acetyl CoA} \xrightarrow{\text{Transferase}} \text{Acetylcholine}$$

- Clinically acetylcholine agonists in combination with acetylcholinesterase inhibitors are used to treat glaucoma (eye disease characterised by high intraocular pressure) and intestinal function after surgery.

*Glutamate*: Is the most important excitatory transmitter in the CNS. It acts on both ionotropic and metabotropic receptors.

- Glutamate is recycled by high affinity transporters into both neurons and glial cells.
- The glial cells convert it into glutamine which then diffuses back into the neuron.
- Mitochondrial glutaminase in the neuron regenerates glutamate for reuse.

$$\text{Glutamine} \xrightarrow{\text{Glutaminase}} \text{Glutamate} + NH_3$$

*γ-amino butyric acid (GABA)* is the major inhibitory transmitter in the brain.

- There are two known GABA receptors. The receptor is the target for several useful therapeutic drugs.
- Benzodiazepines bind to it and cause potentiation of the response to endogenous GABA. These drugs reduce the anxiety and also cause muscle relaxation.
- Barbiturates also bind to GABA receptor and stimulate it directly in the absence of GABA.

It is synthesised from glutamate by the enzyme glutamate decarboxylase.

$$\text{Glutamate} \xrightarrow[B_6PO_4 \;\searrow CO_2]{\text{Glutamate decarboxylase}} \gamma \text{ amino butyric acid (GABA)}$$

### Glycine

Is another inhibitory neurotransmitter and is primarily found in the spinal cord. It blocks the impulses traveling down the cord in the motor neurons to stimulate skeletal muscle. The glycine receptor on the motor neurons is ionotropic and is blocked by strychnine.

### Norepinephrine (Noradrenaline)

Is a major neurotransmitter in the post-ganglionic sympathetic nervous system (*sympathetic nerves arise in the spinal cord and run to the ganglia situated close to the cord, from which the postganglionic nerves are connected to the target tissues).*

- Stimulation of these nerves, is responsible for various features of "fight, flight or fright" response such as:
    - ➢ Stimulation of heart rate
    - ➢ Sweating
    - ➢ Vasoconstriction in the skin
    - ➢ Bronchodilation
- Stimulation of norepinephrine containing neurons in the CNS alters the overall state of alertness or attention.

### Epinephrine (Adrenaline)

Is produced by the adrenal medulla. It is more active than norepinephrine on heart and lungs. It has important stimulatory effects on glycogen metabolism in liver and muscle.

The receptors for norepinephrine and epinephrine are called *adrenoreceptors.* They are divided into α and β types. Epinephrine acts on both types while norepinephrine is more specific for β receptors.

### Dopamine

Is a major inhibitory neurotransmitter in the brain and controls the voluntary movement. It is an intermediate in the synthesis of norepinephrine.

*Note: Norepinephrine, epinephrine and dopamine are together termed as catecholamines. They are derived from the amino acid tyrosine.*

### 5-hydroxytryptamine (5HT)

Is also known as serotonin. Mainly controls the degree of responsiveness of motor neurons in the spinal cord. It is also implicated in the so called vegetative behaviours such as feeding, sexual behaviour and temperature control. It is derived from tryptophan.

### Other Small Molecules

$$\text{Tryptophan} \xrightarrow{\text{Hydroxylase}} \text{5–hydroxytryptophan} \longrightarrow \text{5 HT} \quad (\searrow CO_2)$$

*ATP*: Present in, the synaptic vesicles of sympathetic nerves moves along with norepinephrine. It is responsible for rapid excitatory potentials in the smooth muscle.

*Adenosine:* Inhibitory neurotransmitter in the CNS.

*Histamine:* Present mainly in the hypothalamus and controls the release of pituitary hormones. Also involved in arousal and food intake.

*Nitric oxide:* Excitatory neurotransmitter and is involved in the mechanism of learning.

### Neuropeptides

The first two discovered were pentapeptides and were given the name Enkephalins ("in brain"). They are *"inhibitory neurotransmitters".*

They are termed methionine enkephalin and leucine enkephalin respectively.

Tyr - Gly - Gly - Phe - Met Methionine enkephalin
Tyr - Gly - Gly - Phe - Leu Leucine enkephalin

Enkephalins bind to the same receptors in the brain that binds opiates. Hence they are also termed as *'opioids'*.

- With the discovery of other opioid peptides, a generic name for the group, *endorphins* has been adopted.
- Three opioids, α-endorphin, β-endorphin and γ-endorphin were found to be components of β-lipotropin.

Enkephalins inhibit neuronal activity by decreasing sodium ion conductance at $Na^+$ channels. They also inhibit adenylate cyclase and cAMP formation from ATP necessary for the phosphorylation of membrane and nuclear proteins and in turn for certain functions of the brain.

*Substance P*—contains 11 amino acids and is involved in the transmission of pain.

### *Regulation of Neurotransmitters*

Once the transmitters have served their function they must be removed from the synaptic space.

- Simple diffusion is the major mechanism of removal of neuropeptides.
- Acetylcholine may be destroyed by the enzyme acetylcholinesterase.
- Some of the neurotransmitters may also be taken back into the presynaptic neuron for reuse. This is a major route for removal of catecholamines and amino acids.

## DISORDERS

***Pheochromocytoma:*** It is a condition characterised by the tumour of the adrenal medulla. Increased levels of catecholamines are seen in these patients. This in turn causes hypertension because of the vasoconstrictor action of the catecholamines.

***Parkinson's Disease:*** It is due to the deficient amounts of dopamine. Resting tremors of the hand and other neurological manifestations are characteristic of the disease. Impairment in the function of the basal ganglia (portion of the brain involved in synthesis of dopamine is the principal cause.

# 11 Detoxification

The term 'detoxification' refers to the various biochemical reactions occurring in the body to convert a toxic compound to a less toxic and more soluble form so that it can be easily excreted from the body. Recently the term detoxification is replaced by *Biotransformation* or *Metabolism of Xenobiotics* (Foreign Compounds).

These compounds may be either produced in the body or may gain entry into the body.

Some of the toxic compounds formed by the intestinal bacteria in the body include the following:

- Indole - from tryptophan
- Cadaverine - from lysine
- Tyramine - from tyrosine
- Phenol - from phenylalanine
- Bilirubin - from heme

Some compounds may gain entry into the body by the following means. These are called xenobiotics.

- Drugs
- Food additives
- Cosmetics
- Pollutants

The primary objective of detoxification is to convert the toxic compound to a more polar compound, so that the permeability of the compound through the lipid cell membrane is reduced and it is readily excreted. Most detoxification reactions occur in the liver.

The mechanisms employed in the metabolism of xenobiotics are the ones that are ordinarily used in normal metabolic processes. They are:

1. Oxidation
2. Reduction
3. Hydrolysis
4. Conjugation

The metabolism of xenobiotics, may be divided into two phases which are:

**Phase I :** Reactions include oxidation, reduction and hydrolysis.

**Phase II :** Includes conjugation reactions involving conjugating molecules such as glucuronic acid, amino acids, sulphate, etc.

Generally, detoxification of a compound involves phase I followed by phase II reactions.

A brief consideration of the various mechanisms with examples are given below.

***Oxidation:*** A large number of foreign compounds are detoxified by oxidation. These include—

- Alcohols
- Aldehydes
- Amines
- Aromatic hydrocarbons.

Ethyl alcohol may be completely oxidised in the body to $CO_2$ and $H_2O$.

$$\underset{\text{Ethanol}}{C_2H_5OH} \xrightarrow{\text{Alcohol dehydrogenase}} \underset{\text{Acetaldehyde}}{CH_3CHO} \longrightarrow \underset{\text{Acetic acid}}{CH_3COOH} \xrightarrow{\text{TCA cycle}} CO_2 + H_2O$$

$$\underset{\text{Benzyl alcohol}}{C_6H_5\text{-}CH_2OH} \xrightarrow{(O)} \underset{\text{Benzoic acid}}{C_6H_5COOH}$$

Aldehydes are oxidized to the corresponding acids.

$$\underset{\text{Benzaldehyde}}{C_6H_5CHO} \xrightarrow{(O)} \underset{\text{Benzoic acid}}{C_6H_5COOH}$$

Aliphatic amines are completely oxidised in the body. This is carried out by the enzyme *Monoamine acid oxidase (MAO).*

$$R - CH_2 - NH_2 \xrightarrow[MAO]{} R\text{–}CHO + NH_3$$

$R\text{–}CHO \downarrow CO_2 + H_2O$; $NH_3 \downarrow$ Urea

Aromatic hydrocarbons such as benzene is oxidised to muconic acid which involves opening of the ring.

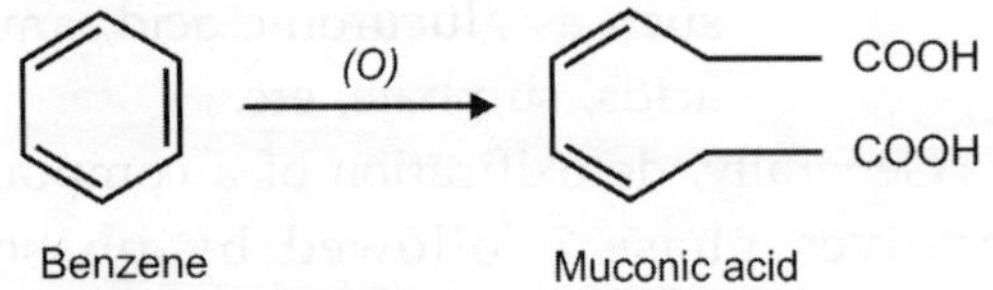

Benzene → Muconic acid

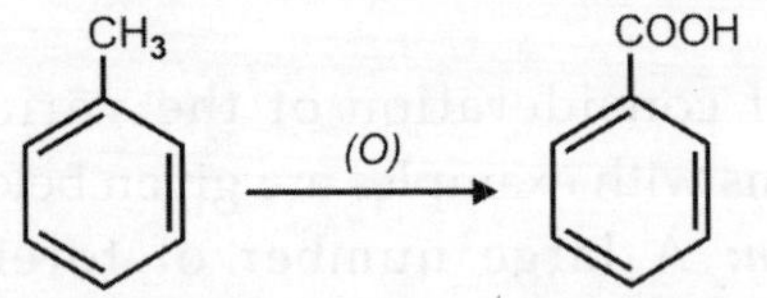

Methyl Benzene → Benzoic acid

- Oxidation usually occurs first and may be followed by conjugation.
- Oxidation of drugs and steroids actually involves hydroxylation and subsequent conjugation with glucuronide or sulphate before their excretion in urine.
- *Hydroxylation* is catalysed by *Mixed Function Oxidase (Mono-oxygenase) System* present in the microsomes of the hepatic cells. Cytochrome $P_{450}$ are heme proteins and is an important component of this system. $P_{450}$ refers to the absorption peak at 450 nm exhibited by the enzyme when exposed to carbon monoxide.

The reactions of cyt $P_{450}$ involves the addition of a hydroxyl group to aliphatic or aromatic compounds and may be represented as follows:

$$R\text{–}H + O_2 + NADPH \longrightarrow R\text{–}OH + H_2O + NADP^+$$

- Cytochrome $P_{450}$ is an inducible enzyme. Its synthesis is increased by the administration of drugs such as phenobarbital. Several drugs are detoxified by the mono-oxygenase system. *For example,*

Meprobromate ⟶ Hydroxy meprobromate
Benzene ⟶ Phenol

*Note: Drugs such as phenobarbital and cholanthrene are detoxified by hydroxylation.*

## Reduction

Is less common and apparently less important than oxidation. Two important examples for reduction are:

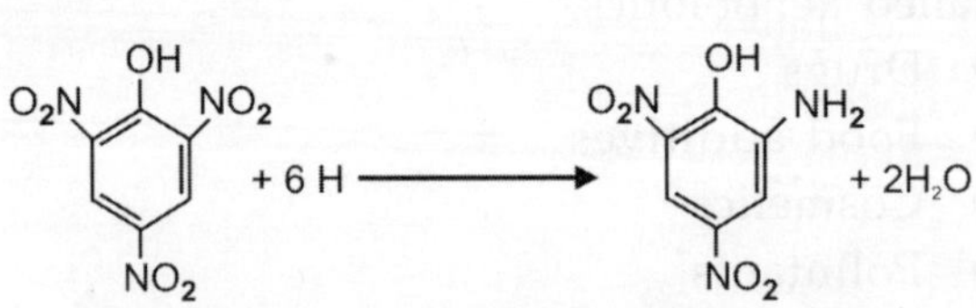

Picric Acid ⟶ Picramic acid
Chloral ⟶ Trichloroethyl alcohol

***Hydrolysis:*** Several compounds are detoxified by hydrolysis. This process involves the hydrolysis of esters, glycosides or amides. Some of the examples are:

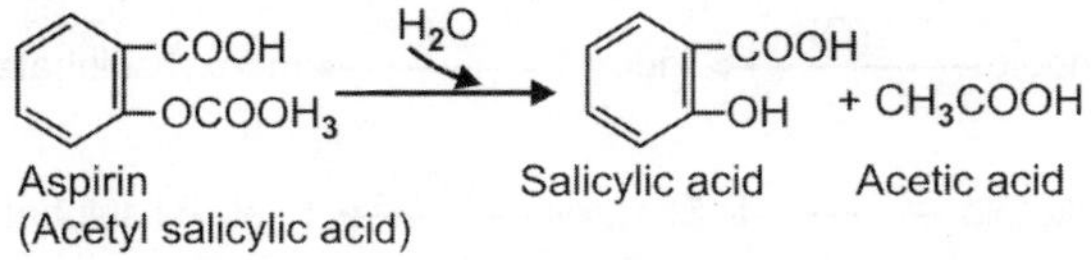


The acetic acid formed is either oxidised or used for the synthesis of physiological compounds and the salicylic acid is excreted after conjugation with glucuronic acid.

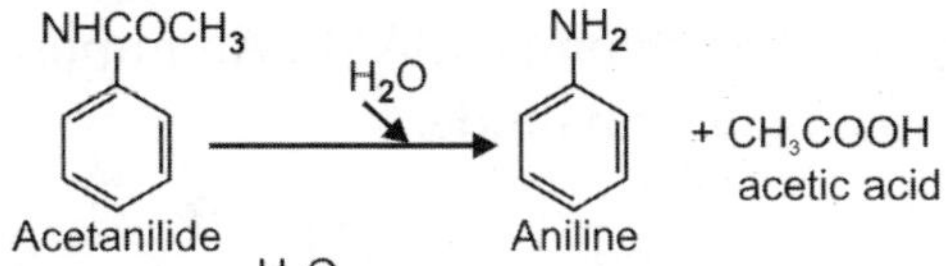


$$\text{Atropine} \xrightarrow{H_2O} \text{Tropic acid + Tropine}$$

$$\text{Procaine} \xrightarrow{H_2O} \text{p-amino benzoic acid + Diethyl amino ethanol}$$

## Conjugation

The chemical combination of one compound with another is referred to as conjugation. Several xenobiotics undergo detoxification by conjugation to produce less toxic and easily excretable compounds. The process of conjugation may occur either directly or after the phase I reaction.

The most common conjugating agents are :

- Glucuronic acid
- Glycine
- Glutathione
- Acetate
- Sulfate

*Glucuronic acid:* Is an oxidation product of glucose. Glucuronic acid participates in detoxification reactions as UDP derivatives. The reaction involves the transfer of glucuronic acid to the compound to be detoxified with the aid of *UDP glucuronyl transferase*. Examples are:

a. Detoxification of Bilirubin :

$$\text{Bilirubin + 2 UDP glucuronic acid} \xrightarrow[\textit{transferase}]{\textit{UDP glucuronyl}} \text{Bilirubin diglucuronide + 2UDP}$$

b. Detoxification of Phenol

$$\text{Phenol + UDP glucuronic acid} \xrightarrow[\textit{transferase}]{\textit{UDP glucuronyl}} \text{Phenyl glucuronide + UDP}$$

*Note:*

1. Products of sex hormone metabolism are excreted as glucuronides. Other compounds excreted using glucuronic acid are sulfa drugs, morphine, salicylic acid, vanillin, etc.
2. The body may use more than one method of detoxification.

*Glycine:* The classic example for use of glycine as conjugating agent, is its conjugation with benzoic acid to form hippuric acid.

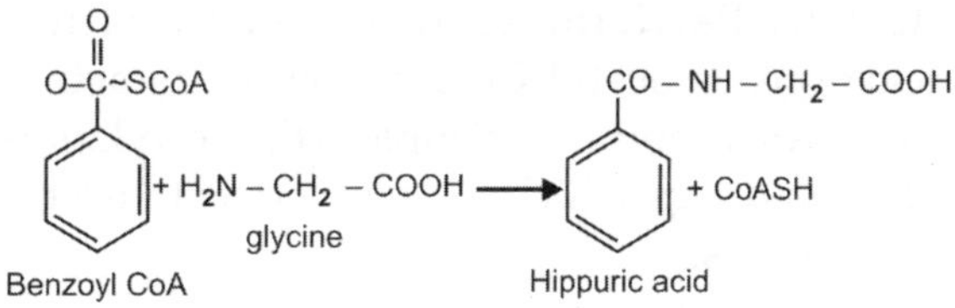


The excretion of hippuric acid was first reported in the urine of cows and horses. Many of the derivatives of benzoic acid are conjugated with glycine in this way.

*Glutathione (GSH):* Compounds such as naphthalene, anthracene, benzyl chloride, bromobenzene, chloroform, etc are detoxified by conjugation with glutathione. The drug is linked to the SH group of cysteine. After conjugation the glutamate and glycine of GSH are removed and an acetyl group is added to the cysteine residue to form mercapturic acid which is readily excreted. *For example,* bromobenzene.

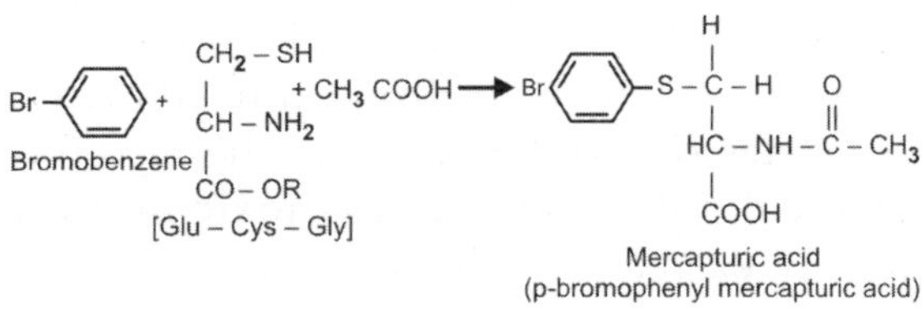

*Glutamine:* Typical example of the employment of glutamine in conjugation reaction is the union of this compound with phenyl acetic acid which is formed from phenylalanine by the minor metabolic pathway.

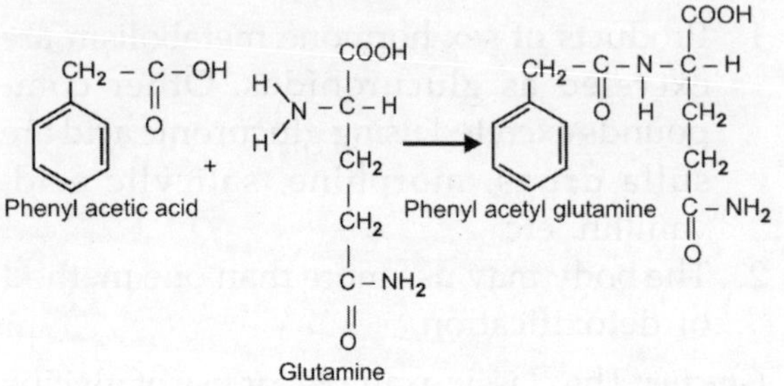


*Note: Large amounts of phenyl acetyl glutamine is excreted in patients with phenylketonuria.*

*Acetate:* Participates in the conjugation reactions as acetyl CoA, which is its active form. An important example is the acetylation of sulfa drugs which are converted to acetyl derivatives.

Isoniazid + Acetyl CoA ⟶ Acetyl Isoniazid + CoA SH

[This occurs after absorption or parenteral administration and the efficiency of the drugs as bacteriostatic agents may thereby be decreased].

### Sulphate

Conjugation by sulphate is achieved by the way of phosphoadenosinephosphosulphate (PAPS) which is also known as *Active Sulphate*. The transfer of $SO_4$ from PAPS to the substance to be detoxified involves the enzyme *sulfotransferase*.

Excretion of indole, skatole and cresol formed by the action of intestinal bacteria on tryptophan involves conjugation with sulphate to form indoxyl and skatoxyl sulphates (*Ethereal Sulphates).* The resulting ethereal sulphates are less toxic and are readily excreted by the kidney.

Indole —(O)⟶ Indoxyl —PAPS⟶ Indoxyl sulphate

Skatole —(O)⟶ Skatoxyl —PAPS⟶ Skatoxyl sulphate

*S-Adenosyl Methionine:* Detoxification of pyridine derivatives occurs by the addition of a methyl group. The methyl group is donated by activated methionine known as S adenosyl methionine (SAM). An important example is nicotinamide.

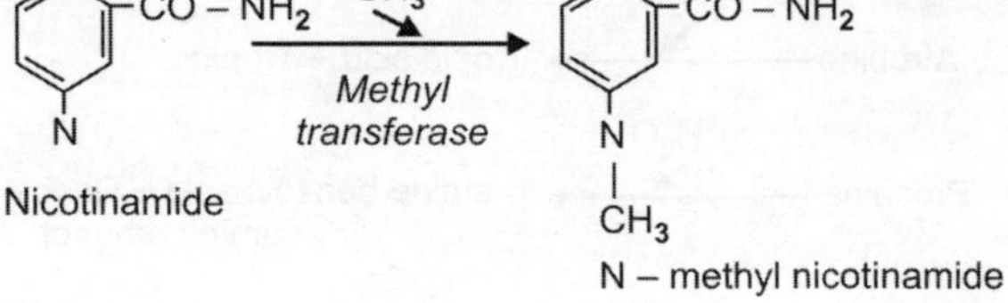


*Detoxification of Cyanide ($CN^-$) :* Is accomplished by the enzyme *rhodonase* which converts the cyanide to thiocyanate. Source of S may be thiosulphate.

HCN + S —Rhodonase⟶ HSCN

*Detoxification of Heavy Metals:* The toxic effects of certain metals such as mercury and cadmium could be overcome by the administration of BAL (British Anti Lewisite).

The various compounds used for conjugation are listed in Table 11.1.

**Table 11.1:** Compounds used for conjugation

| Compound | Substances conjugated |
|---|---|
| Glycine, Glutamine | - Acidic substances |
| Glucuronate | - Acids, alcohols and phenols |
| Acetyl CoA | - Aliphatic and aromatic amines |
| Cysteine | - Aromatic compounds and phenols |
| $SO_4$ | - Steroids and phenols |
| Methyl group | - Pyridine derivatives |

# 12 Water and Electrolyte Balance

Water occupies 75 per cent area of this planet and about 60 to 75 per cent of human body mass. Hence the body may be viewed as a conglomerate of cells which contain fluid and are suspended in fluid. Ions such as $Na^+$, $K^+$, $Ca^{++}$, $Mg^{++}$, $Cl^-$, $HCO_3^-$, $PO_4^{--}$, $SO_4^{--}$ exert osmotic pressure which plays an important role in maintaining normal water distribution between the cells, blood and interstitial fluid. The different organs are connected by a transport system comprising the blood vessels, which pump nutrients into and remove the metabolic waste from the tissues. In this chapter, water and electrolyte metabolism is discussed followed by the mechanisms involved in the homeostasis of plasma pH.

## IMPORTANCE OF WATER

1. Essential constituent of all forms of life—*"Elixir of life"*.
2. In the cell, it is the medium in which all cellular events occur.
3. Required for enzyme action.
4. Required for transport of solutes in the body.
5. Responsible for the three dimensional structures of macrobiomolecules like proteins and nucleic acids.
6. Regulation of body temperature.
7. Provides ions for accelerating biochemical reactions.

## DISTRIBUTION OF WATER IN THE BODY: (FIG. 12.1)

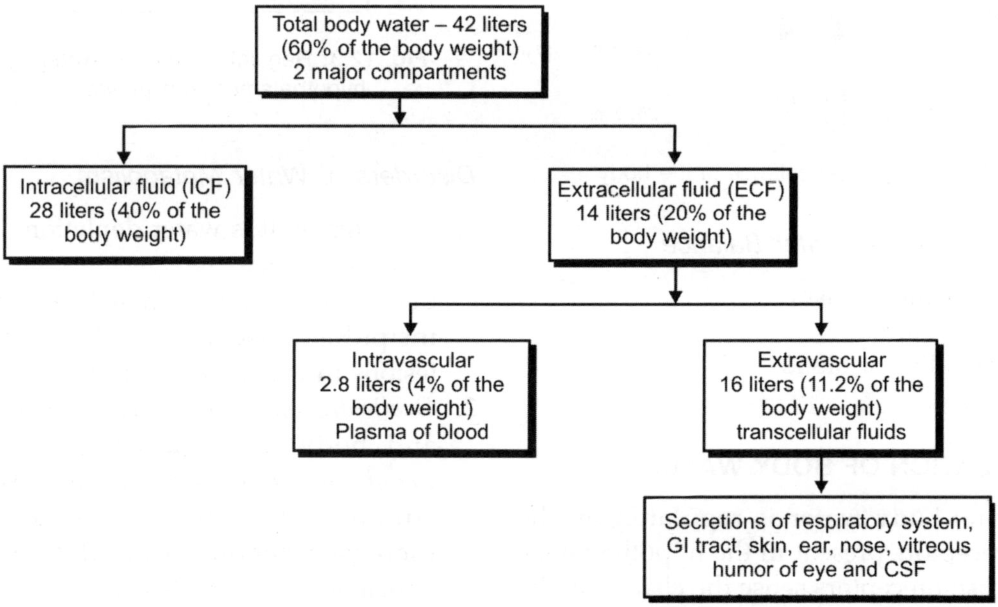


**Fig. 12.1:** Distribution of water in the body

Body water is present in two compartments—extracellular and intracellular compartment.

In the body, water content is highest in the brain (about 90%) and least in the adipose tissue (10%). Since fat is insoluble in water, obese people have low percentage of water (55-65% of their body weight) whereas lean people have higher water content. Hence, females who have more adipose tissue have less water than males.

## Intake and Output of Water

In a normal person intake and output of water is balanced as shown in Figure 12.2.

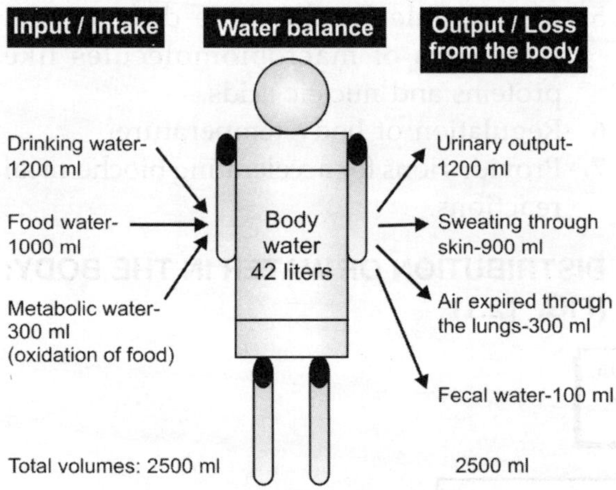


**Fig. 12.2:** Water balance in the body

### *Factors Affecting Water Balance*

1. Rate of metabolism
2. Urine volume
3. Osmolality of blood
4. Climate.

## REGULATION OF BODY WATER

The total body water is regulated by the osmoreceptors present in the hypothalamus. These osmoreceptors sense the change in the plasma osmolality and stimulate the thirst centre as well as the posterior pituitary which releases the antidiuretic hormone (ADH)/vasopressin. The thirst centre increases the intake of water and ADH increases the reabsorption of water at the collecting ducts (Fig. 12.3).

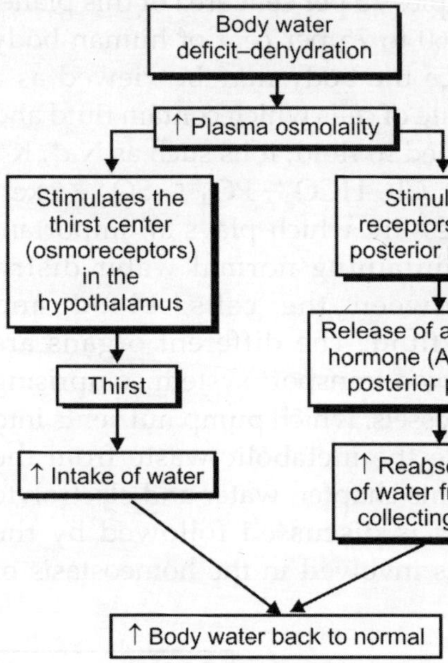


**Fig. 12.3:** Regulation of body water by hypothalamus and pituitary

### *Disorders of Water Metabolism*

1. *Dehydration*: It is water depletion. Plasma volume is decreased.
   *Conditions:* Vomiting, diarrhoea, diabetes insipidus (absence of ADH hormone), lesions in the hypothalamus.
2. *Overhydration*: Increase in the blood volume and body water leading to edema.
   *Conditions:* Water intoxication, excessive administration of intravenous fluids, increased secretion of ADH, protein deficiency, cancer, drugs.

## ELECTROLYTES

Electrolytes are the charged solutes present in body fluids both in ICF and ECF. The osmotic pressure exerted by electrolytes plays an important role in maintaining normal water distribution between the various compartments of the body.

Electrolytes could be inorganic or organic. They could have a positive charge or a negative charge.

### Functions of Electrolytes

Physiological processes dependant on the ECF and ICF electrolyte concentration are:

- Membrane potential
- Neuromuscular excitability
- Nerve impulse transmission
- HCl secretion
- Blood gas transport
- Blood clotting
- Enzyme catalysis
- Bone formation
- Muscle contraction.

### Distribution of the Electrolytes

Concentration of the electrolytes varies in the various compartments is shown in Figure 12.4. $Na^+$ is the major cation and $Cl^-$ is major anion in the ECF. In ICF, $K^+$ is the major cation and proteins and phosphate are the major anions.

### Bicarbonate ($HCO_3^-$)

Bicarbonate is a component (metabolic parameter) of bicarbonate buffer.
*Normal plasma levels*: 22-26 mEq/L
*Functions of bicarbonate:*

- Maintenance of blood pH.

#### *Clinical Importance*

Varies in acid base and electrolyte disturbances.

### Chloride ($Cl^-$)

Chloride is the major extracellular anion.
*Normal plasma levels*: 100-110 mEq/L

#### *Functions*

- Secretion of HCl by parietal cells
- Maintenance of electrical neutrality in erythrocytes
- Maintenance of plasma volume.

#### *Clinical Importance*

Decreased plasma chloride—*Hypochloraemia*

#### *Causes*

- Vomiting and diarrhoea
- Pyloric obstruction
- Cholera
- Ulcerative colitis
- AIDS
- Sunstroke
- Meningitis.

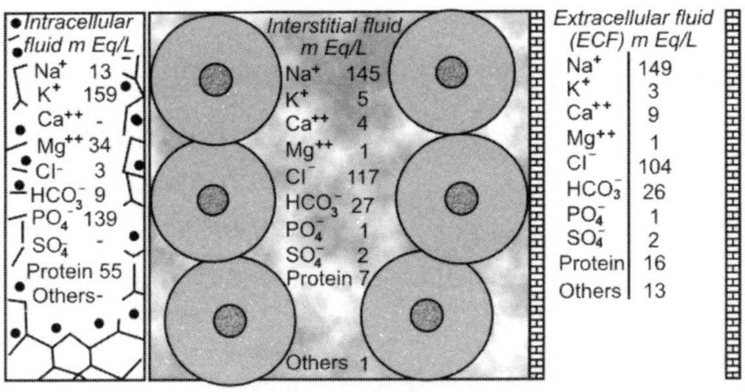


**Fig. 12.4:** Distribution of electrolytes

*Increased plasma chloride—hyperchloraemia*

- Nephritis
- Prostate enlargement
- Cushing's syndrome.

**Phosphate ($PO_4^{--}$)**

*Normal plasma levels*: Adults: 2.5-4.5 mg/ 100 ml
Children: 4.0-6.0 mg/100 ml
*Functions:* Discussed under mineral metabolism.

*Clinical importance*

Discussed under mineral metabolism.

**Sodium ($Na^+$)**

Sodium is the major extracellular cation.
*Normal plasma levels*: 133-146 mEq/L.

*Functions*

- Maintenance of blood volume
- Acid base balance
- Nerve and muscle function
- Absorption of glucose and amino acids in intestine and kidney
- Formation of bile salts
- Activity of $Na^+$ –$K^+$ ATPase.

*Clinical Importance*

Decreased plasma sodium—*hyponatraemia*

*Causes*

- Vomiting and diarrhoea
- Chronic renal failure
- Addison's disease.

*Symptoms*

- Low blood pressure
- Low plasma volume
- Circulatory failure.

Increased plasma sodium—*hypernatraemia*

*Causes*

- Cushing's syndrome
- Prolonged treatment with ACTH, cortisone and sex hormones
- Water retention in the body.

*Symptoms*

- Increased blood pressure—hypertension
- Increased plasma volume.

**Potassium ($K^+$)**

Potassium is the major intracellular cation.
*Normal plasma levels*: 3.8-5.4 mEq/L.

*Functions*

- Role in nerve muscle function
- Maintenance of plasma volume
- Formation of bile salts
- Activity of $Na^+$–$K^+$ ATPase
- Storage of glycogen in muscle and liver 14 mg/kg glycogen
- Growth of tissues.

*Clinical Importance*

Decreased plasma levels—*hypokalaemia.*

*Causes*

- Vomiting and diarrhoea
- Cushing syndrome
- Prolonged use of diuretics
- Treatment with insulin and digitalis.

*Symptoms*

- Muscular weakness
- Tachycardia
- Heart enlargement.

Increased plasma levels—*hyperkalaemia*

*Causes*

- Renal failure
- Addison's disease

- Severe dehydration
- Excess consumption of $K^+$ salts.

*Symptoms*

- Weakness
- Numbness
- Paralysis of extremities
- Low heart sounds
- Cardiac arrest due to collapse of peripheral vasculature.

## Calcium ($Ca^{++}$)

*Normal plasma levels*: 9-11 mg/100 ml

*Functions and clinical importance*: Discussed under mineral metabolism (chapter 8).

All molecules dissolved in the body water contribute to the *osmotic pressure*. Hence osmotic pressure is proportional to the molal concentration of solutes in the solution.

*Osmolarity* is the osmotic pressure exerted by the number of moles per liter of solution. *Osmolality* is the osmotic pressure exerted by the number of moles per kilogram of solvent. One millimole of a substance dissolved in 1 kg of water at 37°C exerts an osmotic pressure of approximately 19 mm of Hg. Under physiologic conditions, the average concentrations of all the osmotically active particles in the ECF is 290 mmol/kg water. And this remains in equilibrium with the ICF.

Crystalloids and water can move easily across the membrane whereas colloidal particles like plasma proteins cannot. Hence proteins are responsible for the *oncotic osmotic pressure (colloidal osmotic pressure)*. This is balanced by the *hydrostatic pressure*, which forces the fluid out of the capillaries (Fig. 12.5). In the arterial end of the capillaries the hydrostatic pressure prevails over the oncotic osmotic pressure. As a result the plasma water and low molecular weight substances filter into the interstitial space. In the venous part of capillaries the oncotic osmotic pressure prevails over the hydrostatic pressure hence fluid is drawn from the extravascular space into the vascular compartment. Reduction in the oncotic pressure leads to extravasation of fluid from the vascular compartment and accumulation in the interstitial space leading to *oedema*.

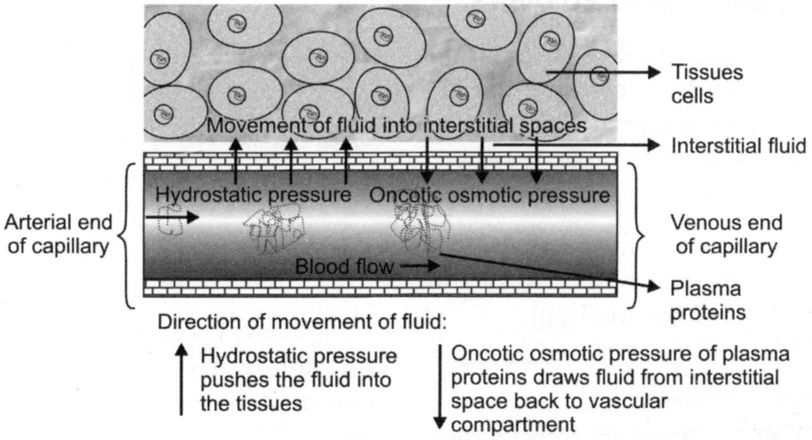


**Fig. 12.5:** Formation of tissue fluid. Normally, Oncotic osmotic pressure = Hydrostatic pressure

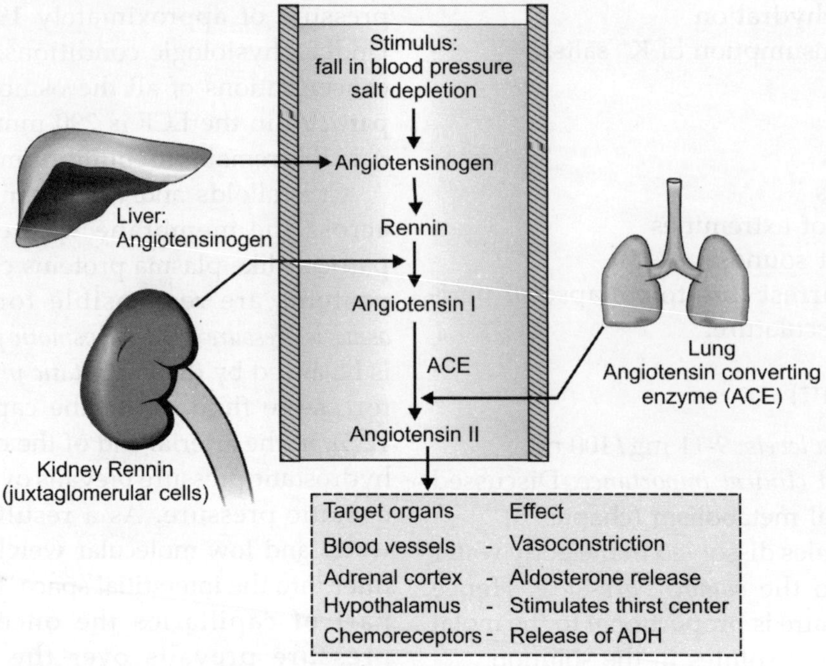


**Fig.12.6:** Rennin angiotensin mechanism

## Regulation of ECF Volume

Regulation of composition, osmolality and volume of ECF is done by the kidneys via

1. Renin angiotensin mechanism (Fig. 12.6).
2. Antidiuretic hormone (Fig. 12.3).

## ACIDS, BASES, SALTS AND BUFFERS

An ***acid*** is a substance capable of donating protons ($H^+$) and thereby increasing the $H^+$ concentration of the medium.

General: $HA \longrightarrow H^+ + A^-$ (conjugate base)

An acid: $HCl \longrightarrow H^+ + Cl^-$

A ***base (alkali)*** is a substance capable of accepting a proton and it decreases the $H^+$ion concentration of the medium.

General: $B + H^+ \longrightarrow BH^+$ (conjugate acid)

A Base: $NH_3 + H^+ \longrightarrow NH_4^+$

*Amphoteric substances*: These are those which can both, *accept as well as donate protons* (Fig. 12.7).

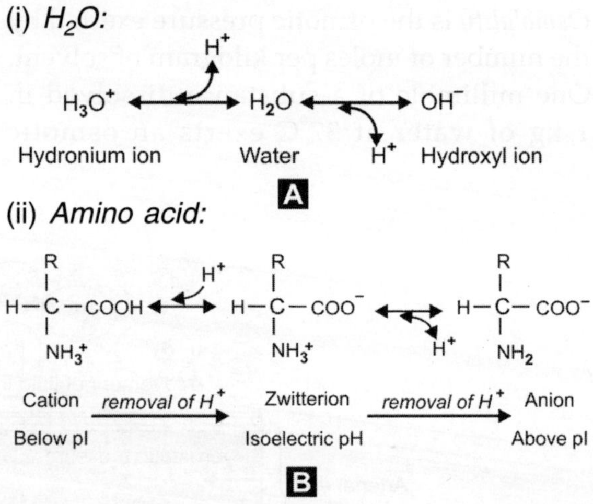


**Fig. 12.7:** Ionization of (A) Water (B) An amino acid

*Strong acids*: These are those, acids which are completely dissociated. *For example,* concentrated sulphuric acid, hydrochloric acid, nitric acid.

$HCl \longrightarrow H^+ + Cl^-$

(Less undissociated acid) (More dissociated acid)

In strong acids concentration of *undissociated acid is less* and *dissociated acid is more.*

*Weak acids*: These are those acids, which are not completely dissociated.
*For example*: Acetic acid

$$CH_3COOH \longrightarrow H^+ + CH_3COO^-$$

(More undissociated acid) (Less dissociated acid)

In weak acids concentration of *undissociated acid is more* and *dissociated acid is less.*

### Dissociation of An Acid

An acid (HA) dissociates into a proton ($H^+$) and its conjugate base ($A^-$).

$$\underset{\text{Acid}}{HA} \longrightarrow \underset{\text{Proton}}{H^+} + \underset{\text{Conjugate base}}{A^-}$$

$$\text{At equilibrium: } \frac{[H^+][A^-]}{[HA]} = K_a \text{ (constant)}$$

$K_a$ is the *dissociation constant* which is temperature dependant.

$$K_a = \frac{[H^+][A^-]}{[HA]}$$

$$\text{Solving for } [H^+] : [H^+] = \frac{K_a[HA]}{[A^-]}$$

Taking –ve logarithm throughout:

$$-\log[H^+] = -\log K_a + \left\{-\log \frac{[HA]}{[A^-]}\right\}$$

$$-\log[H^+] = pH, -\log K_a = pK_a$$

So the equation becomes

$$pH = pK_a - \frac{\log[HA]}{[A^-]}$$

$$pH = pK_a + \frac{\log[A^-]}{[HA]}$$

In general:

$$pH = pK_a + \log \frac{[\text{Conjugate base}]}{[\text{Acid}]}$$

This equation is called as the *Henderson-Hasselbalch equation.*

When the concentration of the conjugate base $[A^-] = [HA]$, the concentration of the undissociated acid (acid is half dissociated), then $[A^-] / [HA] = 1$ and $\log 1 = 0$, $pH = pK_a$

So *$pK_a$ is the pH at which the acid is half dissociated.*

Strong acids will have very low $pK_a$ and weak acids have a higher $pK_a$.

*Isoelectric pH*: It is the pH at which a molecule has equal number of positive and negative charges and has no net charge—*zwitter ion.*

### pH Scale

Sorensen expressed a pH scale for dilute solutions. This is a *logarithmic scale* ranging from 0 to 14 with the centre point being pH 7, which is half way in the scale (Fig. 12.8).

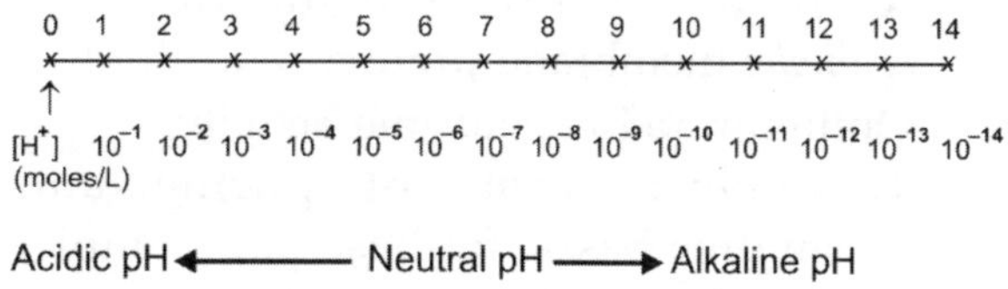


**Fig. 12.8:** Sorensen's pH scale

The $[H^+]$ of $1 \times 10^{-7}$ mol/litre, the pH = 7 ($-\log [H^+] = -\log [10^{-7}] = 7$). This is considered to be neutral. pH of water is found to be 7 and therefore the concentration of $[H^+]$ in water must be $1 \times 10^{-7}$ mol/litre.

If acid is added to water the $[H^+]$ will go above $1 \times 10^{-7}$ mol/litre and the pH will go lower than 7. So lower the pH, higher is the $[H^+]$ concentration or the acidity and higher the pH, lower is the $[H^+]$ concentration (Table 12.1).

**Table 12.1:** pH and $[H^+]$

| | $[H^+]$ | $-log [H^+] = pH$ |
|---|---|---|
| *Water:* | $1 \times 10^{-7}$ mol/litre | - $-log [1 \times 10^{-7}] = 7$ - neutral pH |
| *Acids:* | More than $1 \times 10^{-7}$ mol/litre | $-log$ [more than $1 \times 10^{-7}$] - so pH will be less than 7 |
| *Alkali:* | Less than $1 \times 10^{-7}$ mol/litre | - log [less than $1 \times 10^{-7}$] - so pH will be more than 7 |

If in an acid the $[H^+]$ concentration is $1 \times 10^{-4}$ mol/litre then $pH = -log [1 \times 10^{-4}] = 4$

If in an alkali the $[H^+]$ concentration is $1 \times 10^{-9}$ mol/litre then $pH = -log [1 \times 10^{-9}] = 9$

Hence, For a *difference of pH by 1 unit there is a 10 times difference in the $[H^+]$.*

### *Biological Significance of pH*

- It influences the ionized states of most of the biological molecules and is responsible for their structure and function. *For example:* membrane proteins and lipids.
- Optimum pH for enzyme activity.
- Tautomeric forms of purine and pyrimidine bases, for base pair complementarities in the three-dimensional conformation of DNA is pH dependant.

### *Salts*

Salts are products of a reaction between an acid and a base.

For *example,*

$$\underset{\text{Acid}}{HCl} + \underset{\text{Base}}{NaOH} \longrightarrow \underset{\text{Salt}}{NaCl} + \underset{\text{Water}}{H_2O}$$

Salts *dissociate into their ions*—positively charged *cations* and negatively charged *anions.*

$$\underset{\text{Salt}}{NaCl} \longrightarrow \underset{\text{Cation}}{Na^+} + \underset{\text{Anion}}{Cl^-}$$

Salts can be acidic, alkaline or neutral. This will depend on the acid strength of the cation and alkaline strength of its anion.

Strong acid + Weak base →acidic salt (e.g. $NH_4Cl$)

Weak acid + Strong base →alkaline salt (e.g. sodium acetate)

Strong acid + Strong base →Neutral salt (e.g. NaCl)

### *Conjugate Pairs*

A *conjugate pair* is made of two components

- An acid and its conjugate base $[HA/A^-]$ or
- a base and its conjugate acid $[B/BH^+]$

One member of the conjugate is strong, the other is weak, i.e. if acid HA is strong its conjugate base $A^-$, will be weak.

### *Buffers*

- They can resist a change in pH on addition of either an acid or a base.
- Buffers are conjugate pairs.
- They are made up of any one of the following:
  - Weak acid and its conjugate base for example $H_2CO_3/HCO_3^-$
  - Weak base and its conjugate acid example $NH_3/NH_4^+$
  - Weak acid and its salt for example $CH_3COOH / CH_3COONa$
  - Mixture of two salts of a polybasic acid for example $Na_2HPO_4/NaH_2PO_4$.

### Mechanism of Action of a Buffer

On addition of an acid to a buffer solution, weak acid and salt are formed and on addition of a base, salt and water are formed. Hence there is no drastic change in pH as shown in Figure 12.9.

### *pH of a Buffer Solution is Dependant Upon*

- $pK_a$ of the acid member. (Lower the $pK_a$, lower the pH range where it buffers).
- Ratio of the molar concentration of acid and base components.

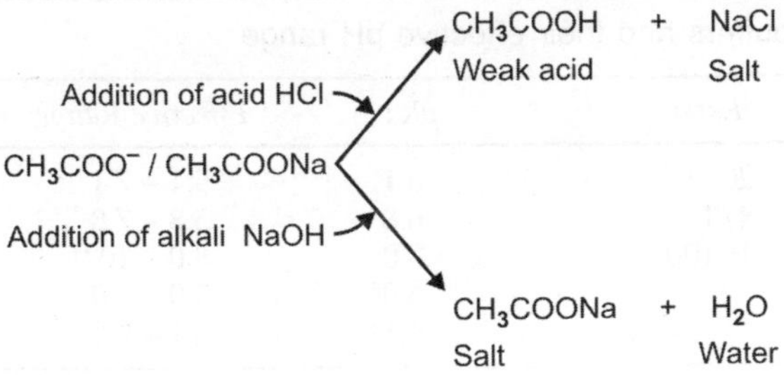


**Fig. 12.9:** Mechanism of action of a buffer

### *Effective Range of pH for a Buffer*

- At $pK_a$ of a buffer, it is most efficient.
- The effective range of pH for a buffer extends ± 1 of its $pK_a$ value.

*Buffering capacity of a buffer*—is the number of equivalents of acid or base to be added to a litre of the buffer solution to change its pH by 1 unit.

It is the ability of a buffer to resist change in pH when an acid or base is added to it.

### *Applications of HH Equation to Buffers*

- Assessing the acid base status of body buffers.
- Predicting the limits of compensation of the body buffers.
- Calculating the pH of a buffer on addition of a known amount of acid or base.
- Calculating the acid/salt ratio of a buffer at known pH of a solution.

## ACID BASE HOMEOSTASIS

Acid base homeostasis can be studied under three headings: (Fig. 12.10)

1. Production of acids.
2. Transport of acids
3. Removal/disposal of acids

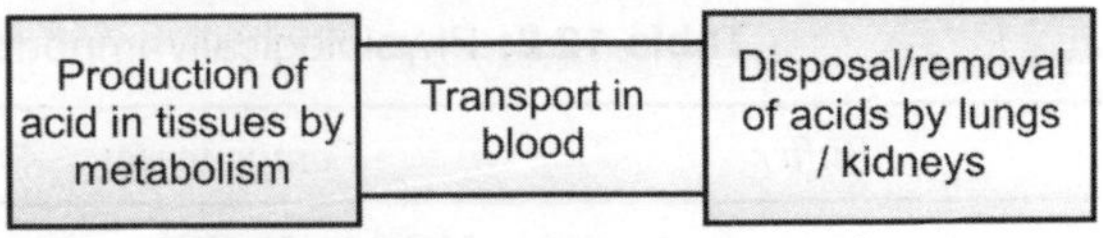


**Fig. 12.10:** Acid base homeostasis

### 1. Production of Acids

During metabolism large quantities of acids are being produced and added to the ECF. The acids produced are of *two types:*

#### *Volatile Acids: Carbon dioxide ($CO_2$)*

- Metabolism in the tissues produces nearly 20,000 mEq of $CO_2$/day (more in the athletes).
- $CO_2$ combines with water to form carbonic acid (weak acid) for transport and then it is eliminated from the body by the lungs.

#### *Non-volatile Acids/Fixed Acids: Such As*

- *Sulphuric acid:* Metabolism of sulphur containing amino acids in proteins.
- *Phosphoric acid:* Metabolism of nucleic acids and phosphoproteins.

These are the major acids produced during metabolism.

- *Lactic acid:* End product of anaerobic glycolysis in contracting muscles and RBC.
- *Keto acids:* Ketone bodies produced from partial oxidation of fatty acids.

These are produced in smaller amounts in normal metabolism.

- About *60–80 mEq of fixed acids are produced /day.*
- These are neutralised by the buffers and excreted by the kidneys.

Production of basic compounds in the body is negligible.

**Table 12.2:** Physiologically important buffers and their effective pH range

| *Buffer* | *Components* | *Ratio* | *pKa* | *Effective Range* |
|---|---|---|---|---|
| Bicarbonate | $HCO_3^-/H_2CO_3$ | 20/1 | 6.1 | 5.1 – 7.1 |
| Phosphate | $Na_2HPO_4/NaH_2PO_4$ | 4/1 | 6.8 | 5.8 – 7.8 |
| Ammonia | $NH_3/NH_4^+$ | 1/100 | 9.0 | 8.0 – 10.0 |
| Haemoglobin | Hb/HHb | | 6.0* | 5.0 – 7.0 |
| Protein buffer | $Pr^-/HPr$ | | 6.1* | 6.1 – 7.1 |

The normal acid challenge is effectively met and pH maintained by the following major mechanisms:

- Buffers
- Respiratory system
- Renal excretion.

## 2. Transport of Acids from Site of Production to Site of Disposal

The first line of defense in the body against an acid load is the *buffers*. These buffers transport the acids produced in the body. The various buffers in the body, their components, $pK_a$ values and effective pH range have been shown in Table 12.2.

$pK_a$ is due to the presence of ionizable imidazole ring of histidine in the protein. Sixteen histidines are present in albumin and thirty eight in haemoglobin/molecule.

### Distribution of Buffers in the Body Compartments

Depending on the concentrations of the components of a buffer system and the pH, the effectiveness of a buffer varies in the various compartments of the body as shown in Table 12.3.

**Table 12.3:** Effective physiological buffers in the body compartments

| *Body compartment* | *Important buffers* |
|---|---|
| Plasma | Bicarbonate, phosphate, protein |
| RBC | Bicarbonate, haemoglobin, protein |
| Kidney | Bicarbonate, phosphate, ammonia |
| Bone | Phosphate, protein |

### Transport and Disposal of Carbon Dioxide

Can be studied as follows:

- Changes at the tissues
- Transport in the blood
- Changes in the lung and disposal
- Role of lungs in acid base homeostasis.

The concentration of respiratory gases in the various body compartments is given in Figure 12.11.

#### *Changes at the Tissues (Fig. 12.12)*

In the arterial blood supplying the tissues, the haemoglobin is in the oxygenated form $HbO_2$. $CO_2$ is produced by metabolism in the tissues. So concentration of $O_2$ is high in the RBC and $CO_2$ is high in the tissues.

1. *Source of $CO_2$ in the tissues* is metabolism. Hence the concentration of $CO_2$ is high in the tissues. It diffuses down concentration gradient into the blood.
2. *Fate of $CO_2$ in the blood*: A small quantity of the $CO_2$ dissolves in plasma and forms carbamino compounds with the proteins present in it. But the major quantity enters the RBCs.

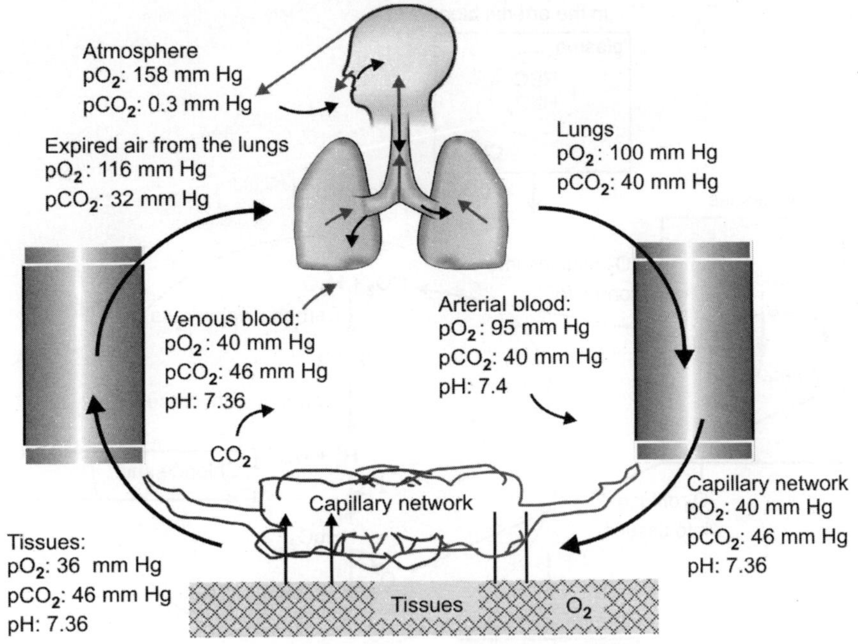


**Fig.12.11:** Concentration of the respiratory gases in the various parts of the circulation *(For colour version see Plate 8)*

3. *In RBCs:*

$$CO_2 + H_2O \xleftrightarrow{\text{Carbonic anhydrase}} H_2CO_3 \text{ (Carbonic acid – a weak acid)} \xleftrightarrow{\text{Carbonic anhydrase}} H^+ + HCO_3^- \text{ (dissociation of acid)}$$

4. *Fate of* $H^+$: It combines with $HbO_2$ to form $HHbO_2$ which dissociates into HHb (protonated Hb) and $O_2$.

$$H^+ + HbO_2 \rightarrow HHbO_2 \rightarrow HHb + O_2 \text{ (released to the tissue)}$$

5. *$O_2$ is released into the tissues* down concentration gradient.
6. *Fate of $HCO_3^-$*: It moves out into the plasma and to maintain electroneutrality, $Cl^-$ moves into the RBC—*chloride shift.*

In venous blood draining the tissues, $CO_2$ is transported in the form of $HCO_3^-$ in the plasma and protonated haemoglobin HHb in the RBC. It is brought to the lungs for disposal without much change in the pH of blood.

### *Changes in the Lungs and $CO_2$ Disposal*

The venous blood coming to the lungs has protonated haemoglobin (HHb) and more chloride ions ($Cl^-$) in the RBCs and more $HCO_3^-$ in the plasma. The concentration of $O_2$ in the lungs is very high Figure 12.12.

1. *$O_2$ diffuses* down concentration gradient *from the lungs into the blood/RBCs.* In the RBCs this $O_2$ combines with HHb to form $HHbO_2$ which dissociates into $HbO_2$ and $H^+$.

$$HHb + O_2 \rightarrow HHbO_2 \rightarrow HbO_2 + H^+$$

2. The $HCO_3^-$ in the plasma moves into the RBCs and the chloride ion moves out into the plasma—*reverse chloride shift.*

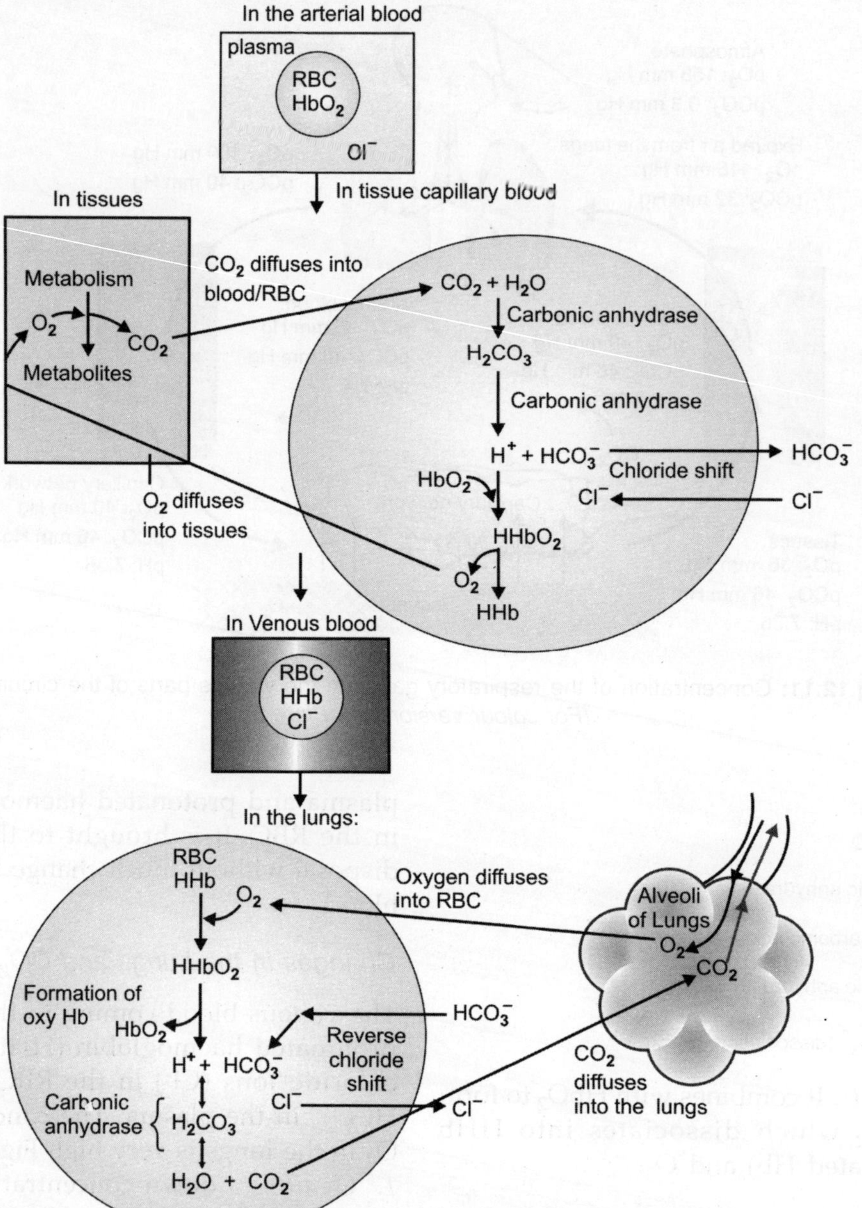


**Fig.12.12:** Transport of respiratory gases

3. *Fate of $HCO_3^-$:*

$$H^+ + HCO_3^-$$
$$\updownarrow \text{ Carbonic anhydrase}$$
$$H_2CO_3 \text{ (carbonic acid - a weak acid)}$$
$$\updownarrow \text{ Carbonic anhydrase}$$
$$CO_2 + H_2O$$

4. $CO_2$, a gas moves down concentration gradient into the alveoli of the lungs and is lost in the air that is expired from the lungs (Fig. 12.12).

## Role of Lungs in Acid Base Homeostasis

The lungs are the second line of defense in the body against an acid load and play an important role in blood pH homeostasis by altering the rate of respiration (exhalation/ retention of $CO_2$). The rate of respiration is controlled by the respiratory centre in the hypothalamus of the brain and the chemoreceptors of the aortic and carotid bodies in the aortic arch. Present in the respiratory center are the chemoreceptors which sense the change in pH of blood. A fall in the blood pH to the acidic side stimulates the chemoreceptors and increases the rate of respiration (hyperventilation). So more $CO_2$ is lost from the lungs and pH is increased to normal.

A rise in the pH to the alkaline side does not stimulate the chemoreceptors and so there is a decrease in the rate of respiration (hypoventilation). So more $CO_2$ is retained in the blood and pH is decreased to normal (Fig. 12.13).

## Role of Kidneys in pH homeostasis

The functional unit of the kidney is called as *nephron*. The parts of a nephron are the glomerulus (tuft of capillaries), Bowman's capsule, proximal convoluted tubule (PCT), loop of Henle (descending arm, ascending arm), distal convoluted tubule (DCT), collecting ducts and ureter. At the glomerulus, arterial blood is filtered (selective dialysis/ultrafiltration). The ultrafiltrate formed in the Bowman's capsule is called as the glomerular filtrate (pH-7.4). The ultrafiltrate passes through the nephron and urine is formed in the collecting ducts. During this passage, the pH can decrease from 7.4 to as low as 4.0 indicating addition of $H^+$ ions (acid). This is called as *acidification of urine* (Table 12.4). The pH of urine varies from 4.4 to 8.0 depending on the amount of acid excreted by the kidneys.

Normally, the kidneys:

1. Eliminate nonvolatile acids (around 50 mEq/day).
2. It conserves bases by minimising its excretion (↑ reabsorption).

The $Na^+$, $Na_2HPO_4$, $HCO_3^-$ filtered at the glomerulus are used for this. Both these processes are based on the formation of $H^+$ and $HCO_3^-$ from $CO_2$ and $H_2O$ within the tubular cells.

### *Mechanism of pH Homeostasis by Kidneys*

*Two mechanisms*:

1. Ion exchange mechanism
2. Buffers—three buffers
   - *bicarbonate*
   - *phosphate*
   - *ammonia*

*Renal ion exchange mechanism (Fig. 12.14).*

*Mechanism*: $Na^+$ from the glomerular filtrate is reabsorbed from the PCT lumen into the PCT cell by a carrier mediated process in exchange for $H^+$ ion, which will move from the cell into the filtrate in the lumen.

*Source of the $H^+$ ion in the PCT cell:* Carbonic anhydrase catalyzed reactions:

$$CO_2 + H_2O \longleftrightarrow H_2CO_3 \longleftrightarrow H^+ + HCO_3^-$$

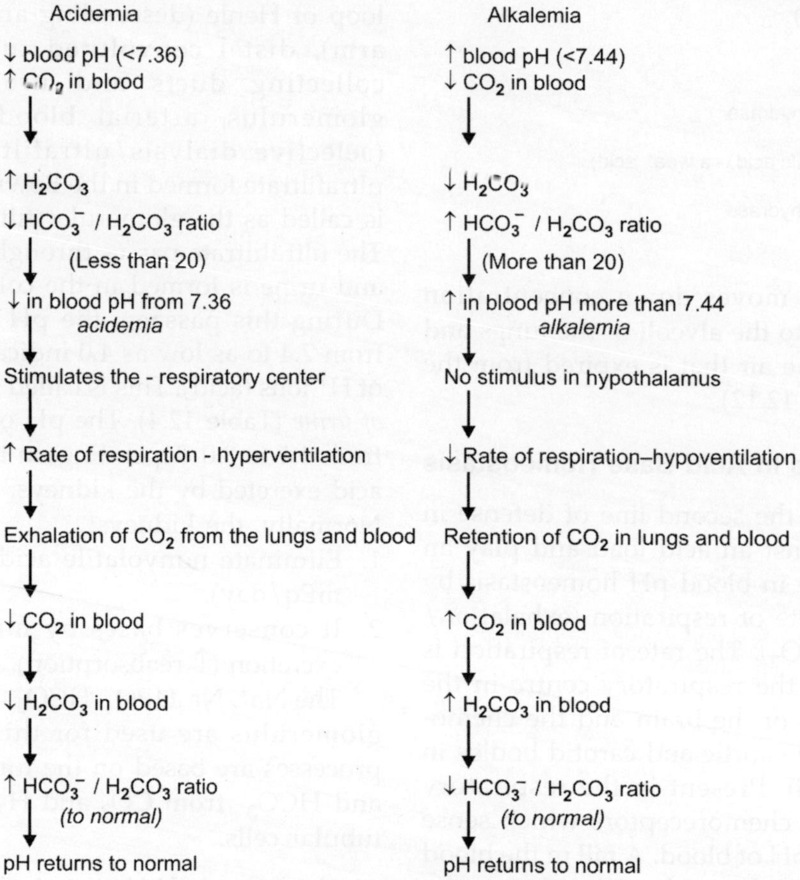


**Fig.12.13:** Role of lungs in pH homeostasis

*Fate of $H^+$*: Exchanged for $Na^+$ from the lumen.

*Fate of $HCO_3^-$* : Enters plasma along with the absorbed $Na^+$ – *restoration of the alkali reserve.*

*Purpose it serves:*

- *Excretion of $H^+$* ions (source from $CO_2$).
- *Reabsorption of $Na^+$* ions filtered from plasma into the glomerular filtrate.
- *Restoration of the $Na^+$ and $HCO_3^-$ in blood* – maintenance of the alkali reserve of the blood.
- *Regulates the pH* when an acid ($H^+$) load has to be eliminated from the blood.

Buffering of the secreted $H^+$ ions is essential for the continuous secretion of $H^+$ ions into the urine. This is done by the buffers as follows:

*Buffering of the filtered bicarbonate:* Glomerular filtrate has 24 mM of bicarbonate/litre and about 3 mM is filtered/minute. About 3.49 mM of $H^+$ are secreted/minute. $HCO_3^-$ buffers the $H^+$ secreted in the tubular fluid to form $H_2CO_3$. Most of the $H^+$ secreted in the tubular fluid is converted to $H_2CO_3$ and reabsorbed back and restored to the blood.

Most of the $HCO_3^-$ mediated buffering of $H^+$ takes place in the PCT. Only 1-20 per

**Table 12.4:** Nephron and urine formation

| *Parts of a nephron* | *pH* | *Changes taking place* |
|---|---|---|
| Glomerulus | 7.4 | Ultra filtration of plasma to form glomerular filtrate in the Bowman's capsule |
| Bowman's capsule | 7.4 | pH of glomerular filtrate is the same as that of plasma. *Contents: organic substances* - urea, uric acid, creatinine and urobilinogen *inorganic substances*: $Na^+$, $HCO_3^-$, $Na_2HPO_4$ and other electrolytes |
| Proximal convoluted tubule | Isotonic ($\leq 7.4$) | Events that occur: • Ion exchange mechanism • Reabsorption and regeneration of $HCO_3^-$ |
| *Loop of Henle:* | | |
| Descending arm | | Reabsorption of water |
| Ascending arm | | Reabsorption of electrolytes |
| Distal convoluted tubule | 6.0 - 6.5 | • Formation of titrable acid ($Na_2HPO_4$) • Buffering by ammonia buffer |
| Collecting ducts | 5.5 – 6.0 | Reabsorption of electrolytes and water (under the influence of ADH). |

cent takes place in the DCT and collecting ducts.

### *Reabsorption and Regeneration of $HCO_3^-$ (Fig. 12.15)*

The bicarbonate from the ultrafiltrate is neutralised by $H^+$ ions released into the proximal tubular lumen by ion exchange mechanism. The $H_2CO_3$ formed will dissociate into $CO_2$ and water. $CO_2$ diffuses into the cells $CO_2$ difference into the cells, the $CO_2$ gives rise to bicarbonate which is restored back to the plasma Fig. 12.16 (reclamation).

### *Excretion of titrable acidity (Phosphate buffer, Fig. 12.16)*

*Site*: Distal convoluted tubule.

*Acids:* Acid phosphate ($NaH_2PO_4$) and small amounts of other acids like lactic acid, acetoacetic acid and β-hydroxy butyric acid

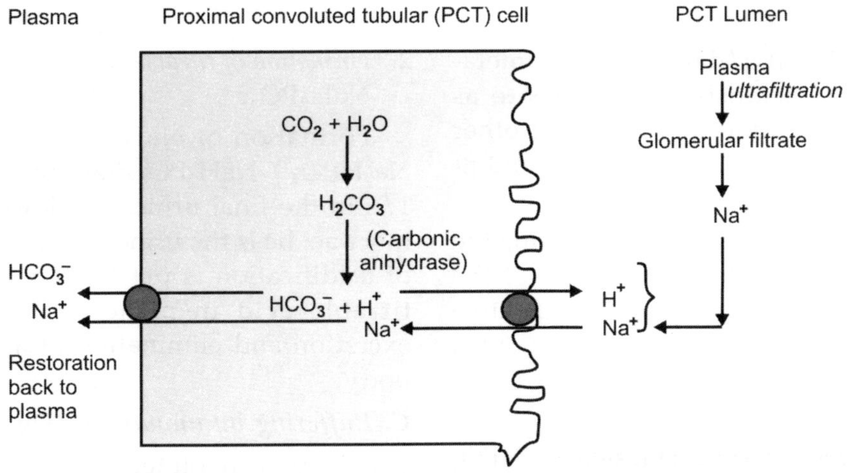


**Fig. 12.14:** Ion exchange mechanism in the PCT cell

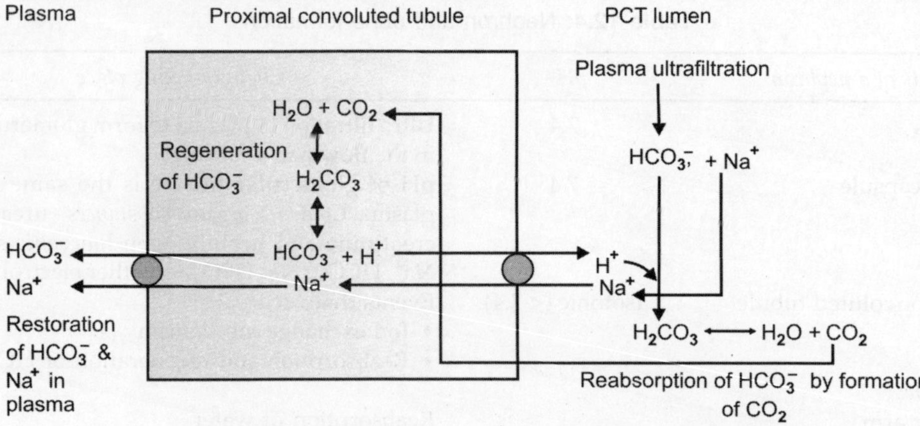


**Fig. 12.15:** Reabsorption and regeneration of $HCO_3^-$

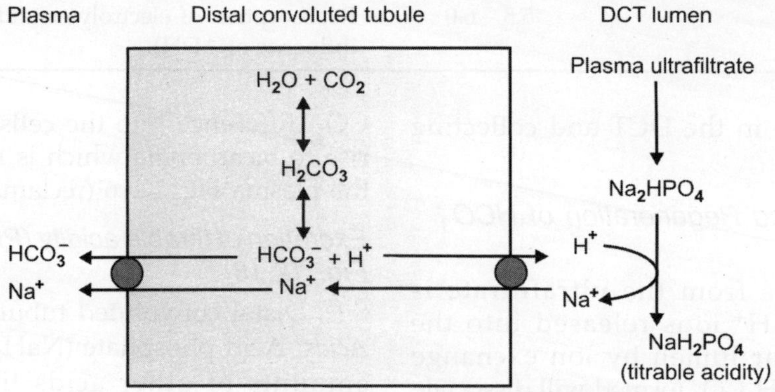


**Fig. 12.16:** Formation of titrable acidity

*Quantity*: 15-45 mEq/day in normal metabolism. During acidosis, it increases to as high as 150 mEq (due to increase in other acids like lactic acid, acetoacetic acid and β-hydroxy butyric acid).

The pH of the tubular filtrate entering the DCT is around 6.9, close to the $pK_a$ of the phosphate buffer 6.8 and falls in the effective range of 5.8–7.8 where it buffers with maximum efficiency.

*Mechanism*

1. *Filtered at the glomerulus*: Plasma $Na_2HPO_4$ into the glomerular filtrate.
2. *Formation of titrable acidity*: $Na_2HPO_4 + H^+$ $NaH_2PO_4$.

Formation of $NaH_2PO_4$ can change the $Na_2HPO_4$ / $NaH_2PO_4$ ratio from 4/1 to 1/9 - 1/6 in the final urine. The lower this ratio, more acidic is the urine. The maximum limit of acidification is pH 4.5. The excretion of titrable acid in urine enables constant excretion and elimination of acid from the body.

***C. Buffering by ammonia: (Fig. 12.17)***

*Site*: Distal convoluted tubule

*Quantity*: 40 – 70 mEq/day

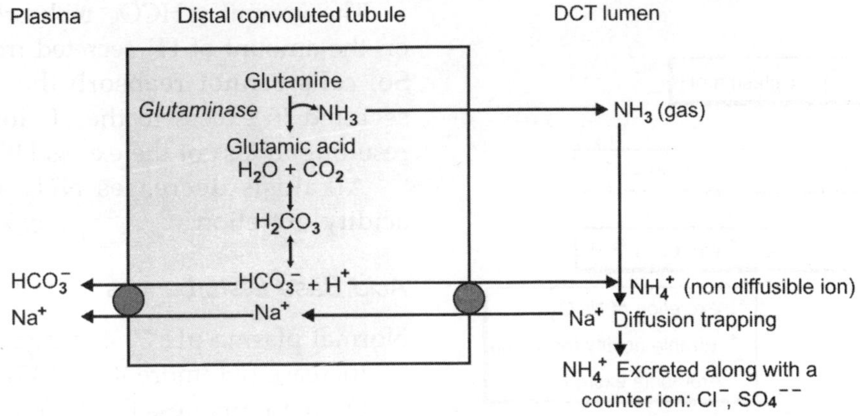


**Fig. 12.17:** Ammonia buffer and diffusion trapping

*Mechanism:*

1. *Source of ammonia:*
   - Glutaminase (60%): Hydrolyses glutamine to ammonia and glutamic acid in renal mitochondria.
   - Amino acid oxidase (40%): Oxidises amino acid to α-keto acid and ammonia
2. Ammonia is a gas and diffuses into the lumen from the DCT cell.
3. *Formation of $NH_4^+$:* In the DCT lumen, ammonia combines with $H^+$.
   $NH_3 + H^+ \rightarrow NH_4^+$
4. *Diffusion trapping*: The tubular membrane is permeable to the ammonia but not to $NH_4^+$. So it is retained in the tubular fluid. This is called as *Diffusion trapping.*

$NH_4^+$ is a weak acid and does not dissociate easily, so its formation effectively decreases the $H^+$ concentration in the lumen enabling further secretion of $H^+$ into the lumen. Conversion of $NH_3$ to $NH_4^+$ decreases the concentration of $NH_3$ so more ammonia will diffuse from the DCT into the lumen. The $H^+$ ions left after buffering with $HCO_3^-$ and $Na_2HPO_4$ are buffered by $NH_3$. So, secretion of $NH_3$ increases when the free $H^+$ ion concentration is high enough to lower the urinary pH below 6.0.

In alkaline urine no free $H^+$ ion is left to be buffered by $NH_3$. So little or no $NH_3$ will be present in urine.

*Elimination of other free acids*

Some $H^+$ ions are accepted by strong conjugate bases such as lactate, acetoacetate, urate, replacing $Na^+$ reabsorbed from their salts. Free weak acids are formed which are excreted without using any base because they have poor ionizability and their elimination changes urinary pH but to a very small extent. During acidosis pH can decrease as low as 4.8 mainly because, concentration of these acids is as high as 150 mmol.

*Role of kidney in pH homeostasis*

Renal correction of an excess of $H^+$ ions in plasma (Fig.12.18) is mainly by 2 mechanisms:

- ↑ excretion of $H^+$ ions
- ↑ reabsorption of $HCO_3^-$

In conditions of acidosis the acid excretion goes as high as 400 mEq /day from 70 mEq. This enhances the activity of glutaminase which increases the excretion of $NH_4^+$. This takes about 3-4 days to set in under conditions of acidosis. And once established, it has high capacity to eliminate acid.

In acidosis

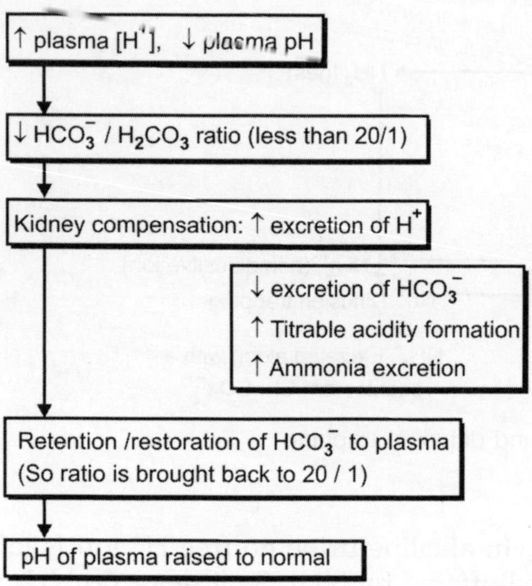


**Fig. 12.18:** Renal correction of an overload of $H^+$ ions.

*Renal correction of alkalaemia (Fig. 12.19)*

Mainly by 2 mechanisms:

- ↓excretion of $H^+$ ions
- ↓reabsorption of $HCO_3^-$

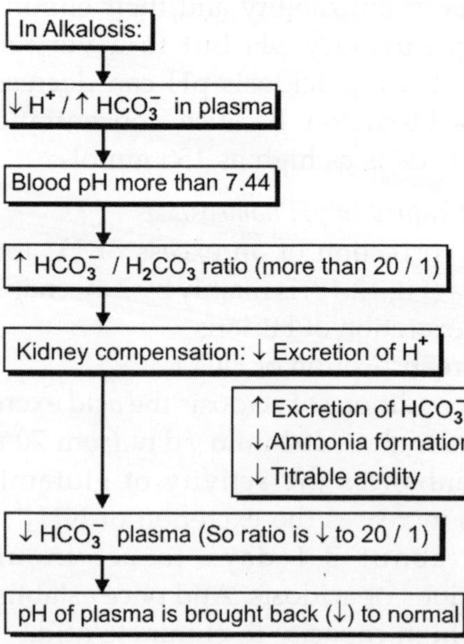


**Fig. 12.19:** Renal correction of alkalosis

The amount of $HCO_3^-$ reabsorbed depends on the amount of $H^+$ secreted from the cells. So, cells cannot reabsorb the $HCO_3^-$ ions secreted in excess to the $H^+$ ions secreted. resulting in loss of the excess $HCO_3^-$ ions.

Alkalosis decreases $NH_3$ and titrable acidity excretion.

### *Acid Base Disturbances*

Normal plasma pH: 7. 4 (range: 7.35 – 7.45)

Increase pH (more than 7.45) = acidaemia

Decrease pH (less than 7.35) = alkalaemia

pH within normal limits is = euphemia

*Acidosis:* It is a clinical state resulting from excess production or accumulation of acids or loss of base from the body.

*Alkalosis:* It is a clinical state resulting from decreased production or loss of acids or accumulation of base in the body.

At pH 7.4: $[H^+]$ = 40 nmol/L (range: 7.35 (45 nmol/L) – 7.45 (35 nmol/L))

*Alkali reserve*: The bicarbonate that remains in plasma after neutralising all acids in plasma is called as alkali reserve.

*Normal value:* 24 mEq/L Range: 22 – 26 mEq/L

In blood, bicarbonate is in association with $Na^+$. So $Na^+$ $HCO_3^-$ converts strong acids to weak acids. pH does not vary much. *For example:*

$$2\,Na^+HCO_3^- + H_2SO_4 \longrightarrow Na_2SO_4 + 2H_2CO_3$$

*Anion Gap:* The sum of all the cations and anions in the ECF is always equal to maintain electrical neutrality. However, there is difference between the measured cations and the measured anions. This is due to the presence of unmeasurable anions like the proteins, sulphate, phosphate and organic acids. This is called as *Anion gap*.

*Normal:* 12 ± 5 mmol / L

*$HCO_3^-/H_2CO_3$ Ratio:* Normal- 20:1

$HCO_3^-$ is called *metabolic parameter* (Since it is regulated by the kidney).

$H_2CO_3$ is called *respiratory parameter* (Since it is regulated by respiration).

Disturbances due to variation in the $HCO_3^-$ are referred to as *metabolic type* of acid. base disturbance and variation in $H_2CO_3$ is referred to as *respiratory disturbance.*

*When pH of blood is altered:*

- Buffers would try to neutralise the change.
- Blood $HCO_3^-$ acts as alkali reserve.
- Respiratory mechanisms would act but compensation does not proceed to completion.
- Renal mechanisms are slower in reacting, but are able to achieve complete compensation.

In a continuous state of acid base disturbance the body—

- Tries to compensate so that pH does not exceed the limits compatible to life.
- The body also tries to remove the primary cause of the imbalance.

*The major types of acid base disturbances are:*

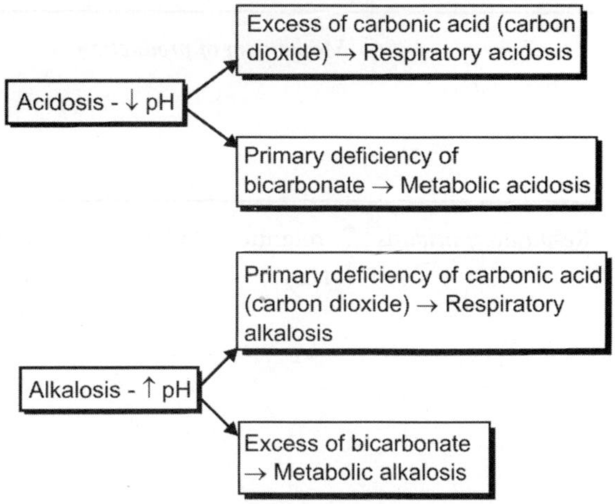


The most important factor in acid-base regulation is the ratio of $HCO_3^-/H_2CO_3$ which has to be maintained since it is the major buffer in plasma and is the first line of defense against any alteration in plasma pH. When there is an imbalance in the acid base status of plasma, due to variation (either ↑ or ↓) in either $HCO_3^-$ or $Pco_2^-$, the abnormality is compensated by varying (↑ or ↓) the other parameter of the ratio in the same direction. That is when there is a fall in plasma pH due to ↑ $Pco_2$, then plasma [$HCO_3^-$] is also ↑ , resulting in a rise in pH. While the $HCO_3^-/H_2CO_3$ ratio remains within normal limits (16/1 to 25/1), corresponding to a plasma pH of 7.3 to 7.5, the situation is said to be *compensated*.

The causes and plasma findings of each of the acid base disturbances is discussed in Table 12.5.

**Table 12.5:** Acid base disturbances

| *Type of disturbances* | *Mechanism of production* | *Causes* | *Changes in plasma Normal Value* pH | Pco$_2$ | HCO$_3^-$ |
|---|---|---|---|---|---|
| | | | 7.35 7.45 | | 22-25 mEq/L |
| *Respiratory acidosis* | ↑ retention of $CO_2$ or ↑ Pco$_2$ | Hypoventilation Seen in lung and CNS diseases, weakness of respiratory muscles, drug overdoses, chronic bronchitis, bronchial asthma, myasthenia gravis, polio | ↓ | ↑ | N or ↓ |
| | | | *Compensated condition* | | |
| | | | 7.3-7.4 | ↑ | ↑ |
| *Metabolic acidosis* | ↑ Production of acids | Diabetic ketoacidosis Lactic acidosis | ↓ | N | ↓↓↓ |
| | ↓Removal of acids | ↑ Renalfailure Drugs-acetazolamide | *Compensated condition* | | |
| | Loss of HCO$_3^-$ | ↑ Diarrhoea, vomiting, intestinal, pancreatic and biliary fistula | | ↓ | ↓ |
| *Respiratory alkalosis* | ↑ removal of $CO_2$ ↓Pco$_2$ | Hypoventilation Anxiety, CNS injury, Salicylate poisoning, fever, septicaemia, encephalitis, cerebral tumours. | ↑ | ↓ | N or ↓ |
| | | | *Compensated condition* | | |
| | | | 7.4 - 7.5 | ↓ | ↓ |
| *Metabolic alkalosis* | ↑ removal of acids | Vomiting, Gastroenteritis, Pyloric stenosis | ↑ | N | ↑↑↑ |
| | | | *Compensated condition* | | |
| | Accumulation of HCO$_3^-$ | Ingestion of excess alkali-antacid therapy, Blood transfusion with citrate, Treatment with NaHCO$_3$. | 7.4 - 7.5 | ↑ | ↑ |
| | | | *Hypokalaemia—retention of $H^+$ and excretion of $K^+$ leading to paradoxic acidosis of urine* | | |

# 13 Muscle and Muscle Contraction

Movement is an important property of the members of the animal kingdom. The tissue which is responsible for movement is the muscle. Muscle is the biochemical transducer which converts potential (chemical) energy to kinetic (mechanical) energy. It is the single largest tissue of the human body.

This chapter deals with the structure and composition of muscle fibre and the mechanism of muscle contraction and the sources of energy for contraction.

In vertebrates there are three types of muscles:

- Skeletal } Striated muscles
- Cardiac }
- Smooth Nonstriated muscle

- The *striated muscles* are composed of multinucleated muscle fibre cells, surrounded by an electrically excitable membrane called the *sarcolemma (Fig. 13.1).*
- An individual cell consists of a bundle of many myofibrils, which are embedded in the cytoplasm of the muscle cell, called the *sarcoplasm,* each cell is about 1μm in diameter.
- Each myofibril is composed of numerous functional units of muscle called the *sarcomere,* which repeats every 2.3μm along the axis of the fibril.
- The presence of sarcomere is responsible for the striated appearance of the muscle fibre.

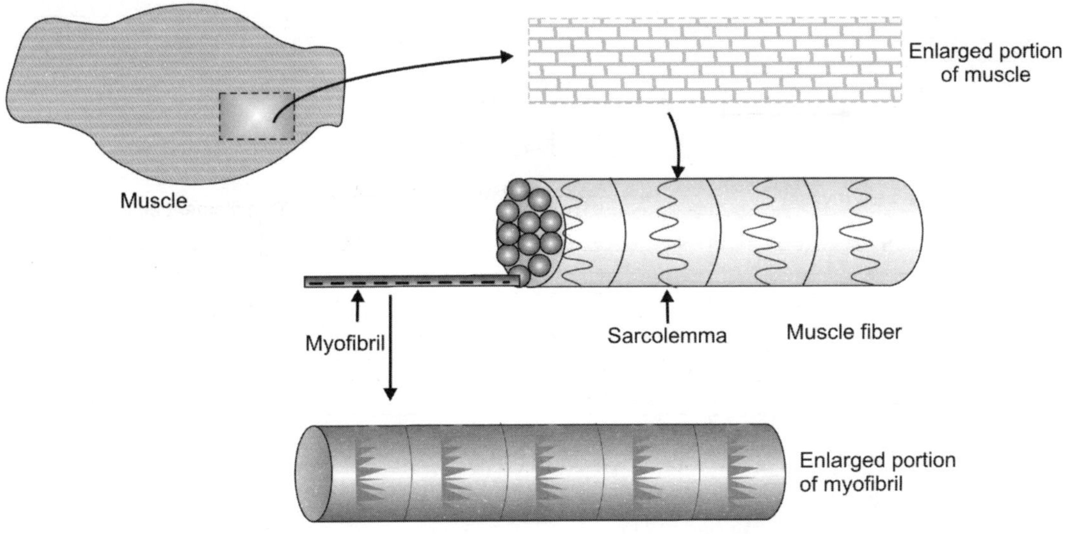


**Fig. 13.1:** Structure of striated muscle

## STRUCTURE OF A SARCOMERE (FIG. 13.2)

Microscopic examination of a sarcomere shows the following features:

- Alternating light and dark bands.
- The light bands of the muscle fibre are termed the *isotropic or I bands,* are made up of thin filaments.
- The dark bands are termed the *anisotropic or A bands,* are made up of thick filaments.
- The central region of the A band appears less dense and this is called as the *H zone.*
- The I band is bisected by a very dense and narrow *Z line,* to which the thin filaments are attached.
- The central portion of the A band is called the *M line,* connecting the thick filaments by a protein called the *M protein.*

## THE THICK FILAMENT

- The thick filament is confined to the A band.
- It is about 16 nm in diameter.
- It is mainly made up of the protein *myosin.*

### Myosin (Fig. 13.3)

- Myosin constitutes about 55% of the muscle protein.
- It has a molecular weight of 4,60,000.
- It is a *hexamer* consisting of six subunits:
  - Two identical heavy chains
  - Two pairs of light chains.

Each Heavy Chain of Myosin has

- A fibrous portion forming intertwined helix with that of the other chain

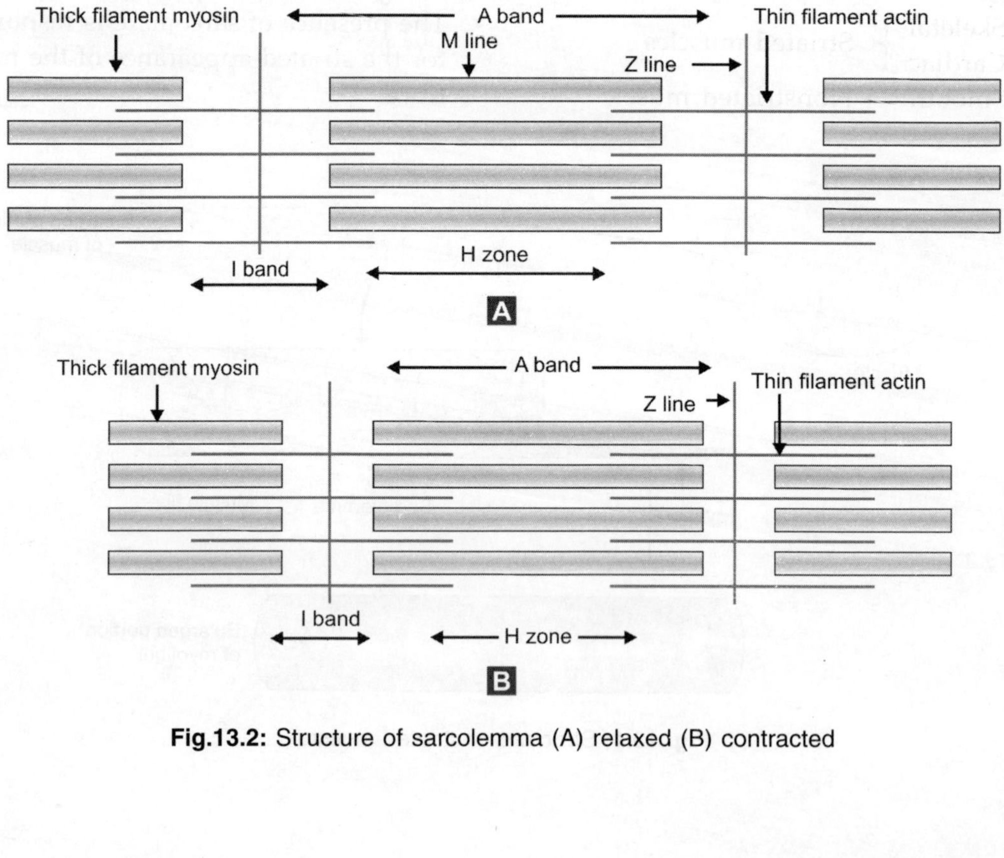


**Fig.13.2:** Structure of sarcolemma (A) relaxed (B) contracted

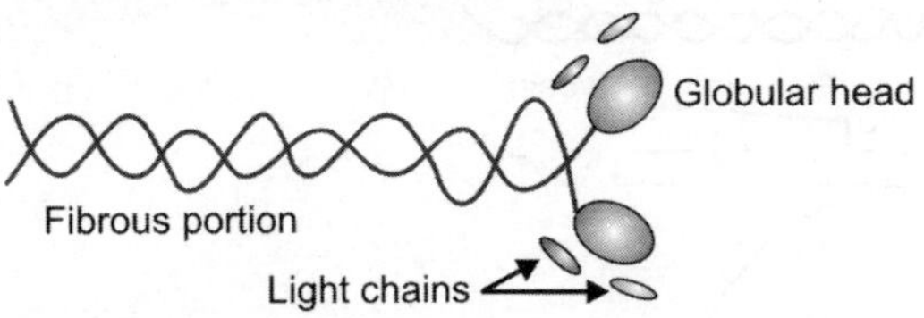


**Fig.13.3:** Structure of myosin molecule

- A globular head attached to one end of the molecule
- To the head portion of the heavy chain, light chains are bound.

*Functions of Myosin Molecule*

1. Myosin molecules assemble to form the thick filament.
2. Has ATPase activity.
3. Binds to actin polymer, the major component of the thin filament during muscle contraction.

## THE THIN FILAMENT (FIG. 13.4)

- The thin filaments lie in the I band, extends into the A band, but not to the H zone of the A band.
- The thin filaments contain the proteins *actin, tropomyosin* and *troponin.*

### Actin

- Major protein of the thin filament.
- In monomeric form it is called the G-actin, because of its globular form.
- Many monomeric G-actin units polymerize into a fibrous form called the *F-actin* or *fibrous actin,* which is a helix made up of actin monomers. Requires $Mg^{++}$ for polymerization.

### Troponin

- It is made up of three polypeptide chains, i.e.
  - troponin *C* (Tn C), binds to calcium

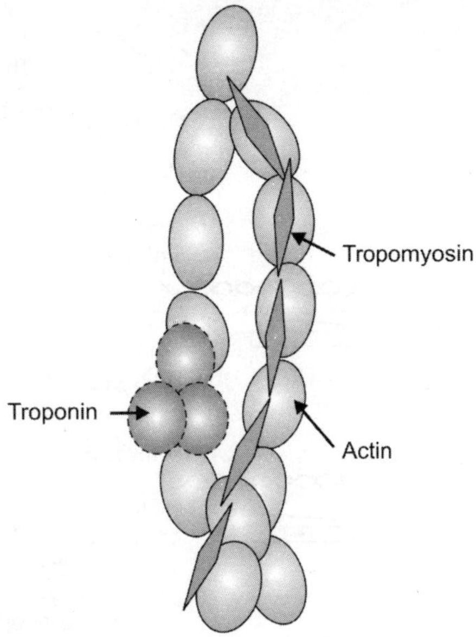


**Fig. 13.4:** Structure of the thin filament

  - troponin *C* (Tn I), binds to actin and prevents binding of actin to myosin and
  - troponin *T* (Tn T), which binds to tropomyosin.
- It is found along the thin filament at regular intervals.

### Tropomyosin

- Is a two stranded molecule.
- Is an alpha helix.
- Lies parallel to the long axis of the thin filament.
- Prevents binding of the thick and thin filaments in the resting muscle.

## MECHANISM OF MUSCLE CONTRACTION (FIG. 13.5)

- Muscle contraction is brought about by the interaction of thick and thin filaments.
- Muscle contraction consists of the cyclic attachment and detachment of the

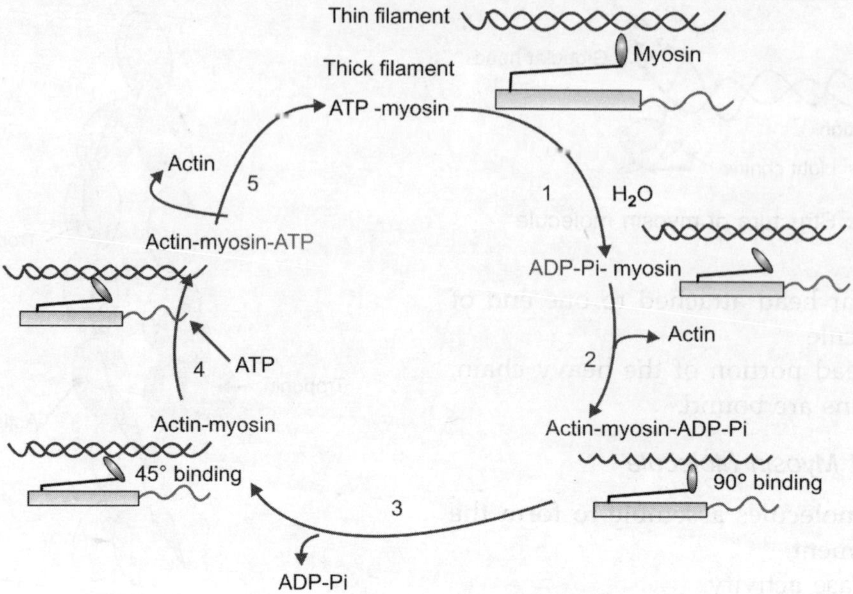


**Fig.13.5:** Mechanism of muscle contraction

globular head portion of myosin to the F–actin filament, followed by a change in the actin-myosin interaction, so that the actin filament is pulled towards the centre of the sarcomere.

- The energy is supplied by hydrolysis of ATP, by the ATPase activity present in the myosin head, which drives the association of myosin with actin.

The Cyclical Events of Muscle Contraction Consist of Five Steps

1. The ATPase of myosin head hydrolyses the ATP that is bound to it, to ADP and Pi. These products remain bound to the myosin head.
2. The myosin head containing ADP and Pi can rotate freely through large angles to locate and bind to F-actin at an angle of 90° with the fibre axis.
3. When myosin binds to actin, ADP and Pi are released and because the conformation with lowest energy for the bond between actin and myosin is 45°, actin is pulled towards the centre of the sarcomere.
4. Now, a new ATP molecule binds to the actomyosin complex.
5. This complex has low affinity for actin. So, actin is released. This is the relaxation step.

The actin on either side of the Z-line has reverse polarity. So, in each sarcomere, the actin is pulled towards the centre, causing the sarcomere to contract. The force developed by the sliding of each actin filament in each sarcomere adds up bringing about muscle contraction. The above cycle keeps repeating to bring about continuous contraction of the muscle.

## REGULATION (FIG. 13.6)

Regulation of muscle contraction is brought about by troponin and tropomyosin through $Ca^{++}$ which is the physiological regulator.

- In resting muscle, $Ca^{++}$ is present within the sarcoplasmic reticulum.
- On stimulation by a nerve impulse, the $Ca^{++}$ from the sarcoplasmic reticulum is

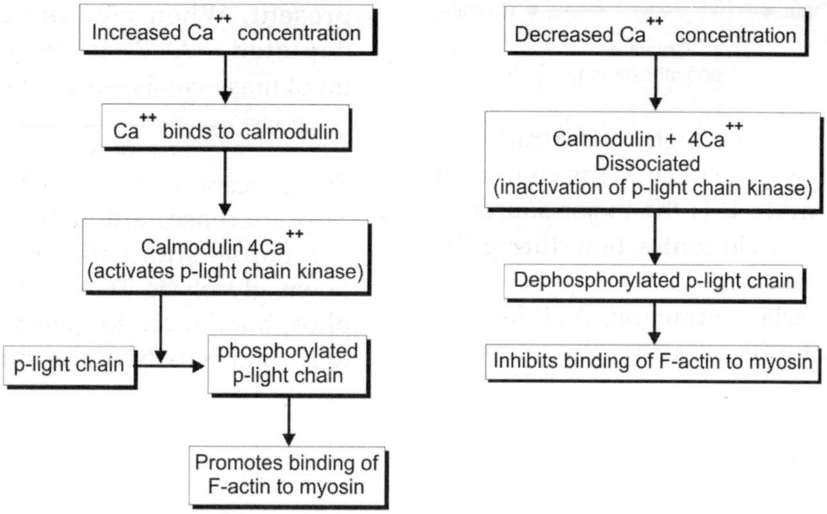


**Fig. 13.6:** Regulation of smooth muscle contraction

released and the cytosolic concentration of Ca$^{++}$ increases by about 10 times (1 μM to 10 μM).

- The troponin C which is the calcium binding subunit has both high and low affinity binding sites for calcium. In resting stage only high affinity sites are occupied by Ca$^{++}$. But, when concentration of Ca$^{++}$ increases in the sarcoplasm, even the low affinity sites are occupied. This results in a conformational change in troponin C which is transmitted to other subunits of troponin and then to tropomyosin.
- This change in conformation shifts the position of tropomyosin on actin, so that the binding sites for myosin head are revealed and favours the binding of myosin to actin.

Smooth muscle differs from striated muscle in that the sarcomeres are not aligned to give a striated appearance and troponin is not present. So, regulation in this case is through a calcium mediated phosphorylation of myosin light chains which are different from that present in striated muscles.

- The smooth muscle myosin contains a light chain, the p-light chain that prevents the binding of myosin head to F-actin.
- The p-light chain must be phosphorylated before it allows F-actin to activate myosin ATPase.
- The phosphorylation of p-light chain commences the attachment—detachment contraction cycle of smooth muscle.

## ENERGY FOR MUSCLE CONTRACTION

- The amount of ATP present in the muscle is sufficient to sustain contractile activity of the muscle for less than one second.
- The reservoir for energy in skeletal muscle is creatine phosphate.
- The phosphate group of creatine phosphate can be transferred to the ADP formed due to hydrolysis of ATP in the presence of creatine phosphokinase, to reform ATP. This is known as *Lohmann's reaction*.

$$\text{Creatine phosphate} + \text{ADP} \xrightleftharpoons{\text{Creatine phosphokinase}} \text{Creatine} + \text{ATP}$$

- Resting muscle has high concentration of creatine phosphate (25 mM) as compared to ATP (4 mM). It is the major source of energy for muscle contraction during the first 4 seconds.
- During muscle contraction, ATP level is high as long as creatine phosphate is present. When creatine phosphate is depleted, ATP is regenerated by myokinase catalysed reaction.

$$2\ \text{ADP} \xrightleftharpoons{\text{Myokinase}} \text{ATP} + \text{AMP}$$

- The concentration of AMP increases.
- This increase, aided by the action of adrenaline stimulates glycogen breakdown, glycolysis, TCA cycle and oxidative phosphorylation, to generate more ATP, to sustain activity of the muscle.

# 14 Connective Tissue

Connective tissue is a system of insoluble protein fibres embedded in a continuous matrix called the extracellular matrix. The cells in the various tissues are embedded in this connective tissue.

## BIOMEDICAL IMPORTANCE

- Components of connective tissue are required to maintain normal tissues.
- Changes occur in these components during old age and diseases causing inflammatory states like *osteoarthritis* and *rheumatoid arthritis.*
- Inherited defects in the enzymes of collagen synthesis result in disorders like *osteogenesis imperfecta* and *Ehlers-Danlos syndrome.*
- Congenital defect in the catabolism of glycosaminoglycans result in a group of disorders called *mucopolysaccharidoses.*
- Alteration in the glycoprotein component of the cell membrane may be responsible for *cancer metastasis.*

## FUNCTIONS

1. Mechanical and supportive
2. Lubrication
3. Elasticity
4. Cellular adhesion and recognition

It consists of three major classes of biomolecules, i.e.

1. Structural proteins:
   - Responsible for mechanical and supportive function and elasticity.
   - Insoluble in nature.
     For example, collagen, elastin.
2. Specialised proteins:
   - Responsible for cellular adhesion and recognition
     For example, fibrillin, fibronectin, laminin.
3. Proteoglycans:
   - Form the ground substance
   - Composed of 90% glycosaminoglycans
   - Associated with small amounts of proteins in tissues.
   - It not only serves as an intercellular binding substance in parenchymatous organs like the liver and muscle, but is distributed throughout the body in cartilage, tendons, ligaments, matrix of bone, kidney, ureters and urethra. It underlies the skin and serves as binding material for blood vessels.

## COLLAGEN

- Major component of connective tissue.
- Most abundant protein in animals and constitutes 25-30% of total weight of proteins.
- At least 19 distinct types of collagen have been identified in mammalian systems.

### Structure of Collagen

Collagen has a triple-helical structure, each triple helix is made up of 3 $\alpha$- chains.

### Primary Structure

1. Each α-chain is made up of about 1000 amino acid residues.
2. The amino acid sequence is unique in that *every third position in* the polypeptide chain is occupied by *a glycine residue.*
3. The sequence of amino acids is represented as *(Gly-X-Y)$_n$*. X and Y can be any amino acid other than glycine.
4. About 100 of the X positions are occupied by proline, while about 100 Y positions are occupied by hydroxyproline. Lysine and hydroxylysine are also present in Y position.
5. Hydroxylysine is present only in collagen.

### Secondary and Tertiary Structure

1. Each α-chain is twisted into a *left handed helix* of 3.3 amino acids per turn.
2. Three of these α-chains are then wound into a right handed super-helix forming rod like molecule.
3. The super helix is *1.4 nm* in diameter and *300 nm* in length.
4. It is absolutely necessary to have glycine in every third position in the α-chain because it is the only amino acid that is small enough to be accommodated in the limited space available down the central core of the triple helix.
5. Collagen is a *glycoprotein*. To some of the hydroxylysine residues galactose or galactosyl glucose are linked through an O-glycosidic bond.
6. There are different types of α-chains with slightly different properties. Three similar or dissimilar α-chains associate to form different types of collagens. (Table 14.1).
7. Many types of collagens are found in small quantities, but are important in determining the physical properties of tissues.

## Synthesis of Collagen

1. Collagen is synthesised on the ribosomes of fibroblasts in a precursor form, i.e. preprocollagen (Fig. 14.1).
2. Once synthesised it undergoes extensive post-translational modifications before becoming part of the mature extracellular collagen fibre.

**Table 14.1:** Types of collagen

| *Type* | *Chain designations* | *Tissue in which found* |
|---|---|---|
| I | $[\alpha 1\,(I)]_2$, $\alpha_2$ (I) | Bone, skin, tendon, scar tissue, heart valve, intestinal and uterine wall. |
| II | $[\alpha 1\,(II)]_3$ | Cartilage, vitreous humour |
| III | $[\alpha 1\,(III)]_3$ | Extensible connective tissues like blood vessels, intestinal and uterine wall, fetal skin. |
| IV | $[\alpha 1\,(IV)]_3$ Or $[\alpha_2\,(IV)]_3$ | Basement membrane, lens capsule |
| V | α1 (V) $\alpha_2$ (V) $\alpha_3$ (V) Or $[\alpha 1\,(V)]_3$ Or $[\alpha 1(V)]_2$ $\alpha_2$ (V) | Found in low amounts in cell surfaces. |

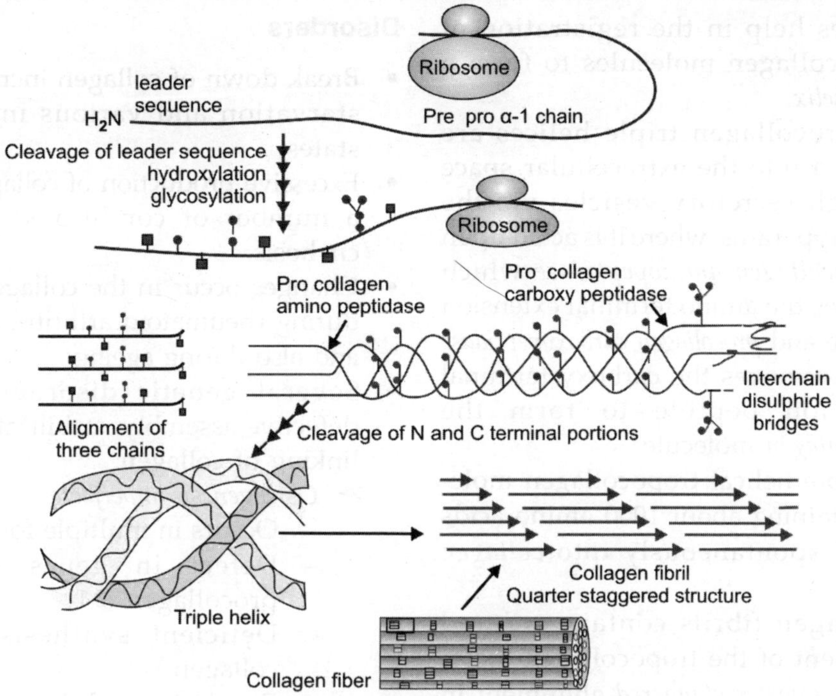


**Fig.14.1:** Synthesis and structure of collagen
*(For colour version see Plate 8)*

a. The *preprocollagen* has a leader sequence at the $-NH_2$ terminal end that directs the polypeptide chain into the vesicular space of endoplasmic reticulum. *This leader sequence is proteolytically cleaved.*

b. In the endoplasmic reticulum, the *proline and lysine* residues at the Y positions of (Gly-X-Y) are *hydroxylated by prolyl hydroxylase and lysyl hydroxylase* respectively.

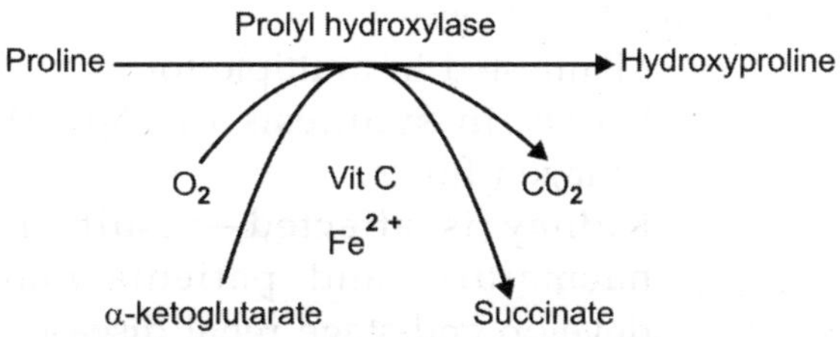


Vitamin C is required for the reaction. The deficiency of vitamin C causes decreased hydroxylation reactions, decreasing the strength of collagen fibres, causing *scurvy.*

c. Some of the hydroxylysine residues are *glycosylated* by addition of galactose or galactosyl glucose residue through an O-glycosidic linkage, which is unique to collagen. There is also addition of oligosaccharide chain to asparagine residues by N-glycosidic linkage.

d. The *procollagen* molecule contains polypeptide extensions of molecular weight 20–35,000 at both amino and carboxy terminal ends. The amino terminal polypeptide forms only intrachain disulphide bonds, while at the carboxy terminal end there is both intra and interchain disulphide bond formation. The interchain disulphide bridges help in the registration of three collagen molecules to form a *triple helix*.

e. The procollagen triple helices are secreted into the extracellular space through secretory vesicles via the Golgi apparatus, where it is acted upon by *procollagen aminopeptidase* which removes the amino terminal extension peptide and *procollagen carboxypeptidase* which removes the carboxy terminal extension peptide to form the *tropocollagen* molecule.

3. These triple helical tropocollagen molecules containing about 1000 amino acids assemble spontaneously into *collagen fibrils*.
4. The collagen fibrils contain ordered arrangement of the tropocollagen molecules in a *quarter staggered* alignment in tendons. In skin it exists as interlacing network of fibres. The arrangement of collagen fibres depends on the function of collagen in that particular tissue. Whatever the arrangement, they always show a cross-striated appearance.
5. The fibroblasts also secrete a glycoprotein called *'fibronectin'* which is present on cell surfaces and blood. The fibronectin binds to the aggregating procollagen fibres and alters the kinetics of fibre formation.
6. The fibrils are further stabilised by the *inter and intrachain crosslinks* between lysine residues aided by the action of *lysyl oxidase*, a copper dependent enzyme.
7. The extent of crosslinking depends on the physiological function and the age of the tissue. Older the collagen, more the extent of crosslinking.

The collagen molecules, once synthesised, have a long half life of several months. They are degraded by the enzyme collagenase.

**Disorders**

- Break down of collagen increases during starvation and various inflammatory states.
- Excessive production of collagen occurs in a number of conditions, e.g. hepatic cirrhosis.
- Changes occur in the collagen molecules during rheumatoid arthritis, osteoarthritis and also during ageing.
- Several genetic disorders result in defective assembly, stabilisation or cross-linking of collagen.

➢ *Osteogenesis imperfecta*
- Occurs in multiple forms
- Defect in genes coding for procollagen α-1
- Deficient synthesis of type I collagen
- Results in brittle bones and skeletal deformities.

➢ *Ehlers-Danlos syndrome:*
- Occurs in multiple forms
- Results in hypermobile joints, skin abnormalities
- Defect may be in:
  - abnormality of synthesis of type III collagen
  - or deficiency of lysyl hydroxylase
  - or deficiency of lysyl oxidase
  - or deficiency of procollagen N-proteinase

➢ *Alport syndrome*
- Manifested in multiple forms.
- Defect in synthesis of type IV collagen fibres.
- Kidney is affected—results in haematuria and patients may develop end-stage renal disease.

In addition to collagen, several proteins occur in the extracellular matrix and perform

**Table 14.2:** Examples of some noncollagenous proteins present in extracellular matrix

| *Protein* | *Function* | *Location* |
|---|---|---|
| Elastin | Provides elasticity to tissues | Lung, large arteries, elastic ligaments, skin, ear cartilage. |
| Fibrillin | Provide scaffold for deposition of elastin | Insoluble microfibrils of lens, periosteum, aorta. |
| Fibronectin | Cell adhesion and migration | Widely distributed in the extracellular matrix, also present in plasma. |
| Laminin | Helps to anchor basal lamina to the cells | Kidneys, muscles |

various functions. Some examples are given in Table 14.2.

## GLYCOPROTEINS

- Proteins that have oligosaccharide chains covalently attached to their backbone (made up of mainly proteins with small amounts of carbohydrates).
- Widely distributed in tissues and have diverse functions, e.g. almost all plasma proteins except albumin are glycoproteins (Table 14.3).
- Only 8 of about 200 monosaccharides found in nature are found in oligosaccharide chains of glycoproteins (Table 14.4).
- There are four types of linkages between polypeptide chains and the sugar residues in glycoproteins (Table 14.5, Fig. 14.2).
- The number of oligosaccharide chains attached to one protein can vary from one to many, with each oligosaccharide chain containing 2, 3 or many monosaccharide units.
- A single glycoprotein may contain oligosaccharides linked by both O-glycosidic and N-glycosidic linkages.

**Table 14.3:** Functions of glycoproteins

| *Function* | *Examples* |
|---|---|
| Structural molecules | Collagen, elastin |
| Enzymes | Proteases, hydrolases, clotting factors |
| Hormones | HCG, TSH |
| Immunologic molecules | Immunoglobulins, blood group substances, interferons |
| Membrane proteins | Glycophorin, receptors. |
| Lubricants | Mucins |
| Folding of certain proteins | Calnexin, calreticulin |

**Table 14.4:** Sugars present in glycoproteins and their activated form required for synthesis

| *Sugar* | *Nucleotide derivative* |
|---|---|
| Galactose | UDP – galactose |
| Glucose | UDP – glucose |
| Mannose | GDP – mannose |
| N – acetyl neuraminic acid (NANA) | CMP – NANA |
| Fucose (6 deoxy-galactose) | GDP – fucose |
| N – acetyl galactosamine | UDP – N acetyl galactosamine |
| N – acetyl glucosamine | UDP – N acetyl glucosamine |
| Xylose | UDP – xylose |

### Synthesis

- Glycoprotein synthesis occurs in two stages.
- Polypeptide chain is synthesised first followed by glycosylation which occurs in the endoplasmic reticulum and Golgi bodies.

**Table14.5:** Types of linkage found in glycoproteins and their site of occurrence

| *Type of linkage* | *Groups involved* | *Examples* |
|---|---|---|
| O-glycosidic | Serine or threonine and N-acetyl galactosamine | Most abundant, mucins. |
| O-glycosidic | Hydroxylysine and galactose | Collagen |
| O-glycosidic | Serine and N-Acetyl glucosamine | Transcription factors, cytosolic proteins |
| N-glycosidic | Asparagine and N-Acetyl galactosamine | Glycoproteins of plasma membrane |
| Glycosyl phosphatidyl inositol anchored (GPI- anchored) | Carboxy-terminal amino acid of protein linked to phosphoryl ethanolamine which in turn is linked to oligosaccharide chain and phosphatidyl inositol | Anchoring of glycoproteins to plasma membrane |

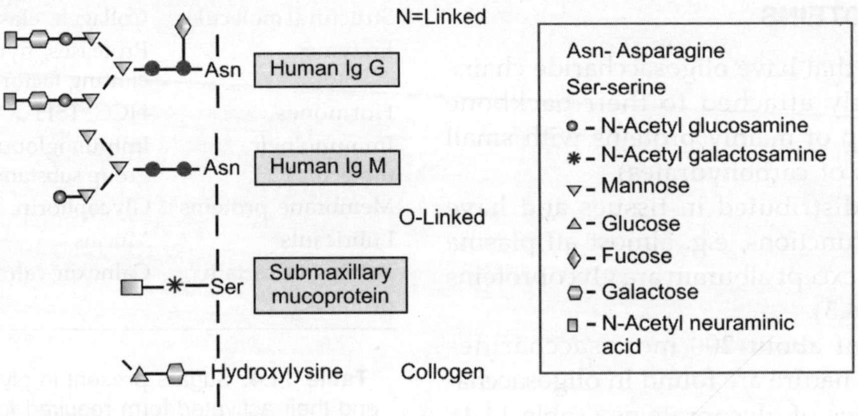


**Fig. 14.2:** Examples of N-linked and O-linked oligosaccharides

- The secretory vesicles containing these synthesised glycoproteins are released by exocytosis.

*O-Glycosylation*

1. During synthesis, each sugar molecule is transferred from its activated nucleotide derivative (Table 14.4) by a specific glucosyl transferase.
2. These glucosyl transferases are found bound to the membranes of endoplasmic reticulum or Golgi bodies.
3. Each one of these enzymes is specific for the sugar transferred and the first one is also specific for the amino acid to which the sugar is linked.
4. The first sugar incorporated during O-glycosylation in most of the cases is N-acetyl galactosamine.
5. The terminal residue is usually NANA.

*N-Glycosylation*

The steps are slightly different from O-glycosylation.

1. The sugar groups are transferred from the nucleotide derivatives by specific glucosyl transferases first to a compound called *dolichol phosphate.*
2. Dolichol phosphate is a lipid on which the pro-oligosaccharide chain is formed.

$$(CH_2{=}\overset{CH_3}{\overset{|}{C}}{-}CH{=}CH)_n{-}CH_2{-}\overset{CH_3}{\overset{|}{C}H}{-}CH_2{-}CH_2{-}O{-}PO_3H_2$$

Dolichol phosphate (n = 15 to 19)

3. The first sugar added is N-acetyl glucosamine, followed by another N-acetyl glucosamine and three mannose residues—*'a pentasaccharide core'*.
4. The oligosaccharide chain is then transferred to the polypeptide chain by *protein-oligosaccharide transferase* present in the endoplasmic reticulum.
5. This is followed by further addition of other monosaccharides and its processing involves removal and modification of sugar residues.
6. There are 3 types of N-linked oligosaccharides (Fig. 14.3).
   i. Complex
   ii. Hybrid
   iii. High mannose

- The oligosaccharide chains of glycoproteins contain biological information which is dependent on the nature, content and sequence of constituent sugars. Removal of one or more sugar residues can result in loss of function of glycoprotein.
- The glycoproteins are anchored to the plasma membrane by glycosyl phosphatidyl inositol structures.
- The carbohydrate containing upper most layer of plasma membrane is known as *'glycocalyx.'*

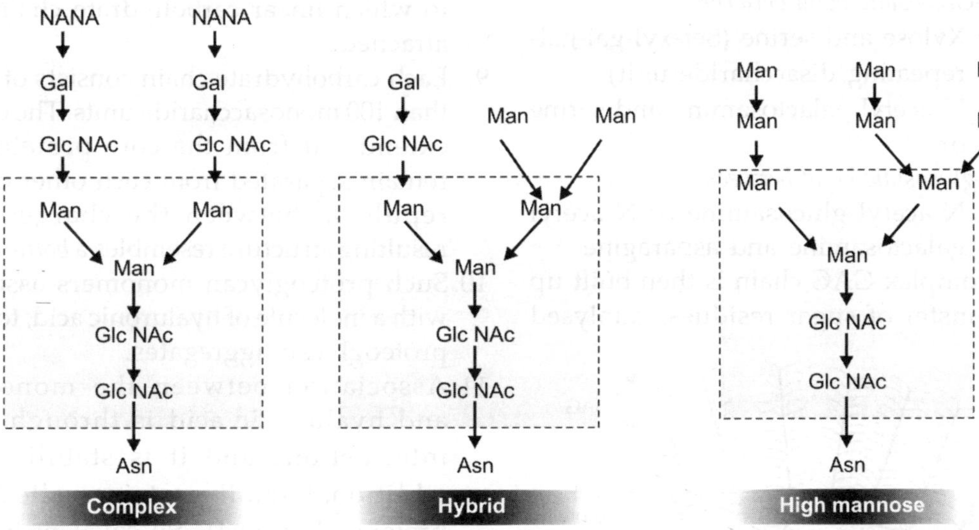


| Asn | – | Asparagine | GlcNAc | – | N-Acetyl glucosamine |
|---|---|---|---|---|---|
| Man | – | Mannose | Gal | – | Galactose |
| NANA | – | N-Acetyl neuraminic acid | | | |

**Fig. 14.3:** Structures of the major types of asparagine-linked oligosaccharides

### Degradation

Is by specific glycosidases present in the lysosomes which remove sugars sequentially from the end of the oligosaccharide chains. Neuraminidase removes the terminal NANA.

## PROTEOGLYCANS (Fig. 14.4)

Proteoglycans are made up of large complexes of negatively charged heteropolysaccharide chains called glycosaminoglycans (90% by weight), associated with small amounts of proteins.

### Glycosaminoglycans (GAG)

Chemistry and functions dealt in Chapter 2.

### Biosynthesis of Proteoglycans

1. Involves attachment of GAG to the protein by
   - *O-Glycosidic bond between:*
     - ➢ Xylose and serine (Ser-xyl-gal-gal-repeating disaccharide unit)
     - ➢ N-acetyl galactosamine and serine or
   - *N-glycosidic bond between:*
     - ➢ N-acetyl glucosamine or N-acetyl galactosamine and asparagine.
2. The complex GAG chain is then built up by transfer of sugar residues, catalysed by specific glucosyl transferases present in Golgi bodies, from their activated nucleotide derivatives.

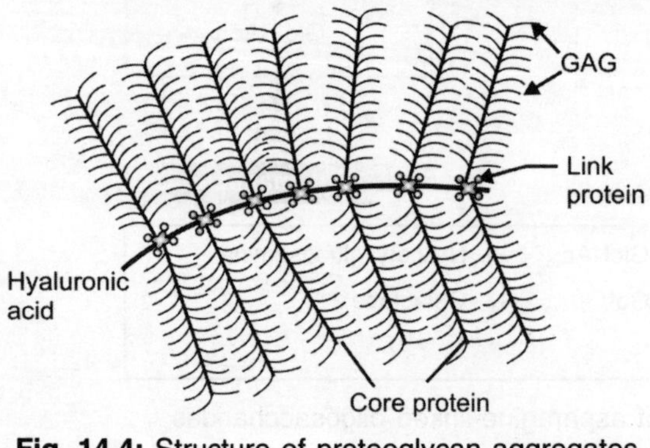


**Fig. 14.4:** Structure of proteoglycan aggregates

3. The glucosyl transferases will catalyse the formation of a single type of glycosidic bond and are sugar specific.
4. Termination of the oligosaccharide chain occurs following sulfation of sugar residues.
5. Sulfation is catalysed by sulfotransferases which transfer '$SO_4$' groups from *3'-phosphoadenosyl 5'-phosphosulfate* (PAPS).
6. Further modifications occur after synthesis, the most important being the epimerisation of glucuronic acid to iduronic acid, its 5–epimer, in heparan and dermatan sulphate, catalysed by a specific epimerase.
7. All glycosaminoglycans except hyaluronic acid are found covalently linked to proteins forming proteoglycan monomers.
8. Each monomer consists of a core protein to which linear carbohydrate chains are attached.
9. Each carbohydrate chain consists of more than 100 monosaccharide units. The chains extend out from the core protein, but remain separated from each other due to repulsion between the charges. The resulting structure resembles a *bottle brush.*
10. Such proteoglycan monomers associate with a molecule of hyaluronic acid, to form proteoglycan aggregates.
11. Association between the monomers and hyaluronic acid is through ionic inter-actions and it is stabilised by additional small proteins called link proteins (Fig. 14.4). No covalent linkage between monomers and hyaluronic acid.

## CATABOLISM

- The core proteins are degraded first followed by oligosaccharides.

**Table 14.6:** Characteristics of some important mucopolysaccharidoses

| Type | Name | Enzyme deficiency | Clinical findings |
|---|---|---|---|
| I | Hurlers' syndrome | α-Iduronidase | Mental retardation +++<br>Skeletal deformity ++<br>Corneal opacity<br>Dermatan sulphate and heparan sulphate in urine |
| II | Hunters' syndrome | Iduronate sulphatase | Mental retardation +<br>Skeletal deformity ++<br>Deafness<br>Heparan sulphate and dermatan sulphate in urine |
| III | Sanfilippo's syndrome | a. Heparan sulfamidase<br>b. N-acetyl glucosaminidase<br>c. N-acetyl transferase<br>d. N-acetyl glucosamine | Mental retardation ++<br>Skeletal deformity +<br>Corneal clouding +<br>Heparan sulphate in urine. |

- The degradation of GAGs occurs by lysosomal hydrolases which are bond specific.
- A number of hydrolases are required.
- The oligosaccharides are degraded sequentially from the last monosaccharide unit.
- Deficiency of any of the hydrolases results in accumulation of the intermediates in lysosomes of various tissues. This group of disorders is termed *'mucopolysaccharidoses.'*

## General Clinical Manifestations of Mucopolysaccharidoses (Table 14.6)

1. Coarse facial hairs
2. Thick skin
3. Corneal opacity, due to accumulation of GAGs.
4. Mental retardation } due to defective cell functions
5. Growth deficiency }
6. Skeletal deformities } due to interaction of accumulated GAG with other proteins like collagen
7. Joint contractures }
8. Herniae
9. Characteristic finding of diagnostic importance is excessive excretion of only one of the GAGs in urine.
10. Can be diagnosed by estimating the hydrolases in plasma.

# 15 Clinical Biochemistry

Any compound present in the living system is under dynamic equilibrium, i.e. the amount of a compound supplied/formed is equal to the amount utilised/taken up, so that, the concentration of the compound is always between a certain range called as the *'reference range'*, under normal conditions. However, during certain diseased conditions, the value of one or a few compounds may vary, making the estimation of it a valuable diagnostic tool. This compound may be a protein, a lipid, a carbohydrate, an amino acid, a hormone, an enzyme or a metabolic end product.

The reference range for a few of such compounds and the clinical conditions in which they are altered is given in Table 15.1.

**Table 15.1:** Reference range and diagnostic importance of some compounds

| *Constituent* | *Reference range* | *Conditions in which increased* | *Conditions in which decreased* |
|---|---|---|---|
| *Albumin* | 3.5 – 5.0 g/dl | —— | Nephrotic syndrome, liver disease, malnutrition, burns, protein losing enteropathy |
| Bicarbonate | 22 – 26 mEq/L | Metabolic alkalosis | Metabolic acidosis |
| Calcium | 9 – 11 mg/dl | Hyperparathyroidism, hypervitaminosis D | Hypoparathyroidism, hypovitaminosis D, rickets, steatorrhoea, |
| Chloride | 96 – 106 mEq/L | Dehydration, respiratory acidosis, Cushing's syndrome | Vomiting, diarrhoea, respiratory alkalosis, Addison's disease |
| Creatine | 0.2 – 0.4 mg/dl | Muscular dystrophy, renal failure | —— |
| Creatinine | 0.8 – 1.5 mg/dl | Renal failure | —— |
| Enzymes: | | | |
| Acid Phosphatase | 2.5 – 12 IU/L | Metastatic prostate cancer | —— |
| Amylase, serum | 80 – 180 Somogyi units/dl or 50-120 IU/L | Acute pancreatitis | —— |

*Contd...*

*Contd...*

| *Constituent* | *Reference range* | *Conditions in which increased* | *Conditions in which decreased* |
|---|---|---|---|
| Alanine transaminase | 3 – 15 IU/L | Liver disease | —— |
| Alkaline phosphatase | 25 – 90 IU/L | Obstructive jaundice, rickets, osteoblastoma | —— |
| Aspartate transaminase | 2 – 20 IU/L | Heart disease, liver disease | —— |
| Creatine phosphokinase | 20 – 80 U/L | Heart disease, muscular dystrophy | —— |
| Lactate dehydrogenase | 60 – 200 IU/L | Heart disease, liver disease | —— |
| Glucose | 70 – 110 mg/dl | Diabetes mellitus hyperadrenalism hyperpituitarism hyperthyroidism | Insulin secreting tumours of the pancreas, insulin overdosage, hypoactivity of the adrenals, pituitary and thyroid |
| Globulins | 2.5 – 3.5 g/dl | Infections, multiple myeloma | —— |
| Lipids: | | | |
| Cholesterol | 150 – 220 mg/dl | Nephrotic syndrome, diabetes mellitus, hypothyroidism, familial hypercholesterolaemia | Hyperthyroidism |
| HDL | 45 – 65 mg/dl | —— | Atherosclerosis, cardiovascular disease* |
| LDL | <165 mg/dl | Atherosclerosis cardiovascular disease* | —— |
| Triacylglycerol | <165 mg/dl | Atherosclerosis cardiovascular disease* | —— |
| pH, blood | 7.35 – 7.45 | Metabolic alkalosis vomiting, Cushing's syndrome, excessive intake of antacids Respiratory alkalosis: hypoxia, hysteria, salicylate poisoning | Metabolic acidosis: Uncontrolled diabetes mellitus, starvation, lactic acidosis Respiratory acidosis: Asthma, pneumonia |
| Phosphate, serum | Adults: 2.5 - 4.5 mg/dl Children: 4-6 mg/dl | Hypoparathyroidism, hypervitaminosis D renal failure | Hyperparathyroidism, rickets, Fanconi syndrome |

*Contd...*

*Contd...*

| *Constituent* | *Reference range* | *Conditions in which increased* | *Conditions in which decreased* |
|---|---|---|---|
| Potassium | 3.5 - 5.0 mEq/L | Renal failure, Addison's disease, diabetic coma, dehydration | Cushing's syndrome, diarrhoea, vomiting, treatment of diabetes mellitus with insulin |
| Sodium | 136 - 145 mEq/L | Cushing's syndrome, dehydration | Diarrhoea, vomiting, chronic renal diseases, Addison's disease, liver cirrhosis, congestive heart failure |
| Urea | 12 - 36 mg/dl | Prerenal: vomiting, diarrhoea, shock, burns, haemorrhage<br>Renal: glomerulonephritis, polycystic kidney, hydronephrosis, malignant hypertension<br>Postrenal: enlargement of prostate, stones in bladder, tumours of bladder and ureter and stricture of urethra | Severe liver disease |
| Uric acid, serum | 3.0 - 7.0 mg/dl | Gout | ——— |

*Risk factor

## ANALYSIS OF URINE (TABLES 15.2 TO 15.4)

The waste products formed in the body are excreted through urine.

### Normal Urine

*Physical characteristics:*

Volume : 1-2 litres
Colour : Pale yellow or straw coloured
Specific gravity : 1.012 – 1.024
pH : 5.5-6.0

**Table 15.2:** Components of normal urine

| *Inorganic constituents:* | *Daily excretion* |
|---|---|
| Chloride (as NaCl) | 8 -15 g |
| Sulphate | 1 g |
| Inorganic phosphate | 1g |
| Calcium | 200 mg |
| Ammonia | 0.5 - 0.8 g |
| *Organic constituents*: | |
| Urea | 20 – 30 g |
| Uric acid | 0.6 –1 g |
| Creatinine | 1-2 g |
| Ethereal sulphate | 100 mg |
| Urobilinogen | traces |

### Analysis of Abnormal Urine

Under certain abnormal conditions, certain other compounds are excreted in the urine in detectable amounts. The physical characteristics may also change. So, analysis of urine is also of diagnostic importance. Since it is a waste product, urine is easily obtained.

**Table 15.3:** Physical characteristics of abnormal urine

| *Physical characteristic* | *Variation* | *Clinical condition* |
|---|---|---|
| Volume | Polyuria (↑) | Diabetes mellitus, diabetes insipidus |
| | Oliguria (↓) | Nephritis, fever, diarrhoea, vomiting |
| | Anuria (nil) | Shock, acute nephritis, mercury poisoning, incompatible blood transfusion, bilateral renal stones |
| Colour | Dark brown or red | Presence of blood as in urinary tract infection, injury during introduction of catheter, incompatible blood transfusion, burns |
| | Yellow | Presence of bilirubin as in hepatic or obstructive jaundice |
| | Black | Presence of melanin |
| | Milky | Presence of lipids or pus/epithelial cells/bacterial cells as in infection |
| pH | Acidic | Fever, diabetic acidosis |
| | Alkaline | Alkali therapy |
| Specific gravity: | Increased | Acute nephritis, fever |
| | Decreased | Diabetes insipidus |

**Table 15.4:** Abnormal constituents present in urine

| *Constituent* | *Pathological conditions in which present* |
|---|---|
| Glucose | Diabetes mellitus, renal glycosuria |
| Lactose | Pregnancy |
| Albumin | Nephrotic syndrome, glomerulonephritis |
| Ketone bodies | Diabetes mellitus, starvation |
| Bile salts and pigments | Hepatic and post hepatic jaundice |
| Blood | Trauma due to introduction of catheter, severe burns, acute glomerulonephritis, incompatible blood transfusion |

# 16 Function Tests

Based on the functions performed by various organs, specific biochemical investigations have been put to use in the laboratory to assess their function. In this chapter the various biochemical tests to evaluate the functioning of the liver and kidney are discussed.

## LIVER FUNCTION TESTS

The liver is the principal organ for the metabolism of carbohydrates, proteins, lipids, porphyrins and bile acids.

- It is capable of synthesising all the plasma proteins, except the immunoglobulins, which are produced by the lymphocytic plasma cell system.
- It is also the major site for storage of iron, glycogen, lipids and vitamins.
- The liver plays an important role in the detoxification of xenobiotics.
- Excretion of endogenous metabolic products such as bilirubin, ammonia, and urea. (Ammonia and urea discussed under protein metabolism).

Abnormalities of bilirubin metabolism are important in the diagnosis of liver diseases. The metabolism of bilirubin is briefly discussed below.

### *Bilirubin Metabolism*

*Bilirubin* is a catabolic product of heme. It is derived mainly from the degradation of haemoglobin of senescent red blood cells (80%) and other heme proteins such as myoglobin, catalase and cytochromes (20%). About 250–350 mg of bilirubin is produced daily.

*Site*: Reticuloendothelial system of the liver and spleen.

*Formation:* The conversion of heme to bilirubin is enzymatic, involving a mixed function oxygenase, known as *heme oxygenase* which is present in the liver microsomes. The green coloured *biliverdin* formed is acted upon by *biliverdin reductase* to yellow coloured pigment bilirubin.

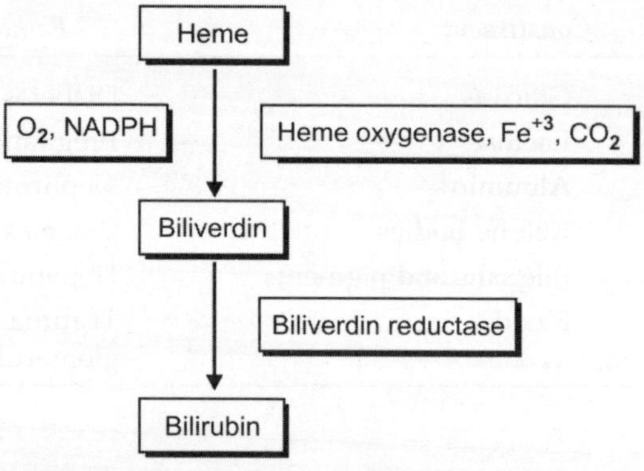


Metabolism of bilirubin involves the following processes (Fig. 16.1).

- Transport to the liver
- Uptake by the hepatocytes
- Conjugation of bilirubin to form bilirubin diglucuronide
- Excretion of conjugated bilirubin into the bile.
- Conversion of bilirubin to urobilinogen in the intestine.

*Transport:* Bilirubin formed is insoluble in aqueous solution (unconjugated bilirubin). Hence it is transported in the plasma, by binding to albumin and forming *albumin bilirubin complex.*

*Uptake:* As the albumin-bilirubin complex enters the liver, bilirubin dissociates and is taken up by the hepatocytes. Within the hepatocyte bilirubin binds to a specific intracellular protein called *ligandin*.

*Conjugation:* The process of conjugation of bilirubin to water soluble bilirubin diglucuronide takes place in the hepatocytes. It involves *UDP glucuronic acid* and the enzyme *UDP glucuronyl transferase.*

$$\begin{array}{ccc}\text{Bilirubin} & \xrightarrow[\text{transferase}]{\text{UDP glucuronyl}} & \text{Bilirubin diglucuronide} \\ + & & + \\ \text{2UDP glucuronic acid} & & \text{2UDP}\end{array}$$

*Excretion:* The water soluble bilirubin diglucuronide (conjugated bilirubin) is readily excreted by way of bile into the intestine. It is an active and rate limiting process.

*Conversion to urobilinogen:* Bilirubin diglucuronides reach the intestine via the bile. In the intestine the glucuronic acid groups are first hydrolysed by *β glucuronid*ase and the bilirubin is reduced by the intestinal bacterial flora to form *biliverdin* (green colour) which is later reduced to urobilinogen.

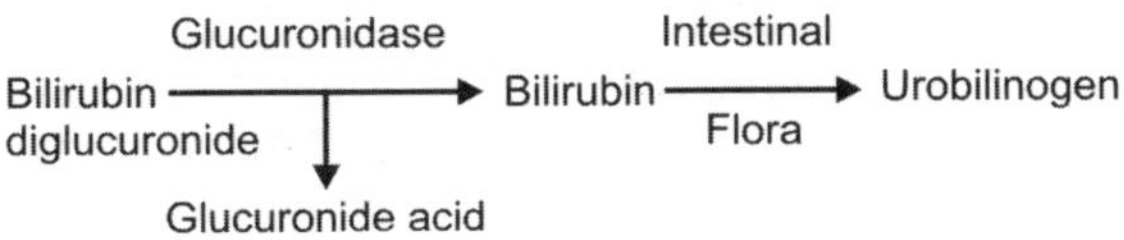


- Some of the urobilinogen is absorbed by way of portal blood which enters the liver (enterohepatic circulation).
- A small amount of urobilinogen (4 mg or less per day) which escapes the circulation is exerted by the kidney. On standing it is oxidized to urobilin a major urinary pigment. Characteristic colour of urine is due to the urobilin.
- A major portion of urobilinogen is reduced further by the intestinal enzymes to *stercobilinogen*. This is autoxidised on standing to brownish coloured pigment and eliminated in the faeces.
- The normal serum bilirubin level is *0.2–0.6 mg/dl.* 2/3 is unconjugated while 1/3 is conjugated.

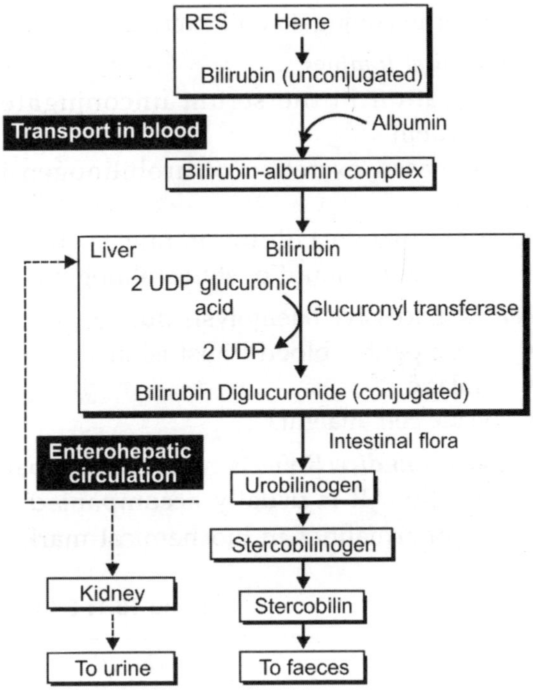


**Fig. 16.1:** Metabolism of bilirubin

*Jaundice:* Is a clinical condition characterised by yellow colouration of the eyes. It is caused by the deposition of bilirubin due to its elevated levels *(hyperbilirubinaemia).* Jaundice is clinically obvious when the plasma bilirubin concentration exceeds 2 mg/dl. Hyperbilirubinaemia occurs when there is an imbalance between production and excretion. Conventionally jaundice is classified into 3 types (Table 16.1).

- Prehepatic : increased production of bilirubin
- Hepatic : impaired hepatic uptake, conjugation or excretion of bilirubin
- Posthepatic : obstruction to biliary drainage.

***Prehepatic jaundice:*** Also referred to as *haemolytic jaundice*. Results from excess production of bilirubin, beyond the ability of the liver to conjugate and excrete.

*Biochemical findings*

- Elevation in the serum unconjugated bilirubin
- Increased excretion of urobilinogen in urine
- Dark brown colour of faeces due to increased amount of stercobilinogen.

*Causes:* Excessive haemolysis due to

- Incompatible blood transfusion
- Malaria
- Sickle cell anaemia

***Hepatic jaundice:*** Reflects generalised hepatic dysfunction. It is usually accompanied by other abnormalities in biochemical markers of hepatocellular dysfunction and varying degrees of intrahepatic obstruction.

*Biochemical findings*

- Increased levels of conjugated and unconjugated bilirubin in the serum.
- Dark yellow coloured urine due to the excessive excretion of bilirubin.
- Increased activities of ALT and AST.
- Slight increase in alkaline posphatase.

*Causes:*

- Viral infections
- Drugs/poisons/toxins, e.g. chloroform, carbon tetrachloride
- Cirrhosis of the liver.

***Posthepatic jaundice:*** Also referred to as *obstructive jaundice, cholestatic jaundice or regurgitation jaundice* and is primarily caused by obstruction to the flow of bile.

*Biochemical findings*

- Elevated levels of conjugated bilirubin and bile acids in the plasma.
- Pale coloured stools caused by absence of faecal bilirubin and urobilin.
- Dark yellow coloured urine due to the presence of water soluble conjugated bilirubin.
- Absence of urobilinogen.

**Table 16.1:** Differential diagnosis of Jaundice

| | *Prehepatic (Hemolytic)* | *Hepatic (Hepatocellular)* | *Posthepatic (Obstructive)* |
|---|---|---|---|
| Serum bilirubin | ↑ <4 mg/dl | ↑ ↑ ↑ upto 30 mg/dl | ↑ ↑ ↑ upto 20 mg/dl |
| Urine bilirubin | Nil | ↑ | ↑↑↑ |
| Urine urobilinogen | ↑↑↑ | ↑ | Nil |
| ALT and AST | Normal | ↑↑ | Normal |
| ALP | Normal | ↑ | ↑↑↑ |
| 5′ Nucleotidase | - | ↑ | ↑↑↑ |
| γ-GGT | - | ↑↑ | ↑↑↑ |

- High increase in alkaline phosphatase

*Causes:* Gallstones, pancreatic tumour.

## Use of Tests in Liver Diseases

The term *"liver function tests"* includes a wide range of tests used in diagnosis of liver disease. Several of them are influenced by factors other than a function of liver, some are an index of liver necrosis rather than of functional impairment. Of the tests described, galactose tolerance and hippuric acid tests are used in conditions in which there is no obstruction to the flow of bile. The bromosulphthalein test, is nearest to measuring a function of the liver, i.e. excretory function. However, LFT is still used for the following:

1. In the diagnosis of sub-clinical liver impairment. It includes:
   - Screening of suspected cases during outbreaks of infective hepatitis.
   - Screening of persons exposed to potentially hepatotoxic agents in industry.
   - In study of alcoholics.
2. For the differential diagnosis of various types of jaundice.
3. To assess the degree of liver failure in known liver disease.
   - prognosis and assessment of extent of residual damage remaining after apparent recovery from liver disease.

## Choice of Liver Function Tests

- Frequently, a combination of tests are employed. No single test would yield a conclusive result.
- A useful routine combination of tests which could be done on a single specimen of clotted blood are—
  - ➢ Bilirubin
  - ➢ ALT and AST
  - ➢ Alkaline phosphatase
  - ➢ Albumin and globulin
  - ➢ Examination of urine for urobilinogen and bilirubin.

If the diagnosis is still not clear galactose tolerance test may be done.

### *Tests to Assess Liver Function*

The liver function tests (LFT) are the biochemical tests commonly employed to assess the functional capacity of the liver. The tests can be classified according to the function of the liver involved.

1. Abnormalities of bile pigment excretion:
   a. Serum bilirubin and van den Bergh's reaction.
   b. Urinary bilirubin
   c. Urinary urobilinogen
2. Synthesis of plasma proteins:
   a. Determination of plasma proteins—albumin, globulin and fibrinogen.
   b. Determination of prothrombin time.
3. Tests based on the role of liver in lipid metabolism:
   a. Determination of serum cholesterol and the ratio of free and ester cholesterol.
4. Determination of serum enzyme activities:
   a. AST and ALT
   b. Alkaline phosphatase
   c. 5′ Nucleotidase
   d. Cholinesterase
5. Tests based on the role of liver in conjugation and excretion.
   a. Bromosulphthalein excretion test
   b. Hippuric acid test.
6. Tests based on the role of liver in carbohydrate metabolism:
   a. Galactose tolerance test
7. Determination of blood ammonia.

*Serum Bilirubin and van den Bergh's Test:*

Normal values

0.2 1.0 mg/dl

Van den Bergh's reagent (Diazo benzene sulphanilic acid) reacts with conjugated bilirubin and gives a purple colour immediately (within 30 sec). This is referred to as *Direct Positive*.

- Addition of methanol dissolves the unconjugated bilirubin which then gives a purple colour (after 30 minutes) and this is referred to as *indirect positive*.
- If the serum contains both unconjugated and conjugated bilirubin in high concentrations purple colour is produced immediately and is further intensified by the addition of alcohol. This type of reaction is known as *biphasic*.

*Note: Normal serum gives a negative Van den Bergh's reaction.*

*Importance:* The Van den Bergh's reaction is highly useful in differentiating the 3 types of jaundice.

Prehepatic (hemolytic)—Indirect positive
Posthepatic (obstructive)—Direct positive
Hepatic—Biphasic

*Urine bilirubin:*

- It can be detected by the Fouchet's test. Fouchet's reagent oxidises yellow bilirubin to green biliverdin.
- *Urine bilirubin is increased in posthepatic and hepatic jaundice* showing an increase in the conjugated bilirubin.
- The conjugated bilirubin being water soluble is excreted in urine in contrast to the unconjugated bilirubin which is insoluble in water and hence not excreted in urine.

1. *Urinary urobilinogen:* Is detected by the Ehrlich's test. A distinct red colour is seen in the presence of increased amounts of urobilinogen.

- Increased amounts are excreted in prehepatic and hepatic jaundice. Normal urine gives a trace of pink colour.
- Normally only traces of urobilinogen is present in urine. However, total absence of urobilinogen in urine is a marker of obstructive jaundice. The various parameters used in differential diagnosis of jaundice is given in Table 16.1.

2. *Synthesis of plasma proteins:* Useful information in liver disease may be obtained by the determination of total proteins, albumin and globulins since they are synthesised in the liver. In liver impairment albumin concentration falls and globulins increase.

*Normal Value*

| | | |
|---|---|---|
| Total protein | - | 6 – 8 g/dl |
| Albumin | - | 3.5 – 5.0 g/dl |
| Globulin | - | 2.5 – 3.5 g/dl |

| | |
|---|---|
| In severe chronic liver failure | - albumin values < 2.5 g/dl and globulin values > 7.0 g/dl are observed |
| Infective hepatitis | - slight increase in globulins |
| Fibrinogen | - reduced in severe acute liver failure. |

*Prothrombin time:* The term *'prothrombin time'* is given to the time required for clotting to take place in fresh citrated plasma to which optimal amounts of thromboplastin and calcium have been added.

Normal value: 10.5–12.5 sec

Determination of prothrombin time is used for the following reasons:

- To decide whether there is a danger of bleeding at operations in biliary tract disease.
- Used to control anticoagulant therapy.

Normal prothrombin time indicates that liver cell function is unimpaired.

3. *Lipid metabolism:* The serum estimation of cholesterol and lipoproteins is done to assess the liver function. However, these parameters are of more significance in cardiac disorders (Refer Chapter 3).
4. *Serum enzymes:* (Refer Chapter 6 for all enzymes and Chapter 4 for transaminases)
   *Transaminases*: ALT and AST
   These are widely used to assess liver function.

   Normal values:
   ALT: 3 – 15 IU/L
   AST: 2 – 20 IU/L
   - The activity of both these enzymes is increased in liver damage. However, ALT is a more sensitive indicator of hepatocellular diseases.

   *γ-Glutamyl transpeptidase:* Normal values: 10 – 15 IU / L.
   - Highly elevated in biliary obstruction and alcoholism.

   *Alkaline phosphatase:* Normal values: 3–13 KA units /dl.
   - Measurement of alkaline phosphatase is widely used in the investigation of the disease of the liver and biliary system.
   - Increased levels are observed in cholestasis [obstructive and hepatocellular jaundice]. Very high values (upto 200 KA units/dl) occur in complete obstruction of the bile ducts.
   - With less complete cholestasis as in stricture of the bile ducts, primary biliary cirrhosis or drug induced cholestatic jaundice the relative increase of alkaline phosphatase is greater than that of bilirubin, thus making it a more sensitive detector.
   - Increase in alkaline phosphatase without any rise in serum bilirubin are seen in primary and secondary carcinoma of the liver.
5. *Nucleotidase:* Normal values: 2 – 15 IU/L.
   - It is more specific than alkaline phosphatase since it is not altered in bone disorders like ALP.
   - *Serum cholinesterase:* Normal values: 2–12 IU/ml.
   - Increase serves as an index of liver cell damage.
   - Activity is *reduced* in poisoning by organophosphorous compounds used as insecticides.
6. *Tests based on conjugation and excretion:* *Hippuric acid synthesis test:* The test depends on the ability of the liver to produce sufficient glycine and conjugate it with an overload of benzoic acid.
   6 g of benzoic acid is given as sodium benzoate orally. The liver removes the benzoic acid by combining it with glycine to form hippuric acid which is excreted in the urine. Amount of hippuric acid excreted during a fixed time is determined.

   Normal health - > than 3.5 g is excreted
   Liver dysfunction - < than 3.5 g is excreted
   Infective hepatitis - < 1gm

*Bromsulphthalein Test (BSP):* This test depends on the excretory function of the liver. The hepatic cells take up BSP, conjugate it with glutathione and the major portion is excreted through the bile with only less than 2% being excreted by the kidney.

A single bolus dose of BSP (5 g/dl) is given intravenously and the serum concentration is measured after 45 min.

In normal people - Retention is less than 5%
Impaired function - Increase in BSP retention is seen.

Increased retention can also result from decreased excretory rate. For example, Dubin-Johnson syndrome.

7. *Galactose tolerance test:* Galactose is more rapidly absorbed from the intestine and its metabolism is entirely hepatic unlike glucose or fructose. The rate of utilisation of galactose is proportional to the functional liver cell mass.
   - 350 mg galactose/kg body weight as 25 to 30% solution is given intravenously within 3 minutes.
   - Blood is taken at 10 min interval for 1 hr.
   - Blood galactose levels are determined.
   - Log of blood galactose values against time are plotted and the $T_{1/2}$ value determined.

     Normal value: 12 ± 2.6 mins.

     Cirrhosis and infective hepatitis—increase in $T_{1/2}$ value.

     ➢ Even when the jaundice has disappeared galactose $T_{1/2}$ may be raised appreciably.
8. *Blood Ammonia:*

   Normal values: < 50 μg/dl
   - Metabolism of ammonia takes place mainly in the liver where it is converted to urea. In liver disease, the ability to remove ammonia coming from the intestine in the portal venous blood may be impaired.
   - Increase in ammonia is found only in more advanced cases of cirrhosis particularly when there are neurological complications.

Tests other than mentioned above are performed when sufficient information is not obtained from routine tests. Besides chemical, serological and radiological tests which are often employed in clinical examination is very essential for a proper diagnosis.

## RENAL FUNCTION TESTS

The kidneys are the vital organs of the body, performing the following important functions:

1. Maintenance of homeostasis: By regulating water, electrolyte and acid base balance in the body.
2. Excretion of metabolic waste products: The end products of protein and nucleic acid metabolism such as urea, creatinine and uric acid are excreted by the kidney.
3. Retention of vital substances: The kidneys reabsorb and retain several vital substances such as glucose, amino acids etc.
4. Hormonal function: The kidneys function as endocrine organs by producing hormones.
   1. *Erythropoietin:* A peptide hormone that stimulates the formation of erythrocytes.
   2. *Calcitriol:* Regulates the calcium absorption from the gut.
   3. *Rennin:* Proteolytic enzyme which stimulates the formation of angiotensin II which in turn leads to aldosterone production. Both these hormones are involved in the regulation of electrolyte balance.

Several tests are employed in the laboratory to assess kidney function. They are broadly divided into four groups:

a. *Analysis of blood:* Estimation of urea, creatinine, protein and electrolyte in serum.
b. *Urine analysis*: Routine urine analysis.
c. Tests which measure glomerular filtration.
d. Tests which measure tubular function.

### 1. Analysis of Analytes in Blood

Estimation of creatinine and urea are the widely used screening tests to assess the kidney function.

*Urea (Fig. 16.2)*

Normal value: 12–36 mg/dl.

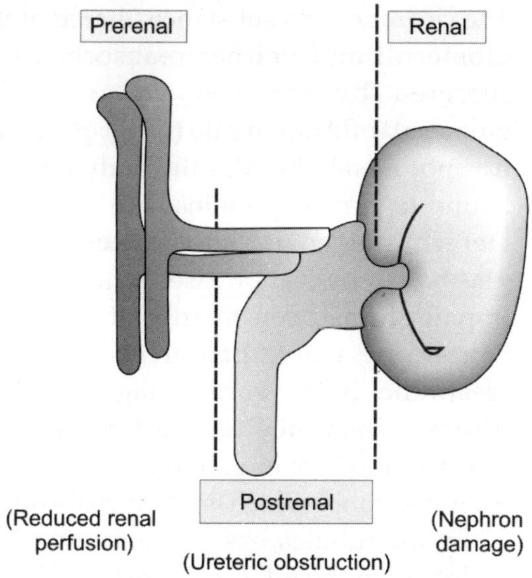


**Fig. 16.2 :** Causes for uraemia

Elevated blood urea (uraemia) is the hall mark of kidney failure. Uraemia is classified into 3 types:

- Prerenal
- Renal
- Postrenal
- *Prerenal*—The kidney fails to receive a proper blood supply. As a result the effective filtration rate in glomerulus is reduced leading to urea retention.
  *Causes:*
  - ➢ Severe and protracted vomiting.
  - ➢ Prolonged diarrhoea
  - ➢ Diabetic coma
  - ➢ Fever
- *Renal:* There is intrinsic damage to the kidney tissue. This may be due to a variety of diseases or a consequence of prolonged prerenal or postrenal problems.
  *Causes:*
  - ➢ Glomerulonephritis
  - ➢ Chronic pyelonephritis
  - ➢ Nephrosclerosis
  - ➢ Mercurial poisoning
- *Postrenal:* Urinary drainage of the kidneys is impaired because of obstruction.
  *Causes:*
  - ➢ Enlargement of prostate
  - ➢ Stones in urinary tract
  - ➢ Tumours of the bladder
  - ➢ Stricture of urethra.

### *Creatinine*

Normal value: 0.8–1.5 mg/dl.
Serum creatinine is elevated in renal failure.

*Note: Urea and creatinine levels rise only when greater than 50% of the renal impairment is seen. They are not useful in early diagnosis of renal disorders but are useful as indication of prognosis or follow up of established renal disorders. Creatinine is a better prognostic index.*

### *Protein*

Albumin levels are decreased in kidney diseases such as glomerulonephritis and nephrotic syndrome due to excessive loss in the urine.

### *Electrolytes*

*Sodium:*

Normal value: 136–145 mEq/L
Hyponatraemia is observed in chronic renal diseases.

*Potassium:*

Normal value: 3.5–5.0 mmol/L
Hyperkalaemia is seen in acute and chronic renal failure.

*Chloride:*

Normal value: 96–106 mEq/L.
Hypochloraemia is observed in renal disease.

## 2. Urine Analysis

Routine urinary examination for physical and chemical constituents is very useful in the assessment of renal function. The physical characteristics that are examined include, volume, pH, specific gravity and osmolality. Chemical tests for abnormal constituents such as proteins, blood, etc are routinely done. (Refer Chapter 14).

*Note: Microalbuminuria is a condition characterised by excretion of small amounts of albumin in urine (< 300 mg/dl) which is not detected by heat and acetic acid test. Normally, 2.5–30 mg/dl of albumin is found in urine. The small increase above this normal range (30–300 mg/dl) in albumin excretion is an indicator of renal impairment in diabetic patients. Since it is an early indication of renal damage it is reversible.*

### 3. Tests of Glomerular Function

Many renal diseases are chronic and the biochemical changes which occur are therefore slow. Hence clearance tests are mainly used.

- *Clearance* is defined as the volume of plasma that would be completely cleared off a substance per minute.
- It represents the volume of plasma (ml) which contains the amount of that substance excreted by the kidney per minute.
- Clearance(C) is expressed as ml/minute, and is calculated by using the formula—

$$C = \frac{UV}{P}$$

U = Concentration of the substance in urine.

V = Volume of urine (in ml) excreted per minute.

P = Concentration of the substance in plasma.

The excretion rate is given by UV. Calculation of clearance is dependent on the mode of excretion of a substance by the kidney.

- The clearance of a substance filtered at the glomeruli and neither reabsorbed nor secreted by the tubules gives the glomerular filtration rate (GFR), provided it is not metabolised in the body and not bound to plasma proteins.
- Determination of substances whose plasma concentration rises when GFR is impaired has been mainly of nitrogen-containing waste products, urea and creatinine. In the earlier stages of renal disease, clearance tests afford a more sensitive index of impaired renal function.
  - ➢ A substance which is partly reabsorbed by the tubules has clearance which is less than the GFR, e.g. urea.
  - ➢ Substances which are not reabsorbed but actually secreted by the tubules have clearances in excess of the GFR, e.g., p-amino hippurate.
  - ➢ Substances which are neither reabsorbed nor secreted by the tubules have a clearance which is almost equal to GFR, e.g. inulin, creatinine.
- Clearance determination is most helpful in the early stages of progressive renal disease while plasma determinations may be more sensitive when renal failure is advanced.

#### *Inulin Clearance Test*

Inulin is a homopolysaccharide of fructose. It is neither absorbed nor secreted by the tubules and is an ideal substance for measuring GFR. Hence inulin clearance is a measure of GFR. The value of GFR as measured by inulin clearance is 125 ml/minute.

The test needs continuous infusion of inulin. Hence this procedure is not used routinely.

### Creatinine Clearance Test

The endogenous creatinine clearance has been widely used as an approximate measure of GFR. Creatinine is not a component of the diet but is formed in the body at a relatively constant rate which is dependent upon the muscle mass. Thus UV will be relatively constant for creatinine.

*Method*

- A 24 hour urine specimen is collected and a sample of blood is taken during the day.
- Creatinine in the serum and urine are determined.
- Clearance is calculated by the ratio as explained earlier.
- V is calculated by dividing the urine volume (ml) by the time of collection (min).

### Clinical Significance

The creatinine clearance averages about 120 ml/min with a range of 95–140 for men and 110 ml per min with a range of 85–125 ml/min for women.

- The creatinine clearance is nearer to the GFR as measured by inulin clearance.
- It varies little with changes in the rate of urine flow.
- Very useful for studying trends in renal function over a long period in the same patient.

*Creatinine coefficient:* Is the urinary creatinine expressed in mg/kg body weight.

Normal range:

20 – 28 →for males

15 – 21 →for females

Elevated values are observed in muscular dystrophy.

### Urea Clearance Test

Urea clearance is defined as the volume (ml) of plasma that would be completely cleared off urea per minute.

Determination of urea clearance over a 24 hr period gives an index of renal function roughly related to GFR. This procedure can be repeated at intervals to follow the progress in a patient.

### Clinical Significance

Normal values:

Maximum urea clearance: 60–95 ml/min, mean = 75 ml/min

Standard clearance : 40–65 ml/min, mean = 54 ml/min.

*Note: The urea clearance in normals averages 75 ml/min if the rate of excretion of urine is > 2 ml/min. For values of V > 2 ml/min, the term **maximum urea clearance** (cm) is used.*

*For values of V < 2 ml/min the term **standard urea clearance** is used. The normal value is 54 ml/min when V = 1 ml/min.*

Urea clearance falls progressively with increasing renal failure.

*CYSTATIN 'C'*

- Offers a more specific and practical measure of monitoring GFR
- Is related to the insufficiency of renal function and is a better marker than creatinine
- Belongs to the super family of cysteine proteinase inhibitors
- Has a molecular weight of 13.26 kDa and is composed of 120 amino acids.

## 4. Tests of Tubular Function

The renal tubules modify the composition of the glomerular filtrate by the process of reabsorption and secretion. The kidney thus exerts its important regulatory function on many body constituents. Damage to the nephrons reduces the functioning tubular mass and seriously affects the regulatory function.

Tests of tubular function are of two types:

a. Tests involving assessment of tubular handling of physiological substances, by imposing an extra load on the tubules to assess functional reserve, which is normally considerable.
b. Tests involving the ability of the tubules to secrete foreign substances injected into the circulation.

One of the earliest signs of impaired tubular function is a decreased ability to secrete either concentrated or diluted urine. This can be readily assessed by determining the specific gravity or osmolality of urine. The specific gravity of urine in normal healthy individuals ranges from 1.012–1.024. With increasing tubular failure, the kidney is still able to concentrate or dilute the urine depending on the water intake to an extent of 1.004–1.040. The range of variation of specific gravity narrows down ultimately balancing at 1.010.

*Concentration test:* It is carried out as follows:

- At 6 pm the patient is given a meal with a good protein content and not more than 200 ml water to drink.
- On the following morning 3 specimens of urine are collected.

8.00 h - Specimen I
9.00 h - Specimen II
10.00 h - Specimen III

Measure the specific gravity or osmolalities of these three specimens.

*Clinical significance:*

- The specific gravity of at least one specimen should exceed 1.022 (osmolality 850 m osmol/Kg.)
- Specific gravity less than 1.022 indicates impaired tubular function.

*Note: Specific gravity may be determined by using an urinometer and osmolality may be determined with an osmometer.*

*Dilution test: It is carried out as follows:*

- The patient is not allowed to take food after midnight.
- At 7:00 AM the bladder is emptied and the patient is given 1200 ml of water to drink within the next half an hour.
- Bladder is emptied hourly for the next 4 hours.
- Volume and specific gravity or osmolality of each of the 4 specimens is determined.

*Clinical Significance*

In a normal person all the water taken is excreted within the 4 hrs.

- The specific gravity of at least one specimen should fall to 1.003 or below.
- With impaired renal function such a low specific gravity is not reached and smaller volumes of urine are passed.
- The test is less sensitive.

## Urine Acidification Test

The kidney plays an important role in maintaining the pH of the body fluids by producing an acid urine. Tubular damage fails to produce an acidic urine.

*Method*

a. Patient is given ammonium chloride (0.1g/kg body weight) in gelatin coated capsules over a period of 1 hr.
b. Empty the bladder an hour later–Discard the specimen. Collect all the urine passed during the next 6 hr.
c. Measure the pH of the urine specimens using a pH meter.

*Clinical Significance*

- In normal persons, at least one specimen shows a pH of 5.3 or less and has an ammonium excretion of 30–90 μmol / min.
- In chronic renal failure, there is systemic metabolic acidosis, but the ability to

secrete a low pH urine is unimpaired, However, the ammonium excretion rate is diminished.

- In renal tubular acidosis, pH of urine fails to reach 5.3 but ammonium excretion may be normal.

### *Phenolsulfonphthalein Test*

Phenolsulfonphthalein (phenol red) is mainly excreted by the tubules and thus gives a measure of their secretory ability.

*Method*

i. Patient is given 500 ml of water to drink.
ii. After 30 min 1 ml of phenol red (accurately measured in 1ml syringe) is injected intravenously.
iii. Bladder is emptied exactly 15 and 120 min later. The amount of dye excreted in each of these samples in a normal person is as given below.

| *Time* | *Dye Excreted* |
|---|---|
| 15 min | 35% |
| 120 min | 6% |

Approximately 70% of the dye is excreted over a period of 2 hr.

*Clinical significance*

The first specimen is the most important.

- With increasing tubular damage percentage of dye excreted is reduced. When the damage is of minor degree the 120 min excretion is normal but the 15 min excretion is reduced.

In general, the assessment of kidney function involves the tests in the following order:

- Routine urine examination.
- Estimation of urea / creatinine.
- Clearance tests.

# 17 Biochemical Techniques

Every molecule present in the body is present in association with large number of similar or dissimilar molecules. A major problem in every area of biochemical research is obtaining sufficient quantities of the substance of interest in a pure form. For this, the substance has to be isolated from its source. This chapter deals with the principles and applications of a few techniques that have been used for purification and estimation of biochemical substances.

## CHROMATOGRAPHY

Is a technique used in biochemistry to separate a substance from a mixture containing similar substances. It is based on differences in their partition in two phases, charge properties, size, specificity or adsorption.

Chromatography usually consists of two phases—a *mobile* and a *stationary* phase. The stationary phase is held on a solid support and the mobile phase, either liquid or gas percolates through the solid support of the stationary phase. Chromatography may also be classified based on the solid support, viz.,

- Paper chromatography, where the support is paper.
- Thin layer chromatography, where the support is finely divided silica gel or alumina which is spread as a thin layer on a glass plate and
- Column chromatography, where solid matrix is filled in a long narrow column.

### Partition Chromatography

It is based on the principle of partition of the compounds to be separated between two solvent phases usually water, *the stationary phase* and an organic solvent which is the *mobile phase*. It can be either paper or thin layer chromatography.

### Paper Chromatography (Fig. 17.1)

The mixture to be separated is applied at a short distance from one end of the paper and the solvent system is allowed to run over the spots. If the solute is more soluble in the mobile phase, it moves faster than the one which is more soluble in water, which is retained and moves slower. The spots can then be visualised by staining with a suitable staining compound. The relative mobility of solutes during paper chromatography is based on their partition coefficient.

*Partition coefficient* is defined as the ratio of the concentration of solute in one phase to its concentration in the other phase.

So, similar solutes which have different partition coefficients will move to different distances.

Along with the mixture, several standard solutions, containing pure solutes are also applied and the movement of the components of the mixture is compared with the movement of the standards and identified by calculating the $R_f$ (Relative Fraction) value.

$R_f$ value is defined as the ratio of the distance moved by the solute to the distance moved by the solvent front. $R_f$ value is different for different solutes. For a given solvent system, the $R_f$ value is characteristic of the compound. From this, it is possible to identify unknown substances by their position in the chromatogram relative to known substances.

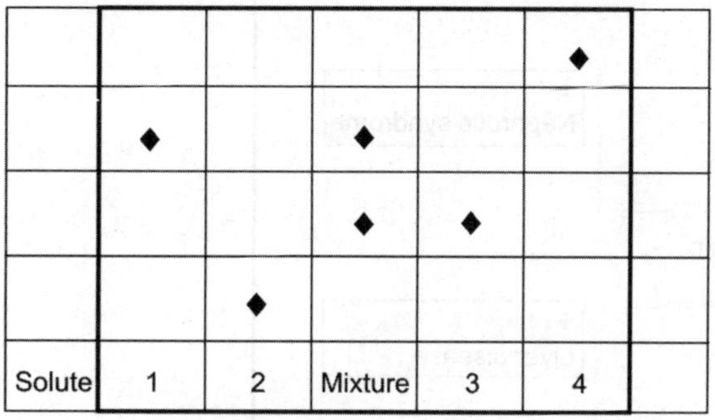


**Fig. 17.1:** Separation by paper chromatography (mixture contains solutes 1 and 3)

*Applications*

- Separation of amino acids.
- Separation of carbohydrates.
  Both used for identification of amino acids and carbohydrates excreted in urine in inherited disorders.
- Separation of lipids.

## ELECTROPHORESIS

This technique of separation is based on the movement of charged particles when subjected to an electric field. Positively charged ions, the cations, move towards cathode (-ve electrode), while negatively charged ions, the anions, move towards the anode (+ve electrode). The rate of migration of different molecules (e.g. proteins) will be different, because the number of charges carried by different molecules vary.

The technique requires an electrophoresis chamber, which consists of two compartments, one for cathode and one for anode and which can be filled with a buffer. It also requires a power pack which can supply either a fixed voltage or current.

The sample is applied on to a solid support which may be paper, agar gel, cellulose acetate or polyacylamide gel on a glass plate and is placed in the electrophoretic chamber. A constant current or voltage is applied and allowed to run for required length of time.

It is then removed from the electrophoretic chamber and stained to visualise the separated bands (Fig. 17.2). Most commonly used stains are Coomassie Brilliant Blue and Amido Schwartz.

*Applications*

1. Separation and purification of plasma proteins, enzymes, lipoproteins, Hb variants, immunoglobulins, isozymes, DNA, RNA.
2. Separation of plasma proteins is performed in clinical lab for diagnostic purposes.

## COLORIMETRY (Fig. 17.3)

Used in biochemical estimations is based on the principle that coloured solutions absorb light.

So, the substance to be measured by colorimetry should be either coloured or must be capable of forming coloured solution on addition of suitable reagents.

White light is a mixture of seven colours of different wavelengths ranging from 340–700 nm (violet to red). Every colour of the VIBGYOR belongs to a certain wavelength of the visible spectrum.

Many substances have the property of absorbing certain wavelengths of light while allowing certain other wavelengths of light to pass through. The solution of the substance will appear to be of the colour which is

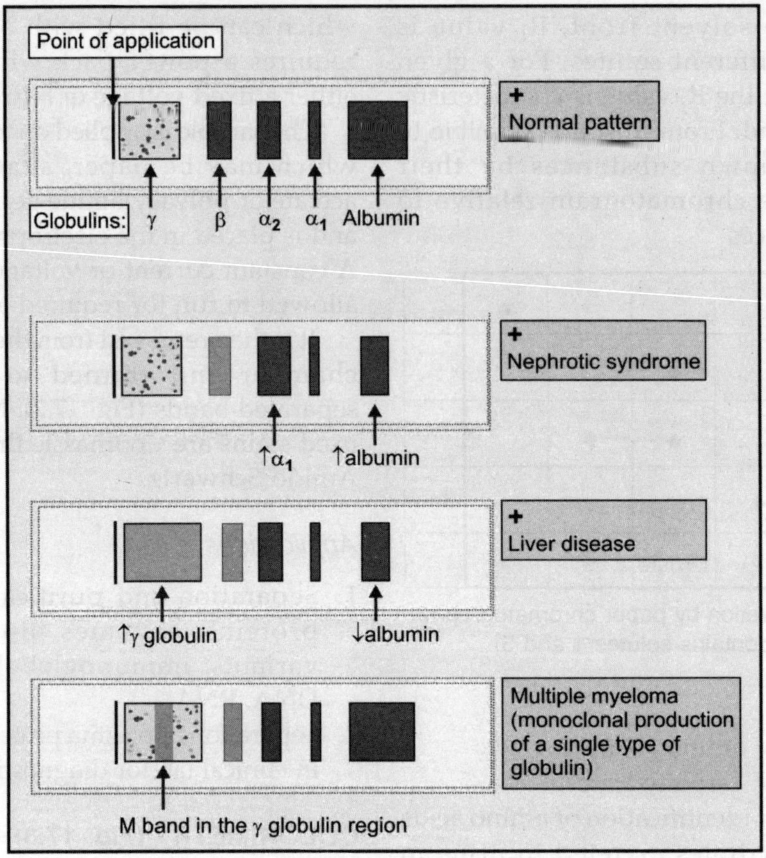


**Fig. 17.2:** Gel electrophoresis pattern of serum

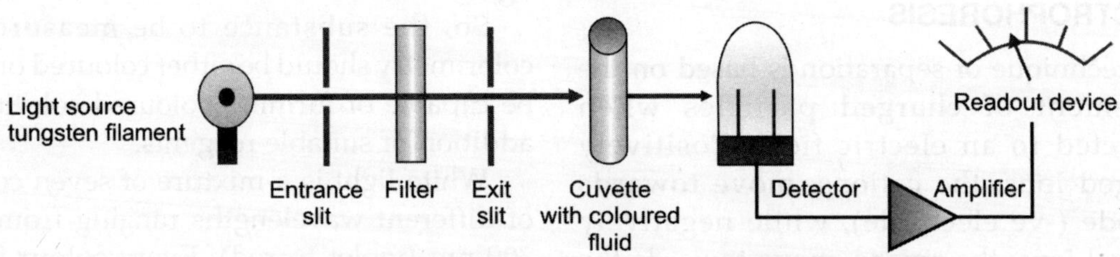


**Fig. 17.3 :** Diagram of the main components of a colorimeter

allowed to pass through or transmitted (e.g., a solution of Hb appears to be red because it absorbs blue-green light and allows red colour to pass through).

The principle of colorimetry is mainly based on two laws:

1. *Beer's law* and
2. *Lambert's law.*

The Beer's law states that the amount of light that is transmitted through a coloured solution decreases exponentially with increase in concentration of the coloured substance. Lambert's law states that the amount of transmitted light decreases exponentially with the increase in depth of the layer of solution that it passes through.

*Transmittance (T)* is the ratio of transmitted light ($I_o$) to that of incident light (I).

$$T = \frac{I_o}{I}$$

*Absorbance (A) or optical density (OD)* is a measure of amount of light absorbed by solution.

The relationship between A and T is exponential can be expressed as

$$A = -\log T$$

By combining the Beer's and Lambert's laws the following equation can be derived.

$A = abc$

Where A is the absorbance.

- a is the molar absorptivity or extinction coefficient, which is a constant for a substance.
- b is the length of light path, which is kept constant in a colorimeter.
- c is the molar concentration of the absorbing substance.

Since both a and b are constants,

$A \propto C$ or

Absorbance or optical density $\propto$ molar concentration of the substance.

The instrument used is called *colorimeter.*

The colorimeter has a source of white light, from which a selected beam falls on a filter which separates the white light into it's component colours. The chosen wavelength is selected by the exit slit and this falls on the cuvette containing the coloured solution (containing the substance to be measured). A certain amount of light, which is proportional to the concentration of the substance present, is absorbed and the remaining light falls on the detector. The signal from the detector is amplified by an amplifier which is connected to a read out device which will give the results as *absorbance.*

A solution containing an unknown amount of a substance is treated in the same way as a set of standards containing known amount of the pure substance. Absorbance of different standards are plotted against concentration to get the *standard or calibration curve* which should be linear (Fig. 17.4).

The amount of the substance present in the unknown can be calculated from the graph.

### Applications

Widely used in estimation of biochemically important substances such as carbohydrates, proteins, amino acids, lipids, and also inorganic compounds like iron, phosphate in blood, serum and other body fluids which is of diagnostic importance.

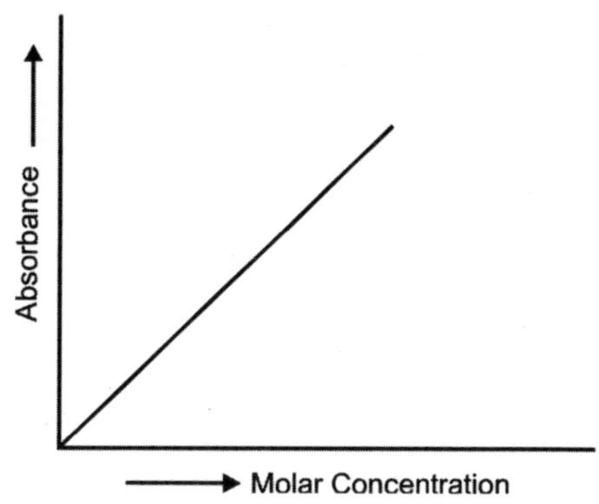


**Fig. 17.4:** Graph showing relationship between absorbance and molar concentration of the substance

## FLAME PHOTOMETRY

In flame photometry, the solution of the mineral (serum) to be determined is passed under carefully controlled conditions as a fine spray into the air supply of a burner for atomization. The flame evaporates and energises the atoms of the mineral. The atoms are excited and the electrons move to the higher energy levels. When they come back to the original level, they emit characteristic radiation and this is measured using the flame photometer. The percentage of atoms energised and therefore the amount of light emitted is proportional to the concentration of the metal in fluid, i.e. serum, urine.

### *Applications*

Estimation of sodium, potassium, calcium and lithium in biologic fluids.

# Appendix

## CHAPTER 1
## CELL BIOLOGY

**10 marks**

1. Draw an eukaryotic cell. Label the parts. Mention the significance of any four cell organelles.

## CHAPTER 2
## CARBOHYDRATES

**8 marks**

1. Classify carbohydrates with suitable examples. Add a note on mucopolysaccharides.
2. What are the normal blood glucose levels during post prandial and fasting conditions? Discuss regulation of blood glucose level.
3. Describe TCA cycle. Indicate the energy yielding steps in this cycle. How is this regulated?
4. Describe gluconeogenesis from pyruvate.
5. Enumerate reactions of glycolysis. Describe the regulation and energetics of glycolysis.
6. Give a brief account of glycogenesis and glycogenolysis. How are the two pathways reciprocally regulated?

**5 marks**

7. Write short essays on:
   a. Hexose monophosphate shunt
   b. Heteropolysaccharides
   c. Formation of pyruvate

**4 marks**

8. How is glucose absorbed in the GI tract?
9. What are the functions of carbohydrates in the body?
10. Write the reactions catalyzed by the key gluconeogenic enzymes.
11. How are carbohydrates digested in the GI tract?

**3 marks**

12. Write a short note on glycogen.

**2 marks**

Write short answers:

13. Disaccharides
14. Two actions of insulin
15. Benedict's test and its clinical applications.
16. Name four biologically important hexoses and their sources.
17. Define a reducing sugar. Give 2 examples.
18. Explain action of amylase and mention the products formed.

## CHAPTER 3
## LIPIDS

**10 marks**

1. Describe the oxidation of fatty acids. Add a note on its energetics.
2. How are lipids classified? Give the composition, properties and functions of triglycerides, lecithin and cephalin.
3. Explain how acetyl CoA formed in mitochondria is utilized in denovo synthesis of fatty acids. Write reactions of de novo synthesis of fatty acids. How is it regulated?

**6 marks**

4. How are lipids digested and absorbed in the GI tract?
5. Name the ketone bodies. How are the ketone bodies formed and utilized? Name two conditions where ketone bodies production increases.

**5 marks**

Write short essays:

6. What is fatty liver? What are the causes? What are lipotropic factors? Name them.
7. Derivatives of cholesterol.

**4 marks**

8. Name the essential fatty acids. What are their functions?
9. How are lipoproteins classified? What are their functions?
10. What is the normal cholesterol level? Name the compounds of biochemical importance formed from cholesterol.

**2 marks**

Write short answers.

11. Surfactant lipids.
12. Polyunsaturated fatty acids
13. LDL
14. Write briefly on lipid peroxidation.

## CHAPTER 4
## PROTEINS

**10 marks**

1. Give an account of the digestion and absorption of proteins.
2. Describe the structure of proteins.

**7 marks**

3. Describe transamination reaction and the formation of ammonia by glutamate dehydrogenase reaction.

**6 marks**

4. Describe urea cycle.
5. Describe amino acids based on their nutritional significance. Give two examples each.
6. Name three aromatic amino acids. Name three specialized products formed from any of them.

**5 marks**

7. Classify amino acids with two examples for each class.
8. Write a note on alkaptonuria. Explain how an alkaptonuria urine sample can show Benedicts test +ve with blood sugar 90 mg / dl.
9. Write reactions catalyzed by SGOT and SGPT. How do they play a role in urea synthesis?
10. Enumerate proteolytic enzymes of GI tract.

**4 marks**

11. What are the functions of proteins in the body?
12. Write briefly on nitrogen balance.
13. Classify proteins giving suitable examples.
14. What are immunoglobulins? Write their structure and function.
15. Write two specialized products formed from tyrosine and glycine.
16. Define digestion. Name two proenzymes.

**3 marks**

17. Enumerate compounds formed from
    (i) glycine (ii) tryptophan (iii) histidine (iv) tyrosine
18. Write short notes on Hartnup's disease.

**2 marks**

19. Give 2 examples each of acidic amino acids and aromatic amino acids.
20. Nonprotein nitrogenous substances.
21. Isoelectric pH of proteins.
22. Write the functions of proteins.

**1 mark**

Name the amino acids from which the following compounds are derived:

23. Dopamine
24. Thyroxine
25. Melatonin
26. Creatine
27. Epinephrine
28. Heme
29. Niacin
30. Histamine
31. Melanin
32. Seratonin
33. GABA

## Chapter 5
## Nucleic Acids

**5 marks**

1. Write a note on purine catabolism and gout.

**4 marks**

2. What is the basic difference between DNA and RNA in structure and function?

**2 marks**

3. What are the functions of DNA?

## Chapter 6
## Enzymes

**10 marks**

1. Define enzymes. Classify them with suitable examples. Explain in detail the physical factors which affect enzyme activity.

**6 marks**

2. Discuss the factors affecting enzyme action.
3. Write short notes on proenzymes. Explain why proenzymes are present.

## Chapter 7
## Vitamins

**10 marks**

1. Describe the sources, daily requirements, chemistry, formation of active forms, functions and deficiency manifestations of vitamin D.
2. Describe the dietary sources, coenzyme forms, biochemical functions, deficiency manifestations and daily requirements of riboflavin.
3. Describe the chemistry, dietary sources, daily requirements biochemical functions and deficiency manifestations of vitamin A.
4. Name any six B-complex vitamins and their coenzyme forms. Explain the biochemical roles of any two of them.

**5 marks**

Write short essay on:

5. Role of vitamin K in blood coagulation.
6. Scurvy
7. Vitamin $B_{12}$
8. Vitamin K
9. Explain $B_{12}$ under (a) absorption (b) transport (c) storage (d) functional role.

**4 marks**

10. Name the deficiency manifestations of the following vitamins.
    a. Thiamin
    b. Niacin
    c. Pyridoxine
11. What are functions of Vitamin C? What is its daily requirement?

**2 marks**

Write short answers.

12. Rickets
13. Pellagra
14. Coenzyme form and functions of pantothenic acid.
15. Coenzyme form and functions of thiamine.
16. Beriberi.
17. Coenzyme form and functions of pyridoxine.
18. Name the fat soluble vitamins.

## CHAPTER 8
## MINERALS

**10 marks**

1. Explain calcium metabolism under (a) source (b) RDA (c) normal blood levels (d) role of vitamin D and PTH in calcium homeostasis (e) factors affecting absorption.
2. Discuss iron metabolism under (a) RDA (b) sources (c) absorption, transport and storage (d) causes of iron deficiency.

**4 marks**

3. Enumerate functions of calcium:
4. Name four trace elements. Give one function for each one of them.
5. What is the normal serum phosphorus level? What is its function?
6. What are the functions of iron in the body?
7. Give one function each for the following minerals
   (a) phosphorus (b) magnesium (c) selenium (d) copper
8. Write a short note on disease caused by iron deficiency and disease of iron excess.

**1 mark**

9. Name two iron containing enzymes.

## CHAPTER 9
## NUTRITION

**6 marks**

1. Calculate the energy requirement of a person with sedentary work on a mixed diet with BMR 30 cal/sq.m/hr. and body surface area of 1.7 $m^2$. Mention the amounts of carbohydrate, protein and fat he has to take to meet his energy requirements.

**5 marks**

2. What are the criteria used to evaluate the biological value (BV) of protein? Give examples of proteins having high BV.

**4 marks**

3. Write briefly on balanced diet.
4. Define BMR. What are the normal values for males and females? Name one condition in which BMR is increased.
5. Name the essential amino acids.
6. Write short notes on SDA.
7. Write short notes on calorific value of foods.
8. Calculate the energy requirements of a student.
9. Write briefly on dietary fibre.
10. Nutritional aspects of carbohydrates.
11. What are the types of PEM and what are their salient features?

**2 marks**

12. Define BMR. How many calories are required for an adult male to maintain BMR?
13. Name factors affecting BMR.

14. Describe two differences between marasmus and kwashiorkar.
15. Nitrogen balance.
16. What is RQ? Mention its value for carbohydrates, proteins and fats. What is its clinical significance?

## Chapter 10
## Hormones and Neurotransmitters

**10 marks**

1. Name the hormones secreted by adrenal gland. Describe their effects on carbohydrate, lipid, protein and electrolyte metabolism.

**2 marks**

2. Name any 4 hormonal disorders.
3. Name a hormone of adrenal medulla and its main function.
4. Write a short note on diabetes insipidus.
5. Name the hormones formed from tyrosine.

## Chapter 11
## Detoxification

**5 marks**

1. Write a short essay on detoxification.

**2 marks**

2. Write a note on detoxification by conjugation.

## Chapter 12
## Water and Electrolyte Balance

**5 marks**

Write short essays:

1. Metabolic acidosis
2. Buffers of blood

**3 marks**

3. What are buffers? Write the importance of bicarbonate buffer system.

**2 marks**

4. Explain renal excretion of acid.

## Chapter 13
## Muscle and Muscle Contraction

**6 marks**

1. Discuss the process of muscle contraction.
2. Describe the roles of calcium and phosphate in muscle contraction.

**4 marks**

3. Name the different sources of energy during muscle contraction.
4. Write short notes on contractile proteins.

**3 marks**

5. Write a short note on creatine phosphate.

## CHAPTER 14
## CONNECTIVE TISSUE

**5 marks**

1. What are glycosaminoglycans ? Mention their functions. Give two examples.

**4 marks**

2. Write briefly on collagen structure and its function.
3. Write briefly on proteoglycans.
4. Write briefly on glycoproteins.
5. Write short note on collagen.

## CHAPTER 15
## CLINICAL BIOCHEMISTRY

**4 marks**

1. Write a note on the importance of serum enzymes in diagnosis of diseases.

**1 mark each**

Clinical/diagnostic importance of—

2. Alkaline phosphatase
3. Creatine phosphokinase
4. Lactate dehydrogenase
5. Amylase
6. SGPT

Name one disease condition each in which the following biochemical parameters are altered:

7. Blood glucose
8. Serum uric acid
9. SGPT
10. Blood urea
11. Acid phosphatase
12. SGOT
13. Cholesterol
14. Serum creatinine
15. Alkaline phosphatase
16. Albumin
17. Creatine phosphokinase

## CHAPTER 16
## FUNCTION TESTS

**12 marks**

1. Explain liver function tests and their diagnostic importance. Mention normal values.

**5 marks**

Short essays:

2. Renal function tests
3. Creatinine clearance and its application
4. Clearance tests

**4 marks**

5. What is jaundice? Classify it.
6. Describe heme degradation.

**3 marks**

7. Mention the importance of estimation of serum enzymes in liver disease.

**2 marks**

Write notes on:

8. Van den Bergh reaction.
9. Clinical significance of Van den Bergh test.
10. Renal function tests.

## Chapter 17
## Biochemical Techniques

**5 marks**

Write short essay on:

1. Paper chromatography
2. Electrophoresis

**2 marks**

3. Write a note on flame photometry.

## Miscellaneous

**4 marks**

Write the names of:

1. S-containing amino acids
2. S-containing lipids
3. S-containing carbohydrate
4. S-containing vitamins.

# Index

**F**



**G**



**H**

**U**



**V**



**W**



**X**



**Z**